INTRODUCTION TO
UNSTEADY THERMOFLUID
MECHANICS

INTRODUCTION TO UNSTEADY THERMOFLUID MECHANICS

Frederick J. Moody

General Electric Company
San Jose, California
and
Mechanical Engineering Department
San Jose State University
San Jose, California

WILEY

A WILEY-INTERSCIENCE PUBLICATION

JOHN WILEY & SONS

New York · Chichester · Brisbane · Toronto · Singapore

Library of Congress Cataloging in Publication Data:
Moody, F. J. (Frederick J.)
    Introduction to unsteady thermofluid mechanics / Frederick J.
  Moody.
        p.   cm.
    "A Wiley-Interscience publication."
    Includes bibliographies and index.
    1. Fluid mechanics. 2. Thermodynamics. I. Title.
  TA357.M64   1989
  620.1'06--dc20
  ISBN 0-471-85705-X   89-33606
                                                     CIP

Printed in the United States of America

10 9 8 7 6 5 4 3 2 1

To Professor William C. Reynolds, whose boundless creativity and enthusiasm turned engineering analysis into an exciting adventure

and

to Professor A. Louis London, whose practical wisdom infused engineering analysis with order, purpose, and dignity.

CONTENTS

9 MULTIDIMENSIONAL INCOMPRESSIBLE BULK AND WATERHAMMER FLOWS

PREFACE

This is a book about unsteady fluid flow and energy transfer for practicing engineers, upper division and graduate students, and engineering and science teachers. The subject has a prominent place in technology, although it seldom is emphasized in modern curricula. Engineers usually discover unsteady thermofluid mechanics when they encounter a surprise system transient in practice. I wrote this introductory book to help reduce the surprise transients and also to fill a gap in structured technology. This stimulating subject indeed has surprises, but it also has unlimited possibilities for application to creative invention, understanding peculiar thermofluid transients, achieving desired thermofluid system responses, and preventing or accommodating undesirable results.

Unsteady thermofluid phenomena involve the time-dependent character of fluid flow and energy transfer. An example of topics includes bulk and propagative flow problems, sound speed, critical flow rate, system pressurization, unsteady condensation or evaporation rates, fluid jets and resulting forces, hot or cold fluid flow in pipes, hydrostatic waves, scale-model design of thermofluid systems, waterhammer, bubble dynamics, submerged structure forces, liquid sloshing in containers, two-fluid interface instability, and fluid-structure interaction.

My objectives for this book are

1. to help others discover this fascinating subject while they acquire an awareness and competent understanding of unsteady thermofluid phenomena;
2. to provide a reference that can be used to help obtain either quick estimates or more detailed solutions of unsteady thermofluid system response;
3. to show how the various phenomena can be formulated from physical principles;
4. to provide example applications that illustrate problem approaches and solution methods;
5. to offer a resource of new, interesting analysis applications that will help make continuing education more enjoyable to engineers and scientists.

The most fascinating, challenging, and rewarding experiences throughout 30 years of engineering practice and teaching have come from problems involving time-dependent fluid flow and energy transfer. The necessity to solve such problems in the energy industry has provided the motivation to revisit undergraduate and graduate course material with the inclusion of unsteady terms involving the time derivatives of mass, momentum, and energy. Although unsteady flow and energy transfer analyses can be found in specialized technical papers, journals, and books, unfamiliarity with the topics made them difficult to locate. Furthermore, many articles were written for experienced specialists, making it necessary to search for more basic treatments. However, the result was always the same: technical growth, increased understanding, and enjoyment in learning how to formulate and solve a new class of practical problems.

It occurred to me in 1975 that problems involving unsteady flow and energy transfer were more common than most engineers realized and that they provided many fulfilling challenges for engineering practice and instruction. Pioneering applications were waiting for someone to wrestle them into a state-of-the-art analytical understanding.

I introduced examples and problems from my own unsteady thermofluid experience into upper division and graduate engineering courses at San Jose State University, where I have been an adjunct professor of engineering since 1971. Moreover, I have included much of the material in numerous advanced engineering analysis courses sponsored by the General Electric Company and offered to engineers with backgrounds ranging everywhere from new hires with no industrial experience to those with years of practice behind them who wanted a "retread." The subject also has been received with enthusiasm during seminars at overseas engineering organizations in Rome, Israel, Japan, Switzerland, Spain, Yugoslavia, and India.

People who attended these classes and seminars included mechanical, civil, electrical, chemical, nuclear, materials science, and aeronautical engineers, math and physics majors, and some who never said what they were, although they attended regularly and claimed that the experience broadened them. Most of them appreciated the introduction of material they never had encountered before. Young engineers were pleased at how the subject helped them to integrate their previous training with unsteady aspects in order to solve current problems which were not part of their curricula. Practicing engineers with 5 to 35 years of experience said that the material was just what they needed because it was fresh, current, and easy to understand, even many years after graduation. Some of this *more seasoned* group also confessed that they no longer feared becoming obsolete. Working selected problems gave some of them a new lease on their careers because it helped them to think in a new territory of problems. Engineering managers attended lectures and seminars on the subject because it revived the scientific part of them and helped them to better appreciate and appropriate the technical talent in their charge.

This book is self-contained and should be easily followed by those with an elementary background in fluid mechanics, physics, and mathematics through calculus and differential equations, with either previous or concurrent exposure to basic heat transfer and thermodynamics.

Graphs, tables, appendices, and solution procedures are included for quick estimates of thermofluid system behavior response. Paper-and-pencil solutions are emphasized, using the included helps when appropriate, and sometimes a pocket calculator. When complicated formulations arise, systematic simplification is described to achieve a level that makes the solution or interpretation less difficult, while often displaying various parameter effects on a system response.

More rigorous formulations include the finer details of a problem, which sometimes are necessary for a given application, ultimately requiring numerical solution. Various problems in this book have been formulated in "cookbook" recipes, or numerically in nondimensional form and graphed to cover a range of parameters for solution estimates. Some rudimentary finite-difference formulations are included for basic solutions which employ simple foreward-difference time advancing, the method of characteristics, over-relaxation techniques, and Runge–Kutta procedures for certain transient behavior. Computational flow charts are given for several rigorous solutions. This coverage is intended to help readers, with introductory computer experience, to write simple programs of their own when a paper-and-pencil solution is not enough. Some experienced programmers will prefer to reformulate problems with modern algorithms. I want to emphasize that this book is not a treatise on computational fluid dynamics and modern numerical solutions. These subjects are adequately treated in other books.

Problems and exercises at the end of each chapter should enhance a study of the material discussed.

Chapters 1–5 introduce various unsteady thermofluid problems involving bulk flow and the propagative behavior of hydrostatic waves. These can be taught in full or in part during a one-semester course for upper division or graduate students.

Selections from all chapters could provide interesting supplemental topics in a traditional course like "Intermediate Fluid Mechanics."

Although Chapter 2 contains numerous graphs and solution helps for problems involving unsteady thermodynamics, it can be reviewed briefly or disregarded without much loss of continuity.

Chapters 6–10 treat unsteady multidimensional flows and uniform systems, waterhammer, large amplitude incompressible and propagative flows, fluid interface instabilities, and miscellaneous topics. These will be appreciated more by those who have been exposed to basic unsteady flows of Chapter 1, and the disturbance and wave propagation coverage of Chapters 3, 4, and 5.

Chapter 6 on system normalization provides powerful analytical tools that consistently inspire those who do creative analyses or experimentation.

Chapters 7, 8, and 9 on waterhammer, large amplitude pressure propagation, and multidimensional flows are among the most useful to fluid engineers.

The variety of the last chapter has appealed to almost everyone.

I received abundant help, suggestions, and encouragement from many engineering co-workers and instructors during the course of writing this book. They are too many to list by name, but they know who they are, and I acknowledge them with heartfelt appreciation. Next-to-last draft chapters were painstakingly read and critiqued by Professor D. C. Wiggert, Professor L. C. Burmeister, Dr. J. Kim, Dr. F. T. Dodge, R. L. Peterson, and Professor E. Elias. Their influence, suggested improvements, and corrections appear on virtually every page. They kindly offered to do this substantial, although nonreimbursable task, and I accepted with deepest thanks and indebtedness. Finally, but actually first of all, it was the multitude of students, both practicing engineers and those working on a degree, whose interest and enthusiasm kept alive the inspiration needed to complete this project. My hope is that they, their contemporaries, and others who follow them, will both enjoy and benefit from the material.

FREDERICK J. MOODY

November, 1989

1 FOUNDATIONS OF UNSTEADY THERMOFLUID MECHANICS

Basic ideas are described in this chapter for use in unsteady thermofluid mechanics. The identification of *propagative* and *bulk* flow system responses is discussed because of its importance in selecting appropriate equations for analysis. Important aspects of equilibrium are reviewed. Summary discussions of the basic principles describing mass and energy conservation and momentum creation are presented with formulations useful for solving problems. Applications include the filling and draining of containers, liquid acceleration forces and expulsion from pipes, waterhammer impact pressure, and missile ejection from a mortar. A short list of unit conversion factors is given in Section A.1 of Appendix A for use in the examples and problems given throughout this book.

1.1 Propagative and Bulk Flows

It is convenient to classify problems according to *propagative* or *bulk* flows. An unbalanced force applied to a fluid produces unsteady motion. The resulting time- and space-dependent fluid properties are analyzed by choosing a system for which the basic principles can be written. Figure 1.1 helps define a *system*, as the word is used in this book. A system here includes an imaginary closed boundary called the control surface (CS), the enclosed control volume (CV), and the contents to be considered in the analysis, which may be particles, mass, momentum, energy, or other properties. Inflows and outflows also may occur across the CS. It is usually convenient to choose a system in which flow properties are continuous in both space

1

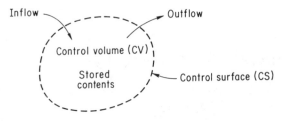

Figure 1.1 System

and time. However, steep gradients or discontinuities in properties may play an important role in a problem; they can be analyzed by careful system selection.

Abrupt property changes are associated with acoustic, shock, and surface waves. Unfortunately, it is possible to choose a system in which unexpected discontinuities may occur; consequently, unrealistic solutions may be obtained. It may be necessary to predict the pressure on a rapidly closing valve as it terminates fluid flow in a long pipe. However, if we choose a system containing all of the fluid in the pipe, we could predict a value of pressure many times higher than that which would occur. The analysis would be wrong because not all fluid particles in the pipe decelerate simultaneously, for the reason that pressure disturbances propagate at finite speed.

Whether or not propagation effects are important in an analysis should be determined before a system is chosen. The procedure involves an estimate of the propagation time t_p, which is compared with the appropriate disturbance time t_d. If propagation occurs in a short time relative to the disturbance time, propagation effects could be of minor importance, appearing only as small variations on the bulk response.

Suppose that the propagation speed in a particular problem is V_p. If L_r is a reference length through which a disturbance propagates, the *propagation time* is defined as

$$t_p = \frac{L_r}{V_p} \tag{1.1}$$

The disturbance time t_d is determined by the disturbance source, and may be the time of valve opening or closure, the acceleration period of a tank containing liquid, or time intervals associated with many other events. If t_p is short relative to t_d, propagation effects usually can be neglected.

Bulk flows are characterized by negligible propagation effects, whereas *propagative flows* include the propagation of disturbances, whether large or small. A criterion for bulk or propagative flow is given as follows:

Criterion for Bulk or Propagative Flows

$$\textit{Bulk flow if } t_p \ll t_d; \textit{ otherwise propagative flow} \tag{1.2}$$

Sometimes the above criterion is difficult to interpret. One may find that $t_p < t_d$, but perhaps t_p is not small enough to neglect propagative effects. Here is a rough guide:

Propagation effects probably are not important in analyses when t_p is less than about $0.1t_d$.

If one includes both propagative and bulk flow effects in any unsteady flow problem, a complete solution will result. However, more completeness often requires more work than a justifiable bulk or propagation flow analyses. Several idealized propagation speeds are listed in Section A.2 of Appendix A.

EXAMPLE 1.1: TRANSPORTING A TANK OF LIQUID

It is necessary to transort a flat tank of liquid at speed V on the conveyor shown in Fig. 1.2. The tank length is L and the undisturbed liquid depth is H. Moreover, the drive motor can be adjusted so that constant acceleration a_c of the conveyor will occur for time t_a, after which acceleration stops and speed V is maintained. It is necessary to estimate the maximum acceleration which can be tolerated without significant spillage.

Anyone who has carried an open tub of water knows how easily it spills over the edge. Acceleration of the tub, caused from side-to-side and back-and-forth motion, creates surface waves whose amplitudes may grow large. The tank in this example is accelerated leftward. The right end pushes the liquid, which rises on the wall and could spill over. The left edge pulls away from the liquid, causing a depression. When constant speed of the conveyor is achieved long enough for waves to settle, the tank no longer exerts a force on the liquid, which then moves as a bulk mass. Disturbance time t_d, during which large waves could be generated, is interpreted as the time of acceleration t_a. If the tank is accelerated slowly enough, small waves created at each end are transmitted back and forth, slowly bringing all liquid into translational motion. Spillage should be negligible if the tank velocity does not increase significantly during any wave travel time over length L. The question to be answered is, how slowly must the tank be accelerated?

Wave motion in a tank is a form of *propagative flow*. Section A.2 of Appendix A shows that hydrostatic wave propagation speed is $V_p = \sqrt{gH}$. Therefore, the corresponding time for wave propagation from one end of the tank to the other is

$$t_p = \frac{L}{V_p} = \frac{L}{\sqrt{gH}}$$

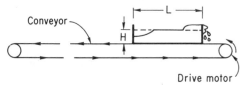

Figure 1.2   Transporting a Tank of Liquid

The criterion of Eq. (1.2) is imposed in the form

$$t_p \ll t_d = t_a = \frac{V}{a_c}$$

Therefore, liquid motion can be treated as one of *bulk flow* without significant spillage if acceleration is restricted by

$$a_c \ll \frac{V}{L} \sqrt{gH}$$

A more rigorous analysis of this problem could be performed with hydrostatic wave theory. However, the criterion of (1.2) is often general enough to make reasonable estimates in problems like this one.

———————

1.2   Quasi-Equilibrium States

Unsteady flow is characterized by local fluid properties which change with time. Thermodynamic properties are functionally related by equations of state which often are based on experiment and may be represented by tables, graphs, or equations. The familiar perfect gas state equation

$$P = \frac{MR_gT}{\mathcal{V}} = \rho R_g T \tag{1.3}$$

contains the constant R_g which is experimentally determined for a particular gas. Thus, when mass M, temperature T, and volume $\mathcal{V}$ are known, pressure P is uniquely determined. Historic experiments leading to Eq. (1.3) were performed at *equilibrium* conditions. Therefore, it is reasonable to question the validity of Eq. (1.3) if any of the variables changes with time during an unsteady flow. Time-dependent P, T, and $\mathcal{V}$ can be measured and continuously recorded in an experiment where the mass M of a known gas is held constant. If the continuous measurements satisfy Eq. (1.3) at all times, the gas is said to be in a continuous state of *quasi-equilibrium*. However, if measured P, T, and $\mathcal{V}$ do not satisfy Eq. (1.3) at some instant, the corresponding state is one of *nonequilibrium*. Since nonequilibrium states of matter are not fully understood, most unsteady thermofluid analyses are based on quasi-equilibrium regions, which permit the assumption of equilibrium states. When nonequilibrium states do exist, as in the interior of a moving shock wave, it is common practice to consider such regions as discontinuities, bounded by fluid at quasi-equilibrium states as indicated by Fig. 1.3.

When a fluid property is disturbed, appropriate use of quasi-equilibrium states depends on the disturbance itself, the size of the system, and the time scale of observation.

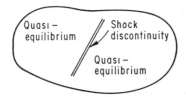

Figure 1.3 Non-equilibrium Shock Viewed as Discontinuity

The *disturbance* may be one of two general types: (1) *recoverable work disturbances*, such as the compression of a gas; and (2) *heating* or *dissipative work disturbances*. Experience shows that a recoverable work disturbance such as compression causes state changes by mechanical propagation. Compression work is done by the piston in Fig. 1.4(*a*), which shows pressure profiles at several times. The compression disturbance travels at acoustic speed, bringing the entire cylinder contents to a new state. Local compression causes rapid local temperature and density changes. The entire system approaches pressure equilibrium after several wave reflections. Experience also shows that heating or dissipative (frictional) work disturbances cause state changes during a time period required for diffusive processes. Figure 1.4(*b*) shows temperature profiles of fluid in a cylinder which is in contact with high temperature at one end and insulated at the other. The entire system approaches temperature equilibrium after the heat diffusion process ends. If both pressure (mechanical) and temperature (thermal) equilibrium are approached during time-dependent changes in a chemically nonreacting system, the system is in a state of quasi-equilibrium. Mechanical stirring, buoyancy-induced fluid circulation, or other mixing processes act to decrease the time required for diffusive processes to reach equilibrium.

The *size* of a system usually is estimated from its greatest dimension. A large system with huge spatial property variations is in a nonequilibrium

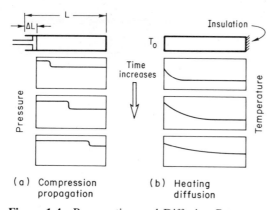

Figure 1.4 Propagation and Diffusion Processes

state, whereas a small interior subsystem can be much closer to equilibrium because the property variations within it are smaller.

The *time scale* is determined from the approximate period of observation, or data sampling. For example, ocean tides observed on an hourly time scale provide almost continuous information, whereas waves arriving at the shore must be observed on a time scale of seconds in order to obtain accurate data.

Fluid in the cylinder of Fig. 1.5 has been disturbed. A general space-dependent property is shown at some instant of time. If we are interested in the state after propagation and diffusion have subsided, we would find the entire system at an equilibrium state. If we were interested in conditions before propagation and diffusion subsided, we would find different states throughout the cylinder and would conclude that the *entire* gas system was in a state of nonequilibrium. However, smaller regions of gas between vertical dashed lines in Fig. 1.5 would approach quasi-equilibrium at average properties, indicated by the horizontal solid lines. Nonequilibrium between adjacent quasi-equilibrium regions may cause the transfer of mass, momentum, or energy, tending to bring the smaller regions into equilibrium with each other. The *maximum* size of each subdivided region should be such that a property varies only a few percent across the region. The *minimum* size must still permit fluid to be treated as a continuum.

The preceding discussion is summarized by stating that when the nature of a fluid disturbance is prescribed, the size of a fluid system which can be assumed in quasi-equilibrium depends on the time scale of observation. Criterion (1.2) can be used to show that if a disturbance propagates at speed V_p, quasi-equilibrium states are closely approached in a system whenever

$$L \ll V_p t \quad \text{quasi-equilibrium} \tag{1.4}$$

where L is the largest system dimension and t is the time period of observation.

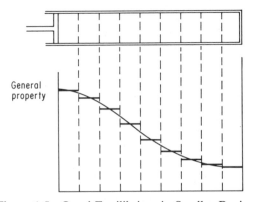

Figure 1.5   Quasi-Equilibrium in Smaller Regions

EXAMPLE 1.2: COMPRESSION EQUILIBRIUM IN A GAS
Air in a cylinder is to be rapidly compressed by a piston. If a pressure reading is to be made 1 s after compression, what length can the cylinder have for all the contained air to be in a quasi-equilibrium state? The acoustic propagation speed is $V_p = 400$ m/s (1312 ft/s).

Equation (1.4) implies that a cylinder length should be much less than

$$L \ll V_p t = (400 \text{ m/s})(1.0 \text{ s}) = 400 \text{ m} \ (1312 \text{ ft})$$

Thus, if L is about 40 m (131 ft) or less, quasi-equilibrium should be approached in 1 s after compression.

If a disturbance diffuses in a system, the extent of diffusion can be estimated by solution of the one-dimensional diffusion equation

$$\frac{\partial \phi}{\partial t} = \alpha \frac{\partial^2 \phi}{\partial x^2}$$

The symbol ϕ designates a diffusing property such as temperature, α is an appropriate diffusion coefficient, and x is the space coordinate. Let initial conditions be described by a uniform ϕ_i in the system. If a disturbance occurs at $x = 0$ and further diffusion cannot occur through the boundary at $x = L$, a full solution shows that quasi-equilibrium states are realized throughout the system if L is approximately restricted by

$$L \ll \sqrt{\frac{\alpha t}{2}} \tag{1.5}$$

The smaller length L of Eqs. (1.4) and (1.5) should be used to evaluate quasi-equilibrium in a given fluid system.

EXAMPLE 1.3: THERMAL EQUILIBRIUM IN A GAS
If gas in a horizontal cylinder is to be heated at one end, what length L would contain gas at quasi-equilibrium states for observations every second? The thermal diffusivity α is 0.09 m²/h (0.97 ft²/h).

If convective circulation is not important and equilibrium is controlled by heat diffusion, Eq. (1.5) yields a length criterion of

$$L \ll \sqrt{\frac{(0.09 \text{ m}^2/\text{h})(1.0 \text{ s})}{2(3600 \text{ s/h})}} = 0.35 \text{ cm} \ (0.138 \text{ in})$$

Thus, a much shorter length of perhaps 0.035 cm (0.014 in) would be appropriate for quasi-equilibrium observations.

It is clear from Examples 1.2 and 1.3 that regions of quasi-equilibrium are large (small) when the disturbance is propagative (diffusive).

1.3 Property Flows and Storage Rates

A *property* is anything that matter has at a given instant of time. Mass, momentum, energy, pressure, volume, and temperature are examples of properties. An *extensive property* depends on the size or *extent* of a system. Mass, volume, and total energy are examples of extensive properties.

An *intensive property* can be observed in a small sample of a larger uniform system. Pressure, temperature, and the ratio of any two extensive properties such as mass/volume (density) and energy/mass (specific energy) are intensive properties.

Suppose that an extensive property Φ is uniform in a system of mass M. The amount of Φ in a large or small mass increment δM is $\delta \Phi$, and the quantity

$$\phi = \frac{\delta \Phi}{\delta M} \tag{1.6}$$

is an intensive property, termed *specific* Φ. A property ϕ generally varies in space and time, and can be expressed functionally as $\phi(x, y, z, t)$. The amount of a given property Φ contained by a control volume (CV) can be written as either a discrete summation of the property increments contained or an integral when passing to the limit. That is,

$$\Phi = \sum_{CV} \delta \Phi = \sum_{CV} \phi \, \delta M \rightarrow \int_{CV} \phi \, dM = \int_{CV} \phi \rho \, d\mathcal{V} \tag{1.7}$$

The amount of mass contained in a CV is simply

$$M = \sum_{CV} \delta M \rightarrow \int_{CV} dM = \int_{CV} \rho \, d\mathcal{V} \tag{1.8}$$

The storage rate of property Φ within a CV is given by its time derivative $d\Phi/dt$, which depends on the flow rates of Φ across the control surface (CS). If $\delta \dot{m}$ is an incremental mass flow rate, the associated flow rate of property Φ is

$$\delta \dot{\Phi} = \phi \, \delta \dot{m} \tag{1.9}$$

Consider an increment of area designated by the vector

$$\delta \mathbf{A} = \delta A \mathbf{n}_a \tag{1.10}$$

where $\mathbf{n}_a$ is a unit vector pointing *outward* from the CS as shown in Fig. 1.6. The fluid velocity relative to the CS is designated by $^{CS}\mathbf{V}$, and the local fluid density is ρ. Thus, the incremental mass flow rate across the CS is

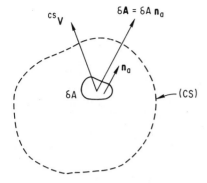

$\delta\mathbf{A} = \delta A\, \mathbf{n}_a$

$^{\text{CS}}\mathbf{V}$

$\mathbf{n}_a$

δA

(CS)

Figure 1.6 Incremental Area on Control Surface

$$\left.\begin{array}{c}\delta\dot{m}_{\text{out}}\\[4pt]\delta\dot{m}_{\text{in}}\end{array}\right\} = \rho\,{}^{\text{CS}}\mathbf{V}\cdot\delta\mathbf{A} \qquad \left\{\begin{array}{l}{}^{\text{CS}}\mathbf{V}\cdot\delta\mathbf{A}>0\\[4pt]{}^{\text{CS}}\mathbf{V}\cdot\delta\mathbf{A}<0\end{array}\right. \tag{1.11}$$

where subscripts out and in designate outflowing and inflowing mass rates. The total mass flow rates crossing a CS are

$$\begin{aligned}\dot{m}_{\text{out}} &= \sum_{\text{CS}}\delta\dot{m}_{\text{out}}\\[6pt]\dot{m}_{\text{in}} &= \sum_{\text{CS}}\delta\dot{m}_{\text{in}}\end{aligned} \tag{1.12}$$

The corresponding incremental property flow rates are

$$\begin{aligned}\delta\dot{\Phi}_{\text{out}} &= \phi_{\text{out}}\,\delta\dot{m}_{\text{out}}\\[6pt]\delta\dot{\Phi}_{\text{in}} &= \phi_{\text{in}}\,\delta\dot{m}_{\text{in}}\end{aligned} \tag{1.13}$$

and the total flow rates across a CS are

$$\begin{aligned}\dot{\Phi}_{\text{out}} &= \sum_{\text{CS}}\phi_{\text{out}}\,\delta\dot{m}_{\text{out}}\\[6pt]\dot{\Phi}_{\text{in}} &= \sum_{\text{CS}}\phi_{\text{in}}\,\delta\dot{m}_{\text{in}}\end{aligned} \tag{1.14}$$

Since the sign of $^{\text{CS}}\mathbf{V}\cdot\delta\mathbf{A}$ in Eq. (1.11) determines whether inflow or outflow is occurring, it is possible to write the difference of outflowing and inflowing mass rates as

$$\dot{m}_{\text{out}} - \dot{m}_{\text{in}} = \sum_{\text{CS}}(\delta\dot{m}_{\text{out}} - \delta\dot{m}_{\text{in}}) \rightarrow \int_{\text{CS}} d\dot{m} = \int_{\text{CS}}\rho\,{}^{\text{CS}}\mathbf{V}\cdot d\mathbf{A} \tag{1.15}$$

Similarly, the difference of general property flow rates is given by

$$\dot{\Phi}_{out} - \dot{\Phi}_{in} = \sum_{CS} [(\phi\,\delta\dot{m})_{out} - (\phi\,\delta\dot{m})_{in}] \rightarrow \int_{CS} \phi\,d\dot{m} = \int_{CS} \phi\rho^{CS}\mathbf{V}\cdot d\mathbf{A}$$
$$(1.16)$$

These relationships are useful in formulating the basic principles describing mass, momentum, and energy conservation.

1.4 Mass Conservation

The principle of mass conservation is formulated for the system of Fig. 1.7, consisting of a CV which is bounded by a CS. Discrete mass units δM are stored inside the CV at any instant, and discrete mass flow rates $\delta\dot{m}$ cross the CS. The CV and CS can be stationary, translating, rotating, or distorting. The basic principle of mass conservation is stated as

  Mass cannot be created or destroyed.

If mass somehow were being created in a system, either outflow would exceed inflow with no change in stored mass, or stored mass would increase with neither outflow nor inflow, or both events would occur simultaneously. However, since the creation of mass violates observation, it follows that the mass conservation principle for a system is described by the word equation

Total mass outflow rate − total mass inflow rate

+ total mass storage rate $= 0$ $\qquad(1.17)$

Equation (1.17) is easy to remember and is general for any system.

Several formulations of the mass conservation principle are useful for working problems. Figure 1.7, Eq. (1.17), and the notation of Section 1.3 yield the formulation

$$\sum_{CS} \delta\dot{m}_{out} - \sum_{CS^-} \delta\dot{m}_{in} + \frac{d}{dt}\sum_{CV} \delta M = 0 \qquad (1.18)$$

which is useful for discrete flow rates crossing a CS. Equations (1.8) and (1.12) with Eq. (1.18) give the alternative form

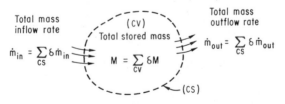

Figure 1.7 Diagram for Mass Conservation Law

$$\dot{m}_{\text{out}} - \dot{m}_{\text{in}} + \frac{dM}{dt} = 0 \tag{1.19}$$

Finally, when the flow properties are piecewise continuous in space, Equations (1.8) and (1.15) are employed in (1.18) to give the integral form as

$$\int_{\text{CS}} \rho\,^{\text{CS}}\mathbf{V}\cdot d\mathbf{A} + \frac{d}{dt}\int_{\text{CV}} \rho\, d\mathcal{V} = 0 \tag{1.20}$$

A differential formulation of the mass conservation principle is readily obtained from Eq. (1.18) and is presented in Chapter 6.

1.5 Liquid Filling and Draining of Containers

Interesting applications of the mass conservation principle come from problems where containers must be filled or drained of liquid. Often the filling or draining time is required in order to ensure compatibility with other aspects of a design or process. A rigid container of horizontal area $A(y)$ at any elevation, containing liquid of density ρ, is shown in Fig. 1.8. Simultaneous mass inflow and outflow rates may be controlled by pumping, pouring, or gravity discharge. The law of mass conservation is given by Eq. (1.19), which applies to any internal shape of the liquid region. It is usually convenient to express the contained mass M in terms of elevation $y_L(t)$, for which the liquid surface must be relatively flat. If the filling or draining time is short compared with the time required for wave propagation across the liquid surface, wave action could cause large surface distortions, making it impossible to relate M to y_L. Propagation velocity of a hydrostatic wave in a liquid of depth y_L is $\sqrt{gy_L}$ (Section A.2 of Appendix A), and propagation time across a liquid surface of horizontal length L is $t_p \approx L/\sqrt{gy_L}$. The filling or draining time (disturbance time) is roughly

$$t_d \approx \frac{M_{\text{full}}}{\dot{m}_{\text{net}}}$$

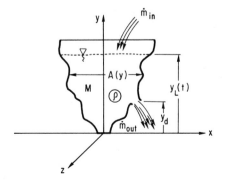

Figure 1.8 Filling and Draining a Container

where $\dot{m}_{net}$ is the approximate net flow rate. The rule of Eq. (1.2) shows that wave action does not have to be considered if

$$\dot{m}_{net} \ll \frac{M_{full}}{L} \sqrt{gy_L} \tag{1.21}$$

Whenever inequality (1.21) is satisfied, the liquid surface will be relatively flat, and dM/dt in Eq. (1.19) can be expressed by

$$\frac{dM}{dt} = \rho A(y_L) \frac{dy_L}{dt} \tag{1.22}$$

If a liquid volume inflow or outflow rate $\dot{V}$ is caused by pumping or pouring, the corresponding mass flow rate is

$$\dot{m} = \rho \dot{V} \tag{1.23}$$

If gravity draining occurs through a drain hole of area A_d, located at elevation y_d, the corresponding mass discharge rate is

$$\dot{m} = \rho C_d A_d \sqrt{2g(y_L - y_d)} \quad \text{(gravity draining)} \tag{1.24}$$

where C_d is a discharge coefficient, ranging from about 0.65 for a sharp-edged hole to 1.0 for a rounded discharge nozzle. The term $y_L - y_d$ is the vertical distance between the discharge hole and the liquid surface.

EXAMPLE 1.4: GRAVITY DRAINING, SPHERICAL TANK
Consider the spherical tank of radius R in Fig. 1.9. The initial liquid surface elevation is y_i, and gravity draining occurs from a hole in the bottom. Determine the time required for complete draining to occur.
 Since $A(y_L) = \pi y_L(2R - y_L)$, Eqs. (1.19), (1.22), and (1.24) yield the problem formulation,

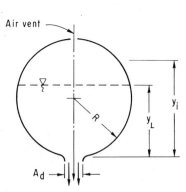

Figure 1.9 Gravity Draining of a Spherical Tank

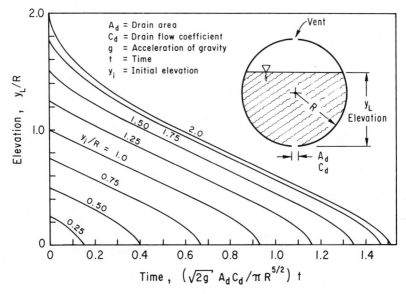

Figure 1.10 Spherical Tank Draining

$$\text{DE:} \qquad \sqrt{y_L}\,(2R - y_L)\,\frac{dy_L}{dt} = -\sqrt{2g}\,\frac{A_d C_d}{\pi} \qquad (1.25)$$

$$\text{IC:} \qquad y_L(0) = y_i \qquad\qquad\qquad\qquad (1.26)$$

for which a solution is

$$\sqrt{\frac{2g}{R}}\,\frac{A_d C_d}{\pi R^2}\,t = \frac{2}{5}\left[\left(\frac{y_L}{R}\right)^{5/2} - \left(\frac{y_i}{R}\right)^{5/2}\right] - \frac{4}{3}\left[\left(\frac{y_L}{R}\right)^{3/2} - \left(\frac{y_i}{R}\right)^{3/2}\right] \qquad (1.27)$$

The time required to drain a spherical tank initially filled to the top is determined by setting $y_i = 2R$ and $y = 0$, which gives

$$\sqrt{\frac{2g}{R}}\,\frac{A_d C_d}{\pi R^2}\,t_{\text{drain}} = \frac{16}{15}\sqrt{2} \qquad (1.28)$$

Figure 1.10 shows the time-dependent liquid elevation for a range of initial values. Similar characteristics are shown in Figs. 1.11, 1.12, and 1.13 for vertical and horizontal cylinders and cones. Figure 1.11(b) is for cases where the inflow rate exceeds the initial draining rate.

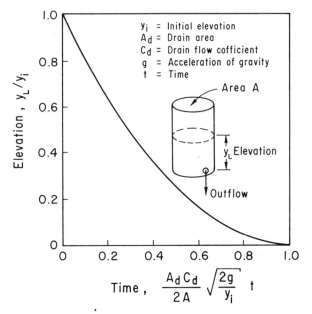

Figure 1.11(a)   Draining a Cylindrical Tank, Vertical

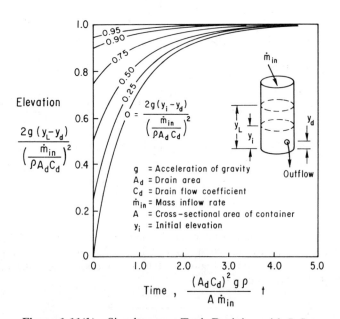

Figure 1.11(b)   Simultaneous Tank Draining with Inflow

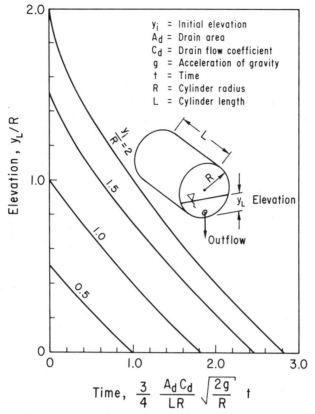

Figure 1.12 Draining a Cylindrical Tank, Horizontal

Draining problems sometimes do have surprises. Consider draining of the two hemispherical tanks shown in Fig. 1.14. Both tanks have the same initial volume of liquid. It seems to many that both hemispheres should require the same time to empty if the drains are identical. The draining time for Case I is obtained from Eq. (1.27) by putting $y_i = R$ and $y_L = 0$, giving

$$\sqrt{\frac{2g}{R}} \frac{A_d C_d}{\pi R^2} t_{\text{drain,I}} = \frac{4}{3} - \frac{2}{5} = \frac{14}{15}$$

A similar analysis for Case II yields

$$\sqrt{\frac{2g}{R}} \frac{A_d C_d}{\pi R^2} t_{\text{drain,II}} = \frac{20}{15}$$

Case II takes 40 percent longer to drain than Case I does!

If a container of liquid is to be quickly emptied by a pump or purged with compressed gas above the liquid surface, inequality (1.21) is more likely to

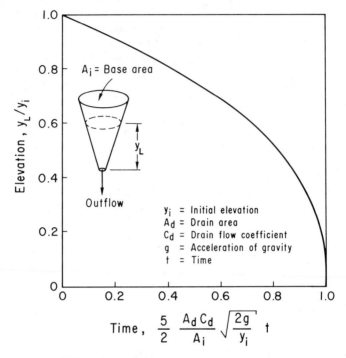

$$\text{Time}, \quad \frac{5}{2} \frac{A_d C_d}{A_i} \sqrt{\frac{2g}{y_i}} \; t$$

Figure 1.13   Draining a Conical Container

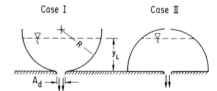

Figure 1.14   Draining Two Hemispheres

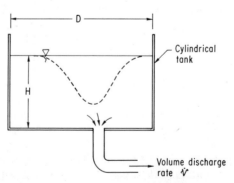

Figure 1.15   Liquid Discharge Choking

be violated. Consequently, the discharge may become choked by drawing the surface downward through the liquid, as indicated in Fig. 1.15.

EXAMPLE 1.5: MAXIMUM GAS PRESSURE, LIQUID PURGE
What maximum gas pressure P_g can be employed for purging a 1.0-m, diameter, 2-m-tall cylindrical tank, half-filled with water, through a 0.10-m diameter nozzle in the bottom into an atmospheric environment without substantial choking of the liquid discharge by gas entrainment?

If the gas pressure P_g is much greater than the hydrostatic head $\rho g y_L / g_0$, the liquid discharge rate is given by

$$\dot{m}_{out} = A_d C_d \sqrt{2 g_0 \rho (P_g - P_\infty)} \tag{1.29}$$

where P_∞ is the ambient pressure. If we put $M_{full} = \rho \pi D^2 y_L / 4$, $L = D = 1.0\,m$, $y_L = 1.0\,m$, $\rho = 1000\,kg/m^3$, $A_d = \pi D_d^2/4$, $D_d = 0.1\,m$, and $C_d = 1.0$ into (1.21), it follows that

$$P_g - P_\infty \ll \rho \, \frac{g}{g_0} \frac{y_L^3}{2} \frac{D^2}{D_d^4 C_d^2} \tag{1.30}$$

$$= \frac{(1000\,kg/m^3)(9.8\,m/s^2)(1^3\,m^3)(1^2\,m^2)}{(1\,kg\text{-}m/N\text{-}s^2)(2)(0.1^4\,m^4)(1^2)}$$

$$\ll 49\,MPa$$

Therefore, if the gas pressure is about one-tenth of this value, or 5 MPa, choking should not occur until the liquid level has dropped considerably. If choking is a serious concern at lower surface levels, smaller y_L should be employed in Eq. (1.30).

1.6 Momentum Creation

The principle of momentum creation is formulated for the system of Fig. 1.16. Discrete mass increments δM within the CV have momentum increments

$$\delta \mathcal{M} = \mathbf{V} \, \delta M \tag{1.31}$$

where $\mathbf{V}$ is velocity relative to the earth. It follows that if a mass flow rate increment $\delta \dot{m}$ is crossing a CS, the associated momentum flow rate is

$$\delta \dot{\mathcal{M}} = \mathbf{V} \, \delta \dot{m} \tag{1.32}$$

A momentum flow rate is out of the system whenever the mass flow rate is outward; it is into the system if the mass flow rate is inward.

An incremental surface force $\delta \mathbf{F}_s$ acting on the CS may include pressure,

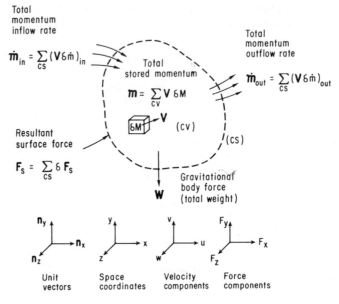

Figure 1.16   General Momentum Principle

shear, or other contact forces. The resultant of all surface force increments is given by

$$F_s = \sum_{CS} \delta F_s \tag{1.33}$$

Also shown is the gravitational body force, or system weight W, given by

$$W = -\left(\frac{g}{g_0} \sum_{CV} \delta M \right) n_y \tag{1.34}$$

Thus, the total force acting on the system of Fig. 1.16 is

$$F = F_s + W \tag{1.35}$$

The basic principle of momentum creation is that

Momentum is created in a system at a rate which is proportional to the net force applied.

Since the creation rate of momentum is the difference of outflow and inflow rates plus the storage rate, the momentum principle for a system is described by the word equation

Total momentum outflow rate − total momentum inflow rate

+ total momentum storage rate = g_0 (net force applied)

$$(1.36)$$

A working formulation, based on the notation of Fig. 1.16, is given by

$$\sum_{cs} (\mathbf{V}\, \delta \dot{m})_{out} - \sum_{cs} (\mathbf{V}\, \delta \dot{m})_{in} + \frac{d}{dt} \sum_{cv} (\mathbf{V}\, \delta M) = g_0 \mathbf{F} \qquad (1.37)$$

An integral formulation, useful for cases with piecewise continuous properties, is obtained from Eqs. (1.37), (1.7) and (1.16), in the form

$$\int_{cs} \mathbf{V}(\rho\,^{cs}\mathbf{V}\cdot d\mathbf{A}) + \frac{d}{dt}\int_{cv} \mathbf{V}\rho\, d\mathcal{V} = g_0 \mathbf{F} \qquad (1.38)$$

Equations (1.37) and (1.38) are vector formulations of the momentum principle, from which appropriate scalar components can be obtained. If a scalar momentum equation is needed for a particular direction with unit vector $\mathbf{n}$, we can write the dot product of $\mathbf{n}$ and either Eq. (1.37) or (1.38). Since $\mathbf{V}\cdot\mathbf{n} = V_n$ and $\mathbf{F}\cdot\mathbf{n} = F_n$, the scalar momentum equation obtained from Eq. (1.38) is

$$\int_{cs} V_n(\rho\,^{cs}\mathbf{V}\cdot d\mathbf{A}) + \frac{d}{dt}\int_{cv} V_n\rho\, d\mathcal{V} = g_0 F_n \qquad (1.39)$$

If $\mathbf{n}$ is either $\mathbf{n}_x$, $\mathbf{n}_y$, or $\mathbf{n}_z$, the velocity and force components would be, respectively u, v, or w, and F_x, F_y, or F_z.

1.7 Surface Forces

If pressure P is exerted uniformly over the area increment δA shown in Fig. 1.17(a), the pressure force magnitude is $P\,\delta A$. A pressure force $\delta \mathbf{F}_p$ acts on a system normal to its surface. The area vector $\delta \mathbf{A} = \delta A \mathbf{n}_A$ is normal to the surface and outward-pointing. The pressure force vector is collinear with $\delta \mathbf{A}$, but of opposite direction,

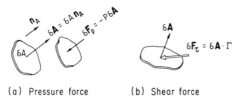

(a) Pressure force   (b) Shear force

Figure 1.17 Surface Forces

$$\delta \mathbf{F}_p = -P \, \delta A \mathbf{n}_A = -P \, \delta \mathbf{A} \qquad (1.40)$$

Shear stress is the *tangent* force per unit surface area shown in Fig. 1.17(*b*). Although an area vector points in the normal direction, shear forces may have any direction normal to the area vector. Thus, shear is *tensorial* in nature. The definition of a shear stress tensor includes information which can be combined with an area vector to obtain the shear force vector. The required tensor properties are included in the *shear stress dyadic*

$$\Gamma = \sum_j \sum_i \mathbf{n}_i \tau_{ij} \mathbf{n}_j \qquad i, j = x, y, z \qquad (1.41)$$

Equation (1.41) is the summation of nine terms, where $\mathbf{n}_i$ and $\mathbf{n}_j$ are unit vectors. The components τ_{ij}, further discussed in Chapter 6, designate the shear stress on an area projection whose normal vector is in the $\mathbf{n}_i$ direction, with an association force component in the $\mathbf{n}_j$ direction. Thus, the shear force $\delta \mathbf{F}_\tau$ on area $\delta \mathbf{A}$ is obtained from

$$\delta \mathbf{F}_\tau = \delta \mathbf{A} \cdot \Gamma \qquad (1.42)$$

which signifies the dot product of area vector $\delta \mathbf{A}$ with the first unit vector $\mathbf{n}_i$ in the dyadic of Eq. (1.41).

Other contact forces on the CS of a fluid system can be expressed as either a pressure or a shear force.

The vector sum of pressure and shear forces, Eqs. (1.40) and (1.42), gives the incremental surface force acting on an area $\delta \mathbf{A}$ as

$$\delta \mathbf{F}_s = \delta \mathbf{F}_p + \delta \mathbf{F}_\tau = -P \, \delta \mathbf{A} + \delta \mathbf{A} \cdot \Gamma \qquad (1.43)$$

from which the total surface force on a fluid system is

$$\mathbf{F}_s = \sum_{CS} \delta \mathbf{F}_s = \sum_{CS} \delta \mathbf{F}_p + \sum_{CS} \delta \mathbf{F}_\tau \qquad (1.44)$$

1.8 Body Forces

Body forces act on every mass particle in a system and may be determined by measuring the net surface force required to prevent system acceleration in a given reference frame.

The earth exerts a downward body force on the object in Fig. 1.18(*a*), whose magnitude is equal to its weight *W*. The weight vector

$$\mathbf{W} = -M \frac{g}{g_0} \mathbf{n}_y \qquad (1.45)$$

is referred to as the *gravitational body force*. The surroundings, which

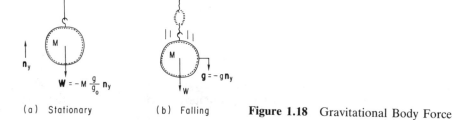

(a) Stationary   (b) Falling   **Figure 1.18** Gravitational Body Force

consist of a rope in this case, exert a vertical upward surface force of magnitude F on the object equal to its weight. That is,

$$\mathbf{F} = -\mathbf{W}$$

and no vertical acceleration relative to earth occurs. Thus, the gravitational body force $\mathbf{W}$ can be determined from an applied force $\mathbf{F}$ which prevents the object from accelerating. Suppose that the rope breaks, thus removing the surface force, as shown in Fig. 1.18(b). The object will fall downward in the direction of its body force $\mathbf{W}$ with acceleration of gravity g. The gravitational acceleration vector is

$$\mathbf{g} = -g\mathbf{n}_y \tag{1.46}$$

When systems are analyzed in reference frames which are not accelerating relative to earth, although they may be moving with constant translational velocity, the gravitational body force $\mathbf{W}$ should be considered in a prediction of fluid motion. Body forces often can be neglected when surface forces are much larger.

1.9 Accelerating Flow in a Pipe, Stationary CV

Suppose it is necessary to predict the horizontal reaction load caused by starting liquid flow in the nonuniform, rigid pipe section of Fig. 1.19(a). The

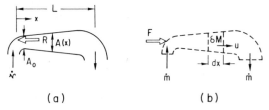

(a)   (b)

Figure 1.19 Accelerating Flow in Nonuniform Pipe

pipe of length L and flow area $A(x)$ is bounded by two $90°$ elbows, and liquid flow begins from rest by starting a pump with a time-dependent volume flow rate

$$\dot{V} = Au = \dot{V}_{max}(1 - e^{-t/\tau}) \qquad (1.47)$$

The time constant τ indicates how fast steady flow is approached. When $t = \tau$, $\dot{V}$ is the fraction $1 - e^{-1}$, or approximately 0.6 of its steady-state value. Therefore, an appropriate disturbance time for the pump start-up is

$$t_d = \tau$$

Whenever reaction loads are required on a flow passage carrying fluid, it is useful to perform a momentum analysis for a CS drawn around that fluid which is in contact with the passage walls being considered, as shown in Fig. 1.19(b). The pipe reaction force R is equal and opposite to the sum of all forces on the CS, denoted by F. If propagation effects are found to be important, the CV can be subdivided into smaller regions. Reference length $L_R = L$ and the acoustic propagation speed $V_p = C$ are employed in Eq. (1.1) to give the propagation time

$$t_p = \frac{L}{C}$$

It is assumed that the pipe length L is short enough so that $t_p \ll t_d$, which makes the problem one of bulk flow. Therefore, the momentum principle can be written for the CV of Fig. 1.19(b) in the horizontal direction without considering propagation effects. Since mass inflow and outflow rates have velocities perpendicular to the horizontal direction, inflow and outflow of horizontal momentum components are zero. An increment of stored mass $\delta M = \rho A(x)\,dx$ is shown whose local velocity is u. It follows from Eq. (1.39) that

$$0 + \frac{d}{dt}\int_0^L u\rho A(x)\,dx = g_0 F$$

The integral depends on the pipe geometry. Since the flow area is given by any arbitrary function $A(x)$, velocity at x is

$$u = \frac{\dot{V}}{A(x)}$$

Thus,

$$F = \frac{\rho L}{g_0}\frac{d\dot{V}}{dt} = R = \frac{\rho L}{g_0}\frac{\dot{V}_{max}}{\tau}e^{-t/\tau} \qquad (1.48)$$

This result has violated the intuition of many engineers. The resulting force does not depend on the cross-sectional variation $A(x)$! Furthermore, Eq. (1.48) shows that the reaction force R goes to zero when steady flow is approached.

EXAMPLE 1.6: INITIAL LIQUID ACCELERATION FORCE

A pump of time constant $\tau = 10.0$ s and steady volume flow rate $\dot{V}_{max} = 1.0 \, m^3/s$ (35.3 ft³/s) begins pumping water of density 1000 kg/m³ (62.4 lbm/ft³) through a flow passage of length $L = 2$ m (6.56 ft) between elbows. Determine the initial longitudinal force exerted on the flow passage.

Equation (1.48) yields the force

$$F = R = \frac{(1000 \text{ kg/m}^3)(2 \text{ m})(1 \text{ m}^3/\text{s})}{(1 \text{ kg-m/N-s}^2)(10 \text{ s})} = 200 \text{ N } (45 \text{ lbf})$$

1.10 Liquid Expulsion from a Uniform Pipe, Distorting CV

Gas or vapor discharge may occur from pipes which are submerged in liquid. The submerged pipes provide a liquid seal normally, requiring the expulsion of a liquid column whenever discharge begins. The time of liquid column expulsion can be an important consideration.

Consider a pipe of uniform flow area A which extends vertically downward into liquid of density ρ a distance L below the surface, as shown in Fig. 1.20. Gas pressure $P(t)$ on the liquid surface in the pipe varies with time. Ambient pressure is P_∞, to which the hydrostatic pressure is added to give pipe discharge pressure as

$$P_e = P_\infty + \rho \frac{g}{g_0} L$$

Expulsive motion of the liquid column is required.

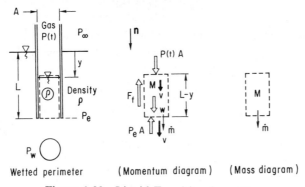

Wetted perimeter   (Momentum diagram)   (Mass diagram)

Figure 1.20   Liquid Expulsion from Pipe

At this point we could determine whether we have bulk or propagation flow. The propagation time would correspond to sonic wave transmission through length L, or

$$t_p = \frac{L}{C}$$

An appropriate problem duration or disturbance time t_d is the time required for full expulsion. However, expulsion time depends on the plug motion, which is not known. Although one could estimate t_d from an approximate analysis, this problem is assumed to be one of bulk flow, for which the expulsion time t_d is obtained and later is compared with t_p. If $t_d \gg t_p$, the bulk flow assumption is justified; otherwise the problem should be reformulated to include propagation effects.

A CS around liquid in the vent decreases in size during expulsion. This example is restricted to a gas-liquid interface which remains flat, although some deformation will occur due to Taylor instability, which is discussed in Chapter 9. Forces acting on the liquid are due to pressure, weight, and wall friction. The momentum formulation in the vertical direction is given by

$$\dot{m}v - 0 + \frac{d}{dt}(Mv) = g_0[P(t)A - P_eA + W - F_f]$$

The left-hand side is simplified by employing mass conservation in the form

$$\dot{m} - 0 + \frac{dM}{dt} = 0$$

Mass and weight of liquid in the pipe are given by

$$M = \rho A(L - y) \quad \text{and} \quad W = M\frac{g}{g_0}$$

The wall surface area in contact with the water column is

$$A_w = P_w(L - y)$$

where P_w is the wetted perimeter. The wall friction force is given by

$$F_f = \tau_w P_w(L - y) = \frac{f_u}{4}\frac{|v|v}{2g_0}\rho P_w(L - y)$$

where f_u is an unsteady friction factor, which usually is smaller than the steady value of Fig. B.1, Appendix B. If we write velocity as $v = dy/dt$, the governing equation for downward motion becomes

$$(L - y)\frac{d^2y}{dt^2} + gy + \left(\frac{f_u P_w}{8A}\right)(L - y)\left(\frac{dy}{dt}\right)^2 = \frac{g_0}{\rho}[P(t) - P_\infty] \quad (1.49)$$

Since the liquid motion starts from rest, appropriate initial conditions are

$$y = 0 \quad \text{and} \quad \frac{dy}{dt} = 0 \qquad \text{at } t = 0 \tag{1.50}$$

Consider the case of a step pressure $P(t) = P_0$. If the right-hand driving pressure term of Eq. (1.49) is large relative to the hydrostatic and friction terms, gy and $(f_u P_w / 8A)(L - y)(dy/dt)^2$ respectively, a solution for $y(t)$ is

$$t = \frac{L}{\sqrt{g_0(P_0 - P_\infty)/\rho}} \sqrt{\frac{\pi}{2}} \ \text{erf} \ \sqrt{\frac{1}{2} \ln \left(\frac{L}{L - y} \right)^2} \tag{1.51}$$

where the error function is defined by

$$\text{erf} \ \eta = \frac{2}{\sqrt{\pi}} \int_0^\eta e^{-z^2} \, dz$$

Equation (1.51) is graphed as a solid line in Fig. 1.21 with velocity $V = dy/dt$. The time for full expulsion at $y = L$ is

$$t_{\text{expulsion}} = \frac{L}{\sqrt{g_0(P_0 - P_\infty)/\rho}} \sqrt{\frac{\pi}{2}} \tag{1.52}$$

The solution for expulsion velocity in Fig. 1.21 approaches an infinite value when y approaches L. This mathematical result comes from a finite pressure force acting on a liquid mass which goes to zero at full expulsion. However, discharging fluid also must accelerate liquid surrounding the submerged pipe. Thus, the total mass being accelerated does not go to zero. The so-called *added mass* effect is included by adding about one radius of fictitious pipe length and can be obtained from methods discussed in Chapter 9.

The problem duration time, based on the preceding solution, is given by

$$t_d = t_{\text{expulsion}}$$

If t_d is compared with the propagation time L/C, it can be determined if the bulk flow analysis was justified. The criterion of Eq. (1.2) shows that $P_0 - P_\infty$ would have to exceed 1000 atm before acoustic propagation effects would be important for *water* expulsion and, probably, for most other liquids.

Another case of liquid expulsion for a step pressure

$$P_0 - P_\infty = \rho \frac{g}{g_0} L$$

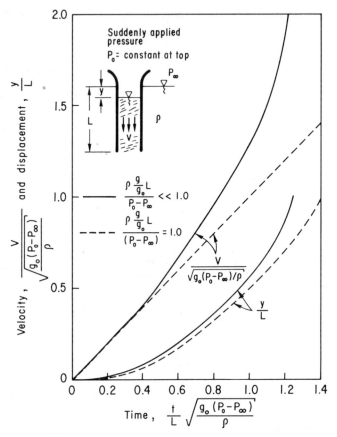

Figure 1.21   Liquid Expulsion, Step Pressure

which is just sufficient to depress the rigid level to the vent discharge, is shown by the dashed line in Fig. 1.21.

One additional liquid expulsion solution for a ramp driving pressure

$$P(t) = P_\infty + Kt \tag{1.53}$$

is shown in Fig. 1.22, where K is a positive constant. The resulting formulation was solved numerically by a Runge-Kutta integration; see Section C.1 of Appendix C.

EXAMPLE 1.7: LIQUID FILLING OR CLEARING OF A PIPE

Figure 1.23 shows two pipes of length L. One initially contains liquid and the other is empty. What takes longer: to expel the liquid column by a sudden pressure increase to P_0 at the left end, or to fill the same pipe with liquid from a reservoir whose pressure is suddenly raised to P_0? What does intuition tell you? See if it agrees with

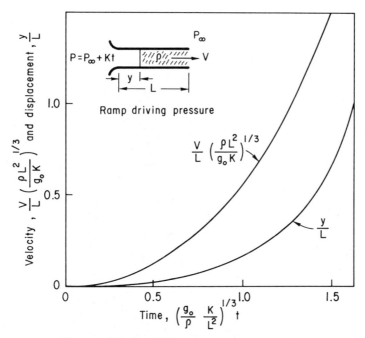

Figure 1.22 Liquid Expulsion, Ramp Pressure

analysis. The expulsion time for clearing the pipe is given by Eq. (1.52). Time for filling the pipe is obtained by writing mass conservation and momentum creation for the CV of Fig. 1.23 as

$$0 - \dot{m} + \frac{dM}{dt} = 0$$

$$0 - \dot{m}u + \frac{d}{dt}(Mu) = g_0(P - P_\infty)A$$

If we put $\dot{m} = \rho A u$ and $M = \rho A x$, and employ the steady Bernoulli equation for liquid flow from the reservoir to the pipe entrance,

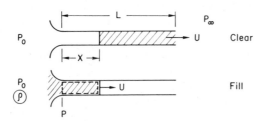

Figure 1.23 Pipe Clearing or Filling with Liquid

$$P = P_0 - \rho \frac{u^2}{2g_0}$$

we obtain the differential equation

$$x \frac{d^2x}{dt^2} + \frac{1}{2} \left(\frac{dx}{dt} \right)^2 = \frac{g_0}{\rho} (P_0 - P_\infty)$$

with initial conditions

$$t = 0, \quad x = 0, \quad \frac{dx}{dt} = \sqrt{\frac{2g_0}{\rho} (P_0 - P_\infty)}$$

The initial dx/dt is a consequence of employing the steady Bernoulli equation, which is consistent with the assumption of a short nozzle entrance relative to pipe length L. The solution for x is

$$x = \sqrt{\frac{2g_0}{\rho} (P_0 - P_\infty)} \, t$$

with filling time

$$t_{\text{fill}} = L \sqrt{\frac{\rho}{2g_0(P_0 - P_\infty)}} \tag{1.54}$$

It follows from Eqs. (1.52) and (1.54) that the ratio of expulsion to filling time is

$$\frac{t_{\text{expulsion}}}{t_{\text{fill}}} = \sqrt{\pi}$$

That is, it takes almost twice as long to expel a full liquid column from the pipe as it does to fill the empty pipe with liquid. Approximately 20 percent of the engineers can trust their intuition on this example!

1.11   Liquid Impact, Moving CV

Liquid of density ρ is flowing at pressure P and velocity u in a uniform, rigid pipe of area A and length L. A sudden disturbance occurs somewhere in the pipe which changes the local velocity and pressure to $u + \Delta u$ and $P + \Delta P$. If either Δu or ΔP are known, it is desirable to determine the other.

Propagation effects dominate this problem, where it is assumed that the disturbance time is about L/C or less. Figure 1.24 shows discontinuities traveling in both directions at speeds S_L and S_R relative to the pipe. Thin CVs are drawn about each discontinuity. Shear forces at the pipe wall and storage inside the CV are negligible. Momentum formulations yield

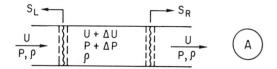

Figure 1.24 Liquid Disturbance

$$\dot{m}\,\Delta u = \begin{cases} -g_0 A\,\Delta P\,, & \text{leftward} \\ +g_0 A\,\Delta P\,, & \text{rightward} \end{cases}$$

The mass flow rates are $\rho A(S_L + u)$ and $\rho A(S_R - u)$, respectively, for the left and right traveling discontinuities. Since $S_L + u$ and $S_R - u$ are equal to the propagation speed C relative to the liquid, we have

$$\Delta P = \begin{cases} -\rho C\,\Delta u/g_0\,, & \text{left traveling} \\ +\rho C\,\Delta u/g_0 & \text{right traveling} \end{cases} \tag{1.55}$$

Equation (1.55) is one form of the waterhammer equation for sudden changes in velocity and pressure. Note that the pipe length L does not influence this result.

EXAMPLE 1.8: WATERHAMMER FORCE, SUDDEN PIPE BREAK

A long rigid pipe of cross-sectional area $A = 0.1\ \text{m}^2$ $(1.08\ \text{ft}^2)$ contains stagnant cold water at pressure $P_0 = 100$ atm (10.1 MPa), Fig. 1.25. The pipe suddenly ruptures. Determine the longitudinal force F required to keep the pipe from moving.

The disturbance travels leftward according to Fig. 1.24, for which $P = P_0$ and $P + \Delta P = P_\infty$. Therefore, $\Delta P = P_\infty - P_0$. Since the initial velocity $u = 0$, it follows from Eq. (1.55) that

$$\Delta u = -\frac{g_0}{\rho C}\,\Delta P = \frac{g_0}{\rho C}\,(P_0 - P_\infty)$$

That is, the right-end pipe break causes liquid discharge toward the right. Let M be the liquid mass in the pipe through which the discontinuity has swept at time t, which now is moving at velocity Δu. The stored momentum is given by $M\,\Delta u$, and the momentum formulation is written for the dotted CS as

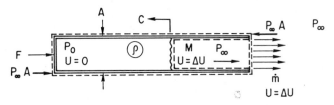

Figure 1.25 Waterhammer Force, Sudden Pipe Break

$$\dot{m}\,\Delta u - 0 + \frac{d}{dt}(M\,\Delta u) = g_0(F + P_\infty A - P_\infty A)$$

Also, since $dM/dt = \rho A(C - \Delta u)$ and $\dot{m} = \rho A\,\Delta u$, it follows that the longitudinal force is

$$F = \frac{1}{g_0}[\rho A(\Delta u)^2 + (\Delta u)\rho A(C - \Delta u)] = (P_0 - P_\infty)A$$

$$= (10\ \text{MPa})(0.1\ \text{m}^2) = 10^6\ \text{N}\ (224{,}800\ \text{lbf})$$

The subject of waterhammer is discussed further in Chapter 7.

1.12 Energy Conservation

The principle of energy conservation is formulated with the help of Fig. 1.26. A fluid increment can store kinetic, gravitational potential, and internal energy,

$$\delta(\text{KE}) = \frac{V^2}{2g_0}\,\delta M \tag{1.56}$$

$$\delta(\text{PE}) = \frac{g}{g_0}\,y\,\delta M \tag{1.57}$$

and

$$\delta U = e\,\delta M \tag{1.58}$$

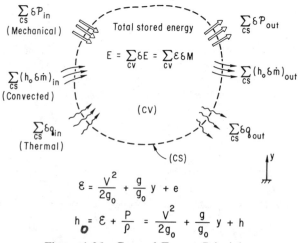

$$\mathcal{E} = \frac{V^2}{2g_0} + \frac{g}{g_0}\,y + e$$

$$h_o = \mathcal{E} + \frac{P}{\rho} = \frac{V^2}{2g_0} + \frac{g}{g_0}\,y + h$$

Figure 1.26 General Energy Principle

Stored energy also may include chemical, electrostatic, electromagnetic, and other forms. However, applications in this text involve a nonreacting, simple compressible substance (SCS), for which the only stored energy forms are given by Eqs. (1.56) to (1.58). Thus, the total stored energy in a mass increment of an SCS is written as

$$\delta E = \mathscr{E} \, \delta M \qquad (1.59)$$

where

$$\mathscr{E} = \frac{V^2}{2g_0} + \frac{g}{g_0} \, y + e \qquad (1.60)$$

Kinetic and gravitational potential energy components depend on motion and position in a given reference frame, but internal thermal energy does not depend on the reference frame. Energy transfer across a CS occurs by work, mass transfer, and heat transfer. Whenever a surface force $\mathbf{F}_s$ moves through a distance $d\mathbf{r}$, the work done is

$$đW_k = \mathbf{F}_s \cdot d\mathbf{r} \qquad (1.61)$$

where $đW_k$ is an inexact differential, whose value depends on the force-displacement path. The power, or rate of work, is

$$\mathscr{P} = \frac{đW_k}{dt} = \mathbf{F}_s \cdot \frac{d\mathbf{r}}{dt} = \mathbf{F}_s \cdot \mathbf{V} \qquad (1.62)$$

A common form of Eq. (1.62) is given in terms of shaft power. If a shaft turns at angular speed $\boldsymbol{\omega}$ with a torque $\mathbf{T} = \mathbf{r} \times \mathbf{F}_s$, the shaft power transmission is

$$\mathscr{P}_{\text{shaft}} = \mathbf{T} \cdot \boldsymbol{\omega} \qquad (1.63)$$

Another form of Eq. (1.62) is obtained for compression or expansion. If a pressure force $P \, \delta A$ moves through a distance $d\mathbf{r}$, the power is expressed in terms of a volume rate of change as

$$\delta \mathscr{P}_c = P \, \delta \mathbf{A} \cdot \frac{d\mathbf{r}}{dt} = P \, \delta \dot{\mathscr{V}} \qquad (1.64)$$

If pressure inside a CV is spatially uniform, the compressive power increments of Eq. (1.64) can be summed over the CS, except where mass transfer occurs, yielding the total compressive power,

$$\mathscr{P}_c = \sum_{\text{CS}} P \, \delta \dot{\mathscr{V}} = P \dot{\mathscr{V}} \qquad (1.65)$$

If a shear force $\delta \mathbf{F}_\tau$ on a CS acts through distance $d\mathbf{r}$, the associated work is $\delta \mathbf{F}_\tau \cdot d\mathbf{r}$. If summed over the CS, the total shear power is

$$\mathcal{P}_\tau = \sum_{CS} \delta \mathbf{F}_\tau \cdot \mathbf{V} = \sum_{CS} \delta A \cdot \mathbf{\Gamma} \cdot \mathbf{V} \qquad (1.66)$$

where $\mathbf{\Gamma}$ is given by Eq. (1.41). Wherever the symbol $\mathcal{P}$ occurs without subscripts, it represents all mechanical power associated with shaft torques, and compression and shear forces on a CS, excluding surface regions across which mass transfer occurs. It should be emphasized that mechanical power modes based on Eq. (1.62) involve surface forces only, *not* body forces.

Whenever a mass flow increment $\delta \dot{m}$ crosses a CS, it carries its stored energy at a rate $\mathcal{E} \, \delta \dot{m}$, plus additional energy it acquires as the system from which it comes expands and performs compressive work on it. If we write

$$\delta \dot{m} = \rho \, \delta \dot{V} \qquad (1.67)$$

the compressive power of Eq. (1.64) is $\delta \mathcal{P}_c = P \, \delta \dot{V} = P \, \delta \dot{m}/\rho$. If $\delta \mathcal{P}_c$ is added to $\mathcal{E} \, \delta \dot{m}$, we get the total power crossing a CS with mass transfer increments:

$$\mathcal{P}_{\dot{m}} = \sum_{CS} h_0 \, \delta \dot{m} \qquad (1.68)$$

where h_0 is the stagnation or total enthalpy per unit mass,

$$h_0 = \mathcal{E} + \frac{P}{\rho} = \frac{V^2}{2g_0} + \frac{g}{g_0} \, y + h \qquad (1.69)$$

and the static enthalpy h is

$$h = e + \frac{P}{\rho} \qquad (1.70)$$

Whenever a temperature difference occurs across a CS, energy is transferred by heating if the boundary is not adiabatic. If $đQ$ is a quantity of energy transfer by heat, its rate is

$$q = \frac{đQ}{dt} \qquad (1.71)$$

If several heat transfer rate increments of δq cross a CS, the total thermal power is expressed as

$$q = \sum_{CS} \delta \left(\frac{đQ}{dt} \right) = \sum_{CS} \delta q \qquad (1.72)$$

where the basic heat transfer forms are

$$\delta q_{\text{conduction}} = -\kappa\, \delta \mathbf{A} \cdot \nabla T$$

$$\delta q_{\text{convection}} = \mathcal{H}\, \delta A (T_{\text{hot}} - T_{\text{cold}}) \tag{1.73}$$

$$\delta q_{\text{radiation}} = \mathcal{F}\sigma\epsilon\, \delta A (T_{\text{hot}}^4 - T_{\text{cold}}^4)$$

The basic principle of energy conservation is given by the statement

Energy cannot be created or destroyed.

Since the rate of creation is the outflow rate minus the inflow rate plus the storage rate, we can write a simple expression of the energy conservation principle by the word equation

Total energy outflow rate − total energy inflow rate

$$+ \text{ total energy storage rate} = 0 \tag{1.74}$$

A working formulation, based on Fig. 1.26, is therefore,

$$\sum_{\text{CS}} (\delta\mathcal{P} + h_0\, \delta\dot{m} + \delta q)_{\text{out}} - \sum_{\text{CS}} (\delta\mathcal{P} + h_0\, \delta\dot{m} + \delta q)_{\text{in}}$$

$$+ \frac{d}{dt} \sum_{\text{CV}} \mathcal{E}\, \delta M = 0 \tag{1.75}$$

An integral formulation can be written with the help of Eqs. (1.7) and (1.16) as

$$\mathcal{P}_{\text{out}} - \mathcal{P}_{\text{in}} + q_{\text{out}} - q_{\text{in}} + \int_{\text{CS}} h_0 \rho\, {}^{\text{CS}}\mathbf{V} \cdot d\mathbf{A} + \frac{d}{dt} \int_{\text{CV}} \mathcal{E}\rho\, d\mathcal{V} = 0 \tag{1.76}$$

Consider cases without mass flow across the CS, negligible kinetic and potential energy terms, and uniform properties within the CV. If we write $(\mathcal{P}_{\text{out}} - \mathcal{P}_{\text{in}})\, dt = đW_{\text{out}}$, $(q_{\text{in}} - q_{\text{out}})\, dt = đQ_{\text{in}}$, and $\mathcal{E}\rho\, d\mathcal{V} = dU$ in Eq. (1.76), it follows that

$$đQ_{\text{in}} = dU + đW_{\text{out}} \tag{1.77}$$

which is a special formulation of the first law of thermodynamics for a fixed-mass, nonflow system. Since we have a nonflow system, the term $đW_{\text{out}}$ corresponds to a differential transfer of energy by compressive work (that is, $đW_c = P\, d\mathcal{V}$) performed by an expanding system. A famous experiment performed by Joule showed that shear work $đW_\tau$ and heat transfer $đQ$ cause equivalent changes in the thermodynamic state of a substance. Therefore, the first law expression of Eq. (1.77) for a fixed-mass system can be written as

$$\eth Q_{\text{in}} = \eth W_{\tau\,\text{in}} = dU + P\,d\mathcal{V} \tag{1.78}$$

Equation (1.78) shows that internal energy of a simple compressible substance, SCS, is changed only by compressive work, heat transfer, or shear work.

1.13 Missile Propulsion from a Tube

A missile of mass M is to be propelled from a cylindrical tube of cross-sectional area A and travel length L, as shown in Fig. 1.27. A fast-burning charge causes rapid release of hot gas into the volume $\mathcal{V}$. The charge can be approximated by an incoming gas mass flow rate $\dot{m}$ with stagnation enthalpy h_0 and a simultaneous heat addition rate q. The missile velocity and displacement are required as functions of time in order to estimate the missile flight trajectory.

CVs are shown dotted around the expanding gas and missile. Energy conservation for the gas is written as

$$P\frac{d\mathcal{V}}{dt} - \dot{m}h_0 - q + \frac{dE_g}{dt} = 0 \tag{1.79}$$

If the kinetic and potential energies of the gas are neglected, E_g is equal to the internal energy (see Table 2.2)

$$E_g = U = \frac{1}{k-1}\,P\mathcal{V} \tag{1.80}$$

where k is the ratio of specific heats c_P/c_V. Moreover, the momentum principle for the missile yields

$$(P - P_\infty)A - F_f - M\frac{g}{g_0} = \frac{M}{g_0}\frac{d^2y}{dt^2} \tag{1.81}$$

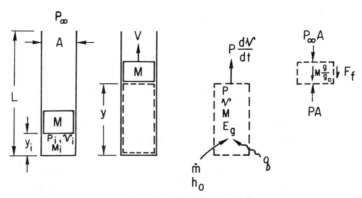

Figure 1.27 Missile Expulsion

where F_f is the wall friction force. Since $V = Ay$, P is eliminated from Eqs. (1.79) and (1.81), giving the differential equation for missile displacement,

$$\frac{1}{k} y \frac{d^3y}{dt^3} + \frac{dy}{dt} \frac{d^2y}{dt^2} + \left(\frac{g_0 A}{M} P_\infty + \frac{g_0}{M} F_f + g \right) \frac{dy}{dt}$$

$$+ \frac{g_0}{Mk} y \frac{dF_f}{dt} = \frac{g_0}{Mk} (k-1)(q + \dot{m} h_0) \tag{1.82}$$

The initial gas region has length y_i, and the missile velocity begins from zero. If the net force expelling the missile rises from a zero value, the initial conditions are

$$t = 0, \quad y = y_i, \quad \frac{dy}{dt} = \frac{d^2y}{dt^2} = 0 \tag{1.83}$$

This analysis is based on a model in which the driving pressure is assumed to be uniform in the gas region. The propagation response time is $t_p = L/C$. If the missile travel time to leave the tube is t_e, the assumption of uniform gas pressure is appropriate only when $t_e \gg t_p$. Otherwise, the analysis would have to be redone, based on nonuniform gas pressure.

EXAMPLE 1.9: MISSILE EJECTION, INSTANT BURN
Consider a case in which the explosive charge used to propel a missile from a mortar tube burns so rapidly that instant gas formation at pressure $P_i = 10\,\text{MPa}$ can be employed to approximate the motion without gas inflow $\dot{m}$ and heat transfer q. Moreover, ambient pressure P_∞, wall friction, and gravity are negligible compared with the driving pressure. The initial gas length is $y_i = 0.1\,\text{m}$ (0.33 ft) in a mortar of area $A = 100\,\text{cm}^2$ (15.5 in^2). If the missile travel length is $L = 1\,\text{m}$ (3.28 ft), what will be the velocity of a 2.0-kg (4.4-lbm) missile at ejection? The gas is approximated by CO_2 at an initial temperature of 1300 K (2340 R) and ratio of specific heats $k = 1.3$.
Equation (1.82) reduces to

$$\frac{1}{k} y \frac{d^3y}{dt^3} + \frac{dy}{dt} \frac{d^2y}{dt^2} = 0$$

Since the pressure force $P_i A$ is suddenly applied, the initial acceleration is $g_0 P_i A/M$. Therefore, initial conditions for this case are

$$t = 0, \quad y = y_i, \quad \frac{dy}{dt} = 0, \quad \frac{d^2y}{dt^2} = \frac{g_0 P_i A}{M}$$

If we substitute $V = dy/dt$, $z = dV/dt$, and $dz/dt = (dz/dy)(dy/dt) = V(dz/dy)$, one integration yields

$$z = \frac{g_0 P_i A}{M} \left(\frac{y_i}{y} \right)^k$$

whereas a second integration gives the missile velocity

$$V = \sqrt{\frac{2Ag_0 y_i P_i}{M(k-1)}} \left[1 - \left(\frac{y_i}{y}\right)^{k-1} \right]$$

It follows that the velocity at ejection, $y = L$, is

$$V(L) = 129 \text{ m/s } (423 \text{ ft/s})$$

It is advisable to determine if this analysis for the assumption of spatially uniform pressure is appropriate, based on the criterion of Eq. (1.2). If the burned gases are largely CO_2, with $k = 1.3$, $R_g = 189$ J/kg-K (35 ft-lbf/lbm-°F), and the temperature is 1300 K (2340 R), the sound propagation speed is

$$C = \sqrt{kg_0 RT} = 565 \text{ m/s } (1853 \text{ ft/s})$$

and the propagation time is

$$t_p = \frac{L}{C} = \frac{1 \text{ m}}{565 \text{ m/s}} = 0.00177 \text{ s}$$

Disturbance time is the time for missile ejection, approximately $t_d = L/[V(L)/2] = (1 \text{ m})/(65 \text{ m/s}) = 0.015$ s. Since t_p is an order of magnitude smaller than t_d, Eq. (1.2) shows that the analysis for spatially uniform pressure is valid.

References

1.1   Byrd, R. B., W. E. Stewart, and E. N. Lightfoot, *Transport Phenomena*, Wiley, New York, 1960.

1.2   Fox, R. W. and A. T. McDonald, *Introduction to Fluid Mechanics*, Wiley, New York, 1978.

1.3   Kahmke, E. *Differentialgleichungen Losungsmethoden and Losungen*, Chelsea, 1959.

1.4   Kreith, F. *Principles of Heat Transfer*, 3rd ed., IEP, 1976.

1.5   Moody, F. J. "Unsteady Condensation and Fluid-Structure Frequency Dependence on Parameters of Vapor Quench Systems," ASME Special Publication PVP-46, *Interactive Fluid-Structural Dynamic Problems in Power Engineering*, 1981.

1.6   Panton, R. L. *Incompressible Flow*, Wiley, New York, 1984.

1.7   Parmakian, J. *Waterhammer Analysis*, Dover, New York, 1963.

1.8   Reynolds, W.C. *Thermodynamics*, 2nd ed., McGraw-Hill, New York, 1975.

1.9   Schlichting, H. *Boundary Layer Theory*, 6th ed., McGraw-Hill, New York, 1968.

1.10 Streeter, V. L. and E. B. Wylie, *Fluid Mechanics*, 6th ed., McGraw-Hill, New York, 1975.

1.11 White, F. M. *Fluid Mechanics*, McGraw-Hill, New York, 1979.

Problems

1.1 A spherical tank of 10 m (32.8 ft) diameter, pressurized with 2000 kg (4400 lbm) of air, suddenly begins to discharge through a small nozzle at an initial rate of 1000 kg/s (2200 lbm/s). Would you recommend a propagation or bulk flow analysis for determining pressure at any arbitrary location in the tank? Acoustic speed in the contained air is about 400 m/s (1312 ft/s).

1.2 A slender instrument line is to be filled with cold water. One end is attached to a tank which will undergo periodic pressure changes every 2.5 s. The other end will be attached to a transducer and control panel which should give a continuous, reasonably accurate pressure reading. Approximately what maximum length of line would you recommend for this application?

1.3 If the instrument line of Problem 1.2 is to contain air, what maximum length would you recommend?

1.4 A small-diameter, 3-m (9.84-ft) long cylinder contains air at 1 atm pressure. One end is sealed, and a piston at the other end must pressurize the contained air to 10 atm. What approximate pressurization rate would you recommend so that pressure of all points in the cylinder will increase uniformly?

1.5 A rocket R travels upward at velocity $V_R = 1000$ m/s (3280 ft/s). Downward discharge gas velocity relative to the rocket is $^R V = 650$ m/s (2132 ft/s). Show that the gas discharge velocity relative to earth is 350 m/s (1148 ft/s) upward.

1.6 The oscillation frequency in rad/s of a spherical gas bubble submerged in liquid of density ρ is approximated from a linearized, bulk flow analysis (Chapter 9) by

$$\omega = \frac{1}{R_\infty} \sqrt{\frac{3kg_0 P_\infty}{\rho}}$$

where k is the gas ratio of specific heats, P_∞ is the average pressure, and R_∞ is the average bubble radius. Use the values $R_\infty = 0.3$ m (0.984 ft), $k = 1.4$, and $P_\infty = 138$ kPa (20 psia), to estimate how large a surrounding water region can be before acoustic effects become important. That is, how far away in the water can pressure measurements be made before acoustic effects noticeably influence the pressure-time history measured?

Answer:

About 12 m (40 ft)

1.7   When a saturated vapor bubble moves into a region of cold liquid, the vapor condensation and interior pressure reduction occur in several milliseconds. The ideal collapse time of a spherical void in liquid, based on instantaneous pressure reduction and a bulk flow analysis, is given by (Section 9.6)

$$\frac{t_{collapse}}{R_i} \sqrt{\frac{g_0 P_\infty}{\rho}} = 0.915$$

where R_i is the initial radius, P_∞ is the ambient pressure, and ρ is the liquid density. A steam bubble collapse occurs in the center of a water tank open at the top with diameter and depth both equal to 1.0 m (3.28 ft). If pressure on the tank bottom is to be calculated to determine the lift force, what bubble size would permit a bulk flow analysis without appreciable acoustic effects? The liquid is standard water at atmospheric pressure.

Answer:

$R_i > 0.024$ m (0.08 ft)

1.8   If a sudden pressure step to P_0 is applied at one end of a water column of length L in a straight, uniform pipe open to pressure $P_\infty = 0$ at the other end, the ideal expulsion time obtained from a bulk flow analysis is given by Eq. (1.52) as

$$\frac{t_{expulsion}}{L} \sqrt{\frac{g_0 P_0}{\rho}} = \sqrt{\frac{\pi}{2}}$$

What maximum value of P_0 can be tolerated before a propagation analysis would be more representative?

Answer:

$P_0 < 350$ MPa (51,250 psi)

1.9   Consider two identical conical containers. Both are positioned vertically, although one is placed vertex up and the other vertex down. Equal size holes exist in the vertex and base of each cone to permit filling and draining. Both cones are filled with liquid and begin to drain at the same time. Show that it requires 8/3 as long for the vertex-up cone to drain as it does for the vertex-down cone.

1.10   Consider a vertical, cylindrical tank of cross-sectional area A_1 initially filled with liquid of density ρ to a level H. Show that the mass discharge rate, for a hole of area a_{d1} and flow coefficient C_{d1} in the bottom, varies with time according to

$$\dot{m} = \alpha - \beta t$$

where α and β are constants, given by

$$\alpha = \rho a_{d1} C_{d1} \sqrt{2gH}$$

$$\beta = \frac{1}{2} \rho \frac{(a_{d1} C_{d1})^2}{A_1} 2g$$

1.11 The mass discharge rate of Problem 1.10 pours into the top of a second vertical cylindrical tank of cross-sectional area A_2 and simultaneously drains from a hole in the bottom of area a_{d2} with flow coefficient C_{d2}. If initially empty, show that liquid elevation y_2 in the tank is obtained from a solution to the problem

DE: $$\frac{dy_2}{dt} + \frac{a_{d2} C_{d2}}{A_2} \sqrt{2gy_2} = \frac{\alpha - \beta t}{A_2 \rho}$$

IC: $$y_2(0) = 0$$

1.12 Show that if the water surface velocity is constant during gravity draining from a hole in the bottom, the tank cross-sectional area must vary with elevation y according to

$$A(y) = \text{constant}\sqrt{y}$$

1.13 A plunger of uniform area A_p moves with speed V_p into a cylinder of larger uniform area A that contains liquid (see Fig. P1.13). Determine

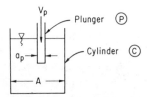

Figure P1.13

the upward velocity of the liquid surface **(a)** relative to an observer on the cylinder, and **(b)** relative to an observer on the plunger.

Answer:

$$^cV = \left(\frac{A_p}{A - A_p}\right)V_p \ ; \quad {}^pV = \left(\frac{A}{A - A_p}\right)V_p$$

1.14 A rigid tank of volume $\mathcal{V}$ contains ideal gas initially at temperature T_1 and pressure P_1. Sonic discharge occurs through an ideal nozzle of area A at a flow rate (Eq. (2.60))

$$\dot{m}_{out} = A\sqrt{\left(\frac{2}{k+1}\right)^{(k+1)/(k-1)} kg_0 P_0 \rho_0}$$

where subscript 0 refers to conditions in the tank, and k is the ratio of specific heats. Use the perfect gas law and the mass conservation principle to show that if gas temperature in the tank remains constant, tank pressure varies with time according to

$$\frac{P_0}{P_i} = \exp\left[-\left(\frac{2}{k+1}\right)^{(k+1)/2(k-1)}\frac{A}{\mathcal{V}}\sqrt{kg_0 R_g T_i}\,t\right]$$

1.15 Suppose that when starting a centrifugal pump, volume flow rate in the pipe system of Fig. P1.15 increases as

$$\dot{\mathcal{V}} = \dot{\mathcal{V}}_{max}(1 - e^{-t/\tau})$$

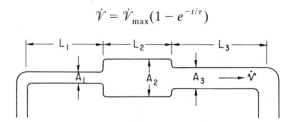

Figure P1.15

Consider a CS around the liquid and show that the pipe reaction force is leftward and varies as

$$F = \frac{\rho \dot{\mathcal{V}}_{max}(L_1 + L_2 + L_3)}{g_0 \tau} e^{-t/\tau}$$

1.16 Consider the pipe of Problem 1.15 with a steady volume flow rate $\dot{\mathcal{V}}$ of liquid whose density is ρ. Assume that one of the pipe sections is open and that liquid is discharging into ambient surroundings. Show that the steady leftward force on the pipe section is $\rho \dot{\mathcal{V}}^2/g_0 A$. Note that for a given volume flow rate, the pipe reaction force is greater for smaller area.

1.17 A container of negligible mass, whose cross-sectional area is A, contains liquid of density ρ at elevation y. Discharge occurs to the left (see Fig. P1.17) through a hole of area a_d and flow coefficient C_d. If

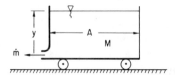

Figure P1.17

the tank is free to move on frictionless wheels, show that

$$\frac{du_c}{dt} = \frac{\dot{m}}{M}\,^c u$$

here u_c is the container velocity and $^c u$ is liquid velocity relative to the container. Express $\dot{m}$ and $^c u$ for gravity draining and show that the tank acceleration remains constant with the value $2ga_d C_d/A$. Then show that if the tank starts from rest with liquid elevation $y = y_i$, its maximum velocity reaches $(8gy_i)^{1/2}$.

1.18 Reservoir pressure can be measured by the use of an attached, clear liquid standpipe as shown in Fig. P1.18. However, if pressure changes

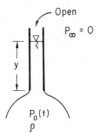

Figure P1.18

are rapid, the observed liquid level may not be dependable. The liquid response is to be studied to determine if it can follow expected pressure transients with reasonable accuracy. The standpipe attachment is ideal so that pressure P is given by the Bernoulli equation $P = P_0(t) - \rho V^2/2g_0$. Write mass conservation and momentum principles for liquid in the standpipe and show that the differential equation for level motion is

$$y\frac{d^2y}{dt^2} + \frac{1}{2}\left(\frac{dy}{dt}\right)^2 + gy = \frac{g_0}{\rho}P_0(t)$$

Verify that for $P_0(t) = $ constant, the steady elevation is $y_{steady} = g_0 P_0/\rho g$. Then assume a step change of $P_0(t)$ from zero to $P_0 = $ constant at $t = 0$, $y = 0$ and show that the initial velocity must be $\sqrt{2g_0 P_0/\rho}$. Write the differential equation in terms of V and y and show that

$$V = \frac{dy}{dt} = \pm\sqrt{\frac{2g_0 P_0}{\rho} - gy}$$

where $+$ refers to upward and $-$ to downward motion. Also show that for the step pressure change the maximum elevation is twice the steady value and requires a time of $(2/g)\sqrt{2g_0 P_0/\rho}$ to reach it.

1.19 A uniform, frictionless pipe is attached to a constant pressure reservoir (see Fig. P1.19). It initially is filled with stagnant water. A

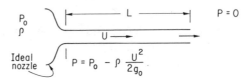

Figure P1.19

valve at the end is opened quickly and provides no restriction to the
flow. Show that the velocity increases according to

$$\frac{u}{\sqrt{2g_0P_0/\rho}} = \tanh\left(\frac{1}{2}\frac{t}{L}\sqrt{\frac{2g_0P_0}{\rho}}\right)$$

1.20 Suppose that the pipe of Problem 1.19 is bent as shown in Fig. P1.20.

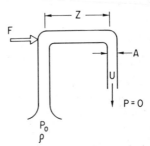

Figure P1.20

Show that the force F required to restrain the pipe from horizontal
motion is given by

$$F = A\frac{Z}{L}P_0\left(1 - \frac{u^2\rho}{2g_0P_0}\right)$$

when the valve is opened. Sketch the force-time behavior.

1.21 In Fig. P1.21 a long column of water traveling at velocity u and zero

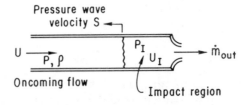

Figure P1.21

pressure in a uniform pipe of area A suddenly arrives at a flow restriction of area a_0. If the restriction flow rate is $\dot{m}_{\text{out}} = a_0\sqrt{2g_0\rho P_I}$, show that the impact pressure must satisfy

$$P_I = \frac{\rho Cu}{g_0} - \frac{C}{g_0}\frac{a_0}{A}\sqrt{2g_0\rho P_I}$$

where $C = u + S = $ sonic speed in the water.

1.22 It is necessary to predict reaction forces on the vertical pipe section in Fig. P1.22 during liquid expulsion. When pressure $P(t)$ increases

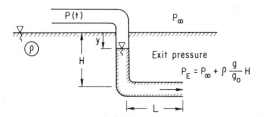

Figure P1.22

above P_∞, liquid will be expelled from the horizontal submerged section. Flow area A is constant. The vertical submergence depth is H, and the horizontal length is L. Assume that the submerged elbow is ideal without pressure losses.

(a) Obtain momentum equations for the horizontal and vertical sections, eliminate elbow pressure, and employ mass conservation to show that the liquid surface displacement is predicted by

$$(H - y + L)\frac{d^2y}{dt^2} + gy = \frac{g_0}{\rho}[P(t) - P_\infty]$$

with initial conditions

$$t = 0, \qquad y = \frac{dy}{dt} = 0$$

(b) Show that the contained liquid exerts a force on the pipe of

$$R = \frac{\rho}{g_0}A\left[(H - y)\frac{d^2y}{dt^2} - \left(\frac{dy}{dt}\right)^2 - g(H - y)\right]$$

where R is upward if positive. Note that if expulsion is not occurring, R is simply the downward weight of contained liquid.

(c) Show that initially, for $P(t) > P_\infty$,

$$R_{\text{initial}} = A\left[\frac{H}{H+L}\left[P(t) - P_\infty\right] - \rho\frac{g}{g_0}H\right]$$

which can be a large upward force proportional to $P(t)$.

(d) Show that when the liquid interface reaches the elbow,

$$R = -\frac{\rho}{g_0}A\left(\frac{dy}{dt}\right)^2$$

which is downward and proportional to velocity squared.

(e) Substitute $V = dy/dt$ into the differential equation of part (a) for the case of $P(t) = P_1 = $ constant to reduce the order by 1 and show that a solution for V^2 in terms of y is

$$V^2 = 2\left[g(H+L) - \frac{g_0}{\rho}(P_1 - P_\infty)\right]\ln\frac{H+L-y}{H+L} + 2gy$$

1.23 A very long pipe of area A is pressurized with liquid to gauge pressure P. If the liquid density is ρ and its sonic speed is C, show that its initial discharge mass flow rate per unit area from one side of a full rupture is given by $g_0 P/C$. If the pipe has an elbow some distance away from the rupture, show that the initial reaction force is PA.

1.24 The rectangular tank of Fig. P1.24 has width L and depth D into the

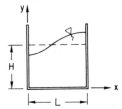

Figure P1.24

page. It is filled with liquid of density ρ to depth H. The liquid undergoes low-amplitude harmonic sloshing in the fundamental mode, which corresponds to the velocity components (Section 9.13)

$$u(x, y, t) = \epsilon\omega_1\frac{e^{\pi y/L} + e^{-\pi y/L}}{e^{\pi H/L} - e^{-\pi H/L}}\sin\frac{\pi x}{L}\sin\omega_1 t$$

$$v(x, y, t) = -\epsilon\omega_1\frac{e^{\pi y/L} - e^{-\pi y/L}}{e^{\pi H/L} - e^{-\pi H/L}}\cos\frac{\pi x}{L}\sin\omega_1 t$$

where ϵ is the slosh amplitude and

$$\omega_1 = \sqrt{\frac{g\pi}{L}\tanh\frac{\pi H}{L}}$$

Show that for small ϵ the horizontal force of liquid on the tank is

$$F_x = -\frac{2\rho D \epsilon \omega_1^2}{g_0} \left(\frac{L}{\pi}\right)^2 \cos \omega_1 t$$

1.25 Water of density 1.0 g/cc (62.4 lbm/ft^3) and 1666 m/s (5464 ft/s) sonic speed initially is at rest in a long cylinder. Suddenly a piston is driven into the pipe at constant velocity of 10 m/s (32.8 ft/s). If initial pressure in the water is zero, show that water pressure on the piston is 1666 N/cm^2 (951 psia).

1.26 A submerged vehicle draws in still water at its front end at mass flow rate $\dot{m}$ and pumps it out at the rear end at speed ^vU relative to the vehicle. If the vehicle mass is M and fluid drag is negligible, show that starting from rest its speed increases according to

$$U_v = {}^vU(1 - e^{-\dot{m}t/M})$$

1.27 A submerged column of liquid is expelled from a horizontal pipe of area A_d into a rectangular region (Fig. P1.27) by suddenly applied air

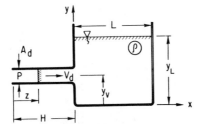

Figure P1.27

pressure P at the left end. Consider a case for which the top liquid surface remains horizontal and both gravity and ambient pressure effects are negligible. Draw a CS around the contained liquid and show that the horizontal rightward force F_x and the downward vertical force F_y are given by

$$F_x = PA_d - \frac{\rho}{g_0} \frac{d}{dt} \int\int\int_V u \, dx \, dy \, dz$$

$$F_y = \frac{\rho}{g_0} \frac{d}{dt} \int\int\int_V v \, dx \, dy \, dz$$

For a horizontal, rising surface verify that

$$\iint u \, dy \, dz = \begin{cases} A_d v_d, & (H-z) < x < 0 \\ A_d v_d \left(1 - \dfrac{x}{L}\right), & 0 < x < L \end{cases}$$

and for a pipe diameter small relative to y_L,

$$\iint v \, dx \, dz = \begin{cases} 0, & 0 < y < y_v \\ A_d v_d, & y_v < y < y_L \end{cases}$$

Then show that

$$F_x = PA_d + \frac{\rho}{g_0} A_d \frac{d}{dt}\left[v_d\left(z - \frac{L}{2}\right)\right]$$

and

$$F_y = \frac{\rho}{g_0} A_d \frac{d}{dt}[v_d(y_L - y_v)]$$

Employ Eq. (1.49) to obtain the differential equation for expulsion,

$$(H - z) \frac{dv_d}{dt} = \frac{g_0}{\rho} P, \qquad z = 0, \quad v_d = 0$$

with $L = H$, $y = z$, $dy/dt = v_d$, $P_\infty = 0$, and negligible friction, and show that

$$\frac{F_x}{PA_d} = 1 + 2 \ln \frac{H}{H-z} + \frac{z - L/2}{H - z}$$

$$\frac{F_y}{PA_d} = 2 \frac{A_d}{A} \ln \frac{H}{H-z} + \frac{y_L - y_v}{H - z}$$

Show that the initial values of F_x and F_y are

$$\frac{F_x}{PA_d} = 1 - \frac{L}{2H} \quad \text{and} \quad \frac{F_y}{PA_d} = \frac{y_L - y_v}{H}$$

1.28 Liquid of density ρ flows at volume rate $\dot{\mathcal{V}}$ in the flow passage (Fig. P1.28) composed of lengths L_1, L_2, and L_3 and flow areas A_1, A_2,

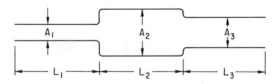

Figure P1.28

and A_3. Assume one-dimensional flow and show that the total fluid kinetic energy is given by

$$E_k = \frac{\rho \dot{V}^2}{2g_0} \left(\frac{L_1}{A_1} + \frac{L_2}{A_2} + \frac{L_3}{A_3} \right)$$

1.29 If the flow passage of Problem 1.28 is turned to a vertical position with section 1 downward, show that the gravitational potential energy of the fluid relative to the lowest end is

$$E_g = \rho \frac{g}{g_0} \left[\frac{V_1}{A_1} \left(\frac{V_1}{2} + V_2 + V_3 \right) + \frac{V_2}{A_2} \left(\frac{V_2}{2} + V_3 \right) + \frac{V_3}{A_3} \frac{V_3}{2} \right]$$

where V_1, V_2, and V_3 are the volumes of each section.

1.30 The pressure difference from inside to outside of a spherical bubble submerged in liquid, caused by surface tension, is given by

$$P - P_\infty = \frac{2\sigma}{R}$$

where R is the bubble radius and σ is the surface tension. Show that the compressive work done to overcome only the surface tension when the bubble radius grows from $R = 0$ to R is given by

$$W_\sigma = \sigma 4\pi R^2$$

Note that this work is proportional to the increase of surface area.

1.31 A water jet of area A_j and velocity V_j relative to earth impinges on a vertical plate which moves at constant speed V in the same direction. Show that maximum mechanical power is transmitted to the plate when its velocity V is $(1/3)V_j$.

1.32 When undisturbed, the liquid level in a rectangular tank is H (see Fig. P1.32). The surface is distorted at some instant of time, where $\eta(x)$ is

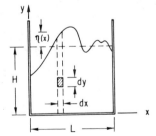

Figure P1.32

displacement from the undisturbed level. The tank has width D into the page and liquid of density ρ. Write the differential gravitational potential energy for the element shown, and verify that the change of liquid potential energy caused by surface distortion is

$$\rho \, \frac{g}{g_0} \, \frac{D}{2} \int_0^L \eta(x)^2 \, dx$$

Use this result to show that if $\eta(x) = b \cos n\pi x / L$, $n = 1, 2, \ldots$, where b is the amplitude, the potential energy is

$$E_p = \frac{D}{4} \, \rho \, \frac{g}{g_0} \, b^2 L$$

which does not depend on the integer n.

1.33 A reaction turbine is like a sophisticated lawn sprinkler. (See Fig. P1.33.) Fluid enters at the center and flows outward through directed

Figure P1.33

nozzles. If the mass flow rate from one jet is $\dot{m}$ with velocity ^nV relative to the nozzle, show that nozzle $V_n = {}^nV/2$ will result in maximum shaft power.

1.34 In Fig. P1.34 liquid and gas of constant densities ρ_L and ρ_g flow in

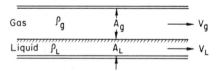

Figure P1.34

separated streams in a flow passage of area A. Respective flow rates of the liquid and gas are $\dot{m}_L$ and $\dot{m}_g$, both of which are constant. The flow areas A_L and A_g can vary but are constrained by $A = A_L + A_g$. Write an expression for the kinetic energy flow rate past a fixed point in the flow passage. Introduce the gas area fraction $\alpha = A_g/A$ and show that the minimum kinetic energy flow rate corresponds to

$$\alpha = \left[1 + \frac{\dot{m}_L}{\dot{m}_g} \left(\frac{\rho_g}{\rho_L} \right)^{2/3} \right]^{-1}$$

Also show that the minimum kinetic energy flow rate corresponds to a velocity ratio

$$\frac{V_g}{V_L} = \left(\frac{\rho_L}{\rho_g}\right)^{1/3}$$

1.35 Ideal gas is charged into a cylinder of area A at a mass flow rate $\dot{m}$ and stagnation enthalpy h_0 (see Fig. P1.35), both of which are

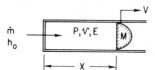

Figure P1.35

constant. Pressure buildup causes a projectile to accelerate in the cylinder. Assume that the contained gas state is spatially uniform (neglect shocks and acoustic waves). Write an energy equation for the gas region, assuming that gas kinetic and potential energies are negligible so that $E = P\mathcal{V}/(k-1)$ obtained from Table 2.2, Chapter 2. Then write a momentum equation for the projectile, assuming zero pressure on its front end. Make use of the relationships $\mathcal{V} = Ax$ and $V = dx/dt$ to show that for an initially stationary projectile, the mathematical formulation for projectile displacement is given by

DE: $$x\,\frac{d^3x}{dt^3} + k\,\frac{dx}{dt}\,\frac{d^2x}{dt^2} = \frac{(k-1)g_0 h_0 \dot{m}}{M}$$

IC: $$t = 0, \quad x = 0, \quad \frac{dx}{dt} = V = 0, \quad \frac{d^2x}{dt^2} = \frac{dV}{dt} = 0$$

1.36 The manometer of area A in Fig. P1.36 contains liquid of density ρ. If

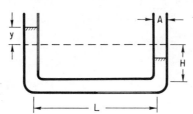

Figure P1.36

the level is displaced and then released, its subsequent motion is required. pressure on each free surface is zero.

(a) Show that for a displacement y the potential energy relative to the horizontal section is

$$\rho A \frac{g}{g_0}\,(H^2 + y^2)$$

(b) Show that the total liquid kinetic energy is

$$\frac{\rho A}{2g_0}(L+2H)\left(\frac{dy}{dt}\right)^2$$

(c) If there are no energy losses, use the condition $dE/dt = 0$ to show that the differential equation describing elevation y is

$$\frac{d^2y}{dt^2} + \frac{2g}{L+2H}\, y = 0$$

1.37 A piston drives a liquid column of density ρ and length L in the flow passage of uniform area A in Fig. P1.37. Motion begins when the

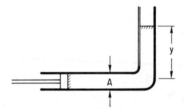

Figure P1.37

vertical elevation y is zero. Assume no pressure losses and write horizontal and vertical momentum equations to express the piston force F in terms of liquid speed V and elevation y. Then show that the total work done on this liquid column after motion starts can be expressed as

$$W_k = \frac{\rho A}{2g_0}(LV^2 + gy^2)$$

1.38 An object of volume $\mathcal{V}$ and density ρ_0 floats in liquid of density ρ. Show that the work required to submerge the object a distance H below the surface is

$$(\rho - \rho_0)\frac{g}{g_0}\,\mathcal{V}H$$

If the object is weightless with $\rho_0 = 0$, show that the work required to submerge the object is equal to that work required to lift the displaced volume of liquid a vertical distance H. If the submerged object is weightless, the work of submergence must appear as a changed energy form in the liquid. Explain which energy form is changed and in what way.

2 UNSTEADY THERMODYNAMICS

Useful thermodynamic formulations are developed in this chapter for predicting the unsteady behavior of thermofluid systems. Selected state properties, the importance of system equilibrium, and stability are discussed. The second law is presented in a form suitable for determining the direction of a process, the most stable state of a system, and the available power loss during system transients. Analyses are included for sound speed and critical flow of gas, liquid-gas, and saturated liquid-vapor mixtures. Formulations are given for predicting vessel blowdown and charging transients. Mass transfer rates by condensation and evaporation also are discussed. Graphs and formulas are presented throughout this chapter for help in estimating solutions for a variety of unsteady thermodynamic problems.

2.1 Systems

The five system categories illustrated in Fig. 2.1 are designated by the terms *general*, *rigid*, *adiabatic*, *fixed mass*, and *isolated*.

A *general system* was introduced in Chapter 1 and may include flows of mass, momentum, and energy across a deformable CS.

A *rigid system* has a nondeformable boundary and does not permit energy transfer by compressive work.

An *adiabatic system* does not permit heat transfer across its boundary.

A *fixed-mass* system contains an identifiable quantity of mass whose magnitude does not change with time.

An *isolated system* has no heat, work, or mass flow across its boundary.

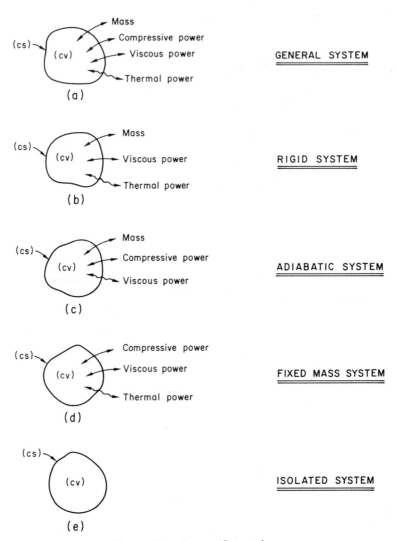

Figure 2.1   System Categories

2.2 Properties

A *property* of matter is any quantity which can be measured or observed at a given instant. *Work* and *heat* are not properties but energy transfer *processes* which change the properties of a system.

Thermodynamic properties are *macroscopic* in nature, although some are conglomerate averages of *microscopic* quantities associated with molecular behavior. Macroscopic pressure, for example, results from the average momentum rate of individual gas molecules striking a wall. Other macroscopic properties include volume, mass, temperature, momentum, and energy.

Properties of matter can be *static, dynamic,* and *thermodynamic.* Static and dynamic properties depend on the *reference frame* of observation and include position, velocity, acceleration, momentum, kinetic energy, and angular velocity. Thermodynamic properties do not depend on the reference frame. If our point of observation was fixed to a moving increment of fluid, we could observe or measure its thermodynamic state properties, but we could not detect its dynamic properties relative to another reference frame.

Some characteristics of matter are called *transport properties,* which usually are considered thermodynamic properties. Transport properties are independent of the reference frame of observation, but are associated with either dynamic or thermodynamic property *gradients.* Dynamic viscosity μ and thermal conductivity κ are transport properties which depend on velocity and temperature gradients, respectively.

Properties also are classified by the terms *extensive* and *intensive.* Extensive properties depend on the amount of a substance and include mass, volume, and energy. Intensive properties do not depend on the amount of a substance. They can be determined from a representative sample, and include pressure, temperature, and all properties which are ratios of two extensive properties, such as density, specific internal energy, or specific enthalpy.

Extensive thermodynamic properties are *single-valued* at any instant. The mass of liquid in a tank, for example, has the value M whether it is resting on the bottom or sloshing actively. However, intensive thermodynamic properties may be either *spatially uniform* or *nonuniform* in a system, such as the temperature in a lake.

Experimental relationships between thermodynamic properties or equations of state such as $P = MR_gT/\mathcal{V}$ are based on systems where the properties are single-valued at the time of observation. It is incorrect to apply an equation of state to a system with nonuniform or multiple values of a property. However, it was mentioned in Section 1.2 that if a system property is nonuniform, the system can be divided into subsystems which are small enough for the contained properties to approach single values at any time. Large or small systems in which properties are uniform are in *self-equilibrium,* although they may not be in equilibrium with adjacent systems. If properties in a system approach spatial uniformity as they change with time and if equilibrium state equations are approximately satisfied, then the process is one of *quasi-equilibrium.* Most analyses of unsteady thermofluid systems can be based upon sufficiently small systems or subsystems, so equilibrium state equations and quasi-equilibrium processes are justified. Cases which may include chemically reacting regions and fluid shocks are treated singularly because the interior states depart significantly from equilibrium.

2.3   Equilibrium Response Time

The equilibrium response time Δt_e is used to designate the approximate time required for a system to achieve a new macroscopic state of equilibrium when it is disturbed. It is determined by disturbance propagation speed and the size of a system. The state of a system is described by its properties, $\phi_1, \phi_2, \ldots,$. Steady state implies that, at a fixed position,

$$\frac{\partial \phi_j}{\partial t} = 0, \qquad j = 1, 2, \ldots \tag{2.1}$$

whereas unsteady state means that local properties may change with time, such that

$$\frac{\partial \phi_j}{\partial t} \neq 0, \qquad j = 1, 2, \ldots \tag{2.2}$$

Whether or not local property changes follow quasi-equilibrium states depends on the equilibrium response time. The concept of an equilibrium response time is useful in determining how finely divided a system must be so that all interior subsystems undergo quasi-equilibrium state changes in response to a disturbance. Compressive work creates pressure disturbances which propagate at acoustic or shock speed. Viscous work and heat create disturbances which propagate at diffusive speeds. Therefore, pressure, shear, and temperature *propagation rates* are needed for a determination of equilibrium response times in a given system. If a disturbance propagation speed is V_p and the characteristic size dimension of a system is L, the equilibrium response time is

$$\Delta t_e \approx \frac{L}{V_p} \tag{2.3}$$

Diffusive disturbances are discussed in Section 2.5. Compressive or decompressive disturbance propagation speeds are given in Section 2.11.

Equation (2.3) can be used to estimate how finely a system should be subdivided, that is, how small a dimension L is required, to analyze unsteady phenomena with equilibrium equations of state.

EXAMPLE 2.1: DISCHARGING AN AIR VESSEL
It is necessary to predict the three-dimensional unsteady pressure disturbances in a cubical vessel of volume $\mathcal{V} = 1.0 \, \text{m}^3$ ($35.28 \, \text{ft}^3$) which is discharging air, $k = 1.4$, $R_g = 287 \, \text{N-m/kg-K}$ ($53.3 \, \text{ft-lbf/lbm-°F}$), from a small nozzle at a rate of $\dot{m} = 10 \, \text{kg/s}$ ($22 \, \text{lbm/s}$). The initial air mass and vessel pressure are $M = 116 \, \text{kg}$ ($255 \, \text{lbm}$) and $P_0 = 10 \, \text{MPa}$ ($1470 \, \text{Psia}$) at a temperature of $T_0 = 300 \, \text{K}$ ($540 \, \text{R}$). Can equilibrium state equations be applied to all the gas in the vessel at any time, or should the vessel be subdivided into smaller volume increments?

The total discharge time or disturbance time would be $t_d = M/\dot{m} = (100 \text{ kg})/(10 \text{ kg/s}) = 10 \text{ s}$. The equilibrium response time Δt_e must be short relative to t_d. If we pick

$$\Delta t_e = 0.1 \text{ s} \ll t_d = 10 \text{ s}$$

and employ the acoustic propagation speed is $\sqrt{kg_0 R_g T} = 347 \text{ m/s}$ (1138 ft/s). Eq. (2.3) gives a characteristic length for equilibrium response of $L \approx C \Delta t_e = (347 \text{ m/s})(0.1 \text{ s}) = 34.7 \text{ m}$ (114 ft). Thus, a volume increment of $L^3 = (34.7)^3 \text{ m}^3 = 41,780 \text{ m}^3$ ($1.47 \times 10^6 \text{ ft}^3$) would permit sufficient equilibrium response to use equilibrium state equations in the vessel. The 1.0-m^3 (35.28 ft^3) vessel does not have to be subdivided, and its unsteady pressure will be uniform at all times during discharge.

If the vessel were a long thin tube, it might have to be subdivided.

2.4 Stability

Stability is used to describe how the state of a system responds to a particular disturbance. There are four classifications of stability: *stable*, *unstable*, *neutrally stable*, and *metastable* states. Figure 2.2 gives several examples to help identify the stability of a system.

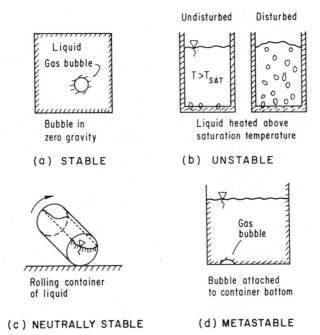

Figure 2.2 Types of Stability

Stable State A disturbance acts on a system, causing forces which restore properties to their initial state when the disturbance is removed. Consider the submerged gas bubble at zero gravity without buoyant rise in Fig. 2.2(*a*). Both the gas and liquid have equal pressures, and surface tension is negligible. If the bubble was squeezed slightly and released, its volume would oscillate because the corresponding pressure disturbance always would act in a direction to restore its undisturbed state.

Unstable State A small disturbance starts a time-dependent change which continues, even with removal of the disturbance, until another stable state is reached. The liquid in Fig. 2.2(*b*) is heated above its saturation temperature by the hot surface. It does not boil because the gas voids at cracks, pits, and surface irregularities are so small that surface tension overcomes the vapor pressure and prevents bubble growth. A slight increase in bubble size or the size of nucleation sites can overcome surface tension and result in rapid boiling.

Neutrally Stable State A disturbance changes the state of a system as long as it is applied. When the disturbance is removed, the system neither returns to its original state nor undergoes unstable behavior. The container of liquid in Fig. 2.2(*c*) can be rolled on the horizontal surface, but will not return to its initial position when the rolling force is removed, nor will it continue motion at an increasing speed.

Metastable State A system state is stable for limited disturbances, but a larger disturbance causes a substantial change to a more stable state. Figure 2.2(*d*) shows a gas bubble attached to the bottom of a liquid container. A sufficiently large disturbance would cause the bubble to break away and rise by buoyancy to the more stable state at the top.

Stability considerations enter unsteady thermofluids in many applications, which include the behavior of fluid jets, vortex shedding, liquid-gas flow patterns, boiling, and the classical instabilities of Mathieu, Taylor, and Helmholtz.

2.5 Transport Properties

Transport properties help in determining the rate at which certain physical quantities are transported from one place to another by diffusive mechanisms. Energy transport occurs as heat transfer if a temperature gradient exists. Velocity gradients in a fluid cause the transport of shear stress and vorticity. A density gradient is associated with mass transport by diffusion.

Basic transport properties include dynamic viscosity μ, thermal conductivity κ, and the mass diffusion coefficient $\mathcal{D}$. Other properties classified as transport properties include the kinematic viscosity $\nu = g_0\mu/\rho$, the thermal diffusivity $\alpha = \kappa/\rho c_p$, and the Prandtl number $\text{Pr} = \mu c_p/\kappa$. Section A.2 in Appendix A lists some representative values of transport properties.

Transport properties correspond to conditions when gradients of temperature, velocity, or mass concentration have reached steady values. *Thermal conductivity* is defined from Fourier's law as

$$\kappa = \frac{-q''}{\partial T/\partial n} \tag{2.4}$$

where q'' is the heat flux and $\partial T/\partial n$ is the temperature gradient in the direction of heat flow. *Dynamic viscosity* for laminar parallel flow is defined by

$$\mu = \frac{\tau}{\partial u/\partial n} \tag{2.5}$$

where τ is the fluid shear stress and $\partial u/\partial n$ is the velocity gradient normal to the flow direction. The *mass diffusion coefficient* is defined by Fick's law in the form

$$\mathcal{D} = -\frac{\dot{m}_i/A}{\partial C_i/\partial n} = -\frac{G_i}{\partial C_i/\partial n} \tag{2.6}$$

where G_i is the mass flux of component i and $\partial C_i/\partial n$ is its concentration gradient. The heat conduction, shear stress, and mass transport properties correspond to steady-state experiments. Transport properties do not depend on the size of a region in a quasi-equilibrium state.

The transport velocity V_p is required in Eq. (1.2) to estimate whether a bulk or propagative analysis is required when diffusive mechanisms dominate an unsteady flow. An estimate of V_p for heat conduction can be obtained by considering the case of a semi-infinite slab at temperature T_∞, whose surface suddenly is raised to T_0. The unsteady temperature profile is given by [2.5]

$$\frac{T - T_\infty}{T_0 - T_\infty} = \frac{2}{\sqrt{\pi}} \int_0^{x/2\sqrt{\alpha t}} \exp(-\eta^2) \, d\eta$$

for which the approximate temperature front penetration distance is

$$x \approx 2\sqrt{\alpha t}$$

Therefore, the *heat conduction penetration velocity* is dx/dt, or

$$V_{p,\text{ heat conduction}} = V_{p,\kappa} = \frac{2\alpha}{x} \tag{2.7}$$

Similar considerations yield the *laminar* and *turbulent shear penetration velocities* normal to the direction of flow:

$$V_{p,\text{ laminar shear}} = V_{p,L} = \frac{2\nu}{x} \tag{2.8}$$

$$V_{p,\text{ turbulent shear}} = V_{p,t} = (0.238)\frac{\nu}{y}\,\text{Re}^{3/4}, \qquad \text{Re} = \frac{U_\infty y}{\nu} \tag{2.9}$$

where U_∞ is the fluid velocity in x and V_p is the shear penetration (or propagation) velocity in the y direction. Equation (2.9) is obtained from the turbulent boundary layer thickness on a flat plate, given by the classical expression $y = (0.38)(U_\infty x/\nu)^{-1/5}x$. Differentiation with respect to t gives dy/dt, which is $V_{p,t}$ and dx/dt, which is U_∞ when we follow a fluid particle.

Friction factors or convection coefficients are available from steady flow correlations. However, applicability of these correlations to unsteady flows depends on how long it takes for the temperature or flow field to develop, relative to the disturbance time.

EXAMPLE 2.2: ACCELERATING FLOW FRICTION FACTOR

A uniform smooth pipe of length $L_p = 40\,\text{m}$ (131 ft) and diameter $D = 0.04\,\text{m}$ (1.57 in) contains stationary water of density $\rho = 1000\,\text{kg/m}^3$ (62.4 lbm/ft^3) and kinematic viscosity $\nu = 1.0 \times 10^{-6}\,\text{m}^2/\text{s}$ (10.8 $\times$ 10^{-6} ft^2/s). A pressure increase of $\Delta P = 1.0\,\text{MPa}$ (147 psia) is suddenly applied at one end, and fluid acceleration begins. Determine if a friction factor f, based on steady flow, can be employed in an analysis of the fluid motion.

The fluid acceleration response time (approximate time for the fluid to reach a new steady flow after a disturbance) should be much longer than the shear response time Δt_ν if a quasi-steady f is justified.

Fluid acceleration response can be estimated by writing a momentum formulation for fluid in the pipe. If a time-averaged friction factor $\bar{f}$ is employed, we obtain the problem

$$\text{DE:} \qquad \frac{dV}{dt} = \frac{g_0}{\rho}\frac{\Delta P}{L_p} - \frac{\bar{f}}{2D}V^2$$

$$\text{IC:} \qquad V(0) = 0$$

for which a solution is given by

$$V(t) = V_{ss}\tanh\left(\frac{1}{2}\frac{t}{\tau}\right)$$

where the steady-state velocity is

$$V_{ss} = \sqrt{\frac{2g_0\,\Delta P D}{\rho \bar{f} L_p}}$$

and

$$\tau = \frac{1}{2}\sqrt{\frac{\rho L_p 2D}{g_0\,\Delta P \bar{f}}}$$

The steady-state Reynolds number is employed with V_{ss} and the smooth pipe friction factor curve of Fig. B.1 of Appendix B. A trial-and-error solution yields for steady state

$$\mathrm{Re} = 5 \times 10^5$$

$$\bar{f} = 0.013$$

$$V_{ss} = 12.4 \, \mathrm{m/s} \quad (40.7 \, \mathrm{ft/s})$$

The flow acceleration response time is

$$\tau = \Delta t_a = 0.25 \, \mathrm{s}$$

Since the steady flow is turbulent, Eq. (2.9) with an approximate distance $y \cong D/2$ is used to estimate the propagation velocity V_p for turbulent shear, and Eq. (2.3) is then employed to obtain the diffusive response time

$$\Delta t_\nu = 0.085 \, \mathrm{s}$$

A steady friction factor will develop in about 0.08 s, before which fluid toward the pipe center will accelerate faster than fluid closer to the wall. A steady $\bar{f}$ reduces the calculated unsteady flow rate below that which would be expected. Although a quasi-steady wall friction factor $\bar{f}$ is not justified for the first 0.08 s, it should give a reasonable estimate of the flow acceleration to steady state.

———————

2.6 State Equations, Simple Compressible Substance (SCS)

Thermodynamic state properties regularily appear in unsteady thermofluid analyses. It is usually necessary to employ one or more state equations to obtain a full solution to problems. The state equations may be algebraic, tabular, graphical, or they may involve machine retrievable properties.

Several common substances used in this book are the *perfect gas*, the *ideal liquid*, and the *two-phase saturated liquid-vapor mixture*. The state properties of real gases and superheated vapors often can be approximated by the perfect gas state equations. The behavior of real liquids can be approximated with ideal incompressible liquid state equations, except when compression is important. Selected thermodynamic properties and state equations are summarized in Tables 2.1 through 2.5. Extensive properties are denoted by capital symbols, intensive properties by lowercase symbols.

2.7 Entropy

Entropy S is a state property which is used in this book to quantify the available energy degradation of a process, the most likely state of a system,

TABLE 2.1 Selected Property Definitions

Name	Definition
Specific heat at constant volume	$c_v = \left(\dfrac{\partial e}{\partial T}\right)_v$
Specific heat at constant pressure	$c_P = \left(\dfrac{\partial h}{\partial T}\right)_P$
Coefficient of volume expansion	$\beta = \dfrac{1}{v}\left(\dfrac{\partial v}{\partial T}\right)_P = -\dfrac{1}{\rho}\left(\dfrac{\partial \rho}{\partial T}\right)_P$
Sound speed	$C = \sqrt{g_0\left(\dfrac{\partial P}{\partial \rho}\right)_s}$
Isothermal compressibility	$K_T = -\dfrac{1}{v}\left(\dfrac{\partial v}{\partial P}\right)_T = \dfrac{1}{\rho}\left(\dfrac{\partial \rho}{\partial P}\right)_T$
Adiabatic or isentropic compressibility	$K_s = -\dfrac{1}{v}\left(\dfrac{\partial v}{\partial P}\right)_s = \dfrac{1}{\rho}\left(\dfrac{\partial \rho}{\partial P}\right)_s$
Modulus of elasticity	$E_m = \rho\left(\dfrac{\partial P}{\partial \rho}\right)_s$
Enthalpy	$H = U + P\mathcal{V}$
Specific enthalpy	$h = e + Pv$

TABLE 2.2 Perfect Gas State Equations

$$P\mathcal{V} = MR_g T \qquad k = \frac{c_p}{c_v}$$

$$P = \rho R_g T \qquad c_p - c_v = R_g$$

$$e = \frac{1}{k-1}\frac{P}{\rho} \qquad h = \frac{k}{k-1}\frac{P}{\rho}$$

$$U = Me = Mc_v T = \frac{1}{k-1}P\mathcal{V}$$

$$H = U + P\mathcal{V} = Mh = Mc_p T = \frac{k}{k-1}P\mathcal{V}$$

$$dS = M\left(c_v\frac{dT}{T} + R\frac{d\mathcal{V}}{\mathcal{V}}\right) = M\,ds$$

$$dS = M\left(c_p\frac{dT}{T} - R_g\frac{dP}{P}\right) = M\,ds$$

Reversible Adiabatic Process:

$$\frac{P}{P_0} = \left(\frac{v_0}{v}\right)^k = \left(\frac{\rho}{\rho_0}\right)^k = \left(\frac{T}{T_0}\right)^{k/(k-1)}$$

TABLE 2.3   Ideal Liquid State Equations

$$v = \frac{1}{\rho} = \text{constant} \qquad ds = c_v \frac{dT}{T} \qquad e = e(T)$$

$$e = c_v T \qquad h = c_p T = e + pv = c_v T + Pv$$

$$dh = c_v \, dT + v \, dP \qquad c_p = c_v$$

TABLE 2.4   Saturated Liquid-Vapor Mixture State Equations[a]

$$\phi_g = \phi_g(P) \quad \text{or} \quad \phi_g(T)$$

$$\phi_f = \phi_f(P) \quad \text{or} \quad \phi_f(T)$$

$$P = P(T)$$

$$M = M_g + M_f \qquad x = \frac{M_g}{M} \qquad 1 - x = \frac{M_f}{M}$$

$$\Phi = \Phi_g + \Phi_f = M_g \phi_g + M_f \phi_f$$

$$\phi = \frac{\Phi}{M} = \phi_g x + \phi_f(1 - x) = \phi_f + x\phi_{fg}$$

$$\phi_{fg} = \phi_g - \phi_f$$

Eliminate x from equations for ϕ_1 and ϕ_2:

$$\phi_1 = \phi_{1f} + \frac{\phi_{1fg}}{\phi_{2fg}} (\phi_2 - \phi_{2f})$$

[a]General thermodynamic property $\Phi = \mathcal{V}, U, H, S$, etc.

TABLE 2.5   Noncondensible Gas-Liquid Mixture State Equations

Same as Table 2.4 except that x remains constant.

and the natural direction of a state change. The entropy differential ds of a uniform, equilibrium system is defined, with reference to Fig. 2.3, as

$$dS = \left(\frac{dQ}{T}\right)_R, \qquad \begin{cases} dQ > 0 & \text{heat inflow} \\ dQ < 0 & \text{heat outflow} \end{cases} \tag{2.10}$$

where T is the absolute temperature. The incremental heat dQ is transferred reversibly. Reversible heat transfer is a limiting case which requires a spatially uniform system temperature without gradients during heat transfer. The isolated system of Fig. 2.4 contains regions 1 and 2 at uniform

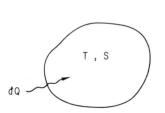

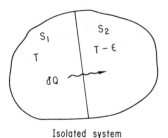

Isolated system

Figure 2.3 Diagram for Definition of Entropy

Figure 2.4 Principle of Entropy Increase

temperatures T and $T - \epsilon$. Total entropy is $S = S_1 + S_2$, from which

$$dS = dS_1 + dS_2 = dQ\left(-\frac{1}{T} + \frac{1}{T - \epsilon}\right)$$

$$= \frac{dQ}{T}\left(\frac{\epsilon}{T} + \cdots\right) > 0 \qquad (2.11)$$

for small ϵ. Equation (2.11) shows that heat transfer in a region of nonuniform temperature causes an increase of entropy. The entropy change is zero if either the temperature is uniform , or if the heat transfer is zero.

The specific entropy is

$$s = \frac{S}{M} \qquad (2.12)$$

and the convected entropy flow rate with a mass flow increment is $s\,\delta\dot{m}$.

The Gibbs equation is useful in determining the entropy change of a system in terms of other properties. The first law expression of Eq. (1.78) for a fixed mass of an SCS relates differential amounts of heat inflow, compressive work outflow, and internal energy increase by

$$dQ_{\text{in}} = dU + P\,d\mathcal{V} \qquad (2.13)$$

If the heat transfer is reversible, Eqs. (2.10) and (2.13) give

$$T\,dS = dU + P\,d\mathcal{V} \qquad (2.14)$$

which is the classical Gibbs equation for an SCS. Another form of Gibbs equation results if the static enthalpy of Table 2.1 is employed, giving

$$T\,dS = dH - \mathcal{V}\,dP \qquad (2.15)$$

Equations (2.14) and (2.15) can be divided by the mass M and written in terms of intensive properties as

$$T \, ds = de + P \, dv = de - \frac{P}{\rho^2} \, d\rho \qquad (2.16)$$

and

$$T \, ds = dh - v \, dP = dh - \frac{1}{\rho} \, dP \qquad (2.17)$$

Thus, the entropy change can be expressed in terms of changes in other properties.

2.8 The Second Law

Equation (2.11) gives one expression of the second law of thermodynamics, which may be stated as

The entropy of an isolated system can only increase or remain constant.

This statement is not especially useful for directly solving problems. A more useful form is

$$\dot{I} \geq 0 \qquad (2.18)$$

where $\dot{I}$ is the irreversibility rate, defined with the help of Fig. 2.5 as

$$\dot{I} = \sum_{\mathrm{cs}} (s \, \delta \dot{m})_{\mathrm{out}} - \sum_{\mathrm{cs}} (s \, \delta \dot{m})_{\mathrm{in}}$$
$$+ \frac{d}{dt} \sum_{\mathrm{cv}} s \, \delta M - \left(\sum_{\mathrm{cs}} \frac{\delta q_{\mathrm{in}}}{T_{\mathrm{in}}} - \sum_{\mathrm{cs}} \frac{\delta q_{\mathrm{out}}}{T_{\mathrm{out}}} \right) \qquad (2.19)$$

The first three summations give the actual rate of creation of entropy in the system. The last two terms in parentheses are the rate of creation of entropy which would occur if the associated processes were reversible. Equation (2.19) applies for any number of entropy flow rates $s \, \delta \dot{m}$ and heat transfer

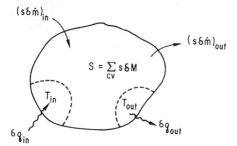

Figure 2.5 Diagram for Irreversibility Rate

rates. An integral formulation is obtained from Eq. (2.19) in the limit, yielding

$$\dot{I} = \int_{cs} s\, d\dot{m} + \int_{cs} \frac{dq}{T} + \frac{d}{dt} \int_{cv} s\, dM$$
$$= \int_{cs} s\rho\,^{cs}\mathbf{V}\cdot d\mathbf{A} + \int_{cs} \frac{q''\cdot d\mathbf{A}}{T} + \frac{d}{dt} \int_{cv} s\rho\, d\mathcal{V} \qquad (2.20)$$

The basic principle expressed by Eq. (2.18) provides a powerful tool in determining the most stable or preferred state of a system, the expected solution when the mathematics of a process give multiple values, and the direction of a process.

The irreversibility rate is related to energy degradation, that is, the loss of available energy which might somehow be extracted as useful work. If T_∞ is the lowest temperature in a system or its immediate surroundings, the rate of available energy degradation can be expressed as the power loss

$$\mathcal{P}_{loss} = T_\infty \dot{I} > 0 \qquad (2.21)$$

Equation (2.21) is useful in determining how the performance of thermofluid systems and processes can be improved.

2.9   Dissipation Without Friction

Consider the oscillation of the spherical gas bubble in liquid, shown in Fig. 2.6. Gravity is neglected. A momentum formulation for the surrounding liquid, later developed in Chapter 9, yields the Rayleigh equation

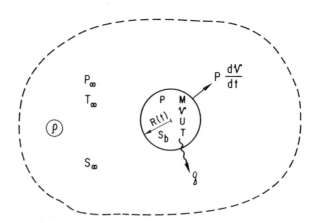

Figure 2.6   Gas Bubble Oscillating in Liquid

$$R\ddot{R} + \frac{3}{2}\dot{R}^2 = \frac{g_0}{\rho}(P - P_\infty) \qquad (2.22)$$

where ρ is the liquid density and P_∞ is the undisturbed, distant pressure. The equilibrium static condition occurs when bubble pressure and temperature correspond to the surroundings; that is, $P = P_\infty$ and $T = T_\infty$ with $R = R_\infty =$ constant. If R is disturbed a slight amount, the gas volume, pressure, and temperature also change. If $P \neq P_\infty$ and $T \neq T_\infty$, R will change with time, based on Eq. (2.22), and heat transfer will occur. If heat transfer q is determined by convection at the bubble wall, we have

$$q = \mathcal{H}A(T - T_\infty), \qquad \mathcal{H} \cong \text{constant} \qquad (2.23)$$

and the energy conservation principle for the bubble gives

$$P\dot{V} + q + \frac{dU}{dt} = 0 \qquad (2.24)$$

If we employ $U = P\mathcal{V}/(k-1)$ and $P\mathcal{V} = MR_g T$ for a perfect gas, with $\mathcal{V} = (4/3)\pi R^3$ and $A = 4\pi R^2$ for a spherical bubble, Eq. (2.24) becomes

$$P\dot{R} + \frac{1}{3k}R\dot{P} + \frac{4\pi\mathcal{H}(k-1)}{3MR_g k}(PR^3 - P_\infty R_\infty^3) = 0 \qquad (2.25)$$

Let the steady radius R_∞ be disturbed initially such that $R(0) = R_\infty + \epsilon$ with the bubble initially at T_∞. The resulting initial pressure disturbance is obtained from Eq. (2.25) with $\dot{P} = \dot{R} = 0$ in the form

$$P(0) = P_\infty\left[\frac{R_\infty}{R(0)}\right]^3 \qquad (2.26)$$

If we linearize Eqs. (2.22) and (2.25) (see Section C.2 of Appendix C) by putting $R(t) = R_\infty + \epsilon R_1(t)$ and $P(t) = P_\infty + \epsilon P_1(t)$, the pressure can be eliminated to obtain the full problem

DE: $\ddot{R}_1 + a\ddot{R}_1 + b\dot{R}_1 + cR_1 = 0$ $\qquad (2.27)$

IC: $t = 0$, $R_1 = 1$, $\dot{R}_1 = 0$, $\ddot{R}_1 = -\dfrac{3g_0 P_\infty}{\rho R_\infty^2}$ $\qquad (2.28)$

where

$$a = \frac{4\pi(k-1)R_\infty^2 \mathscr{H}}{3MR_g} \qquad b = \frac{3g_0 kP_\infty}{\rho R_\infty^2}$$

$$c = \frac{12\pi \mathscr{H} g_0 (k-1)P_\infty}{3MR_g \rho} = a\left(\frac{3P_\infty g_0}{\rho R_\infty^2}\right) \qquad (2.29)$$

An *adiabatic* bubble with $q = 0$ corresponds to $\mathscr{H} = 0$, with the undamped oscillating solution

$$R_1(t) = \frac{1}{k}\cos\left(\sqrt{\frac{3g_0 kP_\infty}{\rho R_\infty^2}}\,t\right) + \frac{k-1}{k} \qquad (2.30)$$

which is plotted in Fig. 2.7. An *isothermal* bubble with $T = T_\infty$ implies instant heat transfer, corresponding to $\mathscr{H} \to \infty$, which yields another undamped oscillatory solution

$$R_1(t) = \cos\left(\sqrt{\frac{3g_0 P_\infty}{\rho R_\infty^2}}\,t\right) \qquad (2.31)$$

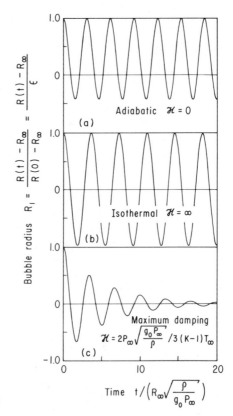

Figure 2.7 Gas Bubble Oscillation

Equation (2.31) also is plotted in Fig. 2.7. The adiabatic and isothermal cases have undamped oscillations with different frequencies. However, any other case with a finite value of $\mathcal{H}$ undergoes a damped oscillation. An example is shown in Fig. 2.7 for

$$\mathcal{H} = \frac{2}{3(k-1)T_\infty} \sqrt{\frac{g_0 P_\infty^3}{\rho}} \tag{2.32}$$

which appears to give maximum damping. It may be surprising to note that this damped bubble oscillation closely resembles damping of a mechanical system by friction. However, there is not frictional dissipation in the formulation of this problem! Neither is energy lost from the bubble-liquid system! The damped oscillation for $0 < \mathcal{H} < \infty$ displays an available power loss, which is a measure of the irreversibility rate.

The amount of damping per cycle of bubble oscillation can be estimated from the available power loss. Consider the isolated system within the dotted boundary of Fig. 2.6, for which $\Sigma_{cv}\, s\, \delta M = S$. Equation (2.19) yields

$$\dot{I} = \frac{dS}{dt}$$

The total entropy is written as the sum of the bubble and liquid entropies, which leads to the irreversibility rate

$$\dot{I} = \frac{dS}{dt} = \frac{dS_b}{dt} + \frac{dS_\infty}{dt}$$

If temperatures of the bubble and liquid systems are uniform, Eq. (2.10) with $q = \dot{d}Q/dt$ yields $dS_b/dt \geq -q/T$ and $dS_\infty/dt \geq q/T_\infty$. It follows from Eq. (2.23) that

$$\dot{I} = \frac{\mathcal{H}A(T - T_\infty)^2}{T_\infty T}$$

The available power loss is based on the lowest temperature such that

$$\mathcal{P}_{\text{loss}} = T_\infty \dot{I} = \frac{\mathcal{H}A T_\infty (T/T_\infty - 1)^2}{T/T_\infty} \tag{2.33}$$

The perfect gas law yields the temperature ratio $T/T_\infty = (P/P_\infty)(R/R_\infty)^3$. Equations (2.26) and (2.22) are used to express T/T_∞ in terms of R_1, for which Eq. (2.33) becomes

$$\mathcal{P}_{\text{loss}} = \mathcal{H}A_\infty T_\infty \epsilon^2 \left(\frac{P_1}{P_\infty} + 3\frac{R_1}{R_\infty} \right)^2 \tag{2.34}$$

The amount of energy initially put into the bubble/water system is $dE = (P - P_\infty)\, d\mathcal{V}$, which corresponds to the work done in changing R_∞ to $R_\infty + \epsilon\, dR_1$. This energy is available to return an equal amount of work to its surroundings. If we put $P - P_\infty = \epsilon P_1$ and $d\mathcal{V} = 4\pi R_\infty^2 \epsilon\, dR_1$, the rate of available energy change dE/dt is diminished by an amount equal to $\mathcal{P}_{\text{loss}}$, giving

$$P_1 4\pi R_\infty^2 \frac{dR_1}{dt} = \mathcal{H} A_\infty T_\infty \left(\frac{P_1}{P_\infty} + 3\, \frac{R_1}{R_\infty} \right)^2 \tag{2.35}$$

Since $p\mathcal{V}^n = P_\infty \mathcal{V}_\infty^n$, where $n = 1$ for isothermal, $n = k$ for adiabatic, and $1 < n < k$ for $\mathcal{H} \neq 0$, it can be shown from $\mathcal{V} = (4/3)\pi R^3$, $A_\infty = 4\pi R_\infty^2$, and the linearization equation $P = P_\infty + \epsilon P_1$ that $P_1/P_\infty = -3nR_1/R_\infty$. Moreover, based on Eqs. (2.30) and (2.31), the circular frequency of oscillation is $\omega = \sqrt{3g_0 n P_\infty / \rho R_\infty^2}$ so that the period of one oscillation is $2\pi/\omega = \tau$. If we put $dt \approx \tau$ and use the value of $\mathcal{H}$ in Eq. (2.32) for maximum damping, Eq. (2.35) can be put into the form

$$\frac{dR_1}{R_1} = -\frac{4\pi(n-1)^2}{n(k-1)\sqrt{3n}} \tag{2.36}$$

Equation (2.36) shows a logarithmic decay of oscillation amplitude. An approximate value of n which matches the damped air bubble oscillation of Fig. 2.7 is 1.19.

2.10   Most Stable Dynamic Properties From Second Law

Sometimes it is desirable to determine the state toward which a system changes for a given set of inflows and outflows. Consider the differential CV of Fig. 2.8 where a fluid stream flows in one direction. Mechanical power $P(\partial \mathcal{V}/\partial t)$ allows expansion of the stream boundary. Heat transfer rates also cross the boundary, and the internal temperature T is uniform. Energy conservation and the second law, Eqs. (2.18) and (2.19), are written as

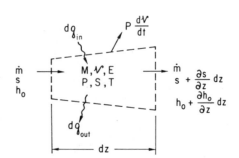

Figure 2.8 Differential Control Volume in Flowing Stream

$$dq_{out} - dq_{in} + \frac{\partial}{\partial z}(\dot{m}h_0)\, dz + P \frac{\partial \mathcal{V}}{\partial t} + \frac{\partial E}{\partial t} = 0 \qquad (2.37)$$

and

$$\frac{\partial(\dot{m}s)}{\partial z}\, dz + \frac{\partial S}{\partial t} \geq \frac{dq_{in} - dq_{out}}{T} \qquad (2.38)$$

If we employ Eqs. (1.60), (1.69), and (2.16), the heat transfer terms can be eliminated from (2.37) and (2.38) to give

$$\frac{\partial}{\partial z}\left[\dot{m}\left(\frac{V^2}{2g_0} + \frac{g}{g_0}\, y \right) \right] dz + \dot{m}v \frac{\partial P}{\partial z}\, dz + \frac{\partial}{\partial t}\left[M\left(\frac{V^2}{2g_0} + \frac{g}{g_0}\, y \right) \right] \leq 0 \qquad (2.39)$$

Although Eq. (2.39) is for one fluid stream, similar equations can be written for other adjacent streams, and summed. If the system is not flowing, the last term of Eq. (2.39) shows that the kinetic and potential energy sum tends toward a minimum. If steady flow conditions occur at any location z, the first two terms are like the Bernoulli head, which approaches a minimum in the flow direction. The last term of Eq. (2.39) shows that if the system is one of fixed mass undergoing slow motion, its potential energy approaches a minimum. One example of this principle is the observation that the lighter of two fluids rises to the top in a mixture.

EXAMPLE 2.3: STABLE LIQUID FLOW IN A CHANNEL
Suppose that a known volume rate of liquid is flowing in a frictionless horizontal flow passage of width w Fig. 2.9 but does not completely fill the cross section. It is necessary to estimate the water depth and velocity.

Equation (2.39) shows that for the steady flow condition at constant surface pressure, the flow rate of kinetic and potential energy approaches a minimum value. One-dimensional flow is assumed at velocity V with the area centroid at elevation $\bar{y}(Y)$, where Y is the surface elevation. The constraint $\dot{V} = A(Y)V = $ constant is employed, and a minimization of $\dot{m}(V^2/2g_0 + g\bar{y}/g_0)$ yields for a rectangular channel

$$Y = \left(\frac{2\dot{V}^2}{gw^2} \right)^{1/3}, \qquad V = \left(\frac{\dot{V}g}{2w} \right)^{1/3} \qquad (2.40)$$

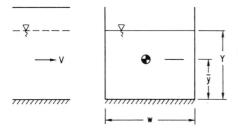

Figure 2.9 Stable Liquid flow in a Channel

This velocity is close to $(\dot{\mathcal{V}}g/w)^{1/3}$, which would prevent a hydrostatic wave from propagating upstream (see Chapter 5).

2.11 Speed of Small Pressure Disturbances

The differential compression dP shown in Fig. 2.10 propagates into undisturbed fluid which is traveling at velocity V. Differential property changes occur across the disturbance, which travels in a direction normal to its plane. The disturbance speed relative to earth, V_D, is to be determined.

A thin CS of area A is drawn around a section of the disturbance. Storage terms are negligible in the thin CV so that the mass, momentum, and energy equations are written as

$$\dot{m} = \rho A(V + V_D) = (\rho + d\rho)A(V_D + V + dV) \qquad (2.41)$$

$$\dot{m}(V + dV) - \dot{m}V = g_0[PA - (P + dP)A] \qquad (2.42)$$

and

$$\dot{m}(h_0 + dh_0) - \dot{m}h_0 + PAV_D - (P + dP)AV_D = 0 \qquad (2.43)$$

The stagnation enthalpy $h_0 = h + V^2/2g_0$ can be employed, and $\dot{m}$ and dV can be eliminated from Eqs. (2.42) and (2.43) to obtain

$$dh - \frac{1}{\rho}\, dP = 0 \qquad (2.44)$$

Comparison with Eq. (2.17) shows that property changes across the differential disturbance occur at constant entropy. Furthermore, if dV is

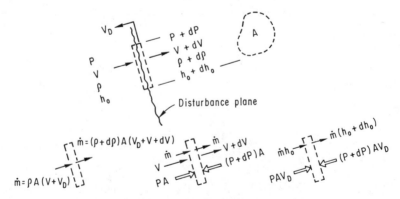

Figure 2.10 Moving Pressure Disturbance

eliminated from Eqs. (2.41) and (2.42), the disturbance propagation speed C relative to the fluid can be expressed in the forms

$$C = V_D + V = \sqrt{g_0 \frac{dP_s}{d\rho_s}} = \sqrt{g_0 \left(\frac{\partial P}{\partial \rho}\right)_s} = v\sqrt{-g_0 \left(\frac{\partial P}{\partial v}\right)_s} \qquad (2.45)$$

Note that the pressure disturbance speed C, called the sound, sonic, or acoustic speed, is a thermodynamic property! Compressible substances can be subjected to a given pressure change ΔP, for which the recorded density change $\Delta \rho$ permits a determination of sound speed from Eq. (2.45), without requiring time and distance measurements! The sound speed C generally is the same for compressive or decompressive disturbances in single-phase substances, although it may be different for two-phase substances like a bubbly mixture of saturated liquid and vapor.

The state equations of Tables 2.1 through 2.5 are used with Eq. (2.45) to obtain expressions for the sound speed in several substances where phase changes are absent. These are summarized below.

SOUND SPEED IN VARIOUS SUBSTANCES

PERFECT GAS

$$C = \sqrt{kg_0 R_g T} = \sqrt{kg_0 \frac{P}{\rho}} \qquad (2.46)$$

IDEAL INCOMPRESSIBLE LIQUID

$$C \rightarrow \infty \qquad (2.47)$$

ORDINARY LIQUID

$$C = \sqrt{\frac{g_0 E_m}{\rho}}$$

$$C_{\text{water}} \approx 1420 \text{ m/s } (4660 \text{ ft/s}) \qquad (2.48)$$

BUBBLY NONCONDENSIBLE GAS-LIQUID MIXTURE OR COMPRESSIVE DISTURBANCE IN SATURATED LIQUID-VAPOR MIXTURE (FIG. 2.13)

$$\frac{C}{C_g} = \frac{(v/v_g)\sqrt{1 - v_L/v_g}}{\sqrt{(v/v_g)[1 - (v_L C_g/v_g C_L)^2] - [v_L/v_g - (v_L C_g/v_g C_L)^2]}} \qquad (2.49)$$

MINIMUM VALUE FROM EQ. (2.49)

$$\frac{C_{\min}}{C_g} \approx 2\sqrt{\frac{v_L}{v_g}} \quad \text{at} \quad \frac{v}{v_g} \approx 2\frac{v_L}{v_g} \qquad (2.50)$$

2.12 Sound Speed, Bubbly, Saturated Liquid-Vapor Mixtures

The state of each phase lies on the saturation dome of Fig. 2.11. A rapid compression drives saturated vapor bubbles momentarily into the super-heated region as shown. The saturated liquid is driven simultaneously into the subcooled region. Since negligible heat transfer will occur between the vapor and liquid during passage of a compressive disturbance, both sub-stances are compressed approximately on isentropic paths. The compressed vapor temperature will exceed that of the compressed liquid, and heat transfer will occur over a period of time to restore mixture temperature equilibrium. However, either the liquid or vapor state may become satu-rated again before the temperature equalize, and vaporization or condensa-tion would occur to ultimately restore saturated equilibrium. All of this takes time, and a compressive disturbance moves as if the vapor was slightly superheated and the liquid was slightly subcooled. Figure 2.12(*a*) gives the compressive sound speeds of saturated steam and water, obtained from Eq. (2.45) where derivatives were taken on the superheated and subcooled sides of the saturation dome. The results of Fig. 2.12(*a*) and Eq. (2.49) were used to obtain the compressive sound speed of a bubbly steam-water mixture, shown in Fig. 2.12(*b*), which gives a reasonable prediction of the data in Fig. 2.13.

A bubbly mixture of liquid and vapor in saturated equilibrium will respond differently to a decompressive disturbance than it will to a compres-sive disturbance. A decompression will cause both saturated liquid and vapor to drop inside the saturation dome. A small amount of liquid will condense from the vapor, whereas a much larger amount of vapor will form simultaneously from the liquid. Vapor formation from the liquid will domi-nate the phase change process, occurring in about 1 ms. This usually is fast enough to assume that a bubbly liquid-vapor mixture is close to saturated

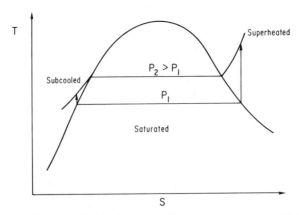

Figure 2.11 Rapid Compression of Saturated Liquid and Vapor

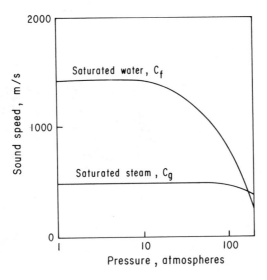

Figure 2.12(a)   Steam and Water Sound Speed, Compressive Disturbances

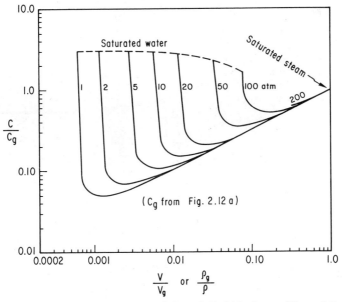

Figure 2.12(b)   Compressive Sound Speed, Bubbly Steam-Water Mixture

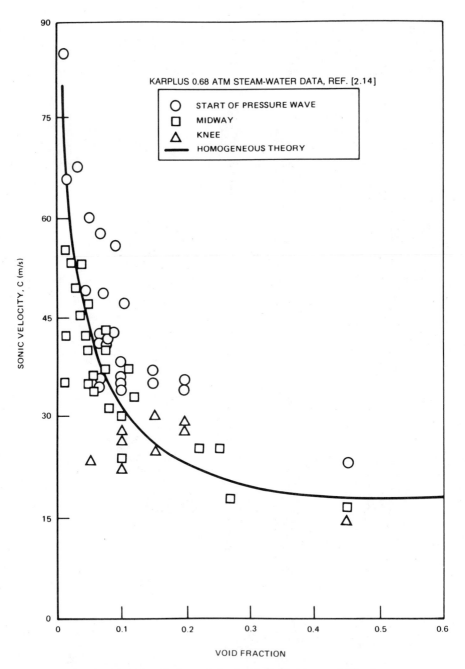

Figure 2.13 Prediction of Sonic Velocity Data of Karplus.

equilibrium during decompression, provided that disturbance times are $1/100$ s or more. The associated sound speed is obtained from Table 2.4 and Eq. (2.45) with $\phi_1 = v$ and $\phi_2 = s$ to obtain

Decompressive Disturbance, Saturated Liquid-Vapor

$$C = v\sqrt{-g_0\left[\frac{dv_f}{dP} - \frac{d(s_f v_{fg}/s_{fg})}{dP} + \frac{d(v_{fg}/s_{fg})}{dP}s\right]^{-1}} \qquad (2.51)$$

Results for saturated steam and water are graphed in Fig. 2.14. Figure 2.15 gives the decompressive sound speed for bubbly steam-water mixtures in saturated equilibrium.

EXAMPLE 2.4: ROUND-TRIP PERIOD OF PRESSURE WAVE
A uniform pipe of length $L = 20$ m (65.6 ft) contains saturated water at 50 atm (735 psia) pressure. One end is attached to a 50-atm reservoir, and a valve at the other end is closed. Sudden partial opening of the valve results in a small decompression which travels through the pipe to the reservoir. The reservoir returns a small compressive disturbance, again pressurizing the pipe to about 50 atm. How much time is required for the disturbance to travel to the vessel and return to the valve?

The decompressive and compressive sound speeds in 50 atm saturated water are respectively 33 m/s (108 ft/s) and 1104 m/s (3621 ft/s) based on Figs. 2.12 through 2.15. Therefore, the initial decompression requires $L/C = (20\text{ m})/(33\text{ m/s}) = 0.61$ s

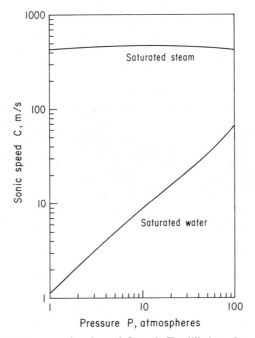

Figure 2.14 Decompressive Sound Speed, Equilibrium Steam and Water

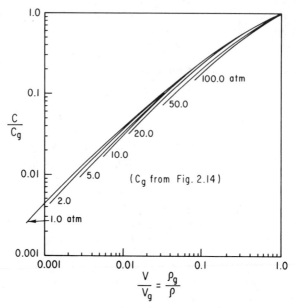

Figure 2.15 Decompressive Sound Speed, Bubbly, Saturated Equilibrium, Steam-Water Mixture

to travel to the reservoir. Although the fluid is in motion after the first wave transit, the return compression requires about $L/C = (20 \text{ m})/(1104 \text{ m/s}) = 0.018 \text{ s}$ to travel back to the valve. The approximate round-trip time, therefore, is 0.628 s.

Increased fluid velocity at each wave transit alters the wave speed relative to the pipe. Both the compressive and decompressive sound speeds should be added algebraically to the fluid velocity.

2.13 Critical Flow

If the fluid in Fig. 2.10 is stationary with $V = 0$, Eq. (2.45) shows that the disturbance propagates at speed $V_D = C$. If the disturbance is stationary, the fluid is traveling at velocity $V = C$, a condition referred to as *critical flow*, at which a small compressive or any size decompressive disturbance cannot travel upstream. Figure 2.16 helps describe critical flow in a nozzle, sometimes called choked flow. An ideal, frictionless, adiabatic converging nozzle connects a reservoir at pressure P_0 to a receiver at pressure P. When $P = P_0$, no flow occurs. If receiver pressure P is slowly reduced, the nozzle flow rate $\dot{m}$ increases and the pressure profile begins to dip as shown. Finally the receiver pressure reaches a value P_c called the critical flow pressure, at which the nozzle pressure profile no longer changes with further reduction of P. When P_c is reached, fluid velocity V in the nozzle discharge plane is

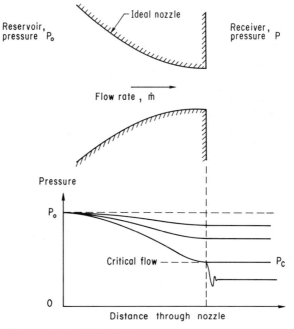

Figure 2.16 Critical Flow Rate in an Ideal Nozzle

equal to the local sound speed C_c, and furthre decompressive disturbances in the receiver cannot propagate into the nozzle. The corresponding critical discharge density is designated by ρ_c, and the critical mass flow rate is

$$\dot{m}_c = \rho_c A C_c$$

Since the sound speed C is a thermodynamic property, it can be expressed functionally as $C = C(P, \rho)$. When critical flow occurs, we have $C_c = C(P_c, \rho_c)$. Therefore, the critical mass flux also is a thermodynamic property which can be written as

$$G_c = \frac{\dot{m}_c}{A} = \rho_c C_c = \rho_c C(P_c, \rho_c) = f(P_c, \rho_c) \tag{2.52}$$

Stagnation properties usually are known in the reservoir. Therefore, a convenient formulation for G_c is obtained by expressing P_c and ρ_c in terms of stagnation conditions. Energy conservation for the stationary, adiabatic nozzle of Fig. 2.16 yields, for negligible storage and potential energy terms,

$$h_0 = h + \frac{V^2}{2g_0}, \qquad dh_0 = 0 = dh + \frac{V \, dV}{g_0}$$

Since the nozzle is adiabatic and frictionless, flow is isentropic, and the Gibbs equation becomes

$$T \, ds = 0 = dh - \frac{1}{\rho} \, dP$$

Elimination of dh gives

$$dh_0 = 0 = \frac{1}{\rho} \, dP + d\left(\frac{V^2}{2g_0}\right)$$

If pressure and velocity are integrated between reservoir conditions $P = P_0$ and $V = 0$ to the critical flow conditions $P = P_c$ and $V = C_c$, the result is

$$C_c^2 = C^2(P_c, \rho_c) = 2g_0 \int_{P_c}^{P_0} \frac{dP}{\rho} \qquad (2.53)$$

The integral is readily obtained since ρ varies only with pressure for isentropic flow; that is,

$$\rho = \rho(P, s), \qquad s = s_0 = \text{constant} \qquad (2.54)$$

Equations (2.52) to (2.54) make it possible to express G_c in terms of stagnation properties P_0 and ρ_0. However, critical flow from a converging nozzle can exist only if the receiver pressure is at or below the critical pressure P_c.

Another method used to obtain the critical flow rate employs the mathematical procedure of maximizing the mass flux with respect to nozzle throat pressure. Velocity $V = G/\rho$ is employed in the steady energy equation $h_0 = h + V^2/2g_0$ where properties h and V are in the exit plane. If ρ and h are expressed functionally as $\rho(P, s)$ and $h(P, s)$, it follows that

$$G = \rho\sqrt{2g_0(h_0 - h)} = \rho(P, s)\sqrt{2g_0[h_0 - h(P, s)]} \qquad (2.55)$$

The assumption of isentropic nozzle flow implies that ρ, h, and G are functions of pressure P only. Therefore, a maximum value of G corresponds to the condition

$$\frac{dG}{dP} = \frac{-g_0(\partial h/\partial P)_s}{\sqrt{2g_0(h_0 - h)}} + \left(\frac{\partial \rho}{\partial P}\right)_s \sqrt{2g_0(h_0 - h)} = 0 \qquad (2.56)$$

The Gibbs equation for constant entropy yields $dh - dP/\rho = 0$, and Eq. (2.45) is substituted into Eq. (2.56) to express pressure at the maximum flow condition, P_{max}, in the form

$$C_{max} = C(P_{max}, s) = \sqrt{2g_0[h_0 - h(P_{max}, s)]}$$

If we integrate $dh - dP/\rho = 0$ from h_0 and P_0 to h and P_{max}, we get

$$C^2_{max} = C^2(P_{max}, s) = 2g_0 \int_{P_{max}}^{P_0} \frac{dP}{\rho} \tag{2.57}$$

Comparison with Eq. (2.53) shows that pressure at maximum flow P_{max} and the critical pressure P_c are the same. Therefore, a maximization of mass flux G from the energy equation yields the same critical flow state as that which corresponds to flow at the local sound speed.

The state equations in Tables 2.1 through 2.5 and acoustic speeds of Sections 2.11 and 2.12 were used to obtain isentropic flow properties for the substances listed below. Recall that steady critical flow can occur only if the critical pressure can be achieved by discharge into a receiver at sufficiently low pressure. Moreover, all pressures are absolute and, therefore, cannot be negative.

CRITICAL FLOW PROPERTIES OF SEVERAL SUBSTANCES

PERFECT GAS

Noncritical from Eq. (2.55) (*Fig.* 2.17)

$$\frac{G}{\sqrt{kg_0 P_0 \rho_0}} = \sqrt{\frac{2}{k-1}\left[\left(\frac{P}{P_0}\right)^{2/k} - \left(\frac{P}{P_0}\right)^{(k+1)/k}\right]} \tag{2.58}$$

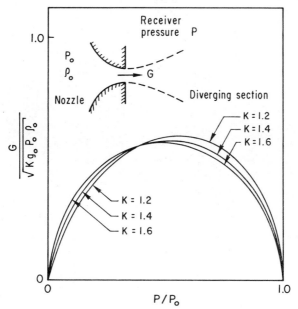

Figure 2.17   Perfect Gas Mass Flux in Ideal Nozzle

Critical Pressure from Eq. (2.57)

$$\frac{P_c}{P_0} = \left(\frac{2}{k+1}\right)^{k/(k-1)}$$

(2.59)

Critical Mass Flux

$$\frac{G_c}{\sqrt{kg_0 P_0 \rho_0}} = \left(\frac{2}{k+1}\right)^{(k+1)/2(k-1)}$$

(2.60)

ORDINARY LIQUID (FIG. 2.18):

Noncritical

$$\frac{G}{\sqrt{2g_0 P_0 \rho}} = \sqrt{1 - \frac{P}{P_0}}$$

(2.61)

Critical Pressure. Unlikely to Achieve in Practice

$$\frac{P_c}{P_0} = 1 - \frac{E_m}{2P_0} \geq 0$$

(2.62)

Critical Mass Flux. Unlikely to Achieve in Practice

$$\frac{G_c}{\sqrt{2g_0 P_0 \rho}} = \sqrt{\frac{E_m}{2P_0}}$$

(2.63)

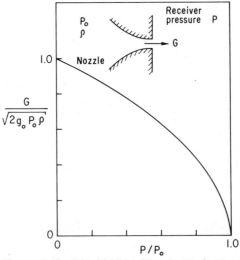

Figure 2.18 Liquid Mass Flux in Ideal Nozzle

INCOMPRESSIBLE LIQUID (FIG. 2.18):

Noncritical. Critical Flow Impossible

$$\frac{G}{\sqrt{2g_0 P_0 \rho}} = \sqrt{1 - \frac{P}{P_0}} \tag{2.64}$$

BUBBLY NONCONDENSIBLE GAS-LIQUID MIXTURE (FIG. 2.19):

Noncritical

$$\frac{G^2 x(k-1)v_{g0}}{2g_0 k P_0} = \frac{1 - (P/P_0)^{(k-1)/k} + [(k-1)/k](1 - P/P_0)b_0}{(P/P_0)^{-1/k} + b_0} \tag{2.65}$$

$$b_0 = \frac{1-x}{x}\frac{v_L}{v_{g0}} \tag{2.66}$$

Critical Pressure

$$b_0^2 + \frac{2}{k}\left[(k+1)\left(\frac{P_c}{P_0}\right)^{-1/k} - \left(\frac{P_c}{P_0}\right)^{-(k+1)/k}\right]b_0$$
$$+ \frac{1}{k-1}\left[(k+1)\left(\frac{P_c}{P_0}\right)^{-2/k} - 2\left(\frac{P_c}{P_0}\right)^{-(k+1)/k}\right] = 0 \tag{2.67}$$

BUBBLY SATURATED LIQUID-VAPOR MIXTURE [FIGS. 2.20(A) AND (B)]:

Equation (2.55) is used directly. See Figs. 2.20(a) and 2.20(b) for G_c and P_c.

Figure 2.17 gives the perfect gas G for several values of k and the full range of P/P_0. Maximum values of G correspond to Eq. (2.58) at the critical pressure P_c of Eq. (2.59). Values of G for $P > P_c$ are subsonic and for $P < P_c$ are supersonic, which occur in steady flow cases only if a diverging section (shown by the dashed line) follows the nozzle throat.

Figure 2.18 gives G in terms of P and P_0 for liquid, which increases with decreasing P/P_0. Liquid critical flow could not occur unless the vessel pressure P_0 was at least half the value of the liquid bulk modulus of elasticity, E_m.

Figure 2.19 gives G for a bubbly noncondensible gas-liquid mixture over the full subsonic range of P/P_0 and values of the parameter b_0. Results correspond to $k = 1.3$, but vary less than 10 percent for $k = 1.4$.

Figures 2.20(a) and 2.20(b) give G_c and P_c for bubbly, homogeneous saturated steam-water mixtures. Equation (2.55) was used with computer-retrievable steam properties to express $G(P_0, h_0; P)$. Stepwise reduction of P from the value P_0 always resulted in a computed maximum value of G.

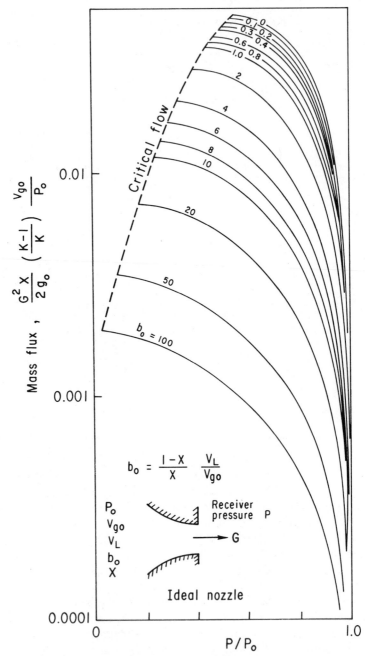

Figure 2.19   Mass Flux, Bubbly Incompressible Liquid–Perfect Gas Mixture in Ideal Nozzle, $k = 1.3$

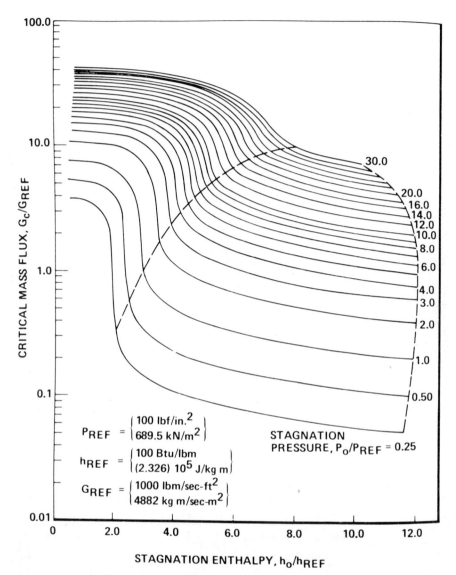

Figure 2.20(a) Homogeneous, Bubbly, Equilibrium, Steam-Water Critical Mass Flux

Figures 2.20(a) and 2.20(b) include the subcooled region where G, at lower values of h_0, approaches the value obtained from Eq. (2.64) for an incompressible liquid. Figure 2.21 shows a comparison of calculated and measured critical flow rates of saturated water.

Maximum values of G for liquid-vapor mixtures are more difficult to determine accurately when h_0 is farther into the subcooled region, because the maximizing condition $dG/dP = 0$ occurs on a steep spike, as indicated in

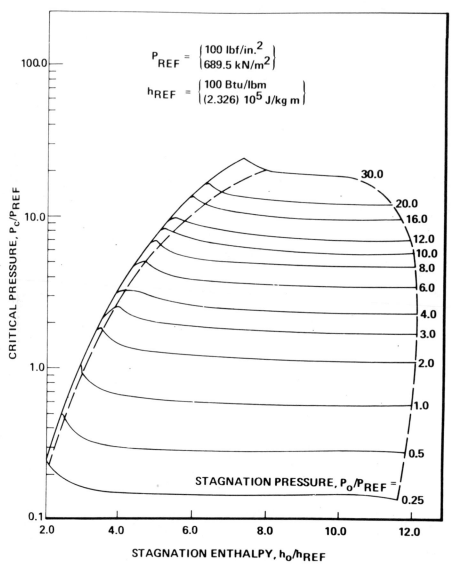

Figure 2.20(*b*)   Homogeneous, Bubbly, Equilibrium, Steam-water Pressure at Critical Flow

Fig. 2.22. However, the pressure P_c at G_{max} is easily distinguished. Since quality x approaches zero at P_c for large subcooling, it was easier to first obtain P_c from $dG/dP = 0$, and then to determine G_{max} from Eq. (2.64) with $P = P_c$. Approximately 1 ms [2.15, 2.16] is required for equilibrium flashing to occur following a decompression, and Bernoulli's equation shows that nonflashing water travels at 150 m/s from a 100-atm stagnation pressure. Therefore, a nozzle should be at least

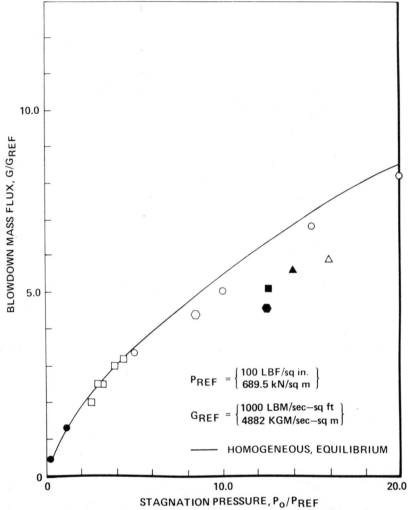

Figure 2.21 Homogeneous Blowdown Rate of Saturated Water [2.9]

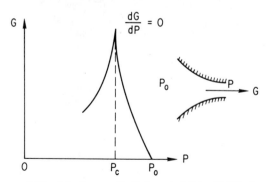

Figure 2.22 Maximum G, Subcooled Liquid Discharge

$$(150 \text{ m/s})(0.001 \text{ s}) = 0.15 \text{ m} \quad (6 \text{ in})$$

or 15 cm long for this equilibrium analysis to apply at 100-atm.

EXAMPLE 2.5: THRUST FROM STEAM-WATER CRITICAL FLOW
A vessel at pressure $P_0 = 7$ MPa (1030 psia) contains saturated water and saturated steam as shown in Fig. 2.23 with $h_{fo} = 1.267 \times 10^6$ J/kg (545 B/lbm) and $h_{go} = 2.772 \times 10^6$ J/kg (1194 B/lbm). Discharge occurs through an ideal nozzle of flow area A into surroundings at pressure $P_\infty = 0.1$ MPa (14.7 psia). The nozzle length exceeds 15 cm, so homogeneous equilibrium discharge occurs. Does saturated water or saturated steam discharge give higher thrust per unit of nozzle area? Many engineers would vote for saturated water because of its higher density and higher critical flow rate.

The momentum principle yields

$$\frac{T}{A} = P_n - P_\infty + \frac{GV_n}{g_0} = P_n - P_\infty + \frac{G^2 v_n}{g_0}$$

where P_n is pressure in the nozzle discharge area and v_n is the discharge specific volume. Figure 2.20(b) gives critical flow pressures of $P_n = P_{fc} = 5.93$ MPa (860 psia) for saturated water and $P_n = P_{gc} = 4.1$ MPa (600 psia) for saturated steam, which are greater than P_∞. Therefore, the discharge rate is critical and Fig. 2.20(a) gives $G_{fc} = 26,900$ kg/s-m^2 (5400 lbm/s-ft^2) and $G_{gc} = 10,270$ kg/s-m^2 (2100 lbm/s-ft^2) for the two cases. Since the nozzle is ideal, isentropic flow yields the specific volume

$$v_n = v_c = v_{fc} + \frac{v_{fgc}}{s_{fgc}}(s_0 - s_{fc})$$

Saturated water discharge corresponds to

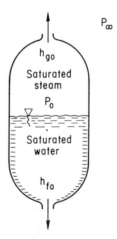

P_∞

h_{go}

Saturated
steam

P_0

Saturated
water

h_{fo}

Figure 2.23   Steam and Water Discharges

$$s_0 = s_f(P_0) = 3.12 \text{ kJ/kg-K} \ (0.7430 \text{ B/lbm-°F}), \qquad P_{fc} = 5.93 \text{ MPa} \ (860 \text{ psia}),$$

$$v_{fc} = 0.0013 \text{ m}^3/\text{kg} \ (0.0211 \text{ ft}^3/\text{lbm}), \qquad v_{fgc} = 0.0317 \text{ m}^3/\text{kg} \ (0.505 \text{ ft}^3/\text{lbm})$$

$$s_{fc} = 3.02 \text{ kJ/kg-K} \ (0.721 \text{ B/lbm-°F}), \qquad s_{fgc} = 2.88 \text{ kJ/kg-K} \ (0.6862 \text{ B/lbm-°F})$$

$$\text{and } v_{fn} = 0.0024 \text{ m}^3/\text{kg} \ (0.037 \text{ ft}^3\text{lbm})$$

for which

$$\left(\frac{T}{A}\right)_f = 7.57 \text{ MPa} \ (1078 \text{ psi})$$

is the saturated water thrust for

$$P_0 = 7 \text{ MPa} \ .$$

Saturated steam discharge corresponds to

$$s_0 = s_g(P_0) = 5.813 \text{ kJ/kg-K} \ (1.3897 \text{ B/lbm-°F}), \qquad P_n = P_{gc} = 4.1 \text{ MPa} \ (600 \text{ psia})$$

$$v_{fc} = 0.00126 \text{ m}^3/\text{kg} \ (0.0201 \text{ ft}^3/\text{lbm}), \qquad v_{fgc} = 0.0473 \text{ m}^3/\text{kg} \ (0.7497 \text{ ft}^3/\text{lbm})$$

$$s_{fc} = 2.81 \text{ kJ/kg-K} \ (0.6720 \text{ B/lbm-°F}), \qquad s_{fgc} = 3.25 \text{ kJ/kg-K} \ (0.7734 \text{ B/lbm-°F})$$

$$\text{and } v_{gn} = 0.045 \text{ m}^3/\text{kg} \ (0.715 \text{ ft}^3/\text{lbm})$$

for which

$$\left(\frac{T}{A}\right)_g = 8.74 \text{ MPa} \ (1272 \text{ psi})$$

is the saturated steam thrust for

$$P_0 = 7 \text{ MPa}$$

It may surprise some that saturated steam gives a higher thrust than saturated water during critical flow discharge!

2.14 Simultaneous Charging and Discharging of a Vessel

Sometimes a fluid region is affected by both mass inflow and outflow, heat transfer, and simultaneous volume expansion. Figure 2.24 portrays such a system. The stored kinetic and potential energy components are negligible. The mass inflow rate has stagnation enthalpy $h_{0,\text{in}}$, determined by the source from which it comes. The mass outflow rate has stagnation enthalpy $h_{0,\text{out}}$, determined by fluid contained near the point of discharge. A homogeneous system will have the same stagnation enthalpy throughout its CV, whereas an equilibrium two-phase system may be discharging liquid, vapor, or mixture at different values of $h_{0,out}$. The compressive power term $P(d\mathcal{V}/dt)$ is required when the system volume changes. Also, if we were analyzing either the liquid or vapor region in an equilibrium system, a volume change would cause compressive energy transfer between regions even though the total system volume remained constant.

Mass and energy conservation equations for the system of Fig. 2.24 are written as

$$\dot{m}_{out} - \dot{m}_{in} + \frac{dM}{dt} = 0 \qquad (2.68a)$$

and

$$\dot{m}_{out}h_{0,out} - \dot{m}_{in}h_{0,in} + q_{out} - q_{in} + P\frac{d\mathcal{V}}{dt} + \frac{dU}{dt} = 0 \qquad (2.68b)$$

The state equations $\mathcal{V} = Mv$ and $U = Me(P, v)$ are employed, and the differential of e is written as

$$de = \left(\frac{\partial e}{\partial P}\right)_v dP + \left(\frac{\partial e}{\partial v}\right)_P dv$$

If we combine Eqs. (2.68a) and (2.68b), we obtain the system pressure rate

$$\frac{dP}{dt}$$

$$= \frac{\dot{m}_{in}(h_{0,in} - f(P)) - \dot{m}_{out}(h_{0,out} - f(P)) + q_{in} - q_{out} - P\dfrac{d\mathcal{V}}{dt} - \left(\dfrac{\partial e}{\partial v}\right)_P \dfrac{d\mathcal{V}}{dt}}{MF\left(P, \dfrac{\mathcal{V}}{M}\right)}$$

$$(2.69)$$

where

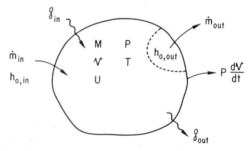

Figure 2.24 Simultaneous Charging and Discharging

$$f(P) = e - v\left(\frac{\partial e}{\partial v}\right)_p \tag{2.70}$$

and

$$F\left(P, \frac{\mathcal{V}}{M}\right) = \left(\frac{\partial e}{\partial P}\right)_v \tag{2.71}$$

The term $F(P, \mathcal{V}/M)$ is positive for all simple compressible substances. Therefore, the sign of the numerator in Eq. (2.69) determines if the pressure rate is positive or negative.

2.15 Pressure Vessel Blowdown

Blowdown is a term used to designate fluid discharge from a pressure vessel into a receiver at lower pressure. One example of blowdown is open-valve gas discharge from tanks, which occurs in many applications including welding, air-operated impact hammering, propane burning, and boiler safety valve operation. Another example of blowdown involves uncontrolled fluid discharge from a pressure vessel through a ruptured pipe. One of the most important blowdown problems is the prediction of time-dependent properties in a pressure vessel during fluid discharge. Most cases of concern involve vessels at sufficiently high pressure to cause critical flow through the discharge opening.

Figure 2.25 shows a rigid vessel of volume $\mathcal{V}$ which is discharging fluid at a mass rate $\dot{m}$ through an ideal nozzle of constant throat area A. The vessel contains uniform fluid of mass M and internal energy U with corresponding pressure P and temperature T at any instant. Heat transfer q may occur between the vessel wall and contained fluid. If we set $\dot{m}_{\text{in}} = 0$, $U = Me$, $h_0 = e + P/\rho$, $\rho = M/\mathcal{V}$, and eliminate $\dot{m}_{\text{out}}$ from Eqs. (2.68a) and (2.68b), we obtain

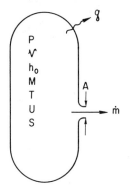

Figure 2.25 Pressure Vessel Blowdown

$$\frac{de}{dt} - \frac{P}{\rho^2}\frac{d\rho}{dt} = -\frac{q_{out} - q_{in}}{\rho \mathcal{V}} \qquad (2.72)$$

Comparison with Eq. (2.16) shows that for an adiabatic vessel wall with $q = 0$, the fluid state during discharge is isentropic. This result is general for any uniform homogeneous fluid.

The discharge mass flow rate at any time during blowdown is taken as the critical flow value such that

$$\dot{m} = G_c A \qquad (2.73)$$

Vessel pressure, temperature, and the mass discharge rate may be required as functions of time for given values of throat area, vessel volume, and initial conditions. Since heat transfer between the fluid and vessel wall depends on many other parameters such as the wall area, thickness, density, thermal conductivity, and heat capacity, it is informative to analyze the bounding cases of adiabatic and isothermal fluid decompression.

Consider the case of *adiabatic blowdown* from a vessel containing perfect gas. Property changes are isentropic, so

$$\frac{M}{\mathcal{V}} = \rho = \rho_i\left(\frac{P}{P_i}\right)^{1/k}, \qquad \text{reversible adiabatic}$$

where subscript i designates the initial condition. Differentiation and substitution of $dM/dt = -\dot{m}$ leads to

$$\left(\frac{P}{P_i}\right)^{(1-k)/k}\frac{d(P/P_i)}{dt} = -\frac{k}{\rho_i}\frac{\dot{m}}{\mathcal{V}} = -\frac{k}{\rho_i}\frac{G_c A}{\mathcal{V}}$$

Equation (2.60) is used to express the critical mass flux of a perfect gas, for which the full problem becomes

DE: $$\left(\frac{P}{P_i}\right)^{(1-3k)/2k}\frac{d(P/P_i)}{dt} = \frac{kA}{\rho_i \mathcal{V}}\sqrt{kg_0 P_i \rho_i}\left(\frac{2}{k+1}\right)^{(k+1)/2(k-1)}$$

IC: $$t = 0, \qquad \frac{P}{P_i} = 1$$

A solution for P/P_i is given by

$$\frac{P}{P_i} = \left[1 + \left(\frac{k-1}{2}\right)\left(\frac{2}{k+1}\right)^{(k+1)/2(k-1)}\sqrt{\frac{kg_0 P_i}{\rho_i}}\frac{At}{\mathcal{V}}\right]^{-2k/(k-1)} \qquad (2.74)$$

Equation (2.74) is plotted in Fig. 2.26 for $k = 1.4$. Note that $\sqrt{kg_0 P_i/\rho_i}$ is the sonic speed in the gas at initial conditions. Also shown for convenience is the temperature

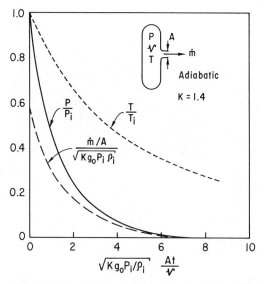

Figure 2.26 Adiabatic Blowdown of Gas Vessel

$$\frac{T}{T_i} = \left(\frac{P}{P_i}\right)^{(k-1)/k}$$

$$(2.75)$$

and the discharging mass flux

$$\frac{\dot{m}/A}{\sqrt{kg_0 P_i \rho_i}} = \left(\frac{2}{k+1}\right)^{(k+1)/2(k-1)}\left(\frac{P}{P_i}\right)^{(k+1)/2k}$$

$$(2.76)$$

Isothermal blowdown of gas corresponds to a vessel gas density

$$\frac{M}{\mathcal{V}} = \rho = \rho_i\left(\frac{P}{P_i}\right), \qquad \text{isothermal}$$

$$(2.77)$$

for which a pressure solution is given by

$$\frac{P}{P_i} = \exp\left[-\left(\frac{2}{k+1}\right)^{(k+1)/2(k-1)}\sqrt{\frac{kg_0 P_i}{\rho_i}}\frac{At}{\mathcal{V}}\right]$$

$$(2.78)$$

Both vessel pressure and the mass discharge rate for the isothermal case are graphed in Fig. 2.27. Cases with heat transfer from vessel walls would lie between the results of Figs. 2.26 and 2.27.

Next, consider the blowdown of a vessel containing a saturated liquid-vapor mixture. An appropriate state equation is obtained from Table 2.4 with $\phi_1 = e = U/M$ and $\phi_2 = v = \mathcal{V}/M$, giving

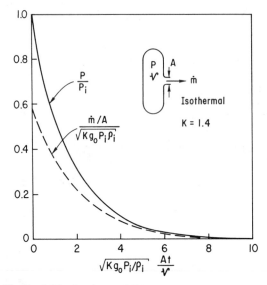

Figure 2.27 Isothermal Blowdown of Gas Vessel

$$\frac{U}{M} = e = e_f(P) + \frac{e_{fg}(P)}{v_{fg}(P)} \left[\frac{\mathcal{V}}{M} - v_f(P) \right] \tag{2.79}$$

If we differentiate Eq. (2.79) with respect to time, and make use of the notation

$$\frac{d\psi(P)}{dt} = \frac{d\psi}{dP}\frac{dP}{dt} = \psi' \frac{dP}{dt}, \qquad \psi = e_f, e_{fg}, v_f, v_{fg}, \ldots \tag{2.80}$$

we may express the pressure rate from Eq. (2.69) in the form

$$\frac{dP}{dt} = \frac{AG_c(P, h_0)}{MF(P, \mathcal{V}/M)} [f(P) - h_0] \tag{2.81}$$

where Eqs. (2.70) and (2.71) become

$$f(P) = e_f - v_f \frac{e_{fg}}{v_{fg}} \tag{2.82}$$

and

$$F\left(P, \frac{\mathcal{V}}{M}\right) = e_f' - \left(v_f \frac{e_{fg}}{v_{fg}}\right)' + \frac{\mathcal{V}}{M}\left(\frac{e_{fg}}{v_{fg}}\right)' \tag{2.83}$$

These functions are plotted in Figs. 2.28 and 2.29. Equation (2.81) was solved numerically to predict pressure decay in a vessel initially filled with

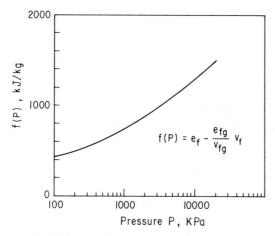

Figure 2.28   Function $f(P)$, Saturated Steam-Water Mixtures

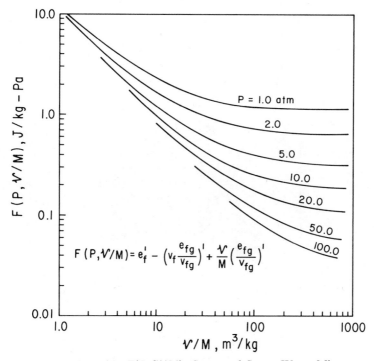

Figure 2.29   Function $F(P, \mathcal{V}/M)$, Saturated Steam-Water Mixtures

saturated water at 68 atm. Figure 2.30 gives the calculated results for pressure and remaining mass. The initial water mass M_i completely filled the vessel, and results for homogeneous critical discharge of saturated water and steam through a nozzle of area A are shown. It is seen that steam blowdown causes faster pressure decay than water blowdown. When the water is expelled and only saturated steam remains, the decompression rate increases. The dimensional time scale At/M_i (m^2-s/kg) in Fig. 2.30 indicates that for a larger blowdown area, the time required to reach various pressures is reduced. Also, if M_i is increased, vessel pressure reduction takes longer. Figure 2.31 shows data from a typical saturated steam-water blowdown, which agrees well with predictive methods presented in this section.

EXAMPLE 2.6: LIQUID-VAPOR BLOWDOWN PRESSURE RATES
An adiabatic pressure vessel contains regions of saturated steam and water at 7 MPa (1030 psia) as in Fig. 2.23, Example 2.5. If fluid is discharging at the critical flow rate through an ideal nozzle of area A, determine the ratio of pressure rates for cases of saturated steam and saturated water discharge. Also determine if it is possible to

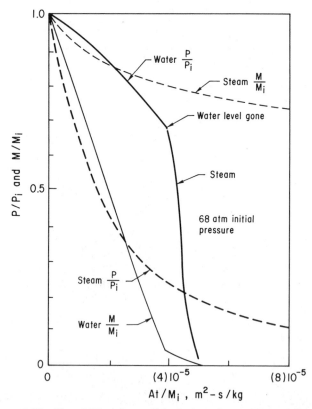

Figure 2.30   Vessel Blowdown Calculation, Steam-Water Mixtures

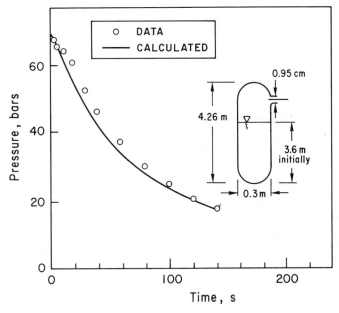

Figure 2.31   Vessel Pressure during Steam Blowdown. [2.17]

discharge a saturated mixture at some stagnation enthalpy from this system and have zero pressure rate.

The pressure rates for steam and water discharge are obtained from Eq. (2.81). Nozzle area A, the contained mass M, and the initial value of $F(P, \mathcal{V}/M)$ are the same for both saturated steam and saturated water discharge. Therefore, the ratio of pressure rates can be written as

$$\frac{(dP/dt)_g}{(dP/dt)_f} = \frac{\dot{P}_g}{\dot{P}_f} = \frac{G_c(P, h_{g0})}{G_c(P, h_{f0})} \frac{f(P) - h_{g0}}{f(P) - h_{f0}}$$

It follows from Fig. 2.20(a) or Example 2.5 that the critical mass fluxes for saturated steam and saturated water at 7 MPa are

$$G_c(P, h_{g0}) = 10,270 \text{ kg/s-m}^2 \ (2100 \text{ lbm/s-ft}^2), \text{ steam}$$

$$G_c(P, h_{f0}) = 26,900 \text{ kg/s} - \text{m}^2 \ (5400 \text{ lbm/s-ft}^2), \text{ water}$$

The terms h_{g0} and h_{f0} could be obtained from steam tables or read from the abscissa of Fig. 2.20(a), giving

$$h_{g0} = 2.772 \times 10^6 \text{ J/kg} \ (1194 \text{ B/lbm})$$

$$h_{f0} = 1.267 \times 10^6 \text{ J/kg} \ (545 \text{ B/lbm})$$

Also, Fig. 2.28 gives

$$f(P) = 1.186 \times 10^6 \text{ J/kg} \quad (511 \text{ B/lbm})$$

from which it follows that

$$\frac{\dot{P}_g}{\dot{P}_f} = 7.47$$

Critical discharge of saturated steam gives about seven times the decompression rate of saturated water from a system at 7 MPa (1030 psia).

Equation (2.81) shows that the vessel pressure rate would be zero during mass discharge if the term $f(P) - h_0$ was zero. The function $f(P)$ is obtained from Eq. (2.82), and h_0 can be expressed in terms of vessel properties as

$$h_0 = h_f + \frac{h_{fg}}{v_{fg}} (v - v_f)$$

Also, since

$$h_f = e_f + Pv_f , \qquad h_g = e_g + Pv_g$$

and

$$h_{fg} = h_g - h_f = e_{fg} + Pv_{fg}$$

it follows that

$$f(P) - h_0 = -\frac{v}{v_{fg}} (e_{fg} + Pv_{fg}) = -v\,\frac{h_{fg}}{v_{fg}}$$

The term vh_{fg}/v_{fg} can never be zero. Therefore, mass cannot be discharged from a saturated adiabatic system without causing decompression.

_____

2.16 Vessel Charging

The process of fluid flow into a receiving vessel is called *vessel charging*. Various design applications require a prediction of the vessel pressure and temperature rates during charging in order to avoid design limits. If we set $\dot{m}_{out} = 0$ in Eqs. (2.68a) and (2.68b) and substitute into Eq. (2.16) with $U = Me$ and $\mathcal{V} = Mv$, the entropy rate of change becomes

$$\frac{ds}{dt} = \frac{1}{MT} [\dot{m}(h_0 - h) - (q_{out} - q_{in})] \qquad (2.84)$$

Equation (2.84) shows that when a vessel is being charged, state changes are not likely to be isentropic as in the case of blowdown from an adiabatic vessel containing a homogeneous mixture. It would be necessary for the bracketed term in Eq. (2.84) to be zero if state changes were isentropic.

Consider the case of charging an adiabatic vessel with perfect gas. Equations (2.68a) and (2.68b) are combined with $q_{in} = q_{out} = 0$, $\dot{m}_{out} = 0$, and $U = p\mathcal{V}/(k-1)$ to give

$$\frac{dP}{dM} = \frac{k-1}{\mathcal{V}} h_{0\infty}$$

Integration from initial conditions $M = M_i$ and $P = P_i$ yields

$$P - P_i = \frac{M_i}{\mathcal{V}}(k-1)h_{0\infty}\left(\frac{M}{M_i} - 1\right) \tag{2.85}$$

It is seen that pressure increases linearly with the contained mass. Vessel temperature is obtained by replacing pressure and enthalpy with the state equations of Table 2.2, giving

$$\frac{T}{T_\infty} = \frac{M_i}{M}\frac{T_i}{T_\infty} + k\left(1 - \frac{M_i}{M}\right) \tag{2.86}$$

When M becomes large relative to M_i,

$$\frac{T}{T_\infty} \to k, \qquad \frac{M}{M_i} \to \infty \tag{2.87}$$

That is, temperature of the contained gas reaches a limit, which is k times the ambient value

EXAMPLE 2.7: GAS CHARGING INTO A VACUUM
An air sample is to be collected from a 20°C (68°F) radioactive environment. It was proposed that a long cylinder should be evacuated. Then a technician should thrust the valved end of the cylinder through a port into the gas environment. While holding the cylinder with one hand, he should operate a mechanism to open the valve with the other. When he hears the woosh of air filling the cylinder, he should close the valve and withdraw the cylinder. Would the technician feel like playing handball that same evening?

This case corresponds to a vessel with an initially contained mass $M_i = 0$, from which Eq. (2.86) gives

$$T = kT_i = 1.4(293 \text{ K}) = 410 \text{ K} = 137°C \text{ (279°F)}$$

This temperature could blister the technician's hands. It is seen than in an adiabatic, evacuated vessel, the temperature immediately rises to kT_i, regardless of how much mass is added. Although actual heat transfer to vessel walls would result in lower temperature, it would be wise to wear insulated gloves before handling a container which is to be charged with gas.

2.17   Charging and Discharging Perfect Gas

The internal energy e of a perfect gas is $Pv/(k-1)$ for which Eq. (2.70) gives $f = 0$, and Eq. (2.69) becomes

$$\frac{dP}{dt} = \frac{\dot{m}_{in}h_{0,in} - \dot{m}_{out}h_{0,out} + q_{in} - q_{out} - P(d\mathcal{V}/dt)}{\mathcal{V}/(k-1)} \tag{2.88}$$

Consider an adiabatic, constant-volume container where the initial pressure and density are P_i and ρ_i. If the charging mass and energy flow rates are constant and discharge occurs through an ideal nozzle of area A_0 to a receiver at P_∞ according to Eq. (2.58), then Eqs. (2.88) and (2.68a) can be written in the forms

$$\frac{dP^*}{dt^*} = k\left[h_{0,in}^* - \frac{1}{G_{in}^*} P^{*3/2}\rho^{*-1/2} \frac{G(P^*)}{\sqrt{kg_0 P_i \rho_i}} \right] \tag{2.89}$$

and

$$\frac{d\rho^*}{dt^*} = 1 - \frac{1}{G_{in}^*} \sqrt{P^* \rho^*} \frac{G(P^*)}{\sqrt{kg_0 P_i \rho_i}} \tag{2.90}$$

where the nondimensional quantities are defined by

$$P^* = \frac{P}{P_i}, \qquad \rho^* = \frac{\rho}{\rho_i}, \qquad t^* = \frac{\dot{m}_{in}}{\rho_i \mathcal{V}} t$$

$$h_{0,in}^* = h_{0,in}\left(\frac{k}{k-1} \frac{P_i}{\rho_i} \right)^{-1}, \qquad G_{in}^* = \frac{\dot{m}_{in}/A_0}{\sqrt{kg_0 P_i \rho_i}} \tag{2.91}$$

If the initial and ambient pressures P_i and P_∞ are equal, we have

$$\frac{G(P^*)}{\sqrt{kg_0 P_i \rho_i}} = \begin{cases} \sqrt{\dfrac{2}{k-1}} (P^{*-2/k} - P^{*-(k+1)/k}), & P^* < \left(\dfrac{k+1}{2}\right)^{k/(k-1)} \\[4mm] \left(\dfrac{2}{k+1}\right)^{(k+1)/2(k-1)}, & P^* \geq \left(\dfrac{k+1}{2}\right)^{k/(k-1)} \end{cases} \tag{2.92}$$

and the nondimensional initial conditions are

$$t^* = 0, \qquad P^* = \rho^* = 1 \tag{2.93}$$

The initial pressure rate with $P^* = 1$ is obtained from Eqs. (2.89) and (2.92) as

$$\left(\frac{dP^*}{dt^*} \right)_{initial} = kh_{0,in}^* \tag{2.94}$$

The maximum or asymptotic values of pressure and density can be obtained by setting both dP^*/dt^* and $d\rho^*/dt^*$ equal to zero, for which Eqs. (2.89) and (2.90) give

$$\sqrt{P^*\rho^*} = G_{\text{in}}^* \frac{\sqrt{kg_0 P_i \rho_i}}{G(P^*)} , \qquad \text{Asymptotic} \qquad (2.95)$$

and

$$\frac{P^*}{\rho^*} = \frac{P}{P_i} \frac{\rho_i}{\rho} = h_{0,\text{in}}\left(\frac{k}{k-1} \frac{P_i}{\rho_i}\right)^{-1} = h_{0,\text{in}}^* , \qquad \text{Asymptotic} \qquad (2.96)$$

The density ρ^* can be eliminated from Eqs. (2.95) and (2.96) to obtain

$$G_{\text{in}}^* \sqrt{h_{0,\text{in}}^*} = P^* \frac{G(P^*)}{\sqrt{kg_0 P_i \rho_i}} \qquad (2.97)$$

Equation (2.97) is used to obtain the asymptotic pressures in Fig. 2.32 for given values of $G_{\text{in}}^* \sqrt{h_{0,\text{in}}^*}$.

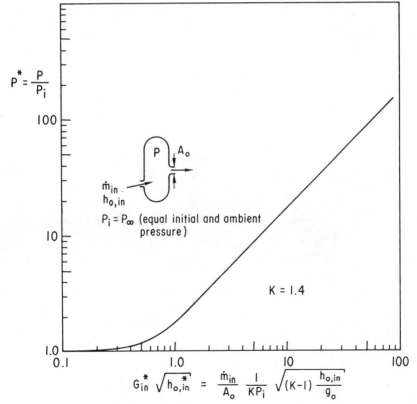

Figure 2.32 Asymptotic Vessel Pressure, Simultaneous Gas Charging and Discharging

EXAMPLE 2.8: VESSEL CHARGING PRESSURE

An adiabatic rigid vessel initially contains atmospheric air at pressure $P_i = P_\infty = 0.101$ MPa (14.7 psia) and density $\rho_i = 1.226$ kg/m^3 (0.076 lbm/ft^3). Air at stagnation enthalpy $h_{0,in} = 577$ kJ/kg (294 B/lbm) begins charging the vessel at a constant mass flow rate of $\dot{m}_{in} = 166$ kg/s (365 lbm/s), while simultaneous discharge occurs through an open nozzle of area $A_0 = 0.01$ m^2 (0.11 ft^2). *Estimate* the pressure-time characteristic in the vessel.

The data and Eq. (2.91) give $h_{0,in}^* = 2.0$ and $G_{in}^* = 40$. Thus, Eq. (2.94) gives the initial nondimensional pressure rate

$$\left(\frac{dP^*}{dt^*}\right)_{initial} = \frac{\mathcal{V}\rho_i}{P_i\dot{m}_{in}}\frac{dP}{dt} = kh_{0,in}^* = (1.4)(2) = 2.8$$

and for $G_{in}^*\sqrt{h_{0,in}^*} = 40\sqrt{2} = 56.6$, Fig. 2.32 gives the asymptotic pressure

$$P_{asymptotic}^* = \frac{P}{P_i} = 98$$

These values permit a sketch of the vessel pressure transient $P^*(t^*)$. Equations (2.89) and (2.90) were solved numerically with a second-order Runge-Kutta integration (Section C.1 of Appendix C), and the nondimensional results are given in Fig. 2.33. The nondimensional time scale of Fig. 2.33 is proportional to vessel volume $\mathcal{V}$. That is, a larger volume decreases the initial pressure rate and requires a longer time for the asymptotic pressure to be reached.

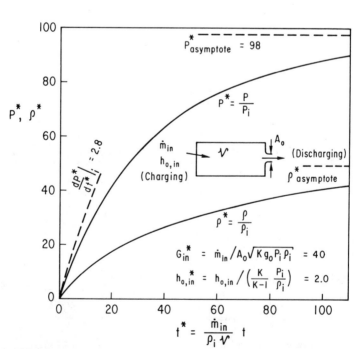

Figure 2.33 Calculated Simultaneous Charging and Discharging of a Gas Vessel

2.18 Charging and Discharging Saturated Mixture

If the system of Fig. 2.24 contains a quasi-equilibrium saturated liquid-vapor mixture, the internal energy is given by

$$e = e_f + \frac{e_{fg}}{v_{fg}} (v - v_f)$$

The pressure rate is obtained from Eq. (2.69) where $f(P)$ and $F(P, \mathcal{V}/M)$ are given by Eqs. (2.82) and (2.83).

Consider a case which includes both inflow and outflow mass rates. If the system volume is fixed and heat transfer is negligible, a rising pressure reaches the maximum value $P_{\max}$, which can be obtained by setting $dP/dt = 0$ in Eq. (2.69). This yields

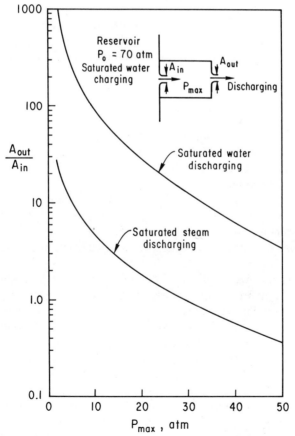

Figure 2.34 Maximum System Pressure, Simultaneous Charging and Discharging Saturated Steam and Water

$$\dot{m}_{in}[h_{0,in} - f(P)] = \dot{m}_{out}[h_{0,out} - f(P)]$$

If both $\dot{m}_{in}$ and $\dot{m}_{out}$ correspond to critical flow, we have

$$\dot{m}_{in} = G_c(P_{0,in}, h_{0,in})A_{in}$$

and

$$\dot{m}_{out} = G_c(P, h_{0,out})A_{out}$$

It follows that when $P = P_{max}$,

$$\frac{A_{out}}{A_{in}} = \frac{G_c(P_{0,in}, h_{0,in})}{G_c(P, h_{0,out})} \frac{h_{0,in} - f(P)}{h_{0,out} - f(P)}, \qquad P = P_{max} \qquad (2.98)$$

If the system is charged at a constant rate with saturated water from a reservoir at pressure $P_{0,in}$, the incoming stagnation enthalpy is $h_f(P_{0,in})$. If saturated water or steam at P_{max} is simultaneously discharging, the outflowing stagnation enthalpy is either $h_f(P_{max})$ or $h_g(P_{max})$, and $f(P)$ corresponds to $f(P_{max})$. The critical mass flux values are determined from Fig. 2.20(a). Values of P_{max} are graphed in Fig. 2.34 for a range of A_{out}/A_{in} when the system is being charged with saturated water from a 70-atm reservoir.

2.19 Vaporization and Condensation

Vaporization is the phase change process of liquid to vapor. Energy must be added to liquid in order to produce vapor at nucleation sites. This is true whether the liquid is heated directly or decompressed below its saturation pressure. When decompression causes vaporization at suspended nucleation sites in the liquid, the process is called *homogeneous boiling* or *flashing*.

Condensation is the process of phase change from vapor to liquid. Vapor condenses when it loses energy by heat transfer to a cool surface. Also, decompression of a saturated vapor causes condensation at nucleation sites in the vapor.

Nucleation sites are small, often microscopic particles or impurities in a fluid, or cavities or protrusions on a surface from which bubbles or droplets can grow during a change of phase. Developments in this section are based on the idealization that neither vaporization nor condensation rates are limited by a lack of nucleation sites.

A simplified model for vaporization or condensation is shown in Fig. 2.35 where pure liquid and vapor regions are separated at an interface. Temperature profiles are shown for heat transfer from vapor to liquid or liquid to vapor. For both temperature profiles, either vaporization or condensation can occur! If net heat transfer is toward the interface, vapor will form. If net heat transfer is away from the interface, liquid will condense.

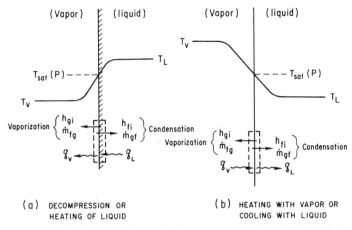

(a) DECOMPRESSION OR HEATING OF LIQUID

(b) HEATING WITH VAPOR OR COOLING WITH LIQUID

Figure 2.35 Vaporization and Condensation at a Liquid-Vapor Interface

Figure 2.35 shows heat transfer rates q_v and q_L from the vapor and liquid sides to the interface. When vaporization occurs, vapor is assumed to form at a saturated interface temperature $T_i = T_{sat}(P_L)$, which corresponds to liquid pressure. This assumption permits the liquid and vapor to be at different pressures if the interface is curved and surface tension plays a role. Moreover, if condensation occurs, liquid is assumed to form at a saturated interface temperature $T_i = T_{sat}(P_v)$, which corresponds to vapor pressure. The mass flow rates $\dot{m}_{fg}$ and $\dot{m}_{gf}$ designate vaporization and condensation rates, respectively.

The energy conservation principle for the dotted CVs in Fig. 2.35(a) and (b) yield

$T_L > T_v$

$$\left.\begin{array}{c} \dot{m}_{fg} \\ \dot{m}_{gf} \end{array}\right\} = \pm \left(\frac{q_L - q_v}{h_{fgi}} \right), \qquad \begin{array}{ll} q_L > q_v & \text{vaporization} \\ q_L < q_v & \text{condensation} \end{array} \qquad (2.99)$$

$T_L < T_v$

$$\left.\begin{array}{c} \dot{m}_{fg} \\ \dot{m}_{gf} \end{array}\right\} = \pm \left(\frac{q_v - q_L}{h_{fgi}} \right), \qquad \begin{array}{ll} q_v > q_L & \text{vaporization} \\ q_v < q_L & \text{condensation} \end{array} \qquad (2.100)$$

The heat transfer rates usually are controlled by conduction or convection mechanisms. If convection dominates in each phase, we may write

$$q_L = \pm \mathcal{H}_L A (T_L - T_i), \qquad \begin{array}{l} T_L > T_v \\ T_L < T_v \end{array} \qquad (2.101)$$

$$q_v = \pm \mathcal{H}_v A (T_i - T_v), \qquad \begin{array}{c} T_L > T_v \\ T_L < T_v \end{array} \qquad (2.102)$$

for which Eqs. (2.99) and (2.100) become

$$\frac{\dot{m}}{A} = \frac{\mathcal{H}_v(T_v - T_i) - \mathcal{H}_L(T_i - T_L)}{h_{fgi}} \begin{cases} >0, & \text{vaporization} \\ <0, & \text{condensation} \end{cases}$$

$$(2.103)$$

Equations (2.99) through (2.103) express the vaporization or condensation mass flux at an interface of vapor and liquid in terms of convective heat transfer rates between the interface and the two phases. If conduction dominates, heat transfer rates in Eqs. (2.101) and (2.102) would be expressed by $-\kappa A \, \partial T/\partial n$ where n is the length coordinate normal to the interface, and κ is the thermal conductivity in the phase being considered.

Sometimes a liquid surface is in contact with a gas mixture. The vapor partial pressure gradient drives vaporization or condensation by mass diffusion according to the general law

$$\frac{\dot{m}}{A} = k_m (P_i - P_v) \begin{cases} >0, & \text{vaporization} \\ <0, & \text{condensation} \end{cases} \qquad (2.104)$$

where P_i is the vapor saturation pressure at the liquid interface temperature, P_v is the vapor partial pressure in the gas mixture, and k_m is a mass transfer coefficient. (See Ref. [2.5] for evaluation of k_m.) Equation (2.104) only applies when the gaseous phase is not pure vapor and consists of two components. When an interface separates pure liquid and vapor, the vaporization and condensation rates are determined by heat transfer according to Eq. (2.103).

2.20 Hot Bubble Response in Saturated Liquid

Consider a case in which superheated vapor bubbles of initial radius R_0 and temperature T_0 are discharged into a tank of saturated liquid. A typical bubble is cooled by heat transfer to the liquid. The already saturated liquid boils at the interface, and vapor enters the bubble. Some bubbling applications require the bubble volume rate of change.

The superheated vapor bubble in Fig. 2.36 is surrounded by saturated liquid. The bubble pressure is assumed to remain equal to that of the surrounding liquid, and liquid inertia is neglected. The CS is a dashed line on the vapor side of the interface. Thus, the heat transfer rate is q_v,

(Saturated liquid)

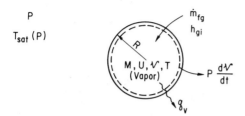

Figure 2.36 Superheated Vapor Bubble in Saturated Liquid

according to Fig. 2.35(b), and the vaporized mass inflow enthalpy corresponds to h_{gi}. Mass and energy conservation are written as

$$-\dot{m}_{fg} + \frac{dM}{dt} = 0 \qquad (2.105)$$

$$q_v + P\frac{d\mathcal{V}}{dt} - \dot{m}_{fg}h_{gi} + \frac{dU}{dt} = 0 \qquad (2.106)$$

The vapor is treated as a perfect gas with $U = P\mathcal{V}/(k-1)$ and $M = P\mathcal{V}/R_gT$. Whenever the perfect gas idealization is employed for vapor, all temperature-dependent thermodynamic properties must be based on the absolute temperature scale. Therefore, the enthalpy h_{gi} must be given by

$$h_{gi} = \frac{k}{k-1}\frac{P}{\rho_{gi}} = c_pT_i$$

and should not be obtained from tables which do not employ an absolute temperature scale. The vaporization enthalpy h_{fg} can be obtained from saturation tables because it is the difference $h_g - h_f$, which does not require an absolute temperature scale.

Since the liquid flowing past a rising bubble is replaced continuously, this problem can be simplified further by taking $h_{gi} \cong h_g$ and assuming that the interface temperature T_i is equal to that of the liquid $T_L = T_f$. This gives $q_L = 0$ from Eq. (2.101). Equations (2.100) and (2.102) are employed for $\dot{m}_{fg}$ and q_v with the geometric relationships $\mathcal{V} = 4\pi R^3/3$ and $A = 4\pi R^2$ to give the full problem for bubble radius as

$$\text{DE:} \qquad \frac{dR^*}{dt^*} = \beta^* - \frac{R^{*3}}{1-\alpha^*(R^{*3}-1)} \qquad (2.107)$$

$$\text{IC:} \qquad R^*(0) = 1 \qquad (2.108)$$

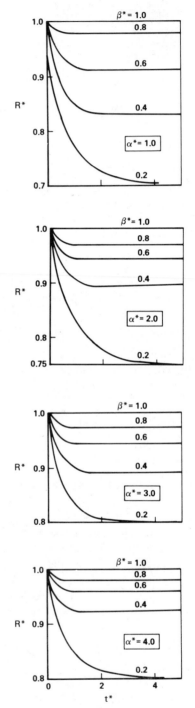

Figure 2.37 Hot Bubble Size Change

where the nondimensional quantities are

$$R^* = \frac{R}{R_0} \qquad t^* = \left[\frac{\mathcal{H}_v T_0 (k-1)(h_{fg} - h_g)}{kPh_{fg}R_0} \right] t$$

$$\alpha^* = \frac{kR_g T_0}{(k-1)(h_{fg} - h_g)} \qquad \beta^* = \frac{T_f}{T_0}, \quad 0 < \beta^* < 1 \tag{2.109}$$

The term $h_{fg} - h_g > 0$ when $h_g = c_P T_i$ is employed for a perfect gas. The asymptotic value of R^* is

$$R_\infty^* \rightarrow \left[\frac{(1 + \alpha^*)\beta^*}{1 + \alpha^*\beta^*} \right]^{1/3} \tag{2.110}$$

Since $0 < \beta^* < 1$, it follows that $0 < R_\infty^* < 1$. That is, even though the vapor mass of a hot bubble in saturated liquid increases, its size shrinks! If this result seems unusual to you; join the majority!

Equation (2.107) was solved for several values of α^* and β^*. These are graphed in Fig. 2.37, which shows that all hot gas bubbles shrink. The temperature decay of a superheated vapor bubble in saturated water was measured by Schmidt [2.3]. Figure 2.38 shows measured temperatures at various elevations after a superheated bubble was released. The solid curve was calculated from Eqs. (2.107) and (2.108) for an estimated heat transfer coefficient of $\mathcal{H} = 300$ W/m²-K (53 B/h-ft²-°F), based on forced convection of air in an enclosed space. Elevation y and time t are related by $y = V_\infty t$, where $V_\infty = 20$ cm/s (7.9 in/s) was employed. The calculation gives reasonable agreement with the data.

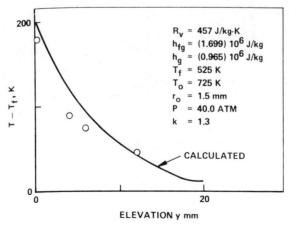

Figure 2.38   Sample Bubble Cooling [2.3]

2.21 Useful Thermodynamic Relationships

It is sometimes necessary to relate the change of a thermodynamic state property to changes in other properties. The thermodynamic properties which occur most often in unsteady thermofluid mechanics are P, ρ, T, e, h, and s. If P and ρ are considered independent, the differentials of T, e, h, and s can be written as

$$d\phi = \left(\frac{\partial\phi}{\partial P}\right)_\rho dP + \left(\frac{\partial\phi}{\partial\rho}\right)_P d\rho, \qquad \phi = T, e, h, s \qquad (2.111)$$

Equation (2.111) can be manipulated to express the partial derivative coefficients of dP and $d\rho$ in terms of tabulated thermodynamic properties. However, the manipulations require relationships which are described below:

1. The mixed second derivatives of continuous functions can be interchanged; that is,

$$\left[\frac{\partial}{\partial\rho}\left(\frac{\partial\phi}{\partial P}\right)_\rho\right]_P = \left[\frac{\partial}{\partial P}\left(\frac{\partial\phi}{\partial\rho}\right)_P\right]_\rho = \frac{\partial^2\phi}{\partial\rho\,\partial P} = \frac{\partial^2\phi}{\partial P\,\partial\rho} \qquad (2.112)$$

2. A constant value of the dependent variable ϕ in Eq. (2.111) yields

$$\left(\frac{\partial P}{\partial\rho}\right)_\phi = -\frac{(\partial\phi/\partial\rho)_P}{(\partial\phi/\partial P)_\rho} = -\left(\frac{\partial\phi}{\partial\rho}\right)_P\left(\frac{\partial P}{\partial\phi}\right)_\rho \qquad (2.113)$$

3. If $\phi_1, \phi_2, \ldots, \phi_n$ are functions of P and ρ, the chain rule for partial differentiation yields

$$\left(\frac{\partial\phi}{\partial P}\right)_\rho = \left(\frac{\partial\phi}{\partial\phi_1}\right)_\rho\left(\frac{\partial\phi_1}{\partial\phi_2}\right)_\rho \cdots \left(\frac{\partial\phi_n}{\partial P}\right)_\rho \qquad (2.114)$$

or

$$\left(\frac{\partial\phi}{\partial\rho}\right)_P = \left(\frac{\partial\phi}{\partial\phi_1}\right)_P\left(\frac{\partial\phi_1}{\partial\phi_2}\right)_P \cdots \left(\frac{\partial\phi_n}{\partial\rho}\right)_P \qquad (2.115)$$

Table 2.1 includes common derivative properties used in thermodynamics. If we employ the Gibbs equation in the forms of Eqs. (2.16) and (2.17) with several properties of Table 2.1 in the functional forms $c_p(p, \rho)$, $c_v(P, \rho)$, $\beta(P, \rho)$ and the sound speed of Eq. (2.45) as $C(P, \rho)$, Eqs. (2.111)–(2.115) can be used to obtain the following thermodynamic property differentials:

IMPORTANT THERMODYNAMIC PROPERTY DIFFERENTIALS

$$dT = \left(\frac{g_0}{\rho\beta C^2} + \frac{\beta T}{\rho c_p}\right) dP - \frac{1}{\beta\rho} d\rho \qquad (2.116)$$

$$de = \left(\frac{g_0 c_p}{\beta \rho C^2}\right) dP + \left(\frac{P}{\rho^2} - \frac{c_p}{\beta \rho}\right) d\rho \tag{2.117}$$

$$dh = \left(\frac{1}{\rho} + \frac{g_0 c_p}{\beta \rho C^2}\right) dP - \left(\frac{c_p}{\beta \rho}\right) d\rho \tag{2.118}$$

$$ds = \left(\frac{g_0 c_p}{\beta \rho T C^2}\right) dP - \left(\frac{c_p}{\beta \rho T}\right) d\rho \tag{2.119}$$

Equations (2.116)–(2.119) can be combined in any suitable way to express a differential in terms of any two others. It should be remembered that P and T are not independent in a multiphase equilibrium system, such as a mixture of saturated liquid and vapor.

EXAMPLE 2.9: ISENTROPIC DECOMPRESSION CHANGES
Fluid pressure is decreased isentropically by a differential amount dP. Determine the resulting changes in ρ, T, e, and h for (1) a perfect gas, (2) an ideal liquid, and (3) a saturated, bubbly liquid-vapor mixture.

Equations (2.116)–(2.119) show that for isentropic pressure changes,

$$d\rho_s = \frac{g_0}{C^2} dP_s$$

for which

$$dT_s = \frac{\beta T}{\rho c_p} dP_s$$

$$de_s = \frac{g_0 P}{\rho^2 C^2} dP_s$$

$$dh_s = \frac{1}{\rho} dP_s$$

If we have a perfect gas, Tables 2.1 and 2.2 give

$$\beta = \frac{1}{T}, \quad c_p = \text{constant}, \quad C = \sqrt{kg_0 \frac{P}{\rho}}$$

$$\rho = \frac{P}{R_g T}, \quad \frac{c_p}{R_g} = \frac{k}{k-1}$$

Thus

PERFECT GAS

$$dT_s = \frac{1}{\rho c_p} dP_s$$

$$de_s = \frac{1}{k\rho} dP_s$$

$$dh_s = \frac{1}{\rho} dP_s$$

If we have an ideal liquid, Tables 2.1 and 2.3 give $c_p = $ constant, $\beta = 0$, and $C \rightarrow \infty$. Thus

IDEAL LIQUID

$$dT_s = de_s = 0$$

$$dh_s = \frac{1}{\rho} \, dP_s$$

Finally, if we have a saturated, bubbly liquid-vapor mixture, Table 2.1 yields $\beta \rightarrow \infty$ because $\partial T = 0$ for $P = $ constant. However, it is seen that

$$\frac{\beta}{c_p} = \frac{1}{v} \frac{(\partial v / \partial T)_p}{(\partial h / \partial T)_p} = \frac{1}{v} \left(\frac{\partial v}{\partial h} \right)_P$$

Since v, h, and P for a saturated mixture are related by (Table 2.4)

$$v = v_f(P) + \frac{v_{fg}(P)}{h_{fg}(P)} \, [h - h_f(P)]$$

we may write

$$\frac{\beta}{c_p} = \frac{1}{v} \left(\frac{\partial v}{\partial h} \right)_P = \frac{1}{v} \frac{v_{fg}}{h_{fg}} = \rho \frac{v_{fg}}{h_{fg}}$$

Thus,

SATURATED, BUBBLY LIQUID-VAPOR MIXTURE

$$dT_s = T v_{fg}(P) \, dP_s$$

$$de_s = \frac{g_0 P}{\rho^2 C^2} \, dP_s$$

$$dh_s = \frac{1}{\rho} \, dP_s$$

$C = $ decompressive sound speed (Section 2.12)

References

2.1   Anderson, J. D., *Modern Compressible Flow*, McGraw-Hill, New York, 1982.

2.2   Bejan, A., *Entropy Generation through Heat and Fluid Flow*, Wiley, New York, 1982.

2.3   Schmidt, H., "Bubble Formation and Heat Transfer during Dispersion of Superheated Steam in Saturated Water-II," *Int. J. Heat Mass Transfer*, **20** (1977), 647–654.

2.4   Huang, F. F., *Engineering Thermodynamics: Fundamentals and Applications*, Macmillan, New York, 1976.

2.5 Kreith, F., *Principles of Heat Transfer*, 3rd ed., IEP, New York, 1976.

2.6 Lahey, R. T., and F. J. Moody, *The Thermal Hydraulics of a Boiling Water Nuclear Reactor*, American Nuclear Society Monograph, 1977.

2.7 Lamb, H., *Hydrodynamics*, 6th ed., Dover, New York, 1945.

2.8 Moody, F. J., *Second Law Thinking, Example Applications in Reactor and Containment Technology*, American Society of Mechanical Engineers Special Publication HTD-33, 1984.

2.9 Moody, F. J., *Maximum Discharge Rate of Liquid/Vapor Mixtures from Vessels*, American Society of Mechanical Engineers Special Publication, *Non-Equilibrium Two-Phase Flows*, 1975.

2.10 Moody, F. J., "Pressure Suppression Containment Thermal-Hydraulics," *Proc. ANS/ASME/NRC International Topical Meeting on Nuclear Reactor Thermal Hydraulics*, 1980, pp. 257 ff.

2.11 Moody, F. J., "Dynamic and Thermal Behavior of Hot Gas Bubbles Discharged into Water," *Nuclear Engineering & Design*, **95** (1986), 47–54.

2.12 Reynolds, W. C., *Thermodynamics*, 2nd ed., McGraw-Hill, New York, 1968.

2.13 Zemansky, M. W., *Heat and Thermodynamics*, 4th ed., McGraw-Hill, New York, 1957.

2.14 Karplus, J. B., *Propagation of Pressure Waves in a Mixture of Water and Steam*, Armour Research Foundation Report ARF 4133-12, 1961.

2.15 Gallagher, E. V., *Water Decompression Experiments and Analysis for Blowdown of Nuclear Reactors*, TID-4500, 1970.

2.16 Edwards, A. R. and T. P. Obrien, "Studies of Phenomena Connected with the Depressurization of Water Reactors," *J. Brit. Nuclear Energy Soc.* **9** (1970).

2.17 General Electric Report NEDO-24708A, Rev. 1, 1980.

2.18 Bejan, A., *Advanced Engineering Thermodynamics*, Wiley, New York, 1988.

Problems

2.1 Small-amplitude waves on a gas-liquid interface in low gravity can be described by a Taylor instability analysis (Section 9.16). Specifically, consider gas of negligible density adjacent to liquid of density ρ between two vertical parallel walls a distance L apart. If the interface surface tension is σ and the body force accelaration b is toward the liquid, the resulting wave profile in two dimensions is given by

$$\eta(x, t) = \sum_{n=1}^{\infty} \epsilon_n \cos \frac{n\pi x}{L} \cos \omega_n t$$

where ϵ_n is the amplitude of the nth harmonic and

$$\omega_n = \sqrt{n\pi \left(\frac{n^2 \pi^2 g_0 \sigma}{L^3 \rho} - \frac{b}{L} \right)}$$

(a) Consider a single harmonic of index n. What value or range of acceleration b will result in stable, neutrally stable, and unstable wave motion?

Answer:

$$b \begin{Bmatrix} < \\ = \\ > \end{Bmatrix} \frac{(n\pi)^2 g_0 \sigma}{L^2 \rho} \qquad \begin{cases} \text{stable} \\ \text{neutrally stable} \\ \text{unstable} \end{cases}$$

(b) Show that the lowest value of b for which the surface becomes unstable is given by

$$b > \frac{\pi^2 g_0 \sigma}{L^2 \rho}, \qquad n = 1 \quad \text{(lowest harmonic)}$$

(c) If b is a known constant value, show that the fastest growing harmonic n corresponds to the next higher integer from the value

$$\frac{L}{\pi} \sqrt{\frac{\rho b}{3 g_0 \sigma}}$$

2.2 Show that the coefficient of volume expansion β and the specific heat at constant pressure c_p for a saturated liquid-vapor mixture are undefined, but that the ratio β / c_p has a finite value

$$\frac{\beta}{c_p} = \frac{1}{v} \frac{v_{fg}}{h_{fg}}$$

2.3 Show that c_p and c_v for an ideal liquid are identical.

2.4 Liquid of density ρ passes through an orifice with a pressure loss given by

$$\Delta P_L = \frac{K_L V^2 \rho}{2 g_0}$$

where K_L is the loss coefficient based on fluid velocity V in the pipe. Neglect heat losses and show that if the average temperature through the orifice is T_∞ and the pipe flow area is A, then the irreversibility rate $\dot{I}$ is given by

$$\dot{I} = \frac{AV}{T_\infty} \Delta P_L = \frac{\dot{V}}{T_\infty} \Delta P_L$$

and the available power loss is

$$\mathscr{P}_{\text{loss}} = \dot{V} \Delta P_L$$

2.5   Liquid flows at a volume flow rate $\dot{V}$ in a horizontal V-shaped channel with an angle θ at the apex. Show that the expected liquid depth z, measured upward from the apex, is given by

$$z = \left(\frac{3\dot{V}^2}{g \tan^2(\theta/2)} \right)^{1/5}$$

Also show that if $\theta = 60°$, then

$$z = \left(\frac{9\dot{V}^2}{g} \right)^{1/5}$$

2.6   Show that if liquid flows at a volume rate $\dot{V}$ in a horizontal, circular pipe of radius R, Fig. P2.6, the expected depth z satisfies the equation

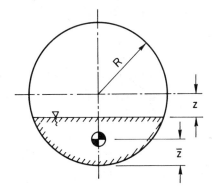

Figure P2.6

$$\frac{\dot{V}^2}{gR^5} = \left[\frac{1}{3}\left(1 - \frac{z^2}{R^2}\right)^{3/2} - \frac{z}{R}\left(\frac{A}{R^2}\right) \right] \frac{A}{R^2}$$

where the liquid cross-sectional area A is geometrically related to z by

$$\frac{A}{R^2} = \frac{\pi}{2} - \frac{z}{R}\sqrt{1 - \frac{z^2}{R^2}} - \sin^{-1}\left(\frac{z}{R}\right)$$

Hint: Verify that for potential energy, the center of gravity of A is at

$$\frac{\bar{z}}{R} = 1 - \frac{2}{3}\frac{(1 - z^2/R^2)^{3/2}}{A/R^2}$$

2.7   Show that the minimum volume flow rate, $\dot{V}_{min}$ in Problem 2.6, which will cause a horizontal circular pipe of radius R to flow full, is given by

$$\dot{V}_{min} = \pi \sqrt{gR^5}$$

2.8   In a bubbly mixture of noncondensible perfect gas and liquid, show that if the gas fraction x is 1.0 then the critical pressure of Eq. (2.67) reduces to that of a perfect gas given by Eq. (2.59).

2.9   Use Eq. (2.49) and the specific volume state equation

$$v = xv_g + (1-x)v_L$$

to show that if the liquid-gas specific volume ratio v_L/v_g is much smaller than the gas mass fraction x, the sound speed in a bubbly mixture of noncondensible gas and liquid can be approximated by

$$C \cong C_g\sqrt{x}, \qquad \frac{v_L}{v_g} \ll x$$

2.10   Verify the minimum sound speed of Eq. (2.50) by first minimizing C/C_g in Eq. (2.49) with respect to v/v_g and then incorporating the fact that for most liquid-gas systems, v_L/v_g and C_g/C_L are small.

2.11   The specific volume of a homogeneous, bubbly gas-liquid mixture is given by

$$v = xv_g + (1-x)v_L$$

If the gas is noncondensible, x is constant. Start with Eq. (2.45) and show that the sound speed in this mixture can be expressed by

$$C = \frac{xv_g + (1-x)v_L}{\sqrt{x(v_g/C_g)^2 + (1-x)(v_L/C_L)^2}}$$

and that the minimum value of C occurs at $x \approx v_L/v_g$.

2.12   Consider a tube filled with many equal plugs of liquid, length L_L, separated by equal plugs of gas, length L_g, with densities ρ_L and ρ_g. Show that the average gas mass fraction of this flow pattern is

$$x = \frac{1}{1 + \rho_L L_L/\rho_g L_g}$$

Sound speed through this plug flow pattern can be defined as the unit of length $L = L_g + L_L$ divided by the sonic transit time $L_g/C_g + L_L/C_L$. Show that the sound speed can be put into the form

$$\frac{C_{\text{plug}}}{C_g} = \left(1 + \frac{1-x}{x}\frac{v_L}{v_g}\right)\left(1 + \frac{1-x}{x}\frac{C_g}{C_L}\frac{v_L}{v_g}\right)^{-1}$$

Compare with the sound speed in a bubbly gas-liquid mixture, obtained in Problem 2.11 and show that, for small values of $C_g v_L / C_L v_g$, sound speed is slower in the bubbly mixture.

2.13 An isolated tunnel of length L and area A contains a burnable hydrogen-air mixture at 1 atm. Ignition occurs at one end of the cylinder, $x = 0$, and the burning front moves through the cylinder at assumed constant speed S. The rate at which mass is burned is designated by $\dot{m}$, and the heat of combustion, or energy released per unit mass burned, is h_c. Since no mass is added by burning, the energy addition rate at the burning front can be treated as if it were a heat transfer rate $q = h_c \dot{m}$ flowing into the burned gas region. Both the unburned air-hydrogen mixture and the burned mixture can be assumed to have the same ratio of specific heats $k = 1.4$.

The unburned gas properties are shown in Fig. P2.13, where M is

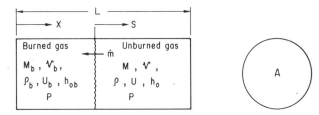

Figure P2.13

the mass of unburned gas, $\mathcal{V}$ is its volume, ρ is its density, U is its internal energy, and h_0 is its stagnation enthalpy. Kinetic and potential energies are negligible. The burned gas has properties designated by the subscript b. Burning is slow enough that both the unburned and burned gases are at uniform pressure P, which increases as burning progresses.

The major objective of this analysis is to determine $P(t)$ while the burning front moves through the cylinder.

(a) Write mass and energy conservation equations for both the burned and unburned gas regions. It will be best to express internal energies in the form $U = P\mathcal{V}/(k-1)$ for both the burned and unburned regions.

(b) Write an energy conservation equation across the burning front to express h_{ob} in terms of h_0 and h_c.

(c) Assume a constant-burning front speed S. Note that $\mathcal{V}_b = ASt$, and eliminate $\mathcal{V}$ and $\mathcal{V}_b$ from the energy conservation equations. Then eliminate $\dot{m}$ to obtain a differential equation for pressure in the cylinder.

(d) If the stagnation enthalpy of the unburned gas is expressed as $h_0 = kP/\rho(k-1)$, it is seen that as P changes, so will h_0. Start with the energy conservation equation and show that the unburned gas is compressed isentropically, for which $P/\rho^k = P_i/\rho_i^k$.

(e) Nondimensionalize the pressure and time and obtain a solution for pressure as a function of time which can be displayed efficiently for an initial pressure P_i and arbitrary heat of combustion h_c.

2.14   An idealized gas cylinder and piston model can be used to illustrate thermal damping (See Fig. P2.14). The model consists of a frictionless

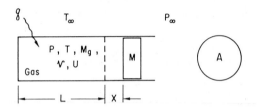

Figure P2.14

piston of mass M in a rigid cylinder of area A which contains a gas column of undisturbed length L, whose mass is M_g. Heat transfer between the gas and surroundings is given by q, which can be in either direction, depending on gas temperature T. The piston is displaced an initial amount x_i and released. The gas is compressed and expanded as the piston oscillates. It will be informative to obtain the differential equation of motion which describes piston displacement x and solve it for various heat transfer parameters. When $x = 0$, gas pressure P is equal to the ambient value P_∞. A linearized analysis (Section C.1 of Appendix C) will be useful in understanding the effect of various parameters on the damping behavior. Several suggestions are offered next to help formulate the problem.

(a) Since only pressure forces act on the piston, show that the piston equation of motion is

$$\ddot{x} = \frac{g_0 A (P - P_\infty)}{M}$$

(b) Show that energy conservation for the gas yields

$$P \frac{d\mathcal{V}}{dt} - q + \frac{dU}{dt} = 0$$

Employ the perfect gas state equations, $P\mathcal{V} = M_g R_g T$, $P_\infty \mathcal{V}_\infty = M_g R_g T_\infty$, and $U = M_g c_v T = P\mathcal{V}/(k-1)$, and the convective heat transfer rate

$$q = \mathcal{H} A_q (T_\infty - T)$$

where $\mathcal{H}$ is an average convection coefficient and A_q is the average cylinder wall area for heat transfer, to show that the energy equation of step (b) can be written as

$$P\frac{d\mathcal{V}}{dt} - \frac{\mathcal{H}A_q P_\infty \mathcal{V}_\infty}{M_g R_g}\left(1 - \frac{P\mathcal{V}}{P_\infty \mathcal{V}_\infty}\right) + \frac{1}{k-1}\left(P\frac{d\mathcal{V}}{dt} + \mathcal{V}\frac{dP}{dt}\right) = 0$$

(c) Write $\mathcal{V} = A(L + x)$ and $\mathcal{V}_\infty = AL$ and eliminate P from steps (a) and (b) to obtain the differential equation of motion

$$\frac{L+x}{k}\frac{M\ddot{x}}{g_0 A} + \left(P_\infty + \frac{M\ddot{x}}{g_0 A}\right)\dot{x}$$
$$- \left(\frac{k-1}{k}\right)\frac{\mathcal{H}A_q T_\infty}{A}\left[1 - \left(1 + \frac{M\ddot{x}}{g_0 A P_\infty}\right)\frac{L+x}{L}\right] = 0$$

(d) Formulate the three required initial conditions for releasing the piston after displacing it an initial amount $x = x_i$. One initial condition can be obtained by putting the initial gas pressure for displacement x_i into step (a).

(e) Nondimensionalize the full problem with the normalized variables

$$x^* = \frac{x}{L} \quad \text{and} \quad t^* = t\sqrt{\frac{g_0 A P_\infty}{ML}}$$

to obtain

$$(1 + x^*)\frac{d^3 x^*}{dt^{*3}} + k\left(1 + \frac{d^2 x^*}{dt^{*2}}\right)\frac{dx^*}{dt^*} + D\left[x^* + \frac{d^2 x^*}{dt^{*2}} + x^*\frac{d^2 x^*}{dt^{*2}}\right]$$
$$= 0$$

where the damping coefficient is

$$D = (k-1)\frac{\mathcal{H}A_q T_\infty}{ALP_\infty}\sqrt{\frac{ML}{g_0 A P_\infty}}$$

(f) Consider a small-amplitude oscillation and show that the adiabatic case $D = 0$ gives an undamped oscillation. Also show that the isothermal case $D \to \infty$ gives another undamped oscillation.

(g) Show that if $0 < D < \infty$, a damped oscillation occurs and estimate the value of D which results in maximum damping.

2.15 A rigid, insulated vessel of volume $\mathcal{V} = 5.0 \, m^3$ ($176 \, ft^3$) is charged with air, $k = 1.4$ and $c_p = 1.0 \, J/gm\text{-}K$ ($0.24 \, B/lbm\text{-}°F$) at a constant critical mass flow rate $\dot{m}_c = 0.1 \, kg/s$ ($0.22 \, lbm/s$) from a reservoir at temperature $T_0 = 300 \, K$ ($540 \, R = 80 \, F$).

(a) Show that the initial pressure rate is $2.4 \, kPa/s$ ($0.35 \, Psi/s$).

(b) If the vessel is initially evacuated, show that the temperature would achieve the value $420 \, K$ ($756 \, R$).

2.16 A rigid, insulated vessel of volume $\mathcal{V}$ contains perfect gas with properties k_1, P_{1i}, and T_{1i}. Another perfect gas with properties k_2 and h_{02} is charged into the vessel at a mass flow rate $\dot{m}$. Derive an expression for the initial pressure rate in the vessel for two cases:

(a) The gases are unmixed, separated by an imaginary flexible envelope, although both are continuously at the same pressure.

(b) The gases are completely mixed with total pressure given by the partial pressure sum.

Answer:

$$\frac{dP}{dt} = \frac{k_1}{k_2}(k_2 - 1) \frac{\dot{m}h_{02}}{\mathcal{V}}, \qquad \text{unmixed}$$

$$\frac{dP}{dt} = (k_1 - 1)\frac{\dot{m}h_{02}}{\mathcal{V}} + \left(\frac{k_2 - k_1}{k_2 - 1}\right)\frac{R_2 T_{1i}}{\mathcal{V}}\dot{m}, \qquad \text{mixed}$$

2.17 Consider a mixture of two perfect gases, identified as gas 1 and gas 2. Total pressure P is the sum of partial pressures P_1 and P_2. The density of each gas present is $\rho_1 = x_1\rho$ and $\rho_2 = x_2\rho$, where x_1 and x_2 are the mass fractions of each gas and ρ is the mixture density. Other properties of gases 1 and 2 are c_{p1}, c_{v1}, R_{g1}, c_{p2}, c_{v2}, and R_{g2}. Start with the equation

$$C = \sqrt{g_0\left(\frac{\partial P}{\partial \rho}\right)_s}$$

and obtain an expression for the mixture sound speed in terms of properties of each gas. Several suggestions follow.

Show that total pressure can be written in the form $P = (x_1 R_{g1} + x_2 R_{g2})\rho T$ and that $P_1/P_2 = x_1 R_{g1}/x_2 R_{g2}$. Note that for an isentropic process of this mixture,

$$ds = x_1 \, ds_1 + x_2 \, ds_2 = 0$$

for which $ds_1/ds_2 = -x_2/x_1$. Let stagnation properties of this mixture be P_0, ρ_0, and T_0, and use the Gibbs equation

$$T \, ds_n = dh_n - \frac{1}{\rho_n} \, dp_n \, , \qquad n = 1, 2$$

to obtain

$$T = T_0 \left(\frac{P}{P_0} \right)^X \, , \qquad X = \frac{x_1 R_{g1} + x_2 R_{g2}}{x_1 c_{p1} + x_2 c_{p2}}$$

Show that for an isentropic process,

$$\frac{P}{P_0} = \left(\frac{\rho}{\rho_0} \right)^Y \, , \qquad Y = \frac{x_1 c_{p1} + x_2 c_{p2}}{x_1 c_{v1} + x_2 c_{v2}}$$

Then show that the sound speed is

$$C = \sqrt{g_0 Y \frac{P}{\rho}}$$

2.18 A valve plunger is driven to shut off flow by the spring-gas cylinder device shown in Fig. P2.18. Chamber 1 in the cylinder is open to the

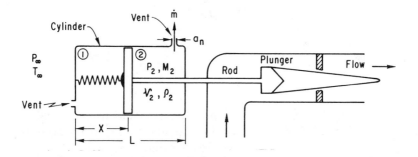

Figure P2.18

atmosphere and contains the spring, which exerts a force F_s on the piston of

$$F_s = F_0 + (F_L - F_0) \frac{x}{L}$$

where F_0 and F_L are the spring forces when piston displacement is $x = 0$ and L, respectively. The plunger rod cross-sectional area is negligible compared with the cylinder area A. Normally the piston is held at position $x = 0$ by compressed air in chamber 2 at initial pressure $P_{20} = F_0/A + P_\infty$, and temperature T_∞. When valve closure is required, chamber 2 is suddenly vented to the atmosphere through an ideal nozzle of flow area a_n. Assume that chamber 1 remains at pressure P_∞, no leakage occurs past the piston or plunger rod, the moving parts have negligible mass, and motion is without friction.

(a) Neglect heat transfer and show that mass and energy conservation, and the perfect gas state equations for chamber 2 yield isentropic state changes so that density and pressure are related by

$$\rho_2 = \rho_{20}\left(\frac{P_2}{P_{20}}\right)^{1/k}$$

(b) Note that the compressed gas force $P_2 A$ must be equal to the sum of atmospheric pressure and spring forces of chamber 1. Moreover, assume that P_2 is sufficiently high that critical flow discharge occurs, based on Eq. (2.60). Show that since $\mathcal{V}_2 = A(L - x)$, pressure P_2 and piston displacement x must satisfy

$$\left\{1 - \frac{F_L}{kP_{20}A}\left[1 - \frac{P_{20}A}{F_L}\left(\frac{P_2}{P_{20}} - \frac{P_\infty}{P_{20}}\right)\right]\left(\frac{P_2}{P_{20}}\right)^{-1}\right\}\frac{d(P_2/P_{20})}{dt}$$

$$= \frac{\dfrac{a_n}{A}\sqrt{kg_o(P_{20}/\rho_{20})/(2/(k+1))^{(k+1)/(k-1)}}}{P_{20}AL/(F_L - F_0)}\left(\frac{P_2}{P_{20}}\right)^{(k-1)/2k}$$

and

$$\frac{x}{L} = \frac{P_{20}A}{F_L - F_0}\left(\frac{P_2}{P_{20}} - \frac{P_\infty}{P_{20}}\right) - \frac{F_0}{F_L - F_0}$$

(c) Determine the initial pressure rate and from this make an estimate of the time required for the piston to travel its full stroke from $x = 0$ to L. Use the following parameters:

$$F_0 = 100 \text{ kN } (22,480 \text{ lbf}) \quad A = 0.2 \text{ m}^2 \ (2.15 \text{ ft}^2)$$
$$F_L = 20 \text{ kN } (4496 \text{ lbf}) \quad a_n = 0.0005 \text{ m}^2 \ (0.0054 \text{ ft}^2)$$
$$L = 0.1 \text{ m } (0.328 \text{ ft}) \quad P_\infty = 101 \text{ kPa } (14.7 \text{ psia})$$
$$T = 300 \text{ K } (540 \text{ R}) \quad k = 1.4$$

Answer:

$$\frac{d(P_2/P_{20})}{dt} = -3.32 \text{ s}^{-1}, \quad t_{\text{stroke}} \approx 0.2 \text{ s}$$

2.19 The critical discharge mass flux G_c of a two-phase saturated homogeneous liquid-vapor mixture depends on the amount of phase change which can occur between the source and the plane of discharge. It was noted in Section 2.13 that if a flow passage from a reservoir was about 15 cm or longer, the saturated homogeneous equilibrium model would be approached and G_c could be obtained from Fig. 2.20. However, if the flow passage is short, the so-called *frozen composition* model, which corresponds to a mixture of incompressible liquid and perfect gas with constant quality x in Fig. 2.19, could be employed for Gc.

A saturated bubbly steam-water mixture of quality $x = 0.2$ and stagnation pressure $P_0 = 7.0\ MPa$ (1020 psia) in a vessel undergoes critical discharge to a low-pressure environment. Compare critical flow rates and critical pressures of the saturated equilibrium and frozen composition models. Use vessel mixture properties:

$$v_{g0} = 0.0278\ \text{m}^3/\text{kg}\quad (0.446\ \text{ft}^3/\text{lbm}) \qquad h_{g0} = 2773\ \text{kJ}/\text{kg}$$
$$(1192\ \text{B}/\text{lbm})$$

$$k = 1.3 \qquad\qquad h_{f0} = 1262\ \text{kJ}/\text{kg}$$
$$(542\ \text{B}/\text{lbm})$$

$$v_L = 0.001348\ \text{m}^3/\text{kg}\ (0.0216\ \text{ft}^3/\text{lbm})$$

Answer:

Saturated Equilibrium Model

$$G_c = 17,575\ \text{kg}/\text{s-m}^2\ (3594\ \text{lbm}/\text{s-ft}^2), \qquad P_c = 4689\ \text{kPa}\ (684\ \text{psia})$$

Frozen Composition Model

$$G_c = 20,628\ \text{kg}/\text{s-m}^2\ (4218\ \text{lbm}/\text{s-ft}^2), \qquad P_c = 3220\ \text{kPa}\ (469\ \text{psia})$$

2.20 Postulated severe accidents in the nuclear energy industry sometimes include the consideration of so-called *steam-explosions*, which involve the discharge of molten core metal particles into a water pool. Immediate quenching of the particles, rapid expansion of the steam formed, and propulsion of the surrounding water mass can occur with explosive violence.

Let a molten metal mass at initial temperature T_0 suddenly be submerged in saturated water at temperature $T_{sat} = 100°C$ and pressure $P_{sat} = 1.0$ atm. Total internal thermal energy of the metal relative to T_{sat} is U_m.

(a) Verify that the steam mass formed, if pressure remained constant at P_{sat}, would be $M_g = U_m/h_{fg}(P_{sat})$.

(b) Show that the corresponding steam expansion would perform mechanical work,

$$W_k = P_{\text{sat}} \, \Delta \mathcal{V}_g = P_{\text{sat}} v_{fg}(P_{\text{sat}}) M_g = \left(\frac{v_{fg}(P_{\text{sat}})}{h_{fg}(P_{\text{sat}})} \, P_{\text{sat}} \right) U_m$$

for which the efficiency of thermal-to-mechanical energy conversion is $\eta = P_{\text{sat}} v_{fg}(P_{\text{sat}})/h_{fg}(P_{\text{sat}})$. Show that for $P_{\text{sat}} = 1.0$ atm, the efficiency is about 7 percent.

3 CONVECTIVE PROPAGATION

Convective problems involve disturbances which are carried with a fluid flow. Two convective analyses described in this chapter provide transient properties of discharging fluid jets and the temperature of fluid streams which exchange thermal energy with a hot or cold pipe wall. The analysis of fluid jets yields the time- and space-dependent cross-sectional jet area and forward velocity for use in determining which structures could become targets intercepted by a jet, and the corresponding impingement force. The interesting phenomenon of incompressible fluid jet shocking is discussed. Another analysis yields the unsteady temperature of hot or cold fluid flows in pipes with wall-to-fluid heat transfer. Formulations are developed for predicting the effects of external temperature baths, electrically heated pipes, and the influence of pipe wall thermal capacity on the discharging fluid temperature. Techniques for obtaining the fluid and pipe wall unsteady temperature profiles are described, which include a fluid path line integration, Laplace transformation, and eigenvalue solutions.

3.1 Incompressible, One-Dimensional Fluid Jets

A fluid jet is formed when discharge occurs from a confining flow passage. Common examples of fluid jets include drinking fountains, drain flows into rivers, and chimney discharges into the atmosphere. Sometimes it is necessary to determine the properties of an unsteady fluid jet in order to properly design structures or other objects in its path to withstand resulting mechanical or thermal loads.

Extensive literature exists for both laminar and turbulent jet penetration and diffusion. The idealized analysis in this chapter is based on an inviscid, unsteady, incompressible, one-dimensional fluid jet and the prediction of its time- and space-dependent velocity and cross-sectional area.

Whether inviscid analysis is justified can be determined by the magnitude of jet boundary diffusion into its surrounding normal to its forward direction. Consider a turbulent jet of initial width D and forward velocity U_∞ in the x direction. Its diffusion rate in the y direction normal to its forward motion proceeds according to the turbulent shear velocity of Eq. (2.9):

$$\frac{dy}{dt} = V_{p,t} = (0.238)\,\frac{\nu}{y}\,\mathrm{Re}^{3/4}, \qquad \mathrm{Re} = \frac{U_\infty y}{\nu} \qquad (3.1)$$

We can follow a jet fluid particle by putting $U_\infty = dx/dt$ and integrating Eq. (3.1) from $x = 0$ and $y = 0$ to x and y to obtain the diffusion penetration length

$$y = 0.38 \left(\frac{\nu}{U_\infty x} \right)^{1/5} x \qquad (3.2)$$

which is simply the boundary layer thickness on a flat plate, used at first to obtain the shear propagation velocity of Eq. (2.9). Equation (3.2) shows that y grows as $x^{4/5}$, and it can be used to determine when inviscid analysis can be applied. A water jet of $D = 1.0\,\mathrm{m}$, velocity $U_\infty = 10\,\mathrm{m/s}$, and $\nu = 4 \times 10^{-7}\,\mathrm{m^2/s}$, submerged in water, would travel $x = 10\,\mathrm{m}$ before the diffusion boundary would grow $8\,\mathrm{cm}$. This amount of diffusion would be small enough in most cases to use the inviscid jet analysis described in this chapter. It is assumed that the starting vortex, resembling a smoke ring at the beginning of discharge, does not strongly affect the jet penetration.

The interaction of a liquid jet with gaseous surroundings would cause interfacial waves with the possible formation and detachment of liquid droplets. Furthermore, a gas jet in surrounding liquid would soon undergo interfacial instability, Chapter 9, and become a column of bubbles. These effects are not considered here.

3.2   One-Dimensional Jets, Basic Formulation

The fluid jet model is shown in Fig. 3.1. A jet of constant density ρ is being discharged from a port at $y = 0$ with a velocity $V(0, t)$, which is a general function of time. The jet cross-sectional area normal to its axis and the average velocity at location y, measured in the direction of flow, and time t are designated $A(y, t)$ and $V(y, t)$. The surrounding fluid pressure $P(y, t)$ is either uniformly constant, or it can be a function of elevation, as in the case of vertical jets submerged in liquid. Although diffusion and shear effects are neglected, the local surrounding fluid pressure acts on the jet boundary.

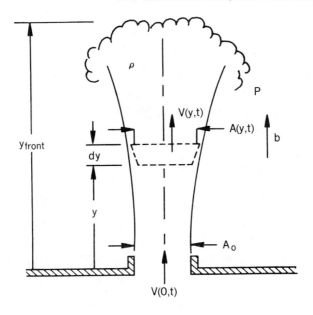

(a) Fluid jet geometry and properties

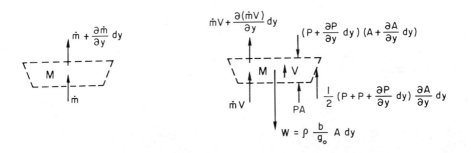

(b) Mass diagram

(c) Momentum diagram

Figure 3.1 One-Dimensional Fluid Jet Model

The mass conservation and momentum creation laws are written from Fig. 3.1(b) and (c) as

$$\frac{\partial \dot{m}}{\partial y} \, dy + \frac{\partial M}{\partial t} = 0 \tag{3.3}$$

and

$$\frac{\partial (\dot{m}V)}{\partial y} \, dy + \frac{\partial (MV)}{\partial t} = -Ag_0 \left(\frac{\partial P}{\partial y} + \rho \frac{b}{g_0} \right) dy \tag{3.4}$$

where b is the vertical body force acceleration (usually that of gravity). When a liquid jet is discharged into gas, the surrounding pressure $P(y, t)$ is essentially uniform, and $\partial P/\partial y = 0$ in Eq. (3.4). When a liquid jet discharges horizontally into the same liquid, $\partial P/\partial y = 0$, and $b = 0$ in the direction of flow, y. The body force acceleration for a vertical jet discharge into the same fluid yields a pressure gradient

$$\frac{\partial P}{\partial y} = -\rho \, \frac{b}{g_0}$$

whether the jet is directed up or down, so the term $\partial P/\partial y + \rho b/g_0$ in Eq. (3.4) vanishes. All cases treated here are summarized as follows:

$$\frac{g_0}{\rho} \frac{\partial P}{\partial y} + b = \alpha = \begin{cases} g, & \text{vertical upward liquid jet in gas} \\[1em] -g, & \text{vertical downward liquid jet in gas} \\[1em] 0, & \text{gas jet submerged in same gas} \\[1em] 0, & \text{liquid jet submerged in same liquid} \\[1em] 0, & \text{fluid jet in zero gravity} \end{cases} \tag{3.5}$$

If we write $\dot{m} = \rho AV$ and $M = \rho A \, dy$, Eqs. (3.3) and (3.4) become

$$\frac{\partial A}{\partial t} + V \frac{\partial A}{\partial y} + A \frac{\partial V}{\partial y} = 0 \tag{3.6}$$

and

$$\frac{\partial V}{\partial t} + V \frac{\partial V}{\partial y} + \alpha = 0 \tag{3.7}$$

where α is determined from Eq. (3.5). Equations (3.6) and (3.7) can be solved for the time and space variation of jet area $A(y, t)$ and the velocity $V(y, t)$.

3.3 Unsteady Fluid Jet Solution

The jet velocity V is written as the function $V = V(y, t)$, for which its total derivative is

$$\frac{dV}{dt} = \frac{\partial V}{\partial t} + \frac{dy}{dt} \frac{\partial V}{\partial y} \tag{3.8}$$

Comparison with Eq. (3.7) leads to the two ordinary differential equations

$$\frac{dy}{dt} = V \tag{3.9}$$

and

$$\frac{dV}{dt} = -\alpha \tag{3.10}$$

Equations (3.9) and (3.10) describe motion of a free fluid particle. Since α is zero or a constant, Eq. (3.10) is first integrated to obtain V, which then is employed in Eq. (3.9) to obtain y. If the initial state of a fluid particle corresponds to its first appearance at $y = 0$ when $t = \tau$, $A = A(0, \tau)$, and $V = V(0, \tau)$, we obtain

$$V = V(0, \tau) - \alpha(t - \tau) \tag{3.11}$$

and

$$y = V(0, \tau)(t - \tau) - \frac{\alpha}{2}(t - \tau)^2 \tag{3.12}$$

Next, the jet area is obtained by writing $A = A(y, t)$, with total derivative

$$\frac{dA}{dt} = \frac{\partial A}{\partial t} + \frac{dy}{dt}\frac{\partial A}{\partial y} \tag{3.13}$$

for substitution into Eq. (3.6). Also, Eq. (3.6) contains $\partial V / \partial y$, which is obtained from Eqs. (3.11) and (3.12) by writing

$$\frac{\partial V}{\partial y} = \left(\frac{\partial V}{\partial y}\right)_t = \frac{(\partial V/\partial \tau)_t}{(\partial y/\partial \tau)_t} = \frac{1}{(t - \tau) - V(0, \tau)(dV(0, \tau)/d\tau + \alpha)^{-1}}$$

$$\tag{3.14}$$

It follows that Eq. (3.6) yields the two ordinary differential equations

$$\frac{dA}{dt} = \frac{A(dV(0, \tau)/d\tau + \alpha)}{V(0, \tau) - (dV(0, \tau)/d\tau + \alpha)(t - \tau)} \tag{3.15}$$

and $dy/dt = V$, which is identical to Eq. (3.9). If we integrate from the initial values $A = A_0$ and $t = \tau$, we get

$$A = \frac{A_0}{1 - (dV(0, \tau)/d\tau + \alpha)(t - \tau)/V(0, \tau)} \tag{3.16}$$

The jet solution given by Eqs. (3.11), (3.12), and (3.16) is easily visualized on the space-time diagram of Fig. 3.2 for the case of fluid jet

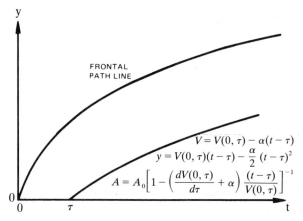

Figure 3.2   Jet Solution, Space-Time Diagram

dischage from elevation $y = 0$, starting at time $t = 0$. The frontal path line may correspond either to the first jet fluid paticle released or to the path of later particles which overtake all particles ahead of them. The overtaking process is a form of shocking.

3.4 Fluid Jet Shocking

Path lines originating at $y = 0$ and different times in Fig. 3.2 may cross each other and create a plane of discontinuity, or a shock. Consider the path lines originating at $t = \tau$ and $t = \tau + d\tau$. The later particle may overtake the earlier particle anywhere throughout the jet length. Equation (3.12) is used to give tha path lines for fluid particles discharged at τ. Another particle discharged at $\tau + d\tau$ follows the path

$$y(\tau + d\tau, t) = V(0, \tau + d\tau)(t - \tau - d\tau) - \frac{\alpha}{2}(t - \tau - d\tau)^2 \qquad (3.17)$$

The later particle overtakes the earlier particle, forming a shock when

$$y(\tau, t) = y(\tau + d\tau, t) \qquad (3.18)$$

It follows that shocking occurs at

$$(t - \tau)_{shock} = \frac{V(0, \tau)}{dV(0, \tau)/d\tau + \alpha} \qquad (3.19)$$

If shocking does occur, $(t - \tau)_{shock} > 0$. Otherwise, no particle overtakes another. Since $V(0, \tau)$ is always positive, Eq. (3.19) shows that a shock will occur only if the denominator also is positive. An upward jet with $\alpha = g$ or a submerged jet with $\alpha = 0$ and an accelerating discharge will result in a shock. A downward jet with $\alpha = -g$ will shock only if the discharge acceleration is greater than g. If we put $t - \tau = (t - \tau)_{\text{shock}}$ into Eqs. (3.11), (3.12), and (3.16), values of the elevation, velocity, and jet area at the shock location are given by

$$y_{\text{shock}} = \left[\frac{V(0, \tau)}{dV(0, \tau)/d\tau + \alpha} \right]^2 \left(\frac{dV(0, \tau)}{d\tau} + \frac{\alpha}{2} \right) \tag{3.20}$$

$$V_{\text{shock}} = \frac{V(0, \tau)[dV(0, \tau)/d\tau]}{dV(0, \tau)/d\tau + \alpha} \tag{3.21}$$

and

$$A_{\text{shock}} \to \infty \tag{3.22}$$

It can be seen that the one-dimensional jet area becomes infinite at a shock, although a two-dimensional analysis would yield an increased, but finite, area.

3.5 Steady Jet Properties

Suppose that a fluid jet has been discharging at constant velocity $V(0, t) = V_i$ from a port of area A_0 at $y = 0$ for a sufficiently long time to establish a steady jet flow. It is desirable to obtain the local jet velocity and area as functions of y. Since each fluid particle in a steady jet originates at $y = 0$ with velocity V_i and area A_0, the quantity $t - \tau$ can be eliminated from Eqs. (3.11), (3.12), and (3.16) to give the steady jet properties

$$V_{\text{steady}} = V_i \sqrt{1 - \frac{2\alpha y}{V_i^2}} \tag{3.23}$$

and

$$A_{\text{steady}} = \frac{A_0}{\sqrt{1 - 2\alpha y/V_i^2}} \tag{3.24}$$

It is seen from Eqs. (3.23) and (3.24) that if $\alpha = 0$, a steady jet has constant velocity and area. If $\alpha = +g$ for a vertical upward liquid jet in gas, Eqs. (3.23) and (3.24) show that the velocity goes to zero and the area becomes large at the front where the radical becomes imaginary. This occurs at the elevation

$$y = y_{\text{front}} = \frac{V_i^2}{2g} \tag{3.25}$$

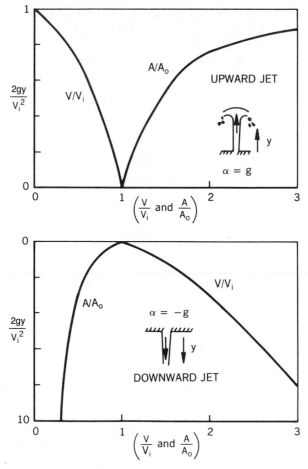

Figure 3.3 Steady Jet Velocity and Area, Upward and Downward Jets

The idealized jet model is based on a spatial coordinate y which always is positive in the direction of jet discharge. If $\alpha = -g$ for a vertical downward liquid jet in gas, the steady velocity increases as the area decreases in y. Steady jet velocity and area profiles are shown in Fig. 3.3 for different values of the constant α.

3.6 Liquid Jet, Step-Ramp Discharge

Let a liquid jet from a port of area A_0 have a discharge velocity described by a step to V_0, followed by a linear acceleration a, such that below the frontal path line of Fig. 3.2 the velocity is given by

$$V(0, t) = V(0, \tau) = V_0 + a\tau \tag{3.26}$$

It follows from Eqs. (3.11), (3.12), and (3.16) that

$$y(\tau, t) = (V_0 + a\tau)(t - \tau) - \frac{\alpha}{2}(t - \tau)^2 \tag{3.27}$$

$$V(\tau, t) = (V_0 + a\tau) - \alpha(t - \tau) \tag{3.28}$$

$$A(\tau, t) = \frac{A_0}{1 - [(a + \alpha)/(V_0 + a\tau)](t - \tau)} \tag{3.29}$$

Equation (3.27) is first solved for τ in terms of y and t, giving

$$\tau = t - \frac{V_0 + at}{\alpha + 2a}\left[1 - \sqrt{1 - \frac{2(\alpha + 2a)y}{(V_0 + at)^2}}\right] \tag{3.30}$$

Substitution into Eqs. (3.28) and (3.29) gives the jet velocity and area in the forms

$$\frac{V}{V_0 + at} = \frac{a + (a + \alpha)\sqrt{1 - \frac{2(\alpha + 2a)y}{(V_0 + at)^2}}}{\alpha + 2a} \tag{3.31}$$

and

$$\frac{A}{A_0} = \frac{a + \alpha + a\sqrt{1 - \frac{2(\alpha + 2a)y}{(V_0 + at)^2}}}{(\alpha + 2a)\sqrt{1 - \frac{2(\alpha + 2a)y}{(V_0 + at)^2}}} \tag{3.32}$$

Possible shock locations are obtained by employing Eqs. (3.26) and (3.30) in (3.20), which gives

$$y_{\text{shock}} = \frac{(V_0 + at)^2}{2(\alpha + 2a)} \tag{3.33}$$

If shocking has not occurred, that is, if no fluid particles have yet overtaken those particles discharged ahead of them, the jet front will correspond to the location of that fluid particle discharged at $\tau = 0$ in Eqs. (3.30) or (3.27), which yields

$$Y_{\text{front}} = V_0 t - \frac{\alpha}{2}t^2, \qquad \text{before shocking} \tag{3.34}$$

The shock location y_{shock} cannot exceed the jet front location y_{front}; that is,

$$y_{\text{shock}} \leq y_{\text{front}} \tag{3.35}$$

It follows from Eqs. (3.33)–(3.35) that for a step-ramp jet discharge, shocking can occur only if

Condition for Shocking in a Step-Ramp Discharge

$$[V_0 - (a + \alpha)t]^2 \leq 0 \tag{3.36}$$

Since the bracketed term is squared, it cannot be less than zero. This means that shocking can only occur at the jet front after time

$$t_{\text{shock}} = \frac{V_0}{a + \alpha}, \qquad a + \alpha > 0, \quad \text{step-ramp discharge} \tag{3.37}$$

The restriction $a + \alpha > 0$ simply implies that the time of shocking cannot be negative. A shock will not occur until a time of $V_0/(a + \alpha)$ has elapsed. The shock will first appear at a location given by putting t_{shock} into either Eq. (3.33) or (3.34), which gives

$$y_{\text{shock appears}} = \frac{\alpha + 2a}{2} \frac{V_0^2}{(a + \alpha)^2}, \qquad a + \alpha > 0 \tag{3.38}$$

Shocking first occurs at the front of a jet with a step-ramp discharge velocity characteristic, thereafter becoming the jet front at a position given by Eq. (3.33). If $V_0 = 0$, shocking will occur immediately at the jet front when discharge begins, provided that $a + \alpha > 0$.

Jet fluid velocity and area at the front before shocking are obtained by putting y_{front} into Eqs. (3.31) and (3.32), giving

$$\frac{V_{\text{front}}}{V_0 + at} = \frac{V_0 - \alpha t}{V_0 + at}, \qquad t < \frac{V_0}{a + \alpha} \quad \text{before shocking} \tag{3.39}$$

and

$$\frac{A}{A_0} = \frac{V_0}{V_0 - (a + \alpha)t}, \qquad t < \frac{V_0}{a + \alpha} \quad \text{before shocking} \tag{3.40}$$

When shocking occurs, the jet front location is obtained from Eq. (3.33) at which $A \to \infty$, and the frontal fluid velocity of Eq. (3.31) is

$$\frac{V_{\text{front}}}{V_0 + at} = \frac{a}{\alpha + 2a}, \qquad t > \frac{V_0}{a + \alpha} \quad \text{after shocking} \tag{3.41}$$

3.7 Submerged Liquid Jet, Step-Ramp Discharge

A submerged port of area A_0 suddenly discharges liquid at the step-ramp velocity described by Eq. (3.26). Equation (3.5) gives $\alpha = 0$ for a submerged jet. The jet velocity and area

$$\frac{V}{V_0 + at} = \frac{1}{2}\left[1 + \sqrt{1 - \frac{4ay}{(V_0 + at)^2}}\right] \tag{3.42}$$

and

$$\frac{A}{A_0} = \frac{1}{2}\left[1 + \frac{1}{\sqrt{1 - 4ay/(V_0 + at)^2}}\right] \tag{3.43}$$

are plotted in Fig. 3.4(a). The jet front progresses according to Eq. (3.34) until shocking occurs at the front at time $t = V_0/a$, after which Eqs. (3.33) and (3.37) show that the shock becomes the jet front. Thus,

$$y_{front} = V_0 t, \qquad t < \frac{V_0}{a} \tag{3.44}$$

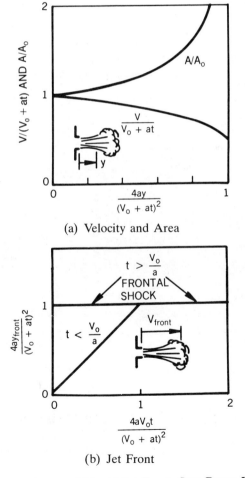

(a) Velocity and Area

(b) Jet Front

Figure 3.4 Submerged Liquid Jet Front, Step-Ramp Discharge

$$y_{\text{shock}} = \frac{(V_0 + at)^2}{4a}, \qquad t > \frac{V_0}{a} \tag{3.45}$$

The jet front is graphed in Fig. 3.4(b). It is expected that for practical applications, a submerged fluid jet shock dissipates the jet so that no substantial effect is imposed on stationary fluid ahead of the shock.

EXAMPLE 3.1: SUBMERGED WATER JET PROPERTIES
A submerged water jet begins discharging from a port of area $A_0 = 0.5\,\text{m}^2$ ($5.4\,\text{ft}^2$) and zero initial velocity, $V_0 = 0$, with a constant acceleration $a = 10\,\text{m/s}^2$ ($32.8\,\text{ft/s}^2$). Determine the unsteady cross-sectional area A and velocity V at a location $y = 5\,\text{m}$ from discharge.

Since $V_0 = 0$, we have $t > V_0/a$, and Fig. 3.4(b) shows that the jet front, corresponding to a frontal shock, arrives at $y = 5\,\text{m}$ (16.4 ft) when $4ay_{\text{front}}/(at)^2 = 1$, or

$$t = \sqrt{\frac{4y_{\text{front}}}{a}} = \sqrt{\frac{(4)(5\,\text{m})}{10\,\text{m/s}^2}} = \sqrt{2}\,\text{s} = 1.414\,\text{s}$$

Equations (3.42) and (3.43) or Fig. 3.4(a) are next employed to obtain the jet velocity and area at $y = 5\,\text{m}$ (16.4 ft) and several times after arrival in the tabulation below. The frontal shock arrives, followed immediately by a diminishing area and increasing liquid jet velocity.

Submerged Water Jet Velocity and Area

t	$\dfrac{4y}{at^2}$	$\dfrac{V}{at}$	$\dfrac{A}{A_0}$	V	A
s	(−)	(−)	(−)	m/s (ft/s)	m² (ft²)
1.414	1.0	0.5	∞	7.0 (23)	∞
2.0	0.5	0.854	1.207	17.08 (56)	0.603 (6.5)
2.5	0.32	0.912	1.106	22.80 (75)	0.555 (6.0)
3.0	0.22	0.941	1.067	28.23 (92)	0.533 (5.7)

If a submerged jet undergoes sudden discharge at constant velocity V_0 and without subsequent acceleration, that is, $a = 0$, Eq. (3.44) gives the jet frontal location as $y_{\text{front}} = V_0 t$ for all times $t < V_0/a = \infty$. Figure 3.4(a) shows that for such a case, since $4ay/(V_0 + at)^2 = 0$, $V/V_0 = A/A_0 = 1.0$. Thus, an idealized jet step discharge at constant V_0 penetrates at constant velocity and uniform area! A children's toy was designed on this principle where a spring-driven piston in a cylinder propelled an invisible plug of air, which could be directed to knock over paper targets across the room.

3.8 Vertical Liquid Jets in Gas

A step-ramp liquid jet discharge velocity of Eq. (3.26) occurs from a port of area A_0. If we employ $\alpha = g$ for a vertical upward jet and $\alpha = -g$ for a vertical downward jet from Eq. (3.5), we obtain the summary results of Eqs. (3.46)–(3.55) based on Eqs. (3.31)–(3.37).

VERTICAL UP AND DOWN LIQUID JETS, STEP-RAMP DISCHARGE

UPWARD (Y IS POSITIVE UP)

$$\frac{V}{V_0 + at} = \frac{a + (a + g)f_u(y, t)}{2a + g} \tag{3.46}$$

$$\frac{A}{A_0} = \frac{a + g + af_u(y, t)}{(2a + g)f_u(y, t)} \tag{3.47}$$

$$f_u(y, t) = \sqrt{1 - \frac{2(2a + g)y}{(V_0 + at)^2}} \tag{3.48}$$

$$y_{\text{front}} = V_0 t - \frac{g}{2} t^2, \qquad t < t_{\text{shock}} = \frac{V_0}{a + g} \tag{3.49}$$

$$y_{\text{front}} = y_{\text{shock}} = \frac{(V_0 + at)^2}{2(2a + g)}, \qquad t > t_{\text{shock}} = \frac{V_0}{a + g} \tag{3.50}$$

DOWNWARD (Y IS POSITIVE DOWN)

$$\frac{V}{V_0 + at} = \frac{a + (a - g)f_d(y, t)}{2a - g} \tag{3.51}$$

$$\frac{A}{A_0} = \frac{a - g + af_d(y, t)}{(2a - g)f_d(y, t)} \tag{3.52}$$

$$f_d(y, t) = \sqrt{1 - \frac{2(2a - g)y}{(V_0 + at)^2}} \tag{3.53}$$

$$y_{\text{front}} = V_0 t + \frac{g}{2} t^2 \tag{3.54}$$

$$y_{\text{front}} = y_{\text{shock}} = \frac{(V_0 + at)^2}{2(2a - g)},$$

$$\text{shock only if } a > g \text{ when } t \geq t_{\text{shock}} = \frac{V_0}{a - g} \tag{3.55}$$

Equations (3.46) and (3.51) give the liquid velocity for upward and downward step-ramp jet discharges. Shocking always occurs in an upward jet, after which the shock becomes the advancing jet front. Shocking in a downward jet can occur only if $a > g$.

3.9   Comparison with Rigorous Two-Dimensional Calculation

Validity of the one-dimensional liquid jet analysis is partly evaluated in this section by comparison with a rigorous two-dimensional calculation. The test problem involves a flat liquid surface which is rising at constant acceleration $a_c = 40 \, \text{m/s}^2$ in a container of rigid vertical walls and cross-sectional area $A_c = 2 \, \text{m}^2$, as shown in Fig. 3.5. The rising liquid surface suddenly encounters an orifice of flow area $A_0 = 1.0 \, \text{m}^2$. If liquid below the orifice continues the same acceleration after impact, it is necessary to determine velocity and elevation of the jet front above the orifice. Liquid velocity at impact is $V_{ci} = 12 \, \text{m/s}$.

Continuity gives the initial velocity through the orifice as

$$V_0 = \frac{A_c}{A_0} V_{ci} = \frac{2 \, \text{m}^2}{1 \, \text{m}^2} (12 \, \text{m/s}) = 24 \, \text{m/s}$$

with corresponding acceleration given by

$$a = \frac{A_c}{A_0} a_c = \frac{2 \, \text{m}^2}{1 \, \text{m}^2} (40 \, \text{m/s}^2) = 80 \, \text{m/s}^2$$

Thus, jet velocity at the orifice corresponds to a vertical upward step-ramp discharge with velocity given in the form

$$V = V_0 + at = 24 + 80t \, \text{m/s} \qquad (t = \text{seconds})$$

Since $\alpha = g$ from Eq. (3.5), the jet front elevation is obtained from Eq. (3.34) as

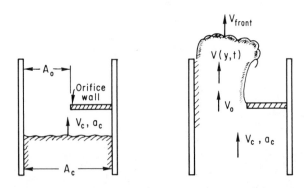

Figure 3.5   Jet from Liquid Surface Impact on Orifice

$$y_{front} = V_0 t - \frac{g}{2} t^2 = 24t - 4.9t^2$$

until a shock forms according to Eq. (3.37) at

$$t_{shock} = \frac{V_0}{a + g} = \frac{24 \text{ m/s}}{80 + 9.8 \text{ m/s}^2} = 0.26 \text{ s}$$

after which Eq. (3.33) gives the jet front elevation as

$$y_{front} = y_{shock} = \frac{(V_0 + at)^2}{2(2a + g)} = \frac{(24 + 80t \text{ m/s})^2}{340 \text{ m/s}^2}$$

The jet velocity at the front is obtained from Eqs. (3.39) and (3.41), giving

$$V_{front} = V_0 - \alpha t = 24 \text{ m/s} - (9.8 \text{ m/s}^2)t$$

$$t < \frac{V_0}{a + \alpha} = \frac{24 \text{ m/s}}{80 + 9.8 \text{ m/s}^2} = 0.26 \text{ s}$$

and

$$V_{front} = (V_0 + at) \frac{a}{2a + g} = (24 + 80t \text{ m/s})(0.47)t > 0.26 \text{ s}$$

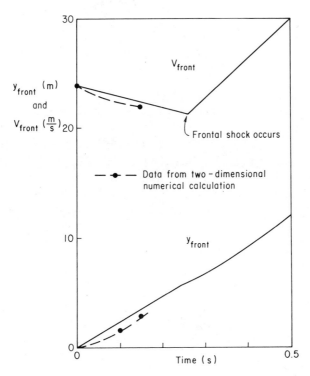

Figure 3.6 Example Computation, Liquid Jet Properties

The jet front and velocity are plotted in Fig. 3.6 for the given sample numbers. These predictions are compared with a rigorous two-dimensional analysis performed by one of the SOLA programs [3.6]. Reasonable agreement is shown to the extent of the numerical solution. The one-dimensional analysis shows that when a shock forms at the jet front, there is an abrupt change in the frontal liquid velocity.

EXAMPLE 3.2: LOCAL ACCELERATION OF UPWARD JET

Two engineers are having an argument about the local acceleration of an upward liquid jet. They have the responsibility of designing braces for use at elevations y_1 and y_2, which suddenly will be submerged by an upward-traveling, accelerating liquid jet. It is known that the unsteady liquid force exerted on a submerged stationary structure is proportional to the local acceleration $(\partial V / \partial t)_y$ (see Chapter 9). The jet discharge in this application corresponds to a step-ramp velocity $V = V_0 + at$. One engineer has a reputation of being overly conservative and wants to design the brace for acceleration a. The other engineer insists that since all liquid particles in the upward jet will have decreasing velocities, acceleration of the jet at various elevations will be substantially less than a. Who is right? Take parameters $V_0 = 24\ \text{m/s}$ (79 ft/s), $a = 80\ \text{m/s}^2$ (262 ft/s²) and structures at $y_1 = 5\ \text{m}$ (16.4 ft) and $y_2 = 10\ \text{m}$ (32.8 ft).

Equation (3.49) shows that a shock occurs at

$$t_{\text{shock}} = \frac{V_0}{a + g} = 0.267\ \text{s}$$

for which the jet front would have risen to

$$y_{\text{front}} = V_0 t - \frac{g}{2} t^2 = 6.058\ \text{m}\quad (19.9\ \text{ft})$$

That is, the jet frontal shock occurs above $y_1 = 5\ m$ and below $y_2 = 10\ m$. Consider the lower structure at $y_1 = 5\ \text{m}$ first, for which Eq. (3.49) is solved for the time of jet arrival, that is,

$$t_{\text{arrival, 1}} = \frac{V_0}{g} \pm \sqrt{\left(\frac{V_0}{g}\right)^2 - \frac{2y_1}{g}} = 0.218\ \text{s}$$

The negative sign was chosen for the upward arrival time. The positive sign would correspond to the jet rising, falling, and arriving at y_1 on the way down, which is not considered in this problem. Furthermore, since the frontal shock forms before the jet front reaches $y_2 = 10\ \text{m}$, Eq. (3.50) with $y_{\text{shock}} = y_2$ is used to determine the corresponding arrival time:

$$t_{\text{arrival, 2}} = \frac{\sqrt{2(2a + g)y_2}}{a} - \frac{V_0}{a} = 0.428\ \text{s}$$

Arrival times $t_{\text{arrival,1}}$ and $t_{\text{arrival,2}}$ determine when the jet front arrives at both structures after which the local acceleration is required.

The upward jet fluid velocity of Eq. (3.46) is differentiated to obtain the local acceleration as

TABLE 3.1 Example Local Jet Acceleration at Two Elevations

Structure at $y_1 = 5$ m		Structure at $y_1 = 10$ m	
t	$\left(\dfrac{\partial V}{\partial t}\right)_y$	t	$\left(\dfrac{\partial V}{\partial t}\right)_y$
s	m/s^2 (ft/s^2)	s	m/s^2 (ft/s^2)
0.218	434 (1424)	0.428	very high
0.5	93 (305)	0.5	140 (459)
1.0	84 (276)	1.0	89 (292)
2.0	81 (266)	2.0	82 (269)
10.0	80 (262)	10.0	80 (262)

$$\left(\frac{\partial V}{\partial t}\right)_y = \frac{a}{2a+g}\left[a + \frac{a+g}{f_u(y,t)}\right]$$

Accelerations calculated at both structures are given in Table 3.1. Neither engineer is ready to believe these results, which show that for an upward step-ramp jet discharge velocity, the acceleration at a stationary structure first *exceeds* the discharge acceleration of $a = 80$ m/s^2, but eventually approaches it.

The results of Example 3.2 seems to violate physical laws! How can a vertical upward liquid jet with constant discharge acceleration have higher acceleration at some fixed elevation? This unusual result is explained by the fact that upward-traveling adjacent fluid mass increments flatten and slow down as they rise, simultaneously moving closer together in the axial direction. But even though the increments slow down, the net result is that a larger velocity difference occurs at a given location in a shorter time. This behavior corresponds to an increased local acceleration.

EXAMPLE 3.3: DOWNWARD JET, IMPINGEMENT VELOCITY
A downward-pointing circular water jet, $\rho = 1$ g/cm^3, is to be discharged through a nozzle of area $A_0 = 1.0$ cm^2 at constant accelerations of $a = g$ and $g/2$ from $V_0 = 0$ starting velocity. The jet will impinge on a horizontal flat plate a distance $y_p = 100$ cm from the nozzle. Determine the impingement force $F = \rho A V^2/g_0$ at the time of arrival for both cases.

Equations (3.51)–(3.55) show that for a downward jet, shocking occurs only if $a > g$. Therefore, this analysis has no shocking. The limiting case of $a = g$ in Eqs. (3.51) and (3.52) gives

$$V = gt \quad \text{and} \quad A = A_0$$

for a nonshocking jet front elevation $y_{front} = gt^2/2$. If we put $y_{front} = y_p$, arrival time of the jet at the flat plate is $t_{arrival} = \sqrt{2y_p/g}$, for which the force is

$$F = 2\rho \frac{g}{g_0} y_p A_0, \qquad a = g$$

$$= 196,000 \text{ dynes} = 1.96 \text{ N}$$

A similar analysis for $a = g/2$ requires a limiting process in Eqs. (3.51) and (3.52). If we put $a = g/2 + \epsilon$ and take the limit as $\epsilon \rightarrow 0$, we obtain

$$V = \frac{gt}{2} + \frac{y}{t}, \qquad A = \left(1 - \frac{2y}{gt^2}\right) A_0, \qquad y_{front} = \frac{g}{2} t^2$$

The arriving impingement force on the flat plate at $y_{front} = y_p$ is

$$F = 0$$

because the jet area is zero at the front.

An impingement force usually is considered as a suddenly applied load, for which dynamic structural response is different than if a load were slowly applied. Both forces predicted in this example would continue to increase in time.

3.10 Heated Liquid Flows in Pipes, Basic Formulation

An incompressible fluid undergoes one-dimensional flow in a rigid pipe and may be heated or cooled by wall heat transfer. Moreover, its entering temperature and flow rate may vary with time. Figure 3.7 shows a simple, frictionless, horizontal pipe of length L and space-dependent cross-sectional area $A(x)$. The pipe is being heated and cooled simultaneously at rates q'_{in} and q'_{out}, where the prime designates wall thermal energy transfer rates per unit length. Since q'_{in} and q'_{out}, generally depend on the local wall and fluid temperatures T_w and T, we have

$$q'_{in} = q'_{in}(x, t, T, T_w)$$

$$q'_{out} = q'_{out}(x, t, T, T_w) \tag{3.56}$$

Axial heat conduction in the fluid is neglected here, although it can play an important role in slow flows and certain liquid metal applications (see Problem 3.20). Fluid of density ρ enters the pipe at velocity $V(0, t)$ and temperature $T(0, t)$. The mass flow rate at the entrance, or at any other location x, is

$$\dot{m}(t) = \rho A(0)V(0, t) = \rho A(x)V(x, t) \tag{3.57}$$

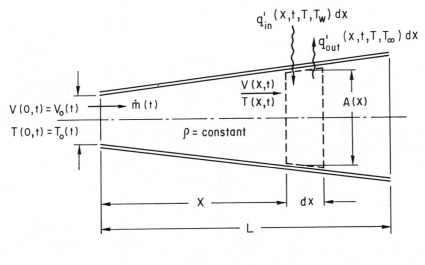

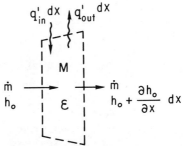

Figure 3.7 Model for Heated or Cooled Liquid Flow in Pipes

The purpose of this section is to formulate a problem from which the time- and space-dependent fluid and pipewall temperatures $T(x, t)$ and $T_w(x, t)$ can be obtained.

Equation (3.57) gives fluid velocity in terms of the entering value $V(0, t)$ as

$$V(x, t) = \frac{A(0)}{A(x)} V(0, t) \qquad (3.58)$$

The energy diagram of Fig. 3.7 yields

$$\dot{m}\, \frac{\partial h_0}{\partial x}\, dx + (q'_{\text{out}} - q'_{\text{in}})\, dx + \frac{\partial}{\partial t} (M\mathscr{E}) = 0 \qquad (3.59)$$

If the substitutions

$$h_0 = h + \frac{g}{g_0} y + \frac{V^2}{2g_0} , \qquad \dot{m} = \rho A V$$

$$\mathscr{E} = e + \frac{g}{g_0} y + \frac{V^2}{2g_0} , \qquad M = \rho A \, dx \qquad (3.60)$$

$$h = c_p T , \qquad e = c_V T$$

are made, and we consider only cases where the kinetic and potential energies are small, Eq. (3.59) yields

$$\frac{\partial e}{\partial t} + V \frac{\partial h}{\partial x} = \frac{q'_{in} - q'_{out}}{\rho A} \qquad (3.61)$$

If we further substitute $h = c_p T$ and $e = c_V T$, and recall from Table 2.3 that for liquid we have $c_p = c_V$, Eq. (3.61) can be written in terms of temperature as

$$\frac{\partial T}{\partial t} + V \frac{\partial T}{\partial x} = \frac{q'_{in}(x, t, T, T_w) - q'_{out}(x, t, T, T_w)}{\rho A(x) c_P} \qquad (3.62)$$

Equation (3.62) is the differential equation which describes time- and space-dependent liquid temperature in a heated or cooled pipe. The velocity V is obtained from Eq. (3.58).

There are many situations where Eq. (3.62) is readily integrated along fluid path lines. Some cases may require a numerical integration, but the process is straightforward. A general path integral solution is discussed next.

3.11   Path Integral Solution

The fluid temperature is written functionally as

$$T = T(x, t) \qquad (3.63)$$

for which the total derivative is

$$\frac{dT}{dt} = \frac{\partial T}{\partial t} + \frac{dx}{dt} \frac{\partial T}{\partial x} \qquad (3.64)$$

Comparison of Eqs. (3.62) and (3.64) suggests that we set

$$\frac{dx}{dt} = V(x, t) \qquad (3.65)$$

which describes a path line traced by a fluid particle. It follows that the temperature must satisfy the ordinary differential equation,

$$\frac{dT}{dt} = \frac{q'_{in}(x, t, T, T_w) - q'_{out}(x, t, T, T_w)}{\rho A(x)c_P} \tag{3.66}$$

The path of any fluid particle is obtained independently of temperature by employing Eq. (3.58) for $V(x, t)$ in Eq. (3.65). Integration for a particle starts at initial time t_i when the particle position is x_i. That is,

$$\int_{x_i}^{x} A(x)\, dx = A(0) \int_{t_i}^{t} V(0, t)\, dt \tag{3.67}$$

for which a general solution gives the fluid particle path

$$x = x(t; x_i, t_i) \tag{3.68}$$

If the solution for x is employed in the right side of Eq. (3.66), the ordinary differential equation for T on the path of Eq. (3.68) reduces to

$$\frac{dT}{dt} = F(t, T, T_w; x_i, t_i) \tag{3.69}$$

If the temperature of a fluid particle at x_i and t_i is given by T_i, an appropriate initial condition for Eq. (3.69) is written as

$$T(x_i, t_i) = T_i \tag{3.70}$$

A graphical display of the path integral solution is given by the time-space diagram of Fig. 3.8. The dividing path line is traced by that liquid particle initially entering the flow passage at time $t = 0$ and corrsponds to

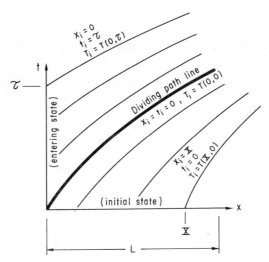

Figure 3.8 Space-Time Diagram, Heated or Cooled Liquid Flow

$$x_i = t_i = 0$$

in Eq. (3.67). All fluid particles initially in the flow passage at time $t = 0$ have path lines *below* the dividing path line, with initial values

$$x_i = X$$
$$t_i = 0 \quad \text{below dividing path line}$$
$$T_i = T(X, 0)$$

All fluid particles entering the flow passage have path lines *above* the dividing path line, and have initial values

$$x_i = 0$$
$$t_i = \tau \quad \text{above dividing path line}$$
$$T_i = T(0, \tau)$$

The solution expressed by a time-space diagram enables one to obtain temperature at any position x and time t by determining which fluid path line passes through x, t and by integrating temperature from the corresponding values of T_i, x_i, and t_i.

The procedure described in this section gives a general path integral solution for obtaining the time- and space-dependent temperature of an incompressible fluid in a heated or cooled flow passage. Specific solutions depend on the nature of heat transfer mechanisms which determine $q'_{in}(x, t, T, T_w)$ or $q'_{out}(x, t, T, T_w)$.

3.12 Cases for Pipes with Negligible Thermal Capacity

Consider a pipe wall of radius R, thickness ΔR, and length ΔL. If its density is ρ_w and its specific heat is c_{vw}, its thermal capacity $C_{t,w}$ is approximately

$$C_{t,w} = \frac{\partial U_w}{\partial T} = \frac{\partial(\rho_w c_{vw} 2\pi R \, \Delta R \, \Delta L T)}{\partial T} = 2\pi \rho_w c_{vw} R \, \Delta R \, \Delta L$$

Similarily, the thermal capacity C_t of a contained plug of fluid with density ρ and specific heat c_v is

$$C_t = \frac{\partial}{\partial T} (\rho c_p \pi R^2 \, \Delta L T) = \pi \rho c_p R^2 \, \Delta L$$

The wall heat capacity can be neglected whenever it is much less than that of the liquid; that is, whenever

$$\frac{C_{t,w}}{C_t} = 2 \frac{\rho_w}{\rho} \frac{c_{vw}}{c_p} \frac{\Delta R}{R} \ll 1 , \qquad \text{negligible pipe thermal capacity}$$

$$(3.71)$$

EXAMPLE 3.4: WATER HEATING DURING PIPE FLOW

A thin-walled heating pipe of radius $R = 0.5$ cm (0.2 in), flow area $A = 0.785$ cm^2 (0.12 in^2), heated perimeter $P_h = 3.14$ cm (1.24 in), and length $L = 2$ m (6.56 ft), raises the temperature of water flowing at velocity $V = 1$ m/s (3.28 ft/s). The pipe has additional properties $\rho_w = 4$g/cm^3 (249 lbm/ft^3), $c_{vw} = 0.42$ kJ/kg-K (0.1 B/lbm-°F), and wall thickness $\Delta R = 0.05$ cm (0.02 in). The water properties are $\rho = 1.0$ g/cm^3 (62.4 lbm/ft^3) and $c_p = 4.2$ kJ/kg-K (1.0 B/lbm-°F). Electric heating coils are suddenly turned on to provide a thermal power input to the liquid according to the ramp function

$$q'(x) = q'_{max} \frac{x}{L}$$

where $q'_{max} = 2000$ W/m (0.58 B/s-ft). The pipe and flowing water initially are at $T_i = 20°C$ (68°F). Determine the liquid temperature at discharge as a function of time.

The criterion of Eq. (3.71) yields

$$\frac{C_{t,w}}{C_t} = 0.08$$

so that an approximate solution for a pipe with negligible thermal capacity seems justified. The electrically heated case is analyzed with Eqs. (3.65) and (3.66), for which

$$\frac{dx}{dt} = V \quad \text{and} \quad \frac{dT}{dt} = \frac{q'_{in}(x)}{\rho A c_p} = \left(\frac{q'_{max}}{L\rho A c_p} \right) x$$

Integration of both equations from x_i, t_i, T_i, to x, t, T yields

$$x - x_i = V(t - t_i)$$

$$T(x, t) = T_0 + \left(\frac{q'_{max}}{L\rho A c_p} \right) \frac{1}{2V} (x^2 - x_i^2)$$

The dividing path line in the t, x diagram of Fig. 3.8 corresponds to

$$x = Vt$$

below which $t_i = 0$, with

$$T(x, t) = T_0 + \left(\frac{q'_{max}}{L\rho A c_p} \right) \left(x - \frac{V}{2} t \right) t , \qquad t < \frac{x}{V}$$

and above which $x_i = 0$, with

TABLE 3.2 Example Water Discharge Temperature, Electrically Heated Pipe with Negligible Thermal Capacity

t(s)	T(L, t) °C	T(L, t) °F
0.0	20.0	68
0.5	20.3	68.5
1.0	24.5	76.1
1.5	25.7	78.3
2.0	26.1	79

$$T(x, t) = T_0 + \left(\frac{q'_{max}}{L\rho A c_p}\right) x^2/2V , \qquad t \geq \frac{x}{V}$$

Discharge temperature corresponds to $x = L$, for which

$$T(L, t) = T_0 + \frac{q'_{max}}{L\rho A c_p} \begin{cases} \left(L - \frac{V}{2} t\right)t , & t < \frac{L}{V} \\ \dfrac{L^2}{2V} , & t \geq \frac{L}{V} \end{cases}$$

It follows that for the given data,

$$\frac{q'_{max}}{L\rho A c_p} = 3.03°C/\text{m-s} \; (0.51 \, F/\text{ft-s})$$

The temperature $T(L, t)$ is tabulated in Table 3.2. The discharge temperature rises from 20°C (68°F) to 26.1°C (79°F) after which it remains constant in time.

Consider a pipe which is not heated electrically, as in Example 3.4. When the condition of Eq. (3.71) is satisfied, the wall becomes a quasi-steady heat transfer boundary at average temperature T_w between the internal fluid at temperature T and the environment at T_∞. If the pipe heat transfer is dominated by convection at the inner and outer surfaces, we have from Fig. 3.9, that

$$q''_{out} = \mathcal{H}(T - T_w) = \mathcal{H}_\infty(T_w - T_\infty)$$

for which

$$q'_{out} = \frac{q''_{out} P_w \, \Delta L}{\Delta L} = \mathcal{H} P_w (T - T_w) = \frac{\mathcal{H} \mathcal{H}_\infty}{\mathcal{H} + \mathcal{H}_\infty} P_w (T - T_\infty) \quad (3.72)$$

where P_w is the wall perimeter, and the wall temperature is

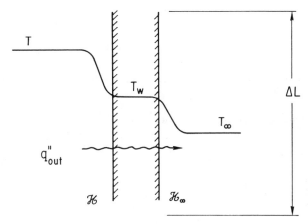

Figure 3.9   Convection-Dominated Heat Transfer

$$T_W = T_\infty + \frac{T - T_\infty}{1 + \mathcal{H}_\infty/\mathcal{H}} \tag{3.73}$$

Consider cases for which the initial pipe and fluid temperatures are equal to the ambient value, and the entrance temperature begins to vary with time. The corresponding initial and boundary conditions are

$$T(x, 0) = T_w(x, 0) = T_\infty \tag{3.74}$$

and

$$T(0, t) - T_\infty = (T_0 - T_\infty)f(t) \tag{3.75}$$

where $f(t)$ is an arbitrary function of time. Equation (3.72) is employed in (3.66) with (3.65) to obtain a uniform pipe solution for conditions of (3.74) and (3.75) in the following form:

Fluid and pipe wall temperatures for incoming temperature variation $(T_0 - T_\infty)f(t)$ with constant ambient temperature T_∞. Thin-walled pipe has negligible thermal capacity.

$$\frac{T(x, t) - T_\infty}{T_0 - T_\infty} = \left(1 + \frac{\mathcal{H}_\infty}{\mathcal{H}}\right) \frac{T_w(x, t) - T_\infty}{T_0 - T_\infty}$$

$$= \left\{ \exp\left[-\left(\frac{\mathcal{H}\mathcal{H}_\infty}{\mathcal{H} + \mathcal{H}_\infty}\right) \frac{P_w}{\rho A c_p} \frac{x}{V} \right] \right\} f\left(t - \frac{x}{V}\right) H_s\left(t - \frac{x}{V}\right) \tag{3.76}$$

The unit step $H_s(t - x/V)$ is 1.0 for all positive values of its argument $t - x/V$, and zero otherwise. Physically, it implies that temperature at any x does not begin to change until the incoming fluid temperature front arrives.

It is seen that the same attenuation envelope is obtained for all incoming fluid temperature functions. Pipe sections further from the entrance are subjected to smaller temperature disturbances.

If fluid initially is flowing in a uniform pipe at temperature $T_0(x, 0)$ which is abruptly submerged in a convective environment at T_∞, the temperature solution is as follows:

Fluid pipe wall temperatures for incoming and initial temperature T_0, and suddenly applied external temperature T_∞. Thin-walled pipe has negligible thermal capacity.

$$\frac{T(x, t) - T_\infty}{T_0 - T_\infty} = \left(1 + \frac{\mathcal{H}_\infty}{\mathcal{H}}\right)\frac{T_w(x, t) - T_\infty}{T_0 - T_\infty}$$

$$= \exp\left\{\left[-\left(\frac{\mathcal{H}\mathcal{H}_\infty}{\mathcal{H} + \mathcal{H}_\infty}\right)\frac{P_w}{\rho A c_p}\right]\begin{bmatrix} t, & t < \dfrac{x}{V} \\ \dfrac{x}{V}, & t \geq \dfrac{x}{V} \end{bmatrix}\right\} \qquad (3.77)$$

EXAMPLE 3.5: LIQUID HEATING FROM A HOT BATH
The same tube and water flow properties of Example 3.4 apply, only instead of electrical heating, the entire tube is suddenly submerged in a hot fluid bath. Heat transfer is controlled by the natural convection bath coefficient $\mathcal{H}_\infty = 50 \text{ W/m}^2\text{-}°C$ (8.8 B/h-ft^2-°F), which is much smaller than the forced convection inside the tube. Determine the required hot bath temperature T_∞ to raise incoming tube water from 20°C (68°F) to 26.1°C (79°F) as in Example 3.4. The transient discharge temperature $T(L, t)$ and wall temperature $T_w(L, t)$ should be determined as well.

The boundary condition of Eq. (3.75) yields $f(t) = 1.0$. Since $\mathcal{H}_\infty \ll \mathcal{H}$, Eq. (3.77) can be employed at $x = L$ to give

$$\frac{T(L, t) - T_\infty}{T_0 - T_\infty} = \exp\left\{-\frac{\mathcal{H}_\infty P_w}{\rho A c_p}\begin{bmatrix} t, & t < \dfrac{L}{V} \\ \dfrac{L}{V}, & t \geq \dfrac{L}{V} \end{bmatrix}\right\}$$

which can be solved for T_∞. That is, for $t > L/V$ at steady state,

$$\frac{\mathcal{H}_\infty P_w L}{\rho A c_p V} = 0.00833$$

for which

$$T_\infty = \frac{T(L, t) - T_0 e^{-0.00833}}{1 - e^{-0.00833}} = 755°C \ (1391°F)$$

A hot bath of 755°C (1391°F) would be required to raise the discharge temperature to 26.1°C (79°F). Since $\mathcal{H}_\infty \ll \mathcal{H}$, the wall and liquid temperature are essentially equal and vary according to

TABLE 3.3 Transient Discharge Water Temperature from Hot Bath Heating Tube Flow

t(s)	$T(L, t) = T_w(L, t)$ °C	°F
0.0	20.0	68
0.5	21.5	71
1.0	23.0	73
1.5	24.6	76
2.0	26.1	79

$$T(L, t) = T_w(L, t) = T_\infty + (T_0 - T_\infty)e^{-0.00416t}$$

$$= 755 - 735\, e^{-0.00416t} \qquad \left(t < \frac{2\,\text{m}}{1\,\text{m/s}} = 2\,\text{s}\right)$$

The discharge temperature $T(L, t)$ is given in Table 3.3. The convective heating design causes the water discharge temperature to increase faster at first and then slower than the electrically heated case given in Table 3.2.

_____

3.13 Thin-Walled Pipe, Thermal Capacity Important

Many practical cases of hot or cold fluid flow in pipes have significant coupling between the pipe wall and fluid temperatures. Here, the thermal capacity criterion of Eq. (3.71) is not satisfied, which means that the pipe has significant thermal capacity. However, the pipe wall is considered thin if its radial temperature gradient is small, so it can be approximated by an average value $T_w(x, t)$ at any axial location. Heat conduction would pass through the pipe wall at the penetration velocity of Eq. (2.7) in a response time of

$$\Delta t_{\kappa,r} = \frac{\Delta R}{V_{p,\kappa}} = \frac{(\Delta R)^2}{2\alpha_w}$$

The pipe longitudinal temperature response corresponds approximately to the fluid transit time L/V. If $\Delta t_{\kappa,r}$ is much smaller than L/V, the thin wall criterion becomes

Thin-Walled Pipe, Uniform Radial Temperature

$$\frac{V(\Delta R)^2}{2L\alpha_w} \ll 1 \tag{3.78}$$

Also, axial heat transfer in a pipe wall is negligible if the fluid transport time L/V is much shorter than the axial conduction response time $\Delta t_{\kappa,L} = L^2/2\alpha_w$. That is,

Negligible Axial Pipe Heat Transfer

$$\frac{2\alpha_w}{VL} \ll 1 \tag{3.79}$$

3.14 Coupled Fluid and Pipe Wall Temperatures

Energy conservation for a thin-walled, nonaxial conducting pipe with substantial thermal capacity, which satisfies the criteria of (3.78) and (3.79), is obtained for Fig. 3.10 as

$$q_{\text{out}_\infty} + q_{\text{out}} - q_{\text{in}} + \frac{dU_w}{dt} = 0$$

If we write the wall internal energy as $U_w = \rho_w A_w c_{vw} T_w\, dx$ and consider convective heating or cooling of the inner and outer walls as $q_{\text{out}_\infty} = \mathcal{H}_\infty P_{w0}(T_w - T_\infty)\, dx$ and $q_{\text{out}} - q_{\text{in}} = \mathcal{H} P_{wi}(T_w - T)\, dx$, it follows that

$$\frac{\partial T_w}{\partial t} + \frac{\mathcal{H} P_{wi}(x)}{\rho_w A_w(x) c_{vw}} \left[(T_w - T) + \frac{\mathcal{H}_\infty}{\mathcal{H}} \frac{P_{w0}(x)}{P_{wi}(x)} (T_w - T_\infty) \right] = 0 \tag{3.80}$$

Equations (3.63)–(3.65) are rewritten for the convective fluid-to-wall heat transfer as

$$\frac{dT}{dt} = \frac{\partial T}{\partial t} + V \frac{\partial T}{\partial x} = \frac{\mathcal{H} P_{wi}(x)(T_w - T)}{\rho A(x) c_p} \quad \text{on} \quad \frac{dx}{dt} = V \tag{3.81}$$

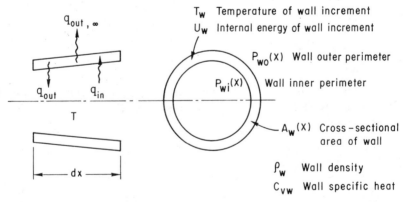

T_W Temperature of wall increment
U_W Internal energy of wall increment

$P_{wo}(x)$ Wall outer perimeter

$P_{wi}(x)$ Wall inner perimeter

$A_W(x)$ Cross-sectional area of wall

ρ_W Wall density
c_{vw} Wall specific heat

Figure 3.10 Energy Conservation, Pipe Wall

Thus, Eqs. (3.80) and (3.81) are two simultaneous first-order partial differential equations for the unsteady fluid and pipe wall temperatures.

3.15   Finite Difference Solution of Coupled Equations

A solution for the fluid temperature of Eq. (3.81) can be obtained from the reduction to two ordinary differential equations given by Eqs. (3.65) or (3.67) and (3.69). The fluid path lines are obtained from integration according to Eq. (3.67), and are shown in Fig. 3.11 for a general flow passage and time-dependent entrance velocity and temperature. If the flow passage has constant area and the entrance velocity is constant, the path lines are straight and parallel. Each path line is identified by the index j, each vertical line by the index i. The crossing of i and j lines identifies mesh points. It is assumed that values of T and T_w are known at each mesh point of the $j-1$ path line before points on the j path line are calculated.

Equation (3.81) is written in a forward difference form between mesh points $i-1$, j and i, j as

$$T_{i,j} = T_{i-1,j} + \frac{\mathscr{H}P_{wi}}{\rho A(x_i)c_p}\,(T_{wi-1,j} - T_{i-1,j})\,\Delta t_{i,i-1} \qquad (3.82)$$

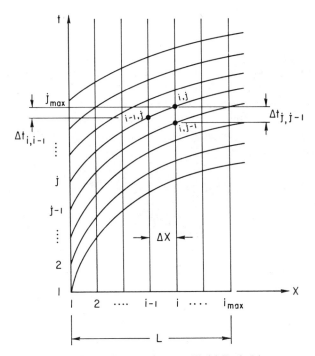

Figure 3.11   Integration on Fluid Path Lines

Thus, $T_{i,j}$ can be calculated at each mesh point on the j path line by a simple marching process from $i = 1$ to $i_{\max}$. Furthermore, Eq. (3.80) is written in forward difference form between mesh points $i, j - 1$ and i, j to give

$$T_{w\,i,\,j} = T_{w\,i,\,j-1} + \frac{\mathcal{H}P_{wi}(x_i)}{\rho_w A_w(x_i)c_{vw}}\left[(T_{w\,i,j-1} - T_{i,j-1})\right.$$
$$\left. + \frac{\mathcal{H}_\infty P_{w0}(x_i)}{\mathcal{H}P_{wi}(x_i)}\,(T_{w\,i,j-1} - T_\infty)\right]\Delta t_{j,\,j-1}$$

$$(3.83)$$

Values of $T_{w\,i,j}$ are obtained on the j path line by marching from $i = 1$ to $i_{\max}$.

The Courant stability criterion

$$\Delta t \le \frac{\Delta x}{V}$$

is employed as is done in the method of characteristics. Reasonable accuracy should be obtained if the computational time step is much less than either the pipe wall thermal response time in Section 3.13 or the fluid convective thermal response time which is obtained from the exponential term in a solution of Eq. (3.81) for a uniform pipe. That is,

$$\Delta t \ll \frac{(\Delta R)^2}{2\alpha} \quad \text{or} \quad \frac{\rho A c_p}{\mathcal{H}P_w}$$

Arbitrary variation of entering fluid temperature can be employed in this method. Also, the initial pipe temperature T_i is assigned to the $j = 1$ path line at mesh points corresponding to $i = 1, 2, \ldots, i_{\max}$.

Figure 3.12 gives a sample computation for an insulated circular pipe and a step change in fluid entrance temperature from T_i to T_0. The variables were normalized according to

$$T^*(x^*, t^*) = \frac{T(x, t) - T_i}{T_0 - T_i} \tag{3.84}$$

$$T_w^*(x^*, t^*) = \frac{T_w(x, t) - T_i}{T_0 - T_i} \tag{3.85}$$

$$x^* = \frac{x}{\tau V}, \qquad t^* = \frac{t}{\tau} \tag{3.86}$$

with

$$\tau = \frac{\rho A c_p}{\mathcal{H}P_{wi}}, \qquad \tau_w = \frac{\rho_w A_w c_{vw}}{\mathcal{H}P_{wi}} \tag{3.87}$$

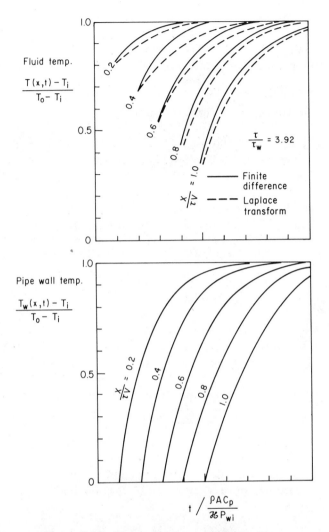

Figure 3.12 Unsteady Temperature Profiles

TABLE 3.4   Example Parameters, Hot Liquid Entering Cold Pipe

$T_0 = 200°C$	$R_i = 0.10$ m
$T_i = 50°C$	$R_0 = 0.112$ m
$\mathcal{H} = 10,000$ W/m^2-°C	$c_p = 4.0$ J/g-°C
$\mathcal{H}_\infty = 0.0$ W/m^2-°C	$V = 1.0$ m/s
$L = 20$ m	$\rho = 1000$ kg/m^3
$\rho_w = 10,000$kg/m^3	$c_{pw} = 0.4$ J/g-°C

Computations were carried out for a pipe length of $x^*_{max} = 1.0$, with 10 space increments of $\Delta x^* = 0.1$ and time steps $\Delta t^* = 0.1$ up to $t^*_{max} = 2.0$. Parameters of Table 3.4 were employed, for which $\tau/\tau_w = 3.92$. It is seen that both the insulated pipe and fluid temperatures approach the entrance value.

3.16 Solution by Laplace Transformation

This section is offered for those who have experience ranging from modest to extensive in the use of Laplace transforms to solve differential equations. The same problem described in Section 3.14 and solved by a finite difference technique in Section 3.15 is used as an example here so that the solutions can be compared.

If T_w is eliminated from Eqs. (3.80) and (3.81), the describing equation for liquid temperature in a constant area, insulated pipe ($\mathcal{H}_\infty = 0$), is given by

$$\frac{\partial}{\partial t}\left(\frac{\partial T}{\partial t} + V \frac{\partial T}{\partial x}\right) + \left(\frac{1}{\tau} + \frac{1}{\tau_w}\right)\frac{\partial T}{\partial t} + \frac{V}{\tau_w}\frac{\partial T}{\partial x} = 0 \qquad (3.88)$$

where τ and τ_w are given by Eq. (3.87). Consider liquid at initial temperature T_i, disturbed suddenly by a time-dependent incoming temperature variation at the entrance. Corresponding initial and boundary conditions are

$$\text{IC:} \qquad T(x, 0) - T_i = \frac{\partial T(x, 0)}{\partial t} = 0 \qquad (3.89)$$

$$\text{BC:} \qquad T(0, t) - T_i = (T_0 - T_i)F_0(t) \qquad (3.90)$$

where $F_0(t)$ is a prescribed disturbance function of time.

The Laplace transform of a function $f(x, t)$ is obtained from

$$\bar{f}(x, s) = \int_0^\infty f(x, t)e^{-st}\,dt \qquad (3.91)$$

and the inversion formula is

$$f(x, t) = \frac{1}{2\pi i}\int_{\gamma - i\infty}^{\gamma + i\infty} \bar{f}(x, s)e^{st}\,ds \qquad (3.92)$$

where the path of complex integration is on a vertical line through γ on the real axis, chosen so that singularities of $\bar{f}(x, s)$ are to the left. The transform operator of Eq. (3.91) is applied to each term in Eqs. (3.88) and (3.90) to obtain the transformed problem,

$$\text{DE:} \qquad \frac{d\bar{T}(x, s)}{dx} + \frac{(1/\tau) + (1/\tau_w) + s}{V[(1/\tau_w) + s]}(s\bar{T} - T_i) = 0 \qquad (3.93)$$

$$\text{BC:} \qquad \bar{T}(0, s) = \frac{T_i}{s} + (T_0 - T_i)\bar{F}_0(s) \tag{3.94}$$

whose solution is

$$\bar{T}(x, s) = \frac{T_i}{s} + (T_0 - T_i)\bar{F}_0(s) \exp\left[\frac{-s[(1/\tau) + (1/\tau_w) + s](x/V)}{(1/\tau_w) + s} \right] \tag{3.95}$$

If $\bar{T}(x, s)$ is substituted into Eq. (3.92) for $\bar{f}(x, s)$, the inversion gives $T(x, t) = f(x, t)$.

The most difficult aspect of the Laplace transformation method is finding the inverse transform when it is not listed in available tables. Various techniques are available for obtaining the inverse transform, including contour integration and numerical inversion. Contour integration is a many-faceted subject and too extensive to include in this text. However, numerical inversion procedures are available and relatively easy to use.

The Laplace Transform Inversion Program found in the IMSL Math Library, Vol. 2, Chapters 3–7, version 1.0, April, 1987, can be employed to obtain inverse Laplace transforms. The program performs complex integration by Simpson's rule with a control parameter for accuracy. The advantage of this program is that the user needs to supply only the transformed variable $\bar{f}(x, s)$, the real constant γ for which all singularities of $\bar{f}(x, s)$ are to the left, the times at which the inverse transform $f(x, t)$ is desired, and the allowable error. The result is a tabulation of the inverse transform at the selected times.

EXAMPLE 3.6: STEP ENTRANCE TEMPERATURE, LAPLACE
The same problem of Sections 3.14 and 3.15 is to be solved in this example, giving the time- and space-dependent liquid temperature in a pipe when the initial liquid at T_i is replaced by incoming liquid at T_0. The Laplace transformation method with parameters of Table 3.4 should be used.

The function $F_0(t) = 1.0$ in Eq. (3.90) for a step change in the incoming temperature, has the transform $\bar{F}_0(s) = 1/s$. The transformed temperature $\bar{T}(x, s) = \bar{f}(x, s)$ of Eq. (3.92) was employed in the IMSL Math Library for the inversion which gave $T(x, t)$. The permissible error specified was 0.005. Results were nondimensionalized with Eqs. (3.84)–(3.87) and are shown as dashed lines in Fig. 3.12. The Laplace transform method with numerical inversion is seen to compare favorably with the finite difference solution.

3.17 Eigenvalue Solution for Thick-Walled Pipes

Hot or cold fluid flow in thick-walled pipes can create undesirable thermal stresses. Thick-walled pipes are characterized by significant radial temperature variation. An eigenvalue solution is described in this section for axisymmetric pipe temperatures, although it can be extended to include angular temperature variation about the pipe axis.

Figure 3.13 shows basic elements of the thick-walled pipe and fluid model. Pipe temperature T_w is described by the general heat conduction equation

$$\frac{\partial T_w}{\partial t} - \alpha_w \nabla^2 T_w = 0$$

written for axisymmetric geometry. It is assumed that the pipe ends have zero axial heat transfer and that the inner and outer walls are in convective environments. The full problem is formulated for temperature scales measured from the ambient T_∞; that is, $T - T_\infty$ and $T_w - T_\infty$ are replaced by T and T_w, respectively. The complete formulation is

Fluid DE

$$\frac{\partial T}{\partial t} + V \frac{\partial T}{\partial x} = \frac{2\mathcal{H}_{in}}{\rho c_v R_{in}} \left[T_w(R_{in}, x, t) - T(x, t) \right] \tag{3.96}$$

Pipe wall DE

$$\frac{\partial T_w}{\partial t} - \alpha_w \left(\frac{\partial^2 T_w}{\partial r^2} + \frac{1}{r} \frac{\partial T_w}{\partial r} + \frac{\partial^2 T_w}{\partial x^2} \right) = 0 \tag{3.97}$$

ICs

$$T(x, 0) = T_w(r, x, 0) = 0 \tag{3.98}$$

Fluid BC

$$T(0, t) = T_0 f(t) \tag{3.99}$$

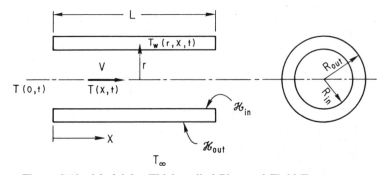

Figure 3.13 Model for Thick-walled Pipe and Fluid Temperature

Pipe End BCs

$$\frac{\partial T_w(r, 0, t)}{\partial x} = \frac{\partial T_w(r, L, t)}{\partial x} = 0 \tag{3.100}$$

Pipe Wall BCs

$$\frac{\partial T_w(R_{in}, x, t)}{\partial r} = \frac{\mathcal{H}_{in}}{\kappa_w} [T_w(R_{in}, x, t) - T(x, t)] \tag{3.101}$$

$$\frac{\partial T_w(R_{out}, x, t)}{\partial r} = -\frac{\mathcal{H}_{out}}{\kappa_w} T_w(R_{out}, x, t) \tag{3.102}$$

A wall temperature solution of the form

$$T_w(r, x, t) = \sum_{n=0}^{\infty} B_n(r, t)\cos\frac{n\pi x}{L} \tag{3.103}$$

satisfies Eq. (3.100), and, by substitution, it also will satisfy (3.97) if

$$\frac{\partial B_n}{\partial t} - \alpha_w \left[\frac{\partial^2 B_n}{\partial r^2} + \frac{1}{r}\frac{\partial B_n}{\partial r} - \left(\frac{n\pi}{L}\right)^2 B_n \right] = 0 \tag{3.104}$$

The right side of Eq. (3.101) is expanded as a cosine series in the form

$$g(x, t) = \frac{\mathcal{H}_{in}}{\kappa_w} [T_w(R_{in}, x, t) - T(x, t)] = \sum_{n=0}^{\infty} A_n(t)\cos\frac{n\pi x}{L} \tag{3.105}$$

for which Fourier analysis yields

$$A_n(t) = \frac{2}{L} \int_0^L g(x, t)\cos\frac{n\pi x}{L} \, dx \tag{3.106}$$

It follows from Eqs. (3.103), (3.101), (3.102), (3.105), and (3.106) that the pipe wall boundary conditions become

$$\frac{\partial B_n(R_{in}, t)}{\partial r} = A_n(t) \tag{3.107}$$

and

$$\frac{\partial B_n(R_{out}, t)}{\partial r} = -\frac{\mathcal{H}_{out}}{\kappa_w} B_n(R_{out}, t) \tag{3.108}$$

Moreover, the initial conditions of Eq. (3.98) yield

$$B_n(r, 0) = 0 \tag{3.109}$$

The function $B_n(r, t)$ can be obtained from a solution of Eqs. (3.104), (3.107), (3.108), and (3.109). However, in order to use the orthogonality property of Eq. (3.104), it is convenient to employ the form

$$B_n(r, t) = \psi_n(r, t) + A_n(t)\left(r - \frac{\kappa_w}{\mathcal{H}_{\text{out}}} - R_{\text{out}}\right) \tag{3.110}$$

which makes both the inner and outer pipe wall boundary conditions homogeneous and yields the problem

DE: $\quad \dfrac{\partial \psi_n}{\partial t} + \alpha_w\left[\dfrac{\partial^2 \psi_n}{\partial r^2} + \dfrac{1}{r}\dfrac{\partial \psi_n}{\partial r} - \left(\dfrac{n\pi}{L}\right)^2 \psi_n\right] = F_n(r, t) \quad$ (3.111)

BC: $\quad \dfrac{\partial \psi_n(R_{\text{in}}, t)}{\partial r} = 0 \tag{3.112}$

$$\dfrac{\partial \psi_n(R_{\text{out}}, t)}{\partial r} + \dfrac{\mathcal{H}_{\text{out}}}{\kappa_w}\psi_n(R_{\text{out}}, t) = 0 \tag{3.113}$$

IC: $\quad \psi_n(r, 0) = 0 \tag{3.114}$

where

$$F_n(r, t) = -\frac{dA_n(t)}{dt}\left(r - \frac{\kappa_w}{\mathcal{H}_{\text{out}}} - R_{\text{out}}\right)$$
$$+ \alpha_w A_n(t)\left[\frac{1}{r} - \left(\frac{n\pi}{L}\right)^2\left(r - \frac{\kappa_w}{\mathcal{H}_{\text{out}}} - R_{\text{out}}\right)\right] \tag{3.115}$$

The related homogeneous form of Eq. (3.111) for $F_n(r, t) = 0$, which also satisfies (3.112) and (3.113), is obtained from the product form $\tau(t)f(r)$, which yields

$$\psi_{n,m}(r, t) = \tau_{n,m}(t)Z(\lambda_m r) \tag{3.116}$$

with

$$Z(\lambda_m r) = J_0(\lambda_m r) + G(\lambda_m)Y_0(\lambda_m r) \tag{3.117}$$

where J and Y are Bessel functions of the first and second kind, respectively. The eigenvalues $\lambda_{n,m}$ and the function $G(\lambda_m)$ are determined from

$$\frac{1}{G(\lambda_m)} = \frac{Y_1(\lambda_m, R_{\text{in}})}{J_1(\lambda_m, R_{\text{in}})}$$
$$= \frac{-\lambda_m Y_1(\lambda_m R_{\text{out}}) + (\mathcal{H}_{\text{out}}/\kappa_w)Y_0(\lambda_m R_{\text{out}})}{\lambda_m J_1(\lambda_m R_{\text{out}}) + (\mathcal{H}_{\text{out}}/\kappa_w)J_0(\lambda_m R_{\text{out}})} \tag{3.118}$$

If we put

$$\psi_n(r, t) = \sum_{m=1}^{\infty} \psi_{n,m}(r, t) \tag{3.119}$$

Eq. (3.111) yields

$$\sum_{m=1}^{\infty} \left\{ \frac{d\tau_{n,m}}{dt} + \alpha_w \left[\lambda_m^2 + \left(\frac{n\pi}{L} \right)^2 \right] \tau_{n,m} \right\} Z(\lambda_m r)$$

$$= F_n(r, t) \tag{3.120}$$

The orthogonality property

$$\int_{R_{in}}^{R_{out}} rZ(\lambda_m r) Z(\lambda_p r) \, dr = \begin{cases} 0, & m \neq p \\ I_{1m}, & m = p \end{cases} \tag{3.121}$$

where I_{1m} is the integral

$$I_{1m} = \int_{R_{in}}^{R_{out}} rZ(\lambda_m r)^2 \, dr \tag{3.122}$$

is used in Eq. (3.120) to give

$$\frac{d\tau_{n,m}}{dt} + \alpha_w \left[\lambda_m^2 + \left(\frac{n\pi}{L} \right)^2 \right] \tau_{n,m} = H_{n,m}(t) \tag{3.123}$$

for which

$$H_{n,m}(t) = -\frac{dA_n(t)}{dt} \frac{I_{2m}}{I_{1m}} + \alpha_w A_n(t) \left[\frac{I_{3m}}{I_{1m}} + \left(\frac{n\pi}{L} \right)^2 \frac{I_{2m}}{I_{1m}} \right] \tag{3.124}$$

$$I_{2m} = \int_{R_{in}}^{R_{out}} r \left(r - \frac{\kappa_w}{\mathcal{H}_{out}} - R_{out} \right) Z(\lambda_m r) \, dr \tag{3.125}$$

and

$$I_{3m} = \int_{R_{in}}^{R_{out}} Z(\lambda_m r) \, dr \tag{3.126}$$

The initial condition of Eq. (3.114) implies that $\tau_{n,m}(0) = 0$, so a solution of (3.123) is

$$\tau_{n,m}(t) = \int_0^t H_{n,m}(\xi) \exp \alpha_w \left[\lambda_m^2 + \left(\frac{n\pi}{L} \right)^2 \right] (\xi - t) \, d\xi \tag{3.127}$$

The solution for pipe wall temperature is expressed from Eqs. (3.103), (3.110), (3.106), (3.105), (3.116), (3.126), (3.117), (3.124), and (3.125) in terms of the unknown fluid temperature $T(x, t)$. However, a solution of Eq.

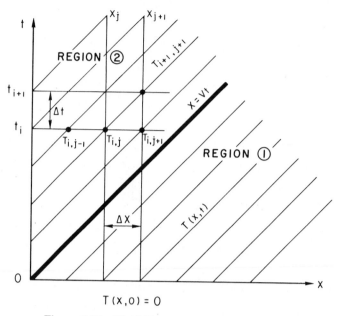

Figure 3.14 Fluid Temperature Solution Plane

(3.96) for $T(x, t)$ is formulated in terms of the initial and boundary conditions of (3.98) and (3.99) as described in Section 3.11. This yields, for the computational mesh of Fig. 3.14,

$$T_{i+1,j+1} = T_{ij} + \frac{2\mathcal{H}_{in}}{\rho c_p R_{in}} [T_w(R_{in}, x_j, t_i) - T_{i,j}] \Delta t \qquad (3.128)$$

It should be remembered that all temperatures are above the ambient value, which has been given the reference value of zero.

The flowchart of Fig. 3.15 gives the computational steps for obtaining the liquid and pipe inside wall temperature for problems with hot or cold liquid flow in thick-walled pipes. Temperatures at other values of r can be obtained by a similar procedure if necessary. The inside pipe wall temperature gradient can be obtained from the boundary condition of Eq. (3.101) when the fluid and pipe inner wall temperature solutions are available.

EXAMPLE 3.7: THICK-WALLED PIPE FLOW TEMPERATURES
Hot liquid enters a cold, short section of thick-walled pipe. Use the data of Table 3.5 to determine temperature profiles of the liquid $T(x, t)$ and pipe inner wall $T_w(R_{in}, x, t)$ at various times, and also the wall temperature gradient at its inner surface. The computational procedure of Figs. 3.14 and 3.15 was employed to obtain the temperature profiles of Figs. 3.16–3.18.

The time and space steps employed in this example were $\Delta t = 3.7$ s and $\Delta x = 1.0$ cm. Summations over eigenvalues obtained from Eq. (3.118) were truncated at 7,

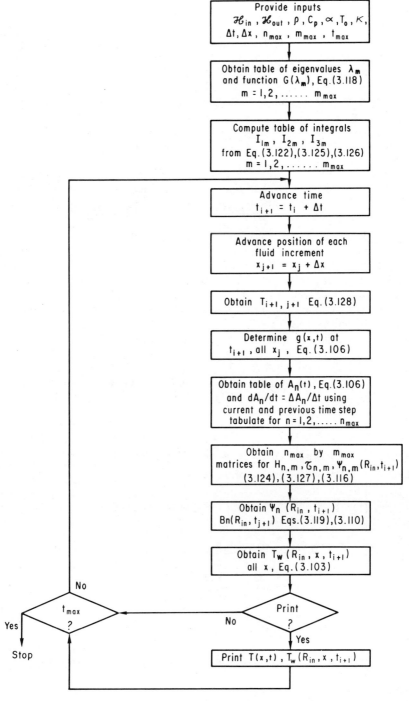

Figure 3.15   Flowchart, Hot or Cold Liquid Flow in Thick-walled Pipes

TABLE 3.5 Thick-Walled Pipe and Liquid Flow Data

$T_0 = 100°C$	Incoming liquid temperature
$f(t) = 1.0$	Function in Eq. (3.99)
$L = 10\,cm$	Pipe length
$R_{in} = 10\,cm$	Pipe inner radius
$R_{out} = 15\,cm$	Pipe outer radius
$\alpha_w = 0.37\,m^2/h$	Pipe thermal diffusivity
$\kappa_w = 294\,W/m\text{-}°C$	Pipe thermal conductivity
$\rho = 0.8\,g/cm^3$	Liquid density
$c_p = 3.4\,J/g\text{-}°C$	Liquid specific heat
$V = 0.27\,cm/s$	Liquid velocity
$\mathcal{H}_{in} = 5880\,W/m^2\text{-}°C$	Liquid-to-pipe convective coefficient
$\mathcal{H}_{out} = 470\,W/m^2\text{-}°C$	Pipe-to-ambient convective coefficient

Ambient temperature assigned zero value.

and the summation involving $g(x, t)$ in Eq. (3.105) was truncated at 47. Figures 3.16–3.18 could be refined with many more terms in both radial and axial summations, but the improvement would be less than 10 percent, based on the magnitude of the last terms considered in the series. More sophisticated time advancement could be employed, although the forward-stepping procedure has been shown to be adequate for this and other types of characteristic problems. Overall, the computational method employed here contains summations which converge rapidly and yield a solution with reasonable accuracy.

Figure 3.17 shows that before $t = 37$ s, a definite liquid temperature front moves through the pipe. The fluid is nonconducting in the axial direction, but moves slowly

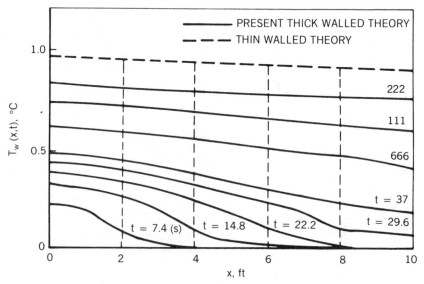

Figure 3.16 Pipe Inner Wall Temperature

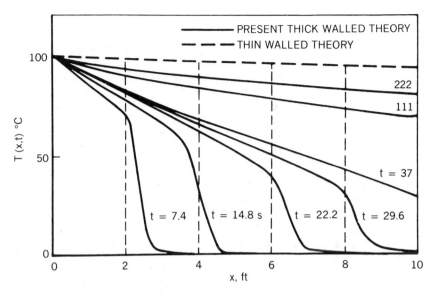

Figure 3.17 Fluid Temperature

enough in this example that axial heat transfer in the pipe actually warms fluid ahead of the front. A refinement of this problem would include axial conduction in the fluid. Figure 3.16 shows the pipe inner wall temperature profile, whose gradient is less steep than that of the liquid, due to the axial heat conduction.

Figure 3.18 shows the temperature gradient for subsequent conservative thermal stress calculations at the pipe inner wall, based on Eq. (3.101) and the results of Figs. 3.16 and 3.17.

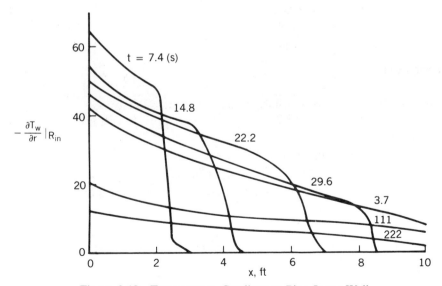

Figure 3.18 Temperature Gradient at Pipe Inner Wall

Fluid and pipe temperatures also were calculated by the method of Section 3.12 for pipes with negligible thermal capacity. Results shown as dashed lines on Figs. 3.16 and 3.17 show abrupt temperature steps moving through the pipe. Clearly, temperatures from the thick-walled method cause small longitudinal temperature gradients, which would yield correspondingly smaller values of thermal stress.

References

3.1   Amos, B. T., and F. J. Moody, "A Procedure for Predicting Temperature Loadings for Thermal Stress Calculations in Thick-Walled Pipes," *Two-Phase Flow and Waterhammer Loads in Vessels, Piping, and Structural Systems*, ASME Special Publication PVP-91, 1984.

3.2   Hildebrand, F. B., *Advanced Calculus for Applications*, 2nd ed., Prentice-Hall, Englewood Cliffs, N.J., 1976.

3.3   Kreith, F., *Principles of Heat Transfer*, 3rd ed., IEP, 1976.

3.4   Landau, L. D., and E. M. Lifshitz, *Fluid Mechanics*, Addison-Wesley, Massachusetts, 1959.

3.5   Moody, F. J. *Approximate Unsteady Fluid Jet Properties from One-Dimensional Theory*, ASME Paper Number 83-FE-25.

3.6   Nichols, B. D., C. W. Hirt, and R. S. Hotchkiss, *SOLA-VOF: A Solution Algorithm for Transient Fluid Flow with Multiple Free Boundaries*, Los Alamos Laboratory Report LA-8355, 1980.

3.7   Rouse, H., *Advanced Mechanics of Fluids*, Wiley, New York, 1959.

3.8   Schlichting, H., *Boundary Layer Theory*, Pergamon, New York, 1955.

3.9   Shapiro, A. H., *The Dynamics and Thermodynamics of Compressible Fluid flow, Vol. I*, Ronald Press, New York, 1953.

3.10   IMSL Math Library, Vol. 2, Chaps. 3–7, version 1.0, April 1987.

Problems

3.1   A liquid jet is discharging at constant velocity V_i. Use the jet properties of Eqs. (3.23) and (3.24) to express the steady velocity and jet area for

(a) a vertical upward liquid jet in gas;

(b) a vertical downward liquid jet in gas.

Discuss what happens to the jet as distance from the source increases.

Answers:

(a)
$$V_{\text{steady}} = v_i\left(1 - \frac{2gy}{V_i^2}\right)^{1/2}$$

$$A_{\text{steady}} = A_0\left(1 - \frac{2gy}{V_i^2}\right)^{-1/2}$$

(b)
$$V_{steady} = V_i\left(1 + \frac{2gy}{V_i^2}\right)^{1/2}$$

$$A_{steady} = A_0\left(1 + \frac{2gy}{V_i^2}\right)^{-1/2}$$

3.2 A liquid jet undergoes discharge at a step velocity $V(0, t) = V_0$. Show that the steady solutions of Eqs. (3.23) and (3.24) agree with the transient solutions given by Eqs. (3.42) and (3.43). Determine how the jet front advances with time. Does a shock form?

3.3 Show from Eqs. (3.39) and (3.41) that the front velocity of a liquid jet just before and after shock formation is continuous, but the jet front acceleration is not continuous.

Answer:

$$\frac{dV_{front}}{dt} = -\alpha \text{ before; } \frac{a^2}{\alpha + 2a} \text{ after}$$

3.4 A water jet discharges into water from an orifice. The discharge velocity increases linearly at acceleration a. How far away from the orifice does shocking begin? Show that the shock is at the front of the jet and express the jet front trajectory $y_{front}(t)$.

Answer:

$y_{shock} = at^2/4$; shock begins immediately at the orifice.

3.5 A water jet of 1000 kg/m^3 (62.4 lbm/ft^3) density discharges into water from an orifice of area $A_0 = 0.1 \text{ m}^2$ (1.07 ft^2) with a step-ramp discharge corresponding to $V_0 = 10 \text{ m/s}$ (32.8 ft/s) and acceleration $a = 5 \text{ m/s}^2$ (16.4 ft/s^2).

(a) Show that the shock forms at 20 m (65.6 ft).

(b) When does the jet arrive at a structure which is 5 m (16.4 ft) from the orifice?

Answer:

0.5 s

(c) If the structure of part (b) is a flat wall which is normal to the jet axis and intercepts the entire jet, determine the initial force exerted by the jet on the wall.

Answer:

$13,000 \text{ N}$ (2922 lbf)

3.6 A liquid jet discharges from an orifice with an accelerating velocity $V(0, \tau) = a\tau$.

(a) For a vertical upward jet show that a shock immediately forms at the front when discharge begins and that the local acceleration after the jet front arrives at elevation y is

$$\left(\frac{\partial V}{\partial t}\right)_y = \frac{a}{2a + g}\left[a + \frac{a + g}{\sqrt{1 - 2(2a + g)y/a^2 t^2}}\right]$$

(b) Show that the frontal shock reaches elevation y at time

$$t = \sqrt{\frac{2(2a + g)y}{a^2}}$$

and that the local acceleration at arrival time is very high.

(c) Show that, as time increases, the local acceleration approaches the discharge value a.

3.7 A liquid jet discharges vertically downward with a velocity $V(0, \tau) = a\tau$ with $a = 2g$.

(a) Show that a shock forms immediately.

(b) Show that the shock location varies according to

$$y_{shock} = (2/3)gt^2$$

(c) Show that the local acceleration is given by

$$\left(\frac{\partial V}{\partial t}\right)_y = \frac{2}{3}g\left[2 + \frac{1}{\sqrt{1 - 3y/2gt^2}}\right]$$

3.8 If a downward-pointing water jet discharges from an orifice of diameter D_0 with velocity $V(0, \tau) = a\tau$, for what value of acceleration a will the emerging jet appear as a uniform diameter liquid column of increasing length?

Answer:

$a = g$

3.9 A vertical upward jet discharges from a hole with a step-ramp velocity $V(0, \tau) = V_0 + a\tau$. Show that the jet will take the form of a liquid column of constant diameter if the discharge acceleration is $a = -g/2$.

3.10 Determine velocity and area functions $V(y, t)$ and $A(y, t)$ for a downward-pointing jet which discharges from a pipe of area A_0 at constant velocity V_0. How do you explain that at any elevation y, neither jet velocity or area vary with time? Does a shock occur?

Answer:

$$V = V_0 \left(1 + \frac{2gy}{V_0^2} \right)^{1/2}$$

$$A = A_0 \left(1 + \frac{2gy}{V_0^2} \right)^{-1/2}$$

3.11 A stainless steel pipe of density $7818 \, kg/m^3$ ($487 \, lbm/ft^3$) and specific heat $460 \, J/kg$-K ($0.11 \, B/lbm$-°F) has a radius $R = 5 \, cm$ ($1.96 \, in$). Approximately what maximum wall thickness will permit an analysis of the unsteady liquid temperature, based on negligible thermal capacity of the wall, for

(a) water flow, $\rho = 1000 \, kg/m^3$ ($62.4 \, lbm/ft^3$), $c_p = 4148 \, j/kg$-K ($1.0 \, B/lbm$-°F)

Answer:

$\Delta R \ll 2.91 \, cm$ ($1.15 \, in$)

(b) mercury flow, $\rho = 13569 \, kg/m^3$ ($846 \, lbm/ft^3$), $c_p = 138 \, J/kg$-K ($0.033 \, B/lbm$-°F)

Answer:

$\Delta R \ll 1.3 \, cm$ ($0.51 \, in$)

3.12 Suppose that a stainless steel pipe of radius $R = 5 \, cm$ ($1.97 \, in$) has a wall thickness $\Delta R = 0.2 \, cm$ ($0.079 \, in$).

(b) What is the maximum liquid velocity in a pipe of length $L = 10 \, m$ ($32.8 \, ft$) for which the radial temperature in the pipe wall can be considered uniform at any location? Stainless steel thermal diffusivity is $0.0387 \, cm^2/s$ ($0.006 \, in^2/s$).

Answer:

$V \ll 19.4 \, m/s$ ($63 \, ft/s$)

(b) How large must the velocity be in order to neglect the pipe axial heat transfer?

Answer:

$V \gg (0.7)10^{-6} \, m/s$ $[(2.5)10^{-6} \, ft/s]$

3.13 Cold water enters and flows in a uniform pipe of length L with negligible thermal capacity. The pipe provides time-dependent but uniform spatial heating throughout its length, given by

$$q'_{in}(x, t) = \frac{Q}{L} (1 + \cos \omega t)$$

where Q/L is constant. Show that a linear temperature gradient can be obtained in the flow passage if the water is pumped so that its velocity varies according to

$$V(t) = V_0(1 + \cos \omega t)$$

where V_0 is constant. If entrance water temperature is T_c, show that the discharge temperature is

$$T(L) = T_c + \frac{Q}{\rho A c_p V_0}$$

3.14 Liquid at $0.0°C$ ($32°F$) temperature flows steadily at 2.0 m/s (6.5 ft/s) in a pipe 10.0 m (32.8 ft) long whose flow area is 0.1 m² (1.07 ft²). Suddenly the entire pipe is heated uniformly at a constant rate so that

$$q'_{in}(x, t) = Kt$$

where t is in seconds and $K = 20,000$ cal/m-s² (24.2 B/ft-s²). Show that for a liquid of $\rho = 1.0$ g/cc (62.4 lbm/ft³) and $c_p = 1.0$ cal/g-°C (1.0 B/lbm-°F) the outlet temperature varies with time according to

$$T(L, t) = \frac{K}{\rho A c_p} \frac{1}{2} t^2 = \left(0.1 \frac{°C}{s^2}\right) t^2, \qquad 0 < t \le 5 \text{ s}$$

$$T(L, t) = \frac{K}{\rho A c_p} \frac{L}{V} \left(t - \frac{L}{2V}\right) = (1.0) \frac{°C}{s} (t - 2.5) \text{ s}, \qquad t > 5 \text{ s}$$

3.15 Consider the problem of incompressible liquid flow in a uniformly heated or cooled flow passage of length L with linearly varying cross section

$$A(x) = A_0 + (A_L - A_0) \frac{x}{L}$$

Mass flow rate $\dot{m}$ is constant. Initially temperature in the passage is $T(x, 0) = 0$, and, at the entrance, $T(0, t) = T_0(t) = 0$.

(a) Employ the nondimensional variables

$$T^* = \frac{T}{q'_{in} L / c_p \dot{m}}, \qquad x^* = \frac{x}{L}, \qquad t^* = \frac{t}{\rho A_0 L / \dot{m}}$$

and show that the mathematical formulation for liquid temperature is given by

DE: $\quad \dfrac{\partial T^*}{\partial t^*} + \dfrac{1}{1+(A_L/A_0-1)x^*}\dfrac{\partial T^*}{\partial x^*} = \dfrac{1}{1+(A_L/A_0-1)x^*}$

BC: $\quad T^*(0,t^*) = 0$

IC: $\quad T^*(x^*,0) = 0$

(b) Show that characteristic lines in the x^*, t^* plane are expressed by

$$t^* - t_i^* = x^* - x_i^* + \left(\frac{A_L}{A_0} - 1\right)\frac{x^{*2} - x_i^{*2}}{2}$$

(c) Show that by writing $dT^*/dt^* = (dT^*/dx^*)(dx^*/dt^*)$ the equation for temperature $T^*(x^*, t^*)$ on the characteristic line which begins at x_i^*, t_i^* is given by

$$T^*(x^*, t^*) = x^* - x_i^*$$

3.16 Liquid of $1000\,\text{kg/m}^3$ ($62.4\,\text{lbm/ft}^3$) density, $c_p = 4.17\,\text{J/g-°C}$ ($1.0\text{B}/$ lbm-°F) and incoming temperature $T_\infty = 20°C$ ($68°F$) enters a uniform circular flow passage of diameter $D = 0.2\,\text{m}$ ($0.656\,\text{ft}$), heated perimeter $P_w = 0.628\,\text{m}$ ($2.06\,\text{ft}$), flow area $A = 0.0314\,\text{m}^2$ ($0.338\,\text{ft}^2$) and length $L = 40\,\text{m}$ ($131.2\,\text{ft}$) at constant velocity $V_0 = 1.0\,\text{m/s}$ (3.28 ft/s). Suddenly the liquid incoming temperature rises to $T_i = 100°C$ ($212\,°F$). The flow passage walls remain at approximately $T_\infty = 20°C$ ($68°F$), and convection heat transfer dominates energy transfer to the walls with $\mathcal{H} = \mathcal{H}_{\text{out}} = 1000\,\text{W/m}^2\text{-°C}$ ($176\,\text{B/h-ft}^2\text{-°F}$). Show that the liquid temperature is given by

$$\frac{T(L,t) - T_\infty}{T_i - T_\infty} = \exp\left[-\frac{\mathcal{H}_{\text{out}}P_w}{\rho A_0 V_0 c_p}x\right], \qquad t > \frac{x}{V_0}$$

and that, at the exit, the normalized temperature above is 0.0 when $t < L/V_0$, and 0.4 when $t > L/V_0$.

3.17 Stationary water in a uniform pipe of area A and length L is initially at the pipe wall temperature T_∞. A pump is started, and water of temperature T_h enters the pipe at $x = 0$ with a time-dependent velocity,

$$V_0(t) = V_\infty(1 - e^{-t/t_p})$$

where V_∞ is the steady velocity and t_p is the flow time constant. Determine the discharging water temperature for a system with the following data:

$T_h = 80°C$ (176°F)

$T_\infty = 20°C$ (68°F)

$\mathcal{H} = 500\ \text{W/m}^2 - \text{K}$ (88 B/h-ft²-°F)

$P_w = 0.314\ \text{m}$ (1.03 ft)

$\rho = 1000\ \text{kg/m}^3$ (62.4 lbm/ft³)

$c_p = 4.17\ \text{J/g-°C}$ (1.0 B/lbm-°F)

$A = 0.00785\ \text{m}^2$ (0.084 ft²)

$t_p = 5\ \text{s}$

$L = 10\ \text{m}$ (32.8 ft)

$V_\infty = 5\ \text{m/s}$ (16.4 ft/s)

Several suggestions follow.

(a) Show that the hot fluid front follows the path

$$\frac{x}{V_\infty t_p} = \frac{t}{t_p} + e^{-t/t_p} - 1$$

on the t, x diagram.

(b) Verify that temperature below the path of part (a) is T_∞ and that, above, path lines correspond to (See Fig. 3.8)

$$\frac{\tau}{t_p} + e^{-\tau/t_p} = \frac{t}{t_p} + e^{-t/t_p} - \frac{x}{V_\infty t_p}$$

with temperature given by

$$\frac{T - T_\infty}{T_h - T_\infty} = \exp\left[-\frac{\mathcal{H} P_h t_p}{\rho A c_p} \left(\frac{t}{t_p} - \frac{\tau}{t_p} \right) \right]$$

(c) Eliminate τ/t_p from the equations in part (b) to obtain $T - T_\infty$ in terms of t/t_p and $x/V_\infty t_p$. Although τ/t_p cannot be obtained explicitly, it can be obtained readily by first graphing the function

$$F\left(\frac{t}{t_p}\right) = \frac{t}{t_p} + e^{-t/t_p}$$

and noting that if t/t_p has the value τ/t_p, the graph of $F(t/t_p)$ yields $F(\tau/t_p)$. Moreover, the path line equation of part (b) can be written as

$$F\left(\frac{\tau}{t_p}\right) = F\left(\frac{t}{t_p}\right) - \frac{x}{V_\infty t_p}$$

Temperature at given values of $x/V_\infty t_p$ and t/t_p can be obtained by entering the graph of F at t/t_p, reading $F(t/t_p)$, calculating $F(\tau/t_p)$ from the last equation, reentering the graph at the calculated $F(\tau/t_p)$, and reading τ/t_p. Then the term $t/t_p - \tau/t_p$ is formulated for use in the temperature equation of part (b) to obtain $T - T_\infty$ at $x/V_\infty t_p$ and t/t_p.

Answer:

$$
\begin{array}{ll}
0 < t < 5.25 & T = 20°C \ (68°F) \\
t = 5.25 \text{ s} & T = 77.0°C \ (170°F) \\
t = 7.50 \text{ s} & T = 77.2°C \ (171°F) \\
t = 10.0 \text{ s} & T = 77.6°C \ (172°F) \\
t = 20.0 \text{ s} & T = 78.8°C \ (174°F)
\end{array}
$$

3.18 A large uniform room of height H and horizontal cross-sectional area A is filled with hot gas at temperature T_∞. A spray of cool water droplets at temperature T_{d0} is introduced at the top and descends as a uniform blanket through the room, as shown in Fig. P3.18. The spray

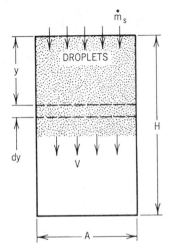

Figure P3.18

mass flow rate is $\dot{m}_s$, and each droplet is assumed to be spherical with radius r and density ρ. Evaporation from the spray is negligible. The only energy transfer rate to the droplets is heat convection. That is, $q_d = \mathcal{H} a_d (T_\infty - T_d)$ is the heat transfer rate to a single droplet of surface area a_d and temperature T_d. It is reasonable to assume that T_∞ changes slowly relative to the time for a droplet to fall a distance H at velocity V.

(a) Show that for a uniform spray droplet region, the total droplet surface area per unit length is

$$
\frac{da}{dy} = \frac{3\dot{m}_s}{r\rho V}
$$

(Hint: Verify that the total droplet introduction rate $\dot{N}$ is $\dot{m}_s/M_d$, where M_d is the mass of a single drop. Then show that the number of drops per unit vertical length is $\dot{N}/V$.)

172   CONVECTIVE PROPAGATION

(b) Use the energy conservation principle for the dotted CV of length dy with a droplet heat transfer per unit vertical length $q' = \mathcal{H}(da/dy)(T_\infty - T_d)$ to show that the describing differential equation for droplet temperature is given by

$$\frac{\partial T_d}{\partial t} + V \frac{\partial T_d}{\partial y} = \frac{3\mathcal{H}}{r\rho c_L}(T_\infty - T_d)$$

where c_L is the droplet specific heat.

(c) Verify that for a constant incoming droplet spray temperature T_{d0}, the solution for $T_d(y, t)$ is

$$T_\infty - T_d = (T_\infty - T_{d0})\exp\left[-\left(\frac{3\mathcal{H}}{r\rho c_L}\right)\frac{y}{V}\right], \qquad 0 < y < Vt, \quad t \leq \frac{H}{V}$$

(d) Write the differential heat transfer to droplets in the section dy and show that the total heat transfer rate to all droplets present at any time is

$$q = \dot{m}_s c_L(T_\infty - T_{d0})\left[1 - \exp\left(\frac{3\mathcal{H}}{r\rho c_L}\right)t\right], \qquad t < \frac{H}{V}$$

with t limited to H/V when $t > H/V$.

(e) Show that after the spray is activated, the hot gas temperature in the room changes according to

$$\frac{T_\infty - T_{d0}}{T_{\infty 0} - T_{d0}} = e^{-(\dot{m}_s c_L/Mc_v)G(t)}$$

where $T_{\infty 0}$ is the initial room temperature, c_v is the gas specific heat at constant volume, M is the mass of gas in the room, and $G(t)$ is given by

$$G(t) = \begin{cases} t - \dfrac{r\rho c_L}{3\mathcal{H}}[1 - e^{-(3\mathcal{H}/r\rho c_L)t}], & t < \dfrac{H}{V} \\ \dfrac{H}{V} + [1 - e^{-(3\mathcal{H}/r\rho c_L V)}]\left(t - \dfrac{H}{V} - \dfrac{r\rho c_L}{3\mathcal{H}}\right), & t > \dfrac{H}{V} \end{cases}$$

3.19 Safety considerations in the nuclear energy field include analyses of postulated severe accidents. One such accident involves a loss of coolant and inability to cool the reactor core, whose temperature rises because of decay heat, supplied by the nuclear fuel. The core and other metallic reactor internals melt and relocate to the bottom of the pressure vessel. Continued lack of cooling causes the vessel to fail, and core debris (corium), in the form of molten liquid metals at temperature T_0, discharges from the vessel onto the concrete floor.

Consider a case where a volume flow rate of molten corium flows horizontally from a location $x = 0$ across the floor at velocity u and depth Y (see Fig. P3.19). Let the corium layer be thin so that its

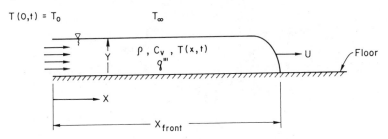

T(0,t) = T₀

T_∞

ρ, C_v, $T(x,t)$

q'''

U

Floor

Y

X

X_{front}

Figure P3.19

temperature $T(x, t)$ is uniform on any vertical plane. Assume that the concrete floor acts like a perfect insulator, and that convective cooling occurs from the top surface at a heat flux $q'' = \mathcal{H}(T - T_\infty)$. Thermal energy is generated in the flowing corium by decay heat, and sometimes also by exothermic chemical reactions between concrete gases released and molten corium metals, resulting in a uniform heat generation rate per unit volume, q'''.

(a) Show that the full problem for $T(x, t)$ is specified as

$$\text{DE:} \qquad \frac{\partial T}{\partial t} + u \frac{\partial T}{\partial x} + \frac{\mathcal{H}}{\rho c_v Y}(T - T_\infty) = \frac{q'''}{\rho c_v}$$

$$\text{IC:} \qquad T(x, 0) = T_\infty$$

$$\text{BC:} \qquad T(0, t) = T_0$$

(b) Show that a solution which satisfies part (a) is

$$T^*(x^*, t^*) = B(1 - e^{-t^*}) + e^{-x^*}\{1 - B[1 - e^{-(t^* - x^*)}]\}H_s(t^* - x^*)$$

$$B = \frac{q''' y}{\mathcal{H}(T_0 - T_\infty)}, \qquad x^* = \frac{\mathcal{H} x}{\rho c_v u Y}$$

$$T^* = \frac{(T - T_\infty)}{(T_0 - T_\infty)}, \qquad t^* = \frac{\mathcal{H} t}{\rho c_v Y}$$

(c) Show that the advancing front temperature changes according to

$$T^*_{front} = e^{-x^*} + B(1 - e^{-x^*})$$

(d) Discuss the effect of parameter B. What conditions could result in solidification of the corium? (liquidus temperature is about

2600 K; solidus temperature is about 2100 K; heat of solidification is $e_{1,s} = 250\,\text{J/gm}$)

(e) If $u = 20\,\text{cm/s}$, $Y = 1.0\,\text{cm}$, $T_0 = 2600\,\text{K}$ liquidus temperature, $T_\infty = 373\,\text{K}$, $\rho = 9000\,\text{kg/m}^3$, $c_v = 0.48\,\text{kJ/kg-K}$, $\mathcal{H} = 500$ W/m^2-K, and $q''' = 1.0\,\text{MW/m}^3$, how far will the corium front travel before it reaches the 2100 K solidus temperature?

3.20 Show that for a case in which axial conduction in the liquid is substantial, Eq. (3.62) becomes

$$\frac{\partial T}{\partial t} + V\frac{\partial T}{\partial x} - \alpha\frac{\partial^2 T}{\partial x^2} = \frac{q'_{in} - q'_{out}}{\rho A c_p}$$

Work out a solution process, and verify it by comparison with separate simplified problems for which solutions can be readily obtained.

3.21 Show that a solution of Eq. (3.81) for the liquid front $x = Vt$ in Example 3.6 is given by

$$\frac{T(x,t) - T_\infty}{T_0 - T_\infty} = \exp\left(-\frac{\mathcal{H}P_w x}{\rho A c_v V}\right)$$

and plot in Fig. 3.12. (Hint: The hot frontal fluid increment always is in contact with a part of the wall boundary at $T_w = T_\infty$.)

4 HYDROSTATIC WAVES, SMALL-AMPLITUDE APPLICATIONS

The objective of this chapter is to introduce hydrostatic wave theory and behavior. Most of the wave action discussed is relatively easy to observe, and can be demonstrated in small table-top studies, an intermediate size wave tank, or a large scale experimental facility. Emphasis is on small-amplitude waves.

Hydrostatic waves can be generated by seismic disturbances which originate on the ocean floor and extend vertically throughout the water depth. The term *hydrostatic* implies that the fluid pressure always is approximated by the local hydrostatic head. Canals, rivers, and other channels which are open to the oceans or other deep bodies of water at one end will experience hydrostatic wave action caused by tides or storms. Moreover, disturbances generated by a moving vertical wall at one end of a wave tank can be approximated by hydrostatic wave propagation. Wave action which is not hydrostatic includes liquid sloshing in containers and surface waves caused by either wind over water or surface vehicles. Nonhydrostatic waves are discussed in Chapter 9.

4.1 One-Dimensional Hydrostatic Waves

Figure 4.1 shows the side and top views of a horizontal channel with vertical sides. The distance $D(x)$ between sides and the rigid channel floor with elevation $B(x)$ above the horizontal reference plane, may vary along the channel. Liquid density is ρ, and the surface elevation above the horizontal is designated by $y(x, t)$. The local horizontal liquid velocity $V(x, t)$ is

175

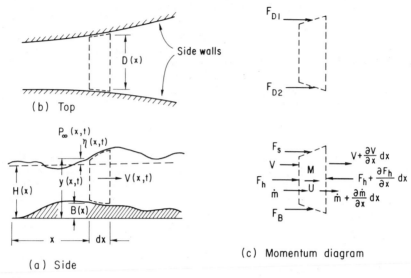

Figure 4.1 Hydrostatic Wave Model

assumed to be uniform at any x location. Also, the vertical component of liquid velocity is considered small relative to V so that vertical momentum is not important. A consequence of uniform horizontal velocity and negligible vertical momentum is that local pressure in the liquid is hydrostatic. Furthermore, it is assumed that the liquid is frictionless.

Consider the momentum diagram of Fig. 4.1(c). The CV is bounded by the vertical channel walls, the channel floor, the free surface at the top, and imaginary vertical planes at x and $x + dx$. If pressure above the liquid surface varies in space and time according to the prescribed function $P_\infty(x, t)$ such as that associated with a moving pressure front over the ocean, then the rightward force F_h on the vertical plane at x is given by pressure at the centroid $P + \rho g(y - B)/2g_0$ times the area $D(y - B)$; that is,

$$F_h = \left[P_\infty + \frac{\rho g(y - B)}{2g_0} \right](y - B)D$$

The corresponding leftward force on the vertical plane at $x + dx$ is $F_h + (\partial F_h/\partial x)\,dx$. The horizontal force F_s exerted on the free surface is the average pressure $(1/2)\{P_\infty + [P_\infty + (\partial P_\infty/\partial x)\,dx]\}$ times the projected area $D(\partial y/\partial x)\,dx$. Second-order differentials are neglected to give

$$F_s = P_\infty D \frac{\partial y}{\partial x}\,dx$$

The force F_B acting on the rigid bottom surface is the average pressure

$$\frac{1}{2}\left[P_B + \left(P_B + \frac{\partial P_B}{\partial x}\,dx\right)\right]$$

times the projected area $-D(dB/dx)\,dx$. Since $P_B = P_\infty + \rho g(y - B)/g_0$,

$$F_B = -\left[P_\infty + \rho g\,\frac{y - B}{g_0}\right]D\,\frac{dB}{dx}\,dx$$

If sections of the vertical channel walls are not parallel to the x direction, they also exert a force on the liquid of

$$F_D = F_{D1} + F_{D2} = \left[P_\infty + \rho g\,\frac{y - B}{2g_0}\right](y - B)\,\frac{dD}{dx}\,dx$$

The total horizontal force acting on the CV is

$$\sum F_x = F_h - \left(F_h + \frac{\partial F_h}{\partial x}\,dx\right) + F_s + F_B + F_D$$

$$= -\left[\frac{\partial}{\partial x}\left(P_\infty + \rho\,\frac{g}{g_0}\,y\right)\right](y - B)D\,dx$$

It follows that the momentum principle for the CV of Fig. 4.1 is given by

$$\frac{\partial(\dot m V)}{\partial x}\,dx + \frac{\partial(MV)}{\partial t} = g_0 \sum F_x = -g_0(y - B)D\,\frac{\partial}{\partial x}\left(P_\infty + \rho\,\frac{g}{g_0}\,y\right)dx$$

Moreover, mass conservation is written as

$$\frac{\partial \dot m}{\partial x}\,dx + \frac{\partial M}{\partial t} = 0$$

If we employ $M = \rho D(y - B)\,dx$ and $\dot m = \rho D(y - B)V$, the mass conservation and momentum equations for hydrostatic wave propagation in a nonuniform channel become

$$\frac{\partial y}{\partial t} + V\,\frac{\partial y}{\partial x} + y\,\frac{\partial V}{\partial x} = \frac{\partial(BV)}{\partial x} - \frac{y - B}{D}\,V\,\frac{dD}{dx} \tag{4.1}$$

and

$$\frac{\partial V}{\partial t} + V\,\frac{\partial V}{\partial x} + g\,\frac{\partial y}{\partial x} = -\frac{g_0}{\rho}\,\frac{\partial P_\infty}{\partial x} \tag{4.2}$$

If a channel is uniform with $B = 0$ and $D = $ constant, the right side of Eq. (4.1) is zero. Equations (4.1) and (4.2) determine the local surface elevation $y = y(x, t)$ and liquid velocity $V = V(x, t)$ in a channel of rectangular cross section with variable wall spacing, an irregular floor elevation, and an unsteady surface pressure.

Let the surface elevation y and liquid velocity V be written as

$$y(x, t) = H(x) + \eta(x, t) \tag{4.3}$$

and

$$V(x, t) = u_\infty(x) + u(x, t) \tag{4.4}$$

where $\eta(x, t)$ is elevation above the undisturbed level $H(x)$, and $u(x, t)$ is the velocity disturbance above an initial steady flow velocity $u_\infty(x)$. Equations (4.1) and (4.2) become

$$\frac{\partial \eta}{\partial t} + (u_\infty + u) \frac{\partial}{\partial x} (H + \eta) + (H + \eta) \frac{\partial}{\partial x} (u_\infty + u)$$
$$= \frac{\partial}{\partial x} [B(u_\infty + u)] - \frac{(H - B + \eta)}{D} (u_\infty + u) \frac{dD}{dx} \tag{4.5a}$$

and

$$\frac{\partial u}{\partial t} + (u_\infty + u) \frac{\partial}{\partial x} (u_\infty + u) + g \frac{\partial}{\partial x} (H + \eta) = -\frac{g_0}{\rho} \frac{\partial P_\infty}{\partial x} \tag{4.5b}$$

When the flow is not disturbed, $\eta = u = 0$, and P_∞ is constant, for which Eqs. (4.5a) and (4.5b) yield the steady flow case,

$$u_\infty \frac{dH}{dx} + H \frac{du_\infty}{dx} = \frac{d}{dx} (Bu_\infty) - \frac{(H - B)}{D} u_\infty \frac{dD}{dx} \tag{4.6a}$$

and

$$u_\infty \frac{du_\infty}{dx} + g \frac{dH}{dx} = 0 \tag{4.6b}$$

A solution of Eqs. (4.6a) and (4.6b) give the undisturbed velocity $u_\infty(x)$ and elevation $H(x)$ in terms of the floor elevation $B(x)$ and channel width $D(x)$. If B and D are uniform constant values, solutions of Eqs. (4.6a) and (4.6b) give constant values of u_∞ and H.

If we subtract the undisturbed flow terms in Eqs. (4.6a) and (4.6b) from (4.5a) and (4.5b), we obtain

$$\frac{\partial \eta}{\partial t} + \frac{\partial}{\partial x} [u_\infty \eta + u(H - B + \eta)] = -\frac{1}{D} [u_\infty \eta + u(H - B + \eta)] \frac{dD}{dx} \tag{4.7a}$$

and

$$\frac{\partial u}{\partial t} + (u_\infty + u) \frac{\partial u}{\partial x} + u \frac{du_\infty}{dx} + g \frac{\partial \eta}{\partial x} = -\frac{g_0}{\rho} \frac{\partial P_\infty}{\partial x} \tag{4.7b}$$

Equations (4.1) and (4.2) generally are used for analysis of large-amplitude waves, whereas Eqs. (4.7a) and (4.7b) are further modified for small-amplitude waves.

4.2 Small-Amplitude Linearization

Small-amplitude waves imply small η and u so that the convective derivative terms are negligible. Formally,

$$\eta \ll H , \qquad u\,\frac{\partial \eta}{\partial x} \approx 0$$

$$\eta\,\frac{\partial u}{\partial x} \approx 0 , \qquad u\,\frac{\partial u}{\partial x} \approx 0 \tag{4.8}$$

It follows that Eqs. (4.7a) and (4.7b) can be written in the linearized forms,

$$\frac{\partial \eta}{\partial t} + u_\infty\,\frac{\partial \eta}{\partial x} + \frac{\partial}{\partial x}\,[(H - B)u] + \eta\,\frac{du_\infty}{dx} = -\frac{1}{D}\,[u_\infty\eta + u(H - B)]\,\frac{dD}{dx} \tag{4.9}$$

and

$$\frac{\partial u}{\partial t} + u_\infty\,\frac{\partial u}{\partial x} + u\,\frac{du_\infty}{dx} + g\,\frac{\partial \eta}{\partial x} = -\frac{g_0}{\rho}\,\frac{\partial P_\infty}{\partial x} \tag{4.10}$$

for hydrostatic waves of small amplitude. Further developments in this chapter are based on Eqs. (4.9) and (4.10).

4.3 Uniform Channels and Surface Pressure

A uniform channel idealization can be used to approximate many applications of hydrostatic wave theory. Reference to Fig. 4.1 shows that a uniform channel is characterized by

$$B(x) = 0 , \qquad D(x) = D = \text{constant} \tag{4.11}$$

for which u_∞ and H are constant. Furthermore, with $\partial P_\infty/\partial x = 0$, Eqs. (4.9) and (4.10) become

$$\frac{\partial \eta}{\partial t} + u_\infty\,\frac{\partial \eta}{\partial x} + H\,\frac{\partial u}{\partial x} = 0 \tag{4.12}$$

$$\frac{\partial u}{\partial t} + u_\infty\,\frac{\partial u}{\partial x} + g\,\frac{\partial \eta}{\partial x} = 0 \tag{4.13}$$

Either η or u can be eliminated from Eqs. (4.12) and (4.13) to obtain the linear wave equation,

$$\frac{\partial^2 \phi}{\partial t^2} + 2u_\infty \frac{\partial^2 \phi}{\partial x \partial t} - (C^2 - u_\infty^2) \frac{\partial^2 \phi}{\partial x^2} = 0 \tag{4.14}$$

$$\text{for } \phi(x, t) = \eta(x, t) \text{ or } u(x, t)$$

It will be shown that $\sqrt{gH}$ is the small-amplitude wave propagation speed relative to the liquid, that is, $\sqrt{gH} = V_p - u_\infty$, designated by

$$C = \sqrt{gH} \tag{4.15}$$

If disturbances are imposed on a nonflowing liquid, u_∞ is set equal to zero in Eq. (4.14).

A general solution for ϕ can be obtained from the function

$$\phi = Ae^{\alpha t + \beta x}$$

where A, α, and β are constants to be determined. Substitution into Eq. (4.14) yields two possible solutions for β, namely $\beta = \pm \alpha/(C \mp u_\infty)$, so

$$\phi = Ae^{\alpha[t \pm x/(C \mp u_\infty)]}$$

Since Eq. (4.14) is linear, any linear combination of solutions also is a solution. We may incorporate a summation of terms for many values of α and the constant A and write

$$\phi = \sum_{n=0}^{\infty} A_n e^{\alpha_n [t + x/(C - u_\infty)]} + \sum_{m=0}^{\infty} A_m e^{\alpha_m [t - x/(C + u_\infty)]}$$

Both summations are seen to be functions of the arguments $t + x/(C - u_\infty)$ or $t - x/(C + u_\infty)$. Therefore, a general solution of Eq. (4.14) is

$$\phi = F_L\left(t + \frac{x}{C - u_\infty}\right) + F_R\left(t - \frac{x}{C + u_\infty}\right)$$

where F_L and F_R are arbitrary functions, twice differentiable with respect to their arguments. If we let ϕ represent the surface elevation η, then

$$\eta(x, t) = F_L\left(t + \frac{x}{C_L}\right) + F_R\left(t - \frac{x}{C_R}\right) \tag{4.16}$$

where C_L and C_R are defined by

$$C_L = C - u_\infty, \qquad C_R = C + u_\infty \tag{4.17}$$

Substitution into eq. (4.12) yields

$$\frac{\partial u}{\partial x} = -\frac{1}{H}\left(F_L' + F_R' + \frac{u_\infty}{C_L}F_L' - \frac{u_\infty}{C_R}F_R'\right)$$

Partial integration with respect to x gives

$$u = -\frac{C}{H}\left[F_L\left(t + \frac{x}{C_L}\right) - F_R\left(t - \frac{x}{C_R}\right)\right] + f_1(t)$$

where $f_1(t)$ is arbitrary. Further substitution of η from Eq. (4.16) into Eq. (4.13) followed by partial integration with respect to t shows that $f_1(t) = 0$, and the corresponding solution for $u(x, t)$ is

$$u(x, t) = -\frac{C}{H}\left[F_L\left(t + \frac{x}{C_L}\right) - F_R\left(t - \frac{x}{C_R}\right)\right] \qquad (4.18)$$

Equations (4.16) and (4.18) are general solutions which satisfy (4.12), (4.13), and also (4.14). If disturbances have been imposed on a stationary stream, then $u_\infty = 0$ in Eq. (4.17). Since Eq. (4.14) is of second order in both x and t, two initial conditions and two boundary conditions are required for a full solution. The meaning of these solutions is described in the next section.

4.4 Meaning of the General Solutions

Equations (4.16) and (4.18) contain the arbitrary functions $F_L(t + x/C_L)$ and $F_R(t - x/C_R)$. The value of each function is determined by its argument. However, a single value of the argument $t + x/C_L$ corresponds to an infinite set of t and x values. The same is true for a single value of the argument $t - x/C_R$. That is, if K_L and K_R are constants, then

$$F_L\left(t + \frac{x}{C_L}\right) = F_L(K_L) = \text{constant}, \qquad t + \frac{x}{C_L} = K_L \qquad (4.19)$$

and

$$F_R\left(t - \frac{x}{C_R}\right) = F_R(K_R) = \text{constant}, \qquad t - \frac{x}{C_R} = K_R \qquad (4.20)$$

Equations (4.16)–(4.20) show that the solutions for $\eta(x, t)$ and $u(x, t)$ can be represented on a *propagation diagram*, which is simply the t, x plane shown in Fig. 4.2 for a uniform channel of length L. The t, x region is bounded by vertical lines at $x = 0$ and $x = L$ and by the initial time $t = 0$. The region extends upward for increasing time. Boundary conditions are required on the $x = 0$ and L lines. Initial conditions also are required only on the boundary $t = 0$.

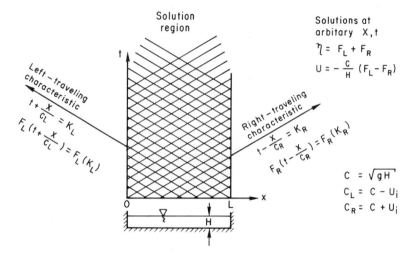

Figure 4.2   Propagation Diagram of General Solution Region

Equation (4.19) gives an infinite family of leftward traveling straight lines with a negative slope $-1/C_L$, for which each line corresponds to a different value of K_L. Likewise, Eq. (4.20) gives an infinite family of rightward traveling straight lines with a positive slope $1/C_R$, each of which corresponds to a different value of K_R. The left and right traveling lines are called left and right traveling *characteristics*, which have been designated by subscripts L and R, respectively.

Each characteristic line represents the propagation path of a disturbance which originates at a boundary of the t, x diagram of Fig. 4.2. The value of the function F_L is constant on each left traveling characteristic, and the value of F_R is constant on each right traveling characteristic. It is important to realize that the values of F_L and F_R do not change where left and right traveling characteristic lines cross each other.

Equations (4.16) and (4.18) show that the surface displacement η and fluid velocity u are obtained anywhere in the t, x plane by employing the sum and difference of F_L and F_R wherever characteristic lines cross. The values of F_L and F_R on each characteristic are determined from either the initial conditions or boundary conditions.

Figure 4.3 helps to show the procedure for determining values of F_L and F_R which propagate on the characteristic lines. Suppose that the initial condition corresponds to an arbitrary flow velocity and surface elevation; that is, for any initial location x_i on the x axis,

$$u(x, 0) = u_i(x_i) \tag{4.21}$$

$$\eta(x, 0) = \eta_i(x_i) \tag{4.22}$$

as noted in Fig. 4.3(a). Equations (4.16) and (4.18) yield

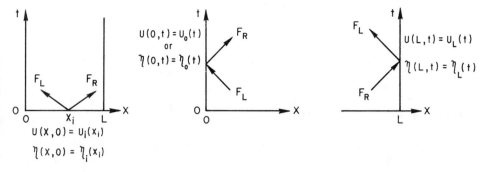

(a) Initial conditions   (b) Left boundary   (c) Right boundary

Figure 4.3   Determination of Functions F_L and F_R

$$\eta(x, 0) = \eta_i(x_i) = F_L + F_R$$

and

$$u(x, 0) = u_i(x_i) = -\frac{C}{H}(F_L - F_R)$$

These equations can be solved simultaneously for F_L and F_R, which originate at $t = 0$. Thus,

$$F_{Li} = \frac{1}{2}\left[\eta_i(x_i) - \frac{H}{C}u_i(x_i)\right] \qquad (4.23)$$

and

$$F_{Ri} = \frac{1}{2}\left[\eta_i(x_i) + \frac{H}{C}u_i(x_i)\right] \qquad (4.24)$$

Next, suppose that the left boundary condition is specified in either of the forms

$$u(0, t) = u_0(t) \quad \text{or} \quad \eta(0, t) = \eta_0(t) \qquad (4.25)$$

as shown in Fig. 4.3(b). The arriving function F_L is known at t, and Eqs. (4.16), (4.18), and (4.25) yield

$$F_R = \eta_0(t) - F_L \quad \text{or} \quad F_R = \frac{H}{C}u_0(t) + F_L \qquad (4.26)$$

Similarily, let the right boundary condition be specified in either of the forms

$$u(L, t) = u_L(t) \quad \text{or} \quad \eta(L, t) = \eta_L(t) \qquad (4.27)$$

as shown in Fig. 4.3(c). The arriving function F_R is known at t, and Eqs. (4.16), (4.18), and (4.27) yield

$$F_L = \eta_L(t) - F_R \quad \text{or} \quad F_L = F_R - \frac{H}{C} u_L(t) \qquad (4.28)$$

4.5   The General Solution in its Crudest Form

A summary of the general solution in its crudest form is given in Fig. 4.4. The known initial conditions make it possible to obtain values of F_L and F_R on all left and right traveling characteristics originating at the line $t = 0$, $0 < x < L$. The left boundary condition and known values of F_L on arriving characteristics yield values of F_R on right traveling characteristics which originate at the left boundary. The right boundary condition and known values of F_R on arriving characteristics yield values of F_L on left traveling characteristics which originate at the right boundary.

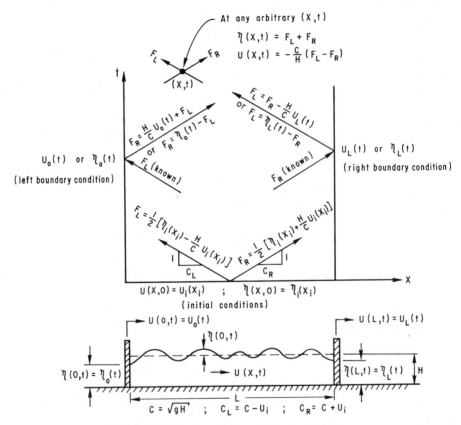

Figure 4.4   Crudest Form of General Solution

The general solution summarized in Fig. 4.4 is referred to as crude because, although it provides solutions for any initial and boundary conditions, it may require numerous characteristic lines to specify η or u at arbitrary values of x and t. The crude solution is most useful for uniform initial conditions and step changes in the boundary conditions.

EXAMPLE 4.1: STEP VELOCITY AT LEFT WALL

A uniform wave channel like that shown in Fig. 4.4 initially contains stagnant liquid, $u_\infty = 0$, of depth H. The right boundary is a stationary wall, and the left boundary is a wall which suddenly begins rightward motion at constant velocity u_0. Determine the resulting wave amplitude, its speed, its reflected amplitude at the right wall, and again at the left moving wall.

The initial and boundary conditions are

$$\text{IC:} \qquad u(x, 0) = \eta(x, 0) = 0$$

$$\text{Left BC:} \qquad u(0, t) = u_0(t) = u_0 = \text{constant}$$

$$\text{Right BC:} \qquad u(L, t) = u_L(t) = 0$$

It is assumed that u_0 is sufficiently small that actual displacement of the left boundary is negligible during the time period of interest. Figure 4.5(a) shows the propagation diagram for this problem. Region i contains only those characteristic lines which originate from the initial conditions, for which Eqs. (4.23) and (4.24) give

$$F_L = F_R = 0$$

Region 1 contains left traveling characteristics with arriving $F_L = 0$, originating from the initial conditions. Figure 4.4 shows that F_R in region 1 should be

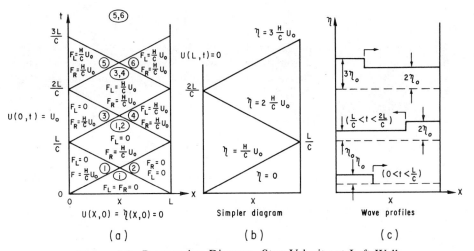

Figure 4.5 Propagation Diagram, Step Velocity at Left Wall

$$F_{R1} = \frac{H}{C} u_0(t) + F_L = \frac{H}{C} u_0 + 0 = \frac{H}{C} u_0$$

Region 2 in Fig. 4.5(a) contains right traveling characteristics with $F_R = 0$, originating from the initial condition. Figure 4.4 shows that F_L in region 2 is given by

$$F_{L2} = F_R - \frac{H}{C} u_L(t) = F_R = 0$$

Region 1, 2 contains values of $F_R = Hu_0/C$ and $F_L = 0$. Region 3 contains $F_L = 0$ from region 2, and Fig. 4.4 yields

$$F_{R3} = \frac{H}{C} u_0 + F_L = \frac{H}{C} u_0$$

Region 4 contains $F_R = Hu_0/C$ from region 1, and Fig. 4.4 yields

$$F_{L4} = F_R - \frac{H}{C} u_L(t) = F_R = \frac{H}{C} u_0$$

This process can be continued, but is stopped at region 5, giving

$$F_{R5} = 2 \frac{H}{C} u_0$$

The amplitude solution $\eta = F_L + F_R$ for any x, t in Fig. 4.5 is

$$\eta_i = \eta_2 = 0$$

$$\eta_1 = \eta_{1,2} = \eta_3 = \frac{H}{C} u_0$$

$$\eta_4 = \eta_{3,4} = \eta_6 = 2 \frac{H}{C} u_0$$

$$\eta_5 = 3 \frac{H}{C} u_0$$
$$\vdots$$

Amplitude results are summarized in Fig. 4.5(b). It is seen that the wave amplitude Hu_0/C propagates rightward from the left moving wall in regions 1, 1, 2, and 3. This wave first arrives at the stationary right wall when $t = L/C$, and the reflected amplitude doubles in regions 4, 3, 4, and 6. Furthermore, the amplitude triples in region 5. Figure 4.5(c) shows the wave profile at several times.

This simple wave behavior can be observed in a shallow pan of water by slowly moving a vertical cardboard sheet.

The wall motion disturbance of Example 4.1 first creates a wave which propagates rightward into undisturbed liquid, before reflection from the right wall. Its shape can be represented by Eq. (4.16) as a one-directional wave with $F_{Li} = 0$ prior to reflection by

$$\eta(x, t) = F_R\left(t - \frac{x}{C_R}\right)$$

Regardless of the geometric shape of the wave, the x, t trajectory of a given disturbance value $\eta = \eta_0$ corresponds to velocity dx/dt, which is the wave propagation speed. Therefore, we have

$$\eta_0 = \text{constant} = F_R\left(t - \frac{x}{C_R}\right), \qquad d\eta_0 = 0 = F'_R\left(t - \frac{x}{C_R}\right)\left(dt - \frac{dx}{C_R}\right)$$

where $F'_R(t - x/C_R)$ is the derivative of F_R with respect to its argument $(t - x/C_R)$. Since F'_R generally is not zero, η_0 moves according to $dx/dt = C_R$, which means that the rightward wave speed is dx/dt. It follows from Eqs. (4.15) and (4.17) that

$$\frac{dx}{dt} = C + u_\infty = C_R = \sqrt{gH} + u_\infty$$

This confirms that $C = \sqrt{gH}$, introduced by Eq. (4.14), is the right traveling wave propagation speed either in stationary liquid with $u_\infty = 0$, or relative to undisturbed liquid which initially is flowing at velocity u_∞. A similar argument for left traveling waves also yields $C = \sqrt{gH}$. When both right and left traveling disturbances exist together, they propagate independently at speeds C_R and C_L, given by Eq. (4.17) for small disturbances. This is not the case for large-amplitude waves, which are discussed in Chapter 5.

Example 4.1 for a step disturbance could have been analyzed with only the triangular regions of Fig. 4.5(b) formed by the sawtooth line which describes motion of the initial disturbance front. If more than one step disturbance occurs at a boundary, the procedure of Fig. 4.4 is easily modified as in the next example.

EXAMPLE 4.2: STEP-UP, STEP-DOWN LEFT WALL VELOCITY
The uniform wave channel of Fig. 4.4 contains nonflowing liquid of depth H. The right boundary is stationary, and the left boundary suddenly begins rightward motion at constant velocity u_0 for a time interval t_0, when it suddenly stops moving and remains stationary. Determine the resulting unsteady wave profile.

The initial and boundary conditions are formally expressed as

$$\text{IC:} \qquad u(x, 0) = \eta(x, 0) = 0$$

$$\text{Left BC:} \qquad u(0, t) = \begin{cases} u_0, & 0 < t < t_0 \\ 0, & t_0 \le t \end{cases}$$

$$\text{Right BC:} \qquad u(L, t) = 0$$

Again, it is assumed that u_0 is sufficiently small that actual displacement of the left boundary is negligible during the time interval of interest. Figure 4.6(a) shows the

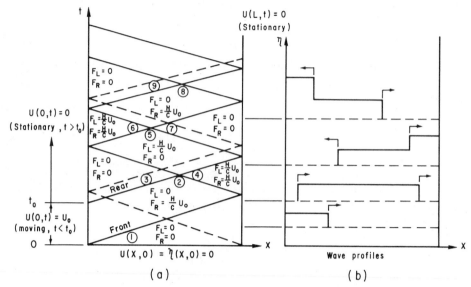

Figure 4.6 Propagation Diagram, Step-Up-Down Wall Velocity

propagation diagram, constructed with the help of Fig. 4.4, but with different regional numbering. The heavy characteristic lines describe motion of the disturbance front and rear.

Left traveling characteristics from region 1 carry values of $F_L = 0$ into regions 2 and 3. Right traveling characteristics generated at the left wall when its velocity is u_0 have $F_R = Hu_0/C$, which propagate through region 2 into region 4, from which $F_L = Hu_0/C$ must propagate leftward to satisfy the stationary right wall boundary condition. The function $F_R = 0$ in regions 3, 5, and 7 is generated at the left wall when it stops moving, and is required to satisfy the stationary wall condition after $t = t_0$. The computational marching process for F_L and F_R in every region is straightforward. Values of the wave profile are readily determined in each region from $\eta = F_L + F_R$, according to Fig. 4.4, and are sketched at selected times in Fig. 4.6(b). The directions of motion shown by arrows are easily determined by noting the slope of the characteristic lines. A positive slope means that the disturbance is propagating rightward at speed C, whereas a negative slope means leftward propagation at speed C.

So far, examples have been given with one disturbed boundary and one stationary boundary. Some interesting wave behavior involves simultaneous motion of both boundaries.

EXAMPLE 4.3: SIMULTANEOUS MOTION OF BOTH END WALLS
The uniform wave channel of Fig. 4.4 contains stationary liquid of depth H. Suddenly the left wall begins to move rightward at constant velocity u_0, and the right

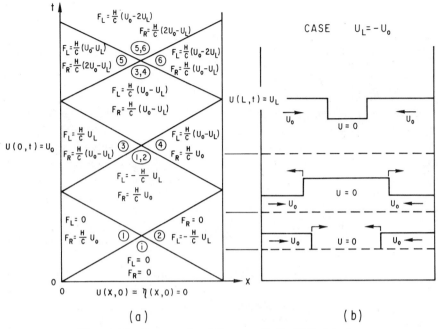

Figure 4.7 Propagation Diagram, Both Walls Moving

wall begins to move rightward at constant velocity u_L. Both u_0 and u_L are small, so that actual wall displacements are negligible compared to the channel length L. Determine the resulting wave profile, and the corresponding liquid velocity.

The initial and boundary conditions are specified formally as

$$\text{IC:} \qquad u(x, 0) = \eta(x, 0) = 0$$

$$\text{Left BC:} \qquad u(0, t) = u_0$$

$$\text{Right BC:} \qquad u(L, t) = u_L$$

The propagation diagram for this problem is shown in Fig. 4.7(a) and was constructed with the help of Fig. 4.4. The wave profile and corresponding velocity are shown in Fig. 4.7(b) for the special case of symmetric wall motion with $u_L = -u_0$.

4.6 Refined Solution, Initial and Boundary Conditions

The procedure developed in this section is offered as one of several tools for solving hydrostatic wave problems in wave tanks of rectangular geometry with any specified boundary and initial conditions. The somewhat tedious development is described, but then it is formulated as a relatively simple procedure which employs Tables 4.1 and 4.2 in the solution of problems. The computational method of characteristics introduced in Chapter 5, Laplace transforms, or other methods also can be employed to obtain the refined solutions to problems discussed here.

The crude solution discussed in Section 4.5 is refined for cases where initial conditions are not uniform and boundary conditions are general functions of time. The same regional numbering scheme of Fig. 4.5(a) is repeated in Fig. 4.8. General initial conditions of Eqs. (4.21) and (4.22) are shown near the $t = 0$ line in Fig. 4.8. Velocity boundary conditions are expressed by Eqs. (4.26) and (4.28), which also are written at $x = 0$ and $x = L$ in Fig. 4.8. Equations (4.16) and (4.18) give solutions for surface elevation $\eta(x, t) = F_L + F_R$ and velocity $u(x, t) = -(C/H)(F_L - F_R)$, which are obtained from the characteristic functions F_L and F_R in all regions of Fig. 4.8. Therefore, it is necessary to express F_L and F_R everywhere in the t, x diagram. Values of F_{Li} and F_{Ri} which originate from the initial conditions are obtained from Eqs. (4.23) and (4.24) as functions of the initial velocity and displacement $u_i(x_i)$ and $\eta_i(x_i)$. These initial condition functions are shown on Fig. 4.8.

Values of F_R originating at the left boundary are obtained from Eq. (4.26) as functions of the boundary condition $u_0(t)$ and the F_L arriving at $x = 0$. Values of F_L originating at the right boundary are obtained from Eq. (4.28) as functions of the boundary condition $u_L(t)$ and the F_R arriving at $x = L$. Both boundary functions F_R and F_L are shown in Fig. 4.8.

If $\eta(x, t)$ or $u(x, t)$ are to be determined at an arbitrary point (x, t), values of F_L and F_R must be obtained for every region of Fig. 4.8 below and including the point (x, t). The procedure for obtaining F_L and F_R is described for two general characteristics which originate at the arbitrary x_i of Fig. 4.8. These characteristics are shown as dotted and dashed lines. When they arrive at a boundary, other characteristic lines originate and travel in the opposite direction. The dotted and dashed sawtooth patterns obtained from the two characteristics originating at x_i are used to express F_L and F_R coming from each boundary. Since the numerical values of F_L and F_R are constant as they travel from one boundary to the other, it is possible to obtain values of F_L and F_R at any point (x, t) if appropriate x_i values are chosen. It would be cumbersome to follow two sawtooth lines from two different values of x_i which intersect at the desired (x, t). Instead, the procedure to be described makes it possible to obtain F_L and F_R at any point (x, t) in any region of the propagation diagram.

Consider the dotted line first on which we have F_{Li}. It arrives at the left boundary $(0, t_1)$, where $t_1 = x_i/C_L$. Then the function F_{R1} originates from $(0, t_1)$, obtained from the left boundary function. The F_{R1} characteristic extends across the diagram and arrives at the right boundary (L, t_4) where $t_4 = x_i/C_L + L/C_R$. The function F_{L4} originates at (L, t_4), obtained from the right boundary function. This process continues and is summarized below up to region 9.

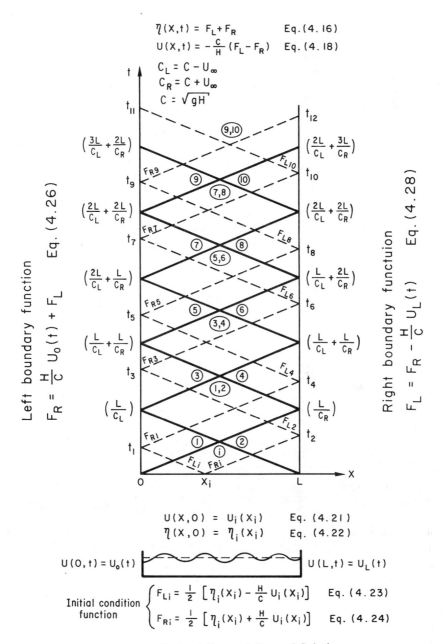

Figure 4.8   Refined Form of General Solution

$$F_{Li}(x_i) = \frac{1}{2} \left[\eta_i(x_i) - \frac{H}{C} u_i(x_i) \right], \qquad t = 0$$

$$F_{R1} = \frac{H}{C} u_0(t_1) + F_{Li}(x_i), \qquad t_1 = \frac{x_i}{C_L}$$

$$F_{L4} = F_{R1} - \frac{H}{C} u_L(t_4), \qquad t_4 = t_1 + \frac{L}{C_R} = \frac{x_i}{C_L} + \frac{L}{C_R}$$

$$F_{R5} = \frac{H}{C} u_0(t_5) + F_{L4}, \qquad t_5 = t_4 + \frac{L}{C_L} = \frac{x_i + L}{C_L} + \frac{L}{C_R}$$

$$F_{L8} = F_{R5} - \frac{H}{C} u_L(t_8), \qquad t_8 = t_5 + \frac{L}{C_R} = \frac{x_i + L}{C_L} + \frac{2L}{C_R}$$

$$F_{R9} = \frac{H}{C} u_0(t_9) + F_{L8}, \qquad t_9 = t_8 = \frac{L}{C_L} = \frac{x_i + 2L}{C_L} + \frac{2L}{C_R}$$
$$\vdots$$

(4.29)

Similarly, we have for the dashed line, which begins with the function F_{Ri},

$$F_{Ri}(x_i) = \frac{1}{2} \left[\eta_i(x_i) - \frac{H}{C} u_i(x_i) \right], \qquad t = 0$$

$$F_{L2} = F_{Ri}(x_i) - \frac{H}{C} u_L(t_2), \qquad t_2 = \frac{L - x_i}{C_R}$$

$$F_{R3} = \frac{H}{C} u_0(t_3) + F_{L2}, \qquad t_3 = t_2 + \frac{L}{C_L} = \frac{L - x_i}{C_R} + \frac{L}{C_L}$$

$$F_{L6} = F_{R3} - \frac{H}{C} u_L(t_6), \qquad t_6 = t_3 + \frac{L}{C_R} = \frac{2L - x_i}{C_R} + \frac{L}{C_L}$$

$$F_{R7} = \frac{H}{C} u_0(t_7) + F_{L6}, \qquad t_7 = t_6 + \frac{L}{C_L} = \frac{2L - x_i}{C_R} + \frac{2L}{C_L}$$

$$F_{L10} = F_{R7} - \frac{H}{C} u_L(t_{10}), \qquad t_{10} = t_7 + \frac{L}{C_R} = \frac{3L - x_i}{C_R} + \frac{2L}{C_L}$$
$$\vdots$$

(4.30)

Both functions F_L and F_R are seen to depend on x_i. However, x_i can be expressed in terms of t and x on each characteristic line. Equations for each segment of the dotted sawtooth line associated with $F_{Li}, F_{R1}, F_{L4}, F_{R5}, \ldots$ are given by

$$F_{Li}: \quad t = \frac{x_i - x}{C_L} \qquad F_{R5}: \quad t = t_5 + \frac{x}{C_R}$$

$$F_{R1}: \quad t = t_1 + \frac{x}{C_R} \qquad F_{L8}: \quad t = t_8 + \frac{L - x}{C_L}$$

$$F_{L4}: \quad t = t_4 + \frac{L - x}{C_L} \qquad F_{R9}: \quad t = t_9 + \frac{x}{C_R}$$
$$\vdots \qquad \qquad \vdots$$

(4.31)

Also, equations for the dashed lines associated with F_{Ri}, F_{L2}, F_{R3}, . . . in Fig. 4.8 can be written as

$$F_{R1}: \quad t = \frac{x - x_i}{C_R} \qquad\qquad F_{L6}: \quad t = t_6 + \frac{L - x}{C_L}$$

$$F_{L2}: \quad t = t_2 + \frac{L - x}{C_L} \qquad F_{R7}: \quad t = t_7 + \frac{x}{C_R} \qquad\qquad (4.32)$$

$$F_{R3}: \quad t = t_3 + \frac{x}{C_R} \qquad F_{L10}: \quad t = t_{10} + \frac{L - x}{C_L}$$

It is seen from Eqs. (4.31) and (4.32) that each arrival time $t_1, t_2, t_3, \ldots$ of a characteristic at a boundary can be expressed in terms of x and t on the new characteristic generated. One example from Eq. (4.32) is that $t_3 = t - x/C_R$, where t and x are any set of points on the F_{R3} characteristic. Moreover, it is possible to eliminate $t_1, t_2, t_3, t_4, t_5, \ldots$ from Eqs. (4.29) and (4.31) or (4.30) and (4.32) in order to obtain the x_i value corresponding to any given point (x, t)! Consider a point on F_{R5} at (x, t) in region 5. Equation (4.29) gives

$$x_i = C_L t_5 - \frac{C_L}{C_R} L - L$$

and from Eq. (4.31), $t_5 = t - x/C_R$ so that

$$x_i = C_L \left(t - \frac{x}{C_R} \right) - \frac{C_L}{C_R} L - L, \qquad x, t \text{ on } F_{R5} \text{ in region 5.} \quad (4.33)$$

A similar process is used to obtain the x_i corresponding to the same point (x, t) in region 5 which lies on an F_{L4} characteristic.

Tables 4.1 and 4.2 were formulated to help obtain F_L and F_R quickly for any (x, t) in the numbered regions of Fig. 4.8. The procedure used to formulate Tables 4.1 and 4.2 is illustrated next for the F_{R5} entry in region 5. It is seen from Eq. (4.29) that

$$F_{R5} = \frac{H}{C} u_0(t_5) + F_{L4}$$

$$= \frac{H}{C} u_0(t_5) - \frac{H}{C} u_L(t_4) + F_{R1}$$

$$= \frac{H}{C} u_0(t_5) - \frac{H}{C} u_L(t_4) + \frac{H}{C} u_0(t_1) + F_{Li}(x_i) \qquad (4.34)$$

Values of the functions u_0 and u_L correspond to the boundary points where they originate, at t_5, t_4, and t_1 in this case. Equation (4.31) gives

$$t_5 = t - \frac{x}{C_R}$$

Table 4.1   The Functions $\hat{F}_R$ and $\hat{F}_L$ for Waves in a Uniform Rectangular Channel (See Fig. 4.8)

$$\hat{F}_{Ri} = \hat{F}_{R2} = 0$$

$$\hat{F}_{R1} = \hat{F}_{R1,2} = \hat{F}_{R4} = \frac{H}{C} u_0(t_0) \qquad\qquad t_0 = t - \frac{x}{C_R}$$

$$\hat{F}_{R3} = \hat{F}_{R3,4} = \hat{F}_{R6} = \hat{F}_{R1} - \frac{H}{C} u_L(t_L) \qquad\qquad t_L = t - \frac{x}{C_R} - \frac{L}{C_L}$$

$$\hat{F}_{R5} = \hat{F}_{R5,6} = \hat{F}_{R8} = \hat{F}_{R3} + \frac{H}{C} u_0(t_0) \qquad\qquad t_0 = t - \frac{x}{C_R} - \frac{L}{C_L} - \frac{L}{C_R}$$

$$\hat{F}_{R7} = \hat{F}_{R7,8} = \hat{F}_{R10} = \hat{F}_{R5} - \frac{H}{C} u_L(t_L) \qquad\qquad t_L = t - \frac{x}{C_R} - \frac{2L}{C_L} - \frac{L}{C_R}$$

$$\hat{F}_{R9} = \hat{F}_{R9,10} = \hat{F}_{R12} = \hat{F}_{R7} + \frac{H}{C} u_0(t_0) \qquad\qquad t_0 = t - \frac{x}{C_R} - \frac{2L}{C_L} - \frac{2L}{C_R}$$

- -

$$\hat{F}_{Li} = \hat{F}_{L1} = 0$$

$$\hat{F}_{L2} = \hat{F}_{L1,2} = \hat{F}_{L3} = -\frac{H}{C} u_L(t_L) \qquad\qquad t_L = t + \frac{x}{C_L} - \frac{L}{C_L}$$

$$\hat{F}_{L4} = \hat{F}_{L3,4} = \hat{F}_{L5} = \hat{F}_{L2} + \frac{H}{C} u_0(t_0) \qquad\qquad t_0 = t + \frac{x}{C_L} - \frac{L}{C_L} - \frac{L}{C_R}$$

$$\hat{F}_{L6} = \hat{F}_{L5,6} = \hat{F}_{L7} = \hat{F}_{L4} - \frac{H}{C} u_L(t_L) \qquad\qquad t_L = t + \frac{x}{C_L} - \frac{2L}{C_L} - \frac{L}{C_R}$$

$$\hat{F}_{L8} = \hat{F}_{L7,8} = \hat{F}_{L9} = \hat{F}_{L6} + \frac{H}{C} u_0(t_0) \qquad\qquad t_0 = t + \frac{x}{C_L} - \frac{2L}{C_L} - \frac{2L}{C_R}$$

$$\hat{F}_{L10} = \hat{F}_{L9,10} = \hat{F}_{L11} = \hat{F}_{L8} - \frac{H}{C} u_L(t_L) \qquad\qquad t_L = t + \frac{x}{C_L} - \frac{3L}{C_L} - \frac{2L}{C_R}$$

$$C_L = C - u_\infty \qquad C_R = C + u_\infty \qquad C = \sqrt{gH}$$

$$\hat{\eta}(x, t) = \hat{F}_L + \hat{F}_R \qquad \hat{u}(x, t) = -\frac{C}{H}(\hat{F}_L - \hat{F}_R)$$

This solution alone corresponds to a case with initially undisturbed liquid $\eta_i = u_i = 0$ and boundary walls at $x = 0$ and $x = L$ with specified velocities $u_0(t)$, $u_L(t)$. Each $\hat{F}_R$ and $\hat{F}_L$ should be evaluated at the t, x values where $\hat{\eta}$ and $\hat{u}$ are desired.

for an arbitrary point (x, t) on F_{R5} in region 5. But t_4 must first be related to the same (x, t) in region 5. This is done by using Eq. (4.29) to write $t_4 = t_5 - L/C_L$, and then putting $t_5 = t - x/C_R$ to obtain

$$t_4 = t - \frac{x}{C_R} - \frac{L}{C_L}$$

similarly,

Table 4.2 The Functions $\tilde{F}_R$ and $\tilde{F}_L$ for Waves in a Uniform Rectangular Channel (See Fig. 4.8)

$\tilde{F}_{Ri}(x_i) = \dfrac{1}{2}\left[\eta_i(x_i) + \dfrac{H}{C}u_i(x_i)\right]$	$x_i = -C_R\left(t - \dfrac{x}{C_R}\right)$
$\tilde{F}_{Li}(x_i) = \dfrac{1}{2}\left[\eta_i(x_i) - \dfrac{H}{C}u_i(x_i)\right]$	$x_i = C_L\left(t + \dfrac{x}{C_L}\right)$
$\tilde{F}_{R2} = \tilde{F}_{Ri}(x_i)$	$x_i = -C_R\left(t - \dfrac{x}{C_R}\right)$
$\tilde{F}_{R1} = \tilde{F}_{R1,2} = \tilde{F}_{R4} = \tilde{F}_{Li}(x_i)$	$x_i = C_L\left(t - \dfrac{x}{C_R}\right)$
$\tilde{F}_{R3} = \tilde{F}_{R3,4} = \tilde{F}_{R6} = \tilde{F}_{Ri}(x_i)$	$x_i = -C_R\left(t - \dfrac{x}{C_R}\right) + \dfrac{2LC}{C_L}$
$\tilde{F}_{R5} = \tilde{F}_{R5,6} = \tilde{F}_{R8} = \tilde{F}_{Li}(x_i)$	$x_i = C_L\left(t - \dfrac{x}{C_R}\right) - \dfrac{2LC}{C_R}$
$\tilde{F}_{R7} = \tilde{F}_{R7,8} = \tilde{F}_{R10} = \tilde{F}_{Ri}(x_i)$	$x_i = -C_R\left(t - \dfrac{x}{C_R}\right) + \dfrac{4LC}{C_L}$
$\tilde{F}_{R9} = \tilde{F}_{R9,10} = \tilde{F}_{R12} = \tilde{F}_{Li}(x_i)$	$x_i = C_L\left(t - \dfrac{x}{C_R}\right) - \dfrac{4LC}{C_R}$
$\tilde{F}_{L1} = \tilde{F}_{Li}(x_i)$	$x_i = C_L\left(t + \dfrac{x}{C_L}\right)$
$\tilde{F}_{L2} = \tilde{F}_{L1,2} = \tilde{F}_{L3} = \tilde{F}_{Ri}(x_i)$	$x_i = -C_R\left(t + \dfrac{x}{C_L}\right) + \dfrac{2LC}{C_L}$
$\tilde{F}_{L4} = \tilde{F}_{L3,4} = \tilde{F}_{L5} = \tilde{F}_{Li}(x_i)$	$x_i = C_L\left(t + \dfrac{x}{C_L}\right) - \dfrac{2LC}{C_R}$
$\tilde{F}_{L6} = \tilde{F}_{L5,6} = \tilde{F}_{L7} = \tilde{F}_{Ri}(x_i)$	$x_i = -C_R\left(t + \dfrac{x}{C_L}\right) + \dfrac{4LC}{C_L}$
$\tilde{F}_{L8} = \tilde{F}_{L7,8} = \tilde{F}_{L9} = \tilde{F}_{Li}(x_i)$	$x_i = C_L\left(t + \dfrac{x}{C_L}\right) - \dfrac{4LC}{C_R}$
$\tilde{F}_{L10} = \tilde{F}_{L9,10} = \tilde{F}_{L11} = \tilde{F}_{Ri}(x_i)$	$x_i = -C_R\left(t + \dfrac{x}{C_L}\right) + \dfrac{6LC}{C_L}$

$$C_L = C - u_\infty \qquad C_R = C + u_\infty \qquad C = \sqrt{gH}$$

$$\tilde{\eta}(x,t) = \tilde{F}_L + \tilde{F}_R \qquad \tilde{u}(x,t) = -\dfrac{C}{H}(\tilde{F}_L - \tilde{F}_R)$$

This solution alone corresponds to a case with stationary boundary walls $u_0 = u_L = 0$ at $x = 0$ and $x = L$, and initially disturbed liquid with $\eta(x,0) = \eta_i(x_i)$ and $u(x,0) = u_i(x_i)$. Each $\tilde{F}_R$ and $\tilde{F}_L$ should be evaluated at the t, x values where $\tilde{\eta}$ and $\tilde{u}$ are desired.

$$t_1 = t_4 - \frac{L}{C_R} = t - \frac{x}{C_R} - \frac{L}{C_L} - \frac{L}{C_R}$$

Thus, t_5, t_4, and t_1 and the x_i expression from Eq. (4.33) are used in (4.34) to write

$$F_{R5} = \hat{F}_{R5} + \tilde{F}_{R5} \tag{4.35}$$

where

$$\hat{F}_{R5} = \frac{H}{C} u_0 \left(t - \frac{x}{C_R} \right) - \frac{H}{C} u_L \left(t - \frac{x}{C_R} - \frac{L}{C_L} \right) + \frac{H}{C} u_0 \left(t - \frac{x}{C_R} - \frac{L}{C_L} - \frac{L}{C_R} \right) \tag{4.36}$$

and

$$\tilde{F}_{R5} = \frac{1}{2} \left[\eta_i(x_i) - \frac{H}{C} u_i(x_i) \right], \ x_i = C_L \left(t - \frac{x}{C_R} \right) - L - \frac{C_L}{C_R} L \tag{4.37}$$

The function F_{R5} has been decomposed into two other functions, which depend either on the boundary conditions, or the initial conditions. The function $\hat{F}_{R5}$ includes only u_0 and u_L, which are specified at the $x = 0$ and $x = L$ boundaries. The function $\tilde{F}_{R5}$ includes only η_i and u_i which are specified initially at $t = 0$. Similarly, functions F_L and F_R in other regions are expressed by separate boundary and initial condition functions. If the liquid initially was undisturbed with $\eta_i(x_i) = u_i(x_i) = 0$, Eq. (4.37) shows that $\tilde{F}_{R5}$ is always zero and can be deleted from the solution for F_{R5} in Eq. (4.35). If the walls are stationary at $x = 0$ and $x = L$, the boundary conditions are $u_0 = u_L = 0$, and since Eq. (4.36) shows that $\hat{F}_{R5}$ is zero, it can be deleted from Eq. (4.35).

Equation (4.35) is a superposition of two solutions. One solution is for initially undisturbed liquid with moving end walls, and the other is for initially disturbed liquid with stationary end walls. Therefore, for the general F_L and F_R functions we may write

$$F_L = \hat{F}_L + \tilde{F}_L \tag{4.38}$$

and

$$F_R = \hat{F}_R + \tilde{F}_R \tag{4.39}$$

Table 4.1 gives $\hat{F}_L$ and $\hat{F}_R$, and Table 4.2 gives $\tilde{F}_L$ and $\tilde{F}_R$. These tables are easily extended by repeating the pattern of the arguments shown. Full solutions for $\eta(x, t)$ and $u(x, t)$ can be expressed as the superposition of solutions for the two cases described; that is,

$$\eta(x, t) = \hat{\eta}(x, t) + \tilde{\eta}(x, t) \tag{4.40}$$

and

$$u(x, t) = \hat{u}(x, t) + \tilde{u}(x, t) \tag{4.41}$$

where

$$\hat{\eta}(x, t) = \hat{F}_L + \hat{F}_R \tag{4.42}$$

$$\hat{u}(x, t) = -\frac{C}{H}(\hat{F}_L - \hat{F}_R) \tag{4.43}$$

and

$$\tilde{\eta}(x, t) = \tilde{F}_L + \tilde{F}_R \tag{4.44}$$

$$\tilde{u}(x, t) = -\frac{C}{H}(\tilde{F}_L - \tilde{F}_R) \tag{4.45}$$

Several examples illustrate the procedure.

Example 4.4: Initial Surface Disturbance
The liquid surface in a wave tank of length L and undisturbed elevation H is initially sloped so that its initial conditions are given by

$$\eta(x,0) = \eta_i(x_i) = a_0\left(1 - \frac{2x_i}{L}\right)$$

$$u(x, 0) = u_i(x) = 0$$

where a_0 is the initial amplitude at the stationary tank end walls. Boundary conditions are

$$u_0(t) = u_L(t) = 0$$

Determine the time-dependent liquid elevation at each end of the tank and in the center.

Since $u_\infty = 0$, we have $C_L = C_R = C$. Since the end walls are stationary, we may use the solution obtained from Table 4.2, given by

$$\eta(x,t) = \tilde{\eta}(x, t) = \tilde{F}_L + \tilde{F}_R$$

It follows from Fig. 4.8 with $C_L = C_R = C$ that elevation at the left wall, $x = 0$, is determined from

$$\eta(0,t) = \tilde{\eta}_1(0, t) = \tilde{F}_{Li} + \tilde{F}_{R1}, \qquad 0 \le t < \frac{L}{C}$$

$$= \tilde{\eta}_3(0, t) = \tilde{F}_{L2} + \tilde{F}_{R3}, \qquad \frac{L}{C} \le t < \frac{2L}{C}$$

$$= \tilde{\eta}_5(0, t) = \tilde{F}_{L4} + \tilde{F}_{R5}, \qquad \frac{2L}{C} \le t < \frac{3L}{C}$$

$$= \tilde{\eta}_7(0, t) = \tilde{F}_{L6} + \tilde{F}_{R7}, \qquad \frac{3L}{C} \le t < \frac{4L}{C}$$
$$\vdots$$

Since $u_\infty = 0$, Table 4.2 and the initial surface distortion $\eta_i(x_i)$ give

$$\tilde{F}_{Ri} = \tilde{F}_{Li} = \frac{1}{2}\,\eta_i(x_i) = \frac{a_0}{2}\left(1 - \frac{2x_i}{L}\right)$$

It follows from further use of Table 4.2 that

$$\tilde{F}_{L1} = \tilde{F}_{Li}(x_i) = \frac{a_0}{2}\left(1 - \frac{2x_i}{L}\right) = \frac{a_0}{2}\left[1 - \frac{2C}{L}\left(t + \frac{x}{C}\right)\right]$$

$$\tilde{F}_{R1} = \tilde{F}_{Li}(x_i) = \frac{a_0}{2}\left[1 - \frac{2C}{L}\left(t - \frac{x}{C}\right)\right]$$

$$\tilde{F}_{L2} = \tilde{F}_{Ri}(x_i) = \frac{a_0}{2}\left(1 - \frac{2x_i}{L}\right) = \frac{a_0}{2}\left[\frac{2C}{L}\left(t + \frac{x}{C}\right) - 3\right]$$
$$\vdots$$

The amplitude at $x = 0$ is

$$\eta(0, t) = (\tilde{F}_L + \tilde{F}_R)|_{x=0} = a_0\left(1 - \frac{2Ct}{L}\right), \qquad 0 \le t < \frac{L}{C}$$

$$= a_0\left(\frac{2Ct}{L} - 3\right), \qquad \frac{L}{C} \le t < \frac{2L}{C}$$

$$= a_0\left(5 - \frac{2Ct}{L}\right), \qquad \frac{2L}{C} \le t < \frac{3L}{C}$$

$$= a_0\left(\frac{2Ct}{L} - 7\right), \qquad \frac{3L}{C} \le t < \frac{4L}{C}$$

The amplitude at $x = L$ is obtained from a similar procedure, which gives

$$\eta(L, t) = a_0\left(\frac{2Ct}{L} - 1\right), \qquad 0 \le t < \frac{L}{C}$$

$$= a_0\left(3 - \frac{2Ct}{L}\right), \qquad \frac{L}{C} \le t < \frac{2L}{C}$$

$$= a_0\left(\frac{2Ct}{L} - 5\right), \qquad \frac{2L}{C} \le t < \frac{3L}{C}$$

$$= a_0\left(7 - \frac{2Ct}{L}\right), \qquad \frac{3L}{C} \le t < \frac{4L}{C}$$

The center elevation $\eta(L/2, t)$ is obtained by putting $x = L/2$ in the functions $\hat{F}_L$ and $\hat{F}_R$ of Table 4.2. It follows that

$$\eta\left(\frac{L}{2}, t\right) = 0, \qquad t > 0$$

The location $x = L/2$ is a pivot point for the surface oscillation, or a *node*. Results for this example are graphed in Fig. 4.9.

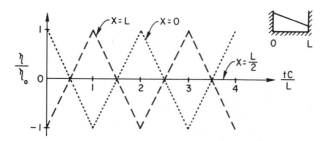

Figure 4.9 Surface Elevation, Initially Sloped Surface

The previous example displays time-dependent surface elevation in a wave tank with rigid end walls and an initially sloped surface. The next example involves an initially stationary liquid and periodic motion of both ends.

EXAMPLE 4.5: SURFACE ELEVATION WITH WALL MOTION
The wave tank of length L initially contains stationary liquid of undisturbed depth H. Left and right periodic wall motions begin simultaneously according to

$$u_0(t) = a_0 \sin \omega_0 t$$
$$u_L(t) = a_L \sin \omega_L t$$

Determine the surface elevation at all x and t up to the time $t = 2L/C$.

The liquid is not flowing as a stream, so $u_\infty = 0$ and $C_L = C_R = C$. Regions on the propagation diagram for the problem are noted in Fig. 4.10(a). The initial conditions are

$$\eta_i(x) = u_i(x) = 0$$

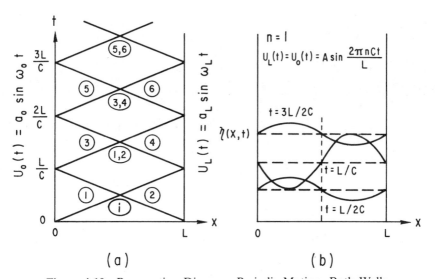

(a)   (b)

Figure 4.10 Propagation Diagram, Periodic Motion, Both Walls

Therefore, Table 4.1 can be used alone to express the surface elevation

$$\eta(x, t) = \hat{\eta}(x, t) = \hat{F}_L + \hat{F}_R$$

Table 4.1 with Fig. 4.10(a) and the specified boundary conditions yield

$$\hat{F}_{Ri} = \hat{F}_{R2} = 0$$

$$\hat{F}_{R1} = \hat{F}_{R1,2} = \hat{F}_{R4} = \frac{H}{C} a_0 \sin \omega_0 \left(t - \frac{x}{C} \right)$$

$$\hat{F}_{R3} = \hat{F}_{R3,4} = \hat{F}_{R6}$$

$$\dot{=} \frac{H}{C} \left[a_0 \sin \omega_0 \left(t - \frac{x}{C} \right) - a_L \sin \omega_L \left(t - \frac{x}{C} - \frac{L}{C} \right) \right]$$

and

$$\hat{F}_{Li} = \hat{F}_{L1} = 0$$

$$\hat{F}_{L2} = \hat{F}_{L1,2} = \hat{F}_{L3} = -\frac{H}{C} a_L \sin \omega_L \left(t + \frac{x}{C} - \frac{L}{C} \right)$$

$$\hat{F}_{L4} = \hat{F}_{L3,4} = \hat{F}_{L5}$$

$$\dot{=} -\frac{H}{C} \left[a_L \sin \omega_L \left(t + \frac{x}{C} - \frac{L}{C} \right) - a_0 \sin \omega_0 \left(t - \frac{x}{C} - \frac{2L}{C} \right) \right]$$

It follows that in each region of Fig. 4.10(a) we have

$$\eta_i(x, t) = 0$$

$$\eta_1(x, t) = \frac{H}{C} a_0 \sin \omega_0 \left(t - \frac{x}{C} \right)$$

$$\eta_2(x, t) = -\frac{H}{C} a_L \sin \omega_L \left(t + \frac{x}{C} - \frac{L}{C} \right)$$

$$\eta_{1,2}(x, t) = \frac{H}{C} \left[-a_L \sin \omega_L \left(t + \frac{x}{C} - \frac{L}{C} \right) + a_0 \sin \omega_0 \left(t - \frac{x}{C} \right) \right]$$

$$\eta_3(x, t) = -\frac{H}{C} \left[a_L \sin \omega_L \left(t + \frac{x}{C} - \frac{L}{C} \right) \right.$$
$$\left. - a_0 \sin \omega_0 \left(t - \frac{x}{C} \right) + a_L \sin \omega_L \left(t - \frac{x}{C} - \frac{L}{C} \right) \right]$$

$$\eta_4(x, t) = -\frac{H}{C} \left[a_L \sin \omega_L \left(t + \frac{x}{C} - \frac{L}{C} \right) \right.$$
$$\left. - a_0 \sin \omega_0 \left(t + \frac{x}{C} - \frac{2L}{C} \right) - a_0 \sin \omega_0 \left(t - \frac{x}{C} \right) \right]$$

$$\eta_{3,4}(x, t) = -\frac{H}{C} \left[a_L \sin \omega_L \left(t + \frac{x}{C} - \frac{L}{C} \right) \right.$$
$$\left. - a_0 \sin \omega_0 \left(t + \frac{x}{C} - \frac{2L}{C} \right) - a_0 \sin \omega_0 \left(t - \frac{x}{C} \right) + a_L \sin \omega_L \left(t - \frac{x}{C} - \frac{L}{C} \right) \right]$$

A disturbance from one end wall requires a time interval L/C to arrive at the opposite wall. If the end wall motion frequencies are integer multiples of C/L; that is, if

$$f_0 = f_L = \frac{nC}{L}, \qquad n = 1, 2, \ldots$$

so that

$$\omega_0 = \omega_L = 2\pi \frac{nC}{L}, \qquad n = 1, 2, \ldots$$

and if the wall motions are in phase, with amplitudes

$$a_0 = a_L = a$$

it follows that

$$\eta_i(x, t) = 0$$

$$\eta_1(x, t) = \frac{H}{C} a \sin 2\pi n \frac{C}{L}\left(t - \frac{x}{C}\right)$$

$$\eta_2(x, t) = -\frac{H}{C} a \sin 2\pi n \frac{C}{L}\left(t + \frac{x}{C} - \frac{L}{C}\right)$$

$$= -\frac{H}{C} a \sin 2\pi n \frac{C}{L}\left(t + \frac{x}{C}\right)$$

$$\eta_{1,2}(x, t) = -2\frac{H}{C} a \sin 2n\pi \frac{x}{L} \cos 2n\pi \frac{Ct}{L}$$

$$\eta_3(x, t) = -\frac{H}{C} a \sin 2\pi n \frac{C}{L}\left(t + \frac{x}{C}\right)$$

$$\eta_4(x, t) = \frac{H}{C} a \sin 2\pi n \frac{C}{L}\left(t - \frac{x}{C}\right)$$

$$\eta_{3,4}(x, t) = 0$$

Several surface profiles are sketched in Fig. 4.10(b) for $n = 1$.

———————

When disturbances are imposed on a steadily flowing stream, separate solutions from Tables 4.1 and 4.2 must be added to obtain a full solution. This is demonstrated in the next example.

EXAMPLE 4.6: WAVES IN A DECELERATING TANK
A rectangular tank of length $L = 3$ m and undisturbed liquid depth $H = 0.2$ m is moving slowly rightward on a conveyor at steady velocity $u_T = u_\infty = 0.1$ m/s (0.328 ft/s), as shown in Fig. 4.11. The conveyor is programmed to slow the tank according to

$$u_T = u_\infty e^{-t/\tau}$$

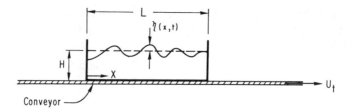

Figure 4.11   Waves in Decelerating Tank

where τ is to be selected so that the surface waves developed do not exceed an amplitude of 0.01 m at the end walls. Determine an appropriate value for τ and express the time-dependent wave amplitude at each wall for one wave round trip. Assume that u_∞ is sufficiently small that the tank end walls do not move appreciably during the time interval considered.

This problem is treated as an initially moving stream with specified velocity boundary conditions at $x = 0$ and L. The justification for neglecting end wall displacement requires that the wall movement

$$x_w = \int_0^t u_T \, dt = u_\infty \tau (1 - e^{-t/\tau})$$

be much smaller than the tank length L during the time required for a wave round trip:

$$\frac{L}{C_R} + \frac{L}{C_L} = \frac{L}{C + u_\infty} + \frac{L}{C - u_\infty} = \frac{2LC}{C^2 - u_\infty^2}$$

That is,

$$x_w = u_\infty \tau \left[1 - \exp\left\{ -\frac{2LC}{\tau(C^2 - u_\infty^2)} \right\} \right] \ll L$$

The appropriate initial and boundary conditions are

$$u(x, 0) = u_\infty \quad \text{and} \quad \eta(x, 0) = 0$$

and

$$u(0, t) = u_0(t) = u(L, t) = u_L(t) = u_\infty e^{-t/\tau}$$

The amplitude during one round trip is obtained from Fig. 4.8, which shows that calculations are required for regions i, 1, 2, 3, and 4. Table 4.2 gives

$$\tilde{F}_{Ri} = \tilde{F}_{R2} = \frac{H}{2C} u_\infty$$

$$\tilde{F}_{Li} = \tilde{F}_{L1} = -\frac{H}{2C} u_\infty$$

$$\tilde{F}_{R1} = \tilde{F}_{R4} = -\frac{H}{2C} u_\infty$$

$$\tilde{F}_{R3} = \frac{H}{2C} u_\infty$$

$$\tilde{F}_{L2} = \tilde{F}_{L3} = \frac{H}{2C} u_\infty$$

$$\tilde{F}_{L4} = -\frac{H}{2C} u_\infty$$

which yields the following $\tilde{\eta}$ solutions for initial flow but zero motion of the end walls:

$$\tilde{\eta}_1 = \tilde{F}_{L1} + \tilde{F}_{R1} = -\frac{H}{C} u_\infty$$

$$\tilde{\eta}_2 = \tilde{F}_{L2} + \tilde{F}_{R2} = \frac{H}{C} u_\infty$$

$$\tilde{\eta}_3 = \tilde{F}_{L3} + \tilde{F}_{R3} = \frac{H}{C} u_\infty$$

$$\tilde{\eta}_4 = \tilde{F}_{L4} + \tilde{F}_{R4} = -\frac{H}{C} u_\infty$$

Table 4.1 yields

$$\hat{F}_{Ri} = \hat{F}_{R2} = 0$$

$$\hat{F}_{R1} = \hat{F}_{R4} = \frac{H}{C} u_\infty e^{-[t+x/C_R]/\tau}$$

$$\hat{F}_{R3} = \frac{H}{C} u_\infty e^{-[t-x/C_R]/\tau}(1 - e^{-L/C_L\tau})$$

$$\hat{F}_{Li} = \hat{F}_{L1} = 0$$

$$\hat{F}_{L2} = \hat{F}_{L3} = -\frac{H}{C} u_\infty e^{-[t+x/C_L-L/C_L]/\tau}$$

$$\hat{F}_{L4} = \frac{H}{C} u_\infty e^{-[t+x/C_L-L/C_L]/\tau}(-1 + e^{-L/C_R\tau})$$

which gives the following solution for end wall deceleration but zero initial fluid motion:

$$\hat{\eta}_1 = \hat{F}_{L1} + \hat{F}_{R1} = \frac{H}{C} u_\infty e^{-[t-x/C_R]/\tau}$$

$$\hat{\eta}_2 = \hat{F}_{L2} + \hat{F}_{R2} = -\frac{H}{C} u_\infty e^{-[t+x/C_L-L/C_L]/\tau}$$

$$\hat{\eta}_3 = \hat{F}_{L3} + \hat{F}_{R3} = \frac{H}{C} u_\infty[-e^{-[t+x/C_L-L/C_L]/\tau} + e^{-[t+x/C_R]/\tau} - e^{-[t-x/C_R-L/C_L]/\tau}]$$

$$\hat{\eta}_4 = \hat{F}_{L4} + \hat{F}_{R4} = \frac{H}{C} u_\infty[-e^{-[t+x/C_L-L/C_L]/\tau} + e^{-[t+x/C_L-2LC/C_LC_R]/\tau} + e^{-[t-x/C_R]/\tau}]$$

Specifically at the left and right ends the solutions obtained from Tables 4.1 and 4.2 are added to determine the wall amplitude. This procedure gives

x = 0

$$\eta_1 = \tilde{\eta}_1 + \hat{\eta}_1 = -\frac{H}{C} u_\infty (1 - e^{-t/\tau}), \qquad 0 \le t < \frac{L}{C_L}$$

$$\eta_3 = \tilde{\eta}_3 + \hat{\eta}_3 = \frac{H}{C} u_\infty [1 + e^{-t/\tau} - 2e^{-[t-L/C_L]/\tau}], \qquad \frac{L}{C_L} \le t < \frac{L}{C_L} + \frac{L}{C_R}$$

and

x = L

$$\eta_2 = \tilde{\eta}_2 + \hat{\eta}_2 = \frac{H}{C} u_\infty (1 - e^{-t/\tau}), \qquad 0 \le t < \frac{L}{C_R}$$

$$\eta_4 = \tilde{\eta}_4 + \hat{\eta}_4 = -\frac{H}{C} u_\infty \left[1 + e^{-t/\tau} - 2e^{-[t-L/C_R]/\tau} \right], \qquad \frac{L}{C_R} < t < \frac{L}{C_L} + \frac{L}{C_R}$$

The maximum amplitudes at the left and right walls are determined next. If amplitudes η_2 and η_4 are differentiated with respect to t at $x = L$, it is found that at $t = L/C_R$, the amplitude growth rates are $d\eta_2/dt > 0$ and $d\eta_4/dt < 0$.. That is, prior to $t = L/C_R$, the amplitude is increasing, after which it decreases. Therefore, the maximum amplitude at $x = L$ is

$$\eta_{2,max} = \eta_2|_{t=L/C_R} = \frac{H}{C} u_\infty (1 - e^{-L/C_R \tau})$$

A similar analysis at $x = 0$ yields

$$\eta_{1,max} = \eta_1|_{t=L/C_L} = -\frac{H}{C} u_\infty (1 - e^{-L/C_L \tau})$$

Thus, τ should be calculated from both of the last two equations. Since increasing the value of τ decreases the maximum amplitude at either end, the larger τ should be used. It is found that, for the numbers given,

$$C = \sqrt{gH} = \sqrt{(9.8 \text{ m/s}^2)(0.2 \text{ m})} = 1.4 \text{ m/s} \ (4.6 \text{ ft/s})$$

$$C_L = C - u_\infty = 1.3 \text{ m/s} \ (4.26 \text{ ft/s}), \qquad C_R = C + u_\infty = 1.5 \text{ m/s} \ (4.9 \text{ ft/s})$$

The amplitude at $x = L$ will be limited to 0.01 m if

$$\tau \ge -\frac{L/C_R}{\ln(1 - \eta_{2,max} C/Hu_\infty)}, \qquad 0 < \frac{\eta_{2,max} C}{Hu_\infty} < 1$$

$$\ge 1.66 \text{ s}$$

whereas at $x = 0$,

$$\tau \geq -\frac{L/C_L}{\ln(1 + \eta_{1,\max}C/Hu_\infty)} , \qquad -1 < \frac{\eta_{1,\max}C}{Hu_\infty} < 0$$

$$\geq 1.92 \ s$$

The value $\tau = 1.92$ s should be employed, for which it can be shown that the tank displacement is 0.17 m. That is, wall movement is much less than the 3 m tank length, which justifies neglecting end wall displacement.

4.7 Outward Propagation Only

There are many cases where a disturbance propagates outward from a source without encountering other disturbances during the time of interest. One example is the action of ocean tides at the mouth of a channel, where wave propagation can occur for long distances inland before substantial reflections occur.

Suppose that a disturbance propagates rightward from a source at $x = 0$. Equations (4.16) and (4.18) with right, but not left traveling disturbances give the general solutions

$$\eta(x, t) = F_R\left(t - \frac{x}{C_R}\right)$$

$$u(x, t) = \frac{C}{H} F_R\left(t - \frac{x}{C_R}\right)$$

(4.46)

If F_R is eliminated, then

$$\eta(x, t) = \frac{H}{C} u(x, t)$$

(4.47)

Equation (4.47) shows that for disturbances traveling in one direction, there is a unique relationship between liquid velocity and surface elevation.

EXAMPLE 4.7: LIQUID ELEVATION ON MOVING WALL
A long wave tank contains stationary liquid of depth $H = 1.0$ m (3.28 ft). If a vertical wall suddenly begins to move at velocity $u = 0.313$ m/s (1.03 ft/s), what is the expected rise in liquid elevation on the wall?
    The propagation speed is

$$C = \sqrt{gH} = 3.13 \ m/s \ \ (10.3 \ ft/s)$$

and Eq. (4.47) yields the elevation

$$\eta = \frac{H}{C} u = 0.1 \ m = 10 \ cm \ \ (3.9 \ in)$$

If a time-dependent disturbance is specified at $x = 0$, for example,

$$u(0, t) = u_0(t) \tag{4.48}$$

it follows that

$$u(0, t) = u_0(t) = \frac{C}{H} F_R(t)$$

from which

$$\eta(0, t) = \frac{H}{C} u_0(t)$$

The velocity and elevation are expressed at any (x, t) by replacing the argument t with $t - x/C_R$; that is,

$$u(x, t) = u_0\left(t - \frac{x}{C_R}\right), \quad \eta(x, t) = \frac{H}{C} u_0\left(t - \frac{x}{C_R}\right), \quad t > \frac{x}{C_R} \tag{4.49}$$

Outward traveling disturbances in a uniform channel repeat the original disturbance at every location x, but at later times. Note that the function $u_0(t - x/C_R)$ in Eq. (4.49) is understood to be zero for $t - x/C_R < 0$.

Any small-amplitude disturbance in a uniform wave tank travels rightward at a single speed C_R. Such a system has no *dispersion*; that is, propagation speed does not depend on the disturbance amplitude or time behavior. However, the wave length does depend on both the propagation speed and the nature of the disturbance.

EXAMPLE 4.8: DETERMINATION OF WAVE LENGTH
The left wall of a wave tank suddenly begins oscillatory motion according to

$$u(0, t) = u_0(t) = b \sin 2\pi f_0 t$$

where b is the maximum velocity and f_0 is the frequency. Consequently, a train of waves propagates into the stationary liquid of undisturbed depth H. Determine the wavelength λ of the waves.

Since there is no initial flow, $C_R = C$. The velocity and surface elevation are obtained from Eq. (4.49) as

$$u(x, t) = b \sin 2\pi f_0\left(t - \frac{x}{C}\right)$$

$$\eta(x, t) = \frac{H}{C} b \sin 2\pi f_0\left(t - \frac{x}{C}\right)$$

The wavelength at a given instant of time corresponds to the length between x and $x + \lambda$, which causes $2\pi f_0(t - x/C)$ to change by 2π; that is,

$$2\pi f_0 \frac{\lambda}{C} = 2\pi \quad \text{or} \quad \lambda = \frac{C}{f_0} \tag{4.50}$$

It is clear that higher frequencies decrease the wavelength being propagated.

4.8 Energy of a Wave in a Uniform Channel

Sometimes it is useful to know how much energy is propagating with a wave. Since the liquid is incompressible and frictional forces are negligible in hydrostatic wave formulations, kinetic and gravitational potential energies are the significant energy forms which propagate. Figure 4.12 shows a region of disturbed surface of length λ propagating to the right in a uniform channel of width D. Gravitational potential energy of the differential section is given by

$$dE_p = \rho \frac{g}{g_0} \frac{(H + \eta)^2}{2} D \, dx$$

The undisturbed potential energy

$$dE_{pi} = \rho \frac{g}{g_0} \frac{H^2}{2} D \, dx$$

is subtracted and the result is integrated over λ to give

$$\Delta E_p = E_p - E_{pi} = \rho \frac{g}{g_0} \frac{D}{2} \int_X^{X+\lambda} (2H\eta + \eta^2) \, dx \tag{4.51}$$

Total kinetic energy in the differential section is

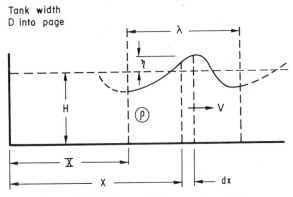

Figure 4.12 Energy Diagram for a Wave

$$dE_k = \frac{\rho D}{2g_0} (H + \eta)V^2 \, dx$$

where V is given by Eq. (4.4). The undisturbed or initial kinetic energy is

$$dE_{ki} = \frac{\rho D}{2g_0} H u_\infty^2 \, dx$$

If the initial value is subtracted from the total and the result is integrated over λ, we obtain

$$\Delta E_k = E_k - E_{ki} = \frac{\rho D}{2g_0} \int_X^{X+\lambda} [(H + n)(u^2 + 2uu_\infty) + \eta u_\infty^2] \, dx \quad (4.52)$$

Equations (4.51) and (4.52) apply to any size wave amplitude. The total energy in a wavelength λ is therefore

$$E_\lambda = \Delta E_p + \Delta E_k$$

$$= \frac{\rho D}{2g_0} \int_X^{X+\lambda} [g(2H\eta + \eta^2) + (H + \eta)(u^2 + 2uu_\infty) + \eta u_\infty^2] \, dx$$

$$(4.53)$$

Consider an initially stationary liquid with $u_\infty = 0$. A right traveling sinusoidal disturbance of wavelength $\lambda = C/f_0$ with velocity

$$u(x, t) = b \sin 2\pi f_0 \left(t - \frac{x}{C} \right) \quad (4.54)$$

and elevation

$$\eta(x, t) = \frac{H}{C} b \sin 2\pi f_0 \left(t - \frac{x}{C} \right) \quad (4.55)$$

contains the total energy

$$E_\lambda = \frac{\rho D b^2 C H}{2g_0 f_0} \quad (4.56)$$

which is equally divided between potential and kinetic forms; that is, Eqs. (4.51) and (4.52) give

$$\Delta E_p = \Delta E_k = \frac{1}{2} \left(\frac{\rho D b^2 C H}{2g_0 f_0} \right) \quad (4.57)$$

EXAMPLE 4.9: ENERGY INPUT TO SINUSOIDAL WAVE

Sinusoidal waves of the form described by Eqs. (4.54) and (4.55) are generated at a moving wall with velocity

$$u_w(t) = u(0, t) = b \sin 2\pi f_0 t$$

for which liquid elevation on the wall is

$$\eta(0, t) = \frac{H}{C} b \sin 2\pi f_0 t$$

The liquid is water in a channel of width $D = 0.5$ m (1.64 ft) with the undisturbed elevation $H = 1.0$ m (3.28 ft). The wave amplitude is to be 0.1 m (0.33 ft) at a frequency of $f_0 = 2$ Hz. Determine the power input from the wall and the energy delivered to the liquid in one cycle of oscillation. Assume no losses in power transmission.

The power delivered by the wall to the liquid is equal to the hydrostatic force times the wall velocity; that is,

$$\mathcal{P} = \left[\rho \frac{g}{g_0} \frac{H + \eta}{2} \right] D(H + \eta) u_w$$

The energy delivered in one period of oscillation $t = 1/f_0$ is given by

$$E = \int_0^{1/f_0} \mathcal{P} \, dt = \frac{\rho D g}{2g_0} \int_0^{1/f_0} (H + \eta)^2 u_w \, dt$$

Substitution for η and u_w yields

$$E = \frac{\rho D b^2 C H}{2 g_0 f_0}$$

which is the same energy carried by a wave in Eq. (4.56). Since

$$C = \sqrt{gH} = 3.13 \text{ m/s } (10.3 \text{ ft/s})$$

and the amplitude of η is

$$\frac{Hb}{C} = 0.1 \text{ m } (0.33 \text{ ft})$$

we have

$$b = 0.313 \text{ m/s } (1.03 \text{ ft/s})$$

for which the power is

$$\mathcal{P} = \left\{ \begin{matrix} 766 \text{ W} \\ 0.73 \text{ B/s} \end{matrix} \right\} [1.0 + 0.1 \sin 2\pi(2)t]^2 \sin 2\pi(2)t$$

Although the power varies with time over a cycle of oscillation, the maximum power required to move the wall is $(1.1)^2(766) = 927$ W. The energy delivered to the liquid in one cycle of oscillation is 38 J (28 ft/bs).

4.9 Wave Transmission at a Channel Discontinuity

Figure 4.13 shows a rectangular wave channel of liquid depth H_I and width D_I, which has an abrupt transition to another channel of liquid depth H_{II} and width D_{II}. The liquid initially is flowing rightward at steady velocity u_{iI}

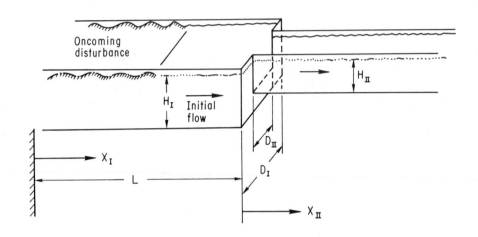

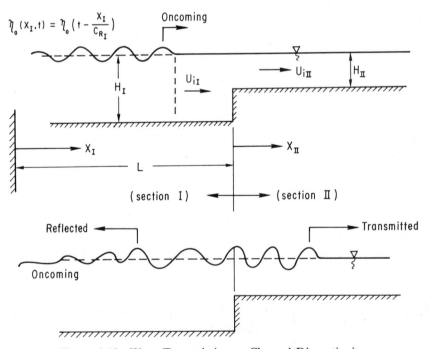

Figure 4.13 Wave Transmission at Channel Discontinuity

in section I. Since the same volume rate flows in each section, the initial velocity in section II is

$$u_{i\text{II}} = \frac{H_\text{I} D_\text{I}}{H_\text{II} D_\text{II}} \, u_{i\text{I}}$$

Figure 4.13 shows the right traveling disturbance $\eta_0(x_I, t)$ coming from the left in section I. The propagation diagram of Fig. 4.14 shows that when η_0 reaches the channel discontinuity at arrival time t_a, a right traveling disturbance is transmitted into section II and a left traveling disturbance is reflected back into section I. The transmitted and reflected properties are to be determined.

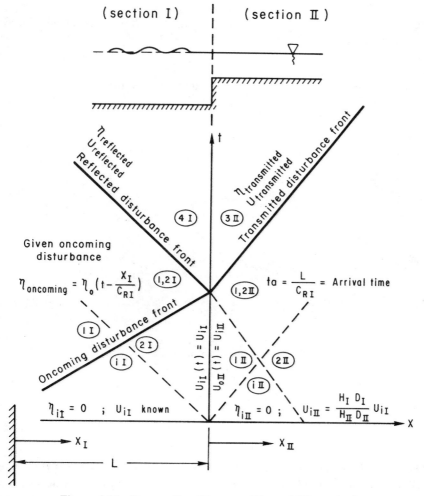

Figure 4.14   Propagation Diagram, Channel Discontinuity

The propagation diagram of Fig. 4.14 is shown as two regions corresponding to channel sections I and II, separated by a common boundary at the discontinuity. The oncoming disturbance is known in Section I, but the reflected and transmitted disturbances are unknown. Boundary conditions which describe continuous elevation and volume flow rate across the discontinuity are given by

$$\eta_I\big|_{x_I=L} = \eta_{II}\big|_{x_{II}=0} \tag{4.58}$$

and

$$H_I D_I u_I\big|_{x_I=L} = H_{II} D_{II} u_{II}\big|_{x_{II}=0} \tag{4.59}$$

Figures 4.8, 4.14, Tables 4.1 and 4.2, and Eqs. (4.40)–(4.45) were employed with the boundary conditions (4.58) and (4.59) to obtain Eqs. (4.60)–(4.66) for the transmitted and reflected waves.

HYDROSTATIC WAVES AT A CHANNEL DISCONTINUITY (FIG. **4.14**)

TRANSMITTED

$$\eta_{transmitted} = \frac{2D_I C_I}{D_I C_I + D_{II} C_{II}} \, \eta_{oncoming} \tag{4.60}$$

$$u_{transmitted} = \frac{2C_{II} D_I C_I}{H_{II}(D_I C_I + D_{II} C_{II})} \, \eta_{oncoming} - \frac{H_I D_I}{H_{II} D_{II}} u_{iI} \tag{4.61}$$

$$\text{with } \eta_{oncoming} = \eta_0\left(t - \frac{L}{C_{RI}} - \frac{X_{II}}{C_{RII}}\right) \tag{4.62}$$

$$C_I = \sqrt{gH_I}, \qquad C_{II} = \sqrt{gH_{II}} \tag{4.63}$$

REFLECTED

$$\eta_{reflected} = \frac{2D_I C_I}{D_I C_I + D_{II} C_{II}} \, \eta_{oncoming} \tag{4.64}$$

$$u_{reflected} = \frac{2C_I D_{II} C_{II}}{H_I(D_I C_I + D_{II} C_{II})} \, \eta_{oncoming} - u_{iI} \tag{4.65}$$

$$\text{with } \eta_{oncoming} = \eta_0\left(t - \frac{L}{C_{RI}} - \frac{(L - X_I)}{C_{LI}}\right) \tag{4.66}$$

It is seen from Eqs. (4.60) and (4.63) that if the oncoming waves arrive from a channel of great width or depth relative to the channel where transmission occurs, the maximum transmitted wave amplitude is twice the oncoming value; that is, $\eta_{transmitted} \to 2\eta_{oncoming}$. Moreover, if waves arriving from a small channel enter a large channel, virtually no disturbance will be transmitted, that is, $\eta_{transmitted} \to 0$.

EXAMPLE 4.10: OCEAN WAVE TRANSMISSION AT A CHANNEL

An inland channel waterway has an average width of 20 m (65.6 ft) and an average depth of 10 m (32.8 ft), but the shore width is many times greater than the channel width. Ocean waves of amplitude $a_0 = 1.0$ m (3.28 ft) arive at the channel with frequency $f = 0.2$ Hz. Determine the transmitted wave properties in the channel.

The ocean is designated as section I with

$$H_I = 10 \text{ m} \quad (32.8 \text{ ft}) \quad \text{and} \quad D_I \to \infty$$

The channel is designated as section II with

$$H_{II} = 10 \text{ m} \quad (32.8 \text{ ft}) \quad \text{and} \quad D_{II} = 20 \text{ m} \quad (65.6 \text{ ft})$$

The arriving waves are approximated by the sinusoidal function

$$\eta_0\left(t - \frac{x_I}{C_I}\right) = a_0 \sin 2\pi f\left(t - \frac{x_I}{C_I}\right)$$

with $a_0 = 1.0$ m (3.28 ft) and $f = 0.2$ Hz. Equation (4.60) gives the transmitted wave elevation as

$$\eta_{transmitted} = \begin{Bmatrix} 2.0 \text{ m} \\ 6.56 \text{ ft} \end{Bmatrix} \sin 0.4\pi\left(t - \frac{x_{II}}{C_{II}} - \frac{L}{C_I}\right)$$

where L is arbitrary. If the channel water initially is not flowing, use of $C_{II} = \sqrt{gH_{II}}$ with Eq. (4.61) gives the velocity

$$u_{transmitted} = (2.0 \text{ m/s}) \sin 0.4\pi\left(t - \frac{x_{II}}{C_{II}} - \frac{L}{C_I}\right)$$

These results show that the wave amplitude doubles and the channel velocity amplitude is 2 m/s (6.56 ft/s). Since $H_I = H_{II} = 10$ m (32.8 ft), the propagation speeds in both the ocean and channel are equal; that is, $C_I = C_{II} = 10$ m/s (32.8 ft/s).

If there are successive channel discontinuities, the transmitted wave amplitude is determined by successive use of Eq. (4.60) at each discontinuity.

EXAMPLE 4.11: WAVE GROWTH, SEVERAL DISCONTINUITIES

Figure 4.15(a) shows a channel of uniform width D, which contains a step decrease in depth from $H_I = 3$ m to $H_{II} = 1$ m. Figure 4.15(b) shows the same channel with two successive equal steps, where $H_{I,II} = 2$ m, giving the same overall depth change

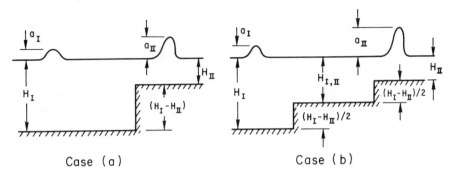

Case (a)   Case (b)

Figure 4.15 Wave Amplitude Increase in One and Two Steps

as shown in Fig. 4.15(a). If an oncoming wave in section I has amplitude a_I, what will be the transmitted amplitude for each case?

Equation (4.60) gives the amplitude multiplier for case (a) as

$$\frac{\eta_{II}}{\eta_I} = \frac{2C_I}{C_I + C_{II}} = \frac{2}{1 + \sqrt{H_{II}/H_I}} = 1.268$$

The first step of Fig. 4.15(b) yields

$$\frac{\eta_{I,II}}{\eta_I} = \frac{2}{1 + \sqrt{H_{I,II}/H_I}} = \frac{2}{1 + \sqrt{2/3}}$$

For the second step, we have

$$\frac{\eta_{II}}{\eta_{I,II}} = \frac{2}{1 + \sqrt{H_{II}/H_{I,II}}} = \frac{2}{1 + \sqrt{1/2}}$$

The overall amplitude multiplier in Fig. 4.15(b) is

$$\frac{\eta_{II}}{\eta_I} = \frac{\eta_{II}}{\eta_{I,II}} \frac{\eta_{I,II}}{\eta_I} = \frac{2}{1 + \sqrt{2/3}} \frac{2}{1 + \sqrt{1/2}} = 1.29$$

It is interesting that the amplitude multiplication for two successive steps is greater than it is for a single step. Many more steps of smaller size will cause a more dramatic amplitude multiplication.

There are conditions which could be useful in various experiments where it is desirable to transmit a wave through a channel area discontinuity with a negligible amplitude change. If the oncoming wave is a step with

$$\eta_0 = N_0 = \text{constant}$$

then, according to Eq. (4.60), the transmitted wave amplitude will be equal to the oncoming amplitude if

$$\frac{D_{\mathrm{I}}}{D_{\mathrm{II}}} = \frac{C_{\mathrm{II}}}{C_{\mathrm{I}}} = \sqrt{\frac{H_{\mathrm{II}}}{H_{\mathrm{I}}}} \qquad (4.67)$$

Thus, an oncoming and transmitted wave step will be the same at a channel area discontinuity if the width and depth vary according to Eq. (4.67).

If the oncoming wave is sinusoidal, the condition of Eq. (4.67) causes the transmitted and oncoming amplitudes to be equal, but the wavelength and propagation speed in section II will differ from that in section I.

4.10   Other Methods for Solving Small Amplitude Wave Flows

The general solution for small-amplitude hydrostatic waves in uniform channels given in Section 4.6 applies to nearly all initial and boundary conditions which could be specified. However, other solution methods can be employed.

The eigenvalue method requires a transformation of the dependent variables η or u of Eq. (4.14) to render the wall boundary conditions homogeneous. Then the separation of variables procedure, the orthogonality property of Eq. (4.14), and the initial conditions can be employed to develop a full solution. Computations involve evaluation of truncated infinite series and often require numerical evaluation of integrals.

Linear transformations, such as the Laplace or Fourier transforms, also can be used to obtain a solution of hydrostatic waves in uniform channels. Sometimes the inverse transform, which accomodates the specified end wall boundary conditions, is difficult to obtain. (However, see Section 3.16 for discussion of a numerical procedure for the inversion of Laplace transforms.)

The solution method of Section 4.6 suffers none of the evaluation difficulties which may arise in eigenvalue or linear transform methods. If the initial and boundary conditions are written in analytical form, the solution is straightforward and involves a continuous sequential addition of boundary and initial conditions.

4.11   Subcritical, Critical, and Supercritical Flows

The initial flow velocity u_∞ in a uniform channel is either subcritical ($u_\infty < C$), critical ($u_\infty = C$), or supercritical ($u_\infty > C$). Equation (4.14) is written for each of these cases as

$$\frac{\partial^2 \phi}{\partial t^2} + 2u_\infty \frac{\partial^2 \phi}{\partial x \partial t} - (C^2 - u_\infty^2) \frac{\partial^2 \phi}{\partial x^2} = 0, \qquad u_\infty < C \qquad (4.68)$$

$$\frac{\partial^2 \phi}{\partial t^2} + 2C \frac{\partial^2 \phi}{\partial x \partial t} = 0 , \qquad u_\infty = C \tag{4.69}$$

$$\frac{\partial^2 \phi}{\partial t^2} + 2u_\infty \frac{\partial^2 \phi}{\partial x \partial t} + (u_\infty^2 - C^2) \frac{\partial^2 \phi}{\partial x^2} = 0 , \qquad u_\infty > C \tag{4.70}$$

where ϕ is either η or u. The subcritical case described by Eq. (4.68) already has been discussed, but is included here for comparison. The method of Section 4.3 is employed to obtain the solutions

$$\eta(x, t) = F_L\left(t + \frac{x}{C - u_\infty}\right) + F_R\left(t - \frac{x}{C + u_\infty}\right)$$

$$u(x, t) = -\frac{C}{H}\left[F_L\left(t + \frac{x}{C - u_\infty}\right) - F_R\left(t - \frac{x}{C + u_\infty}\right)\right] \qquad u_\infty < C \tag{4.71}$$

$$\eta(x, t) = F_c(x) + F_{Rc}\left(t - \frac{x}{2C}\right)$$

$$u(x, t) = -\frac{C}{H}\left[F_c(x) - F_{RC}\left(t - \frac{x}{2C}\right)\right], \qquad u_\infty = C \tag{4.72}$$

$$\eta(x, t) = F_{sL}\left(t - \frac{x}{u_\infty - C}\right) + F_{sR}\left(t - \frac{x}{u_\infty + C}\right)$$

$$u(x, t) = -\frac{C}{H}\left[F_{sL}\left(t - \frac{x}{u_\infty - C}\right) - F_{sR}\left(t - \frac{x}{u_\infty + C}\right)\right] \qquad u_\infty > C \tag{4.73}$$

A propagation diagram for each case is sketched in Fig. 4.16, showing the propagation paths of a local disturbance at x_d. The disturbance head and tail move in opposite directions in subcritical flow. If the flow is critical, the disturbance cannot propagate opposite to the direction of flow. The critical flow case can be seen as a standing hydraulic jump, such as that stationary wave which appears on the surface of a stream where it passes over a submerged rock. If the flow is supercricital, both the head and tail propagate in the direction of flow, and the tail lags behind the head. The oncoming fluid is not disturbed by a local disturbance in critical or supercritical flows. Figure 4.16 is drawn for an abrupt disturbance at $x = x_d$. If the disturbance is a functiòn of time, a succession of similarly sloped characteristics with different F functions would originate at various t and $x = x_d$.

Figure 4.17 shows propagation diagrams for subcritical, critical, and supercritical flows with right traveling disturbances generated at $x = 0$. If the initial conditions are

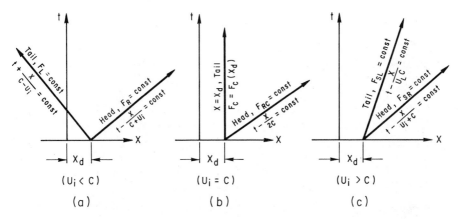

Figure 4.16 Propagation. (*a*) Subcritical, (*b*) Critical, (*c*) Supercritical

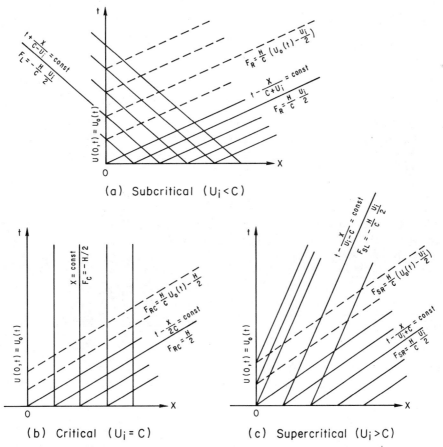

Figure 4.17 Propagation Diagrams, Disturbance at $x = 0$. (*a*) Subcritical ($u_i < c$). (*b*) Critical ($u_i = c$). (*c*) Supercritical ($u_i > c$)

$$u(x,0) = u_\infty , \qquad \eta(x,0) = 0 \tag{4.74}$$

then the characteristic lines originating from $t = 0$ carry the functions F_L and F_R according to

$$u_\infty < C , \quad F_L = -\frac{H}{C}\frac{u_\infty}{2} , \quad F_R = \frac{H}{C}\frac{u_\infty}{2} , \qquad t = 0$$

$$u_\infty = C , \quad F_c = -\frac{H}{2} , \qquad F_{Rc} = \frac{H}{2} , \qquad t = 0 \tag{4.75}$$

$$u_\infty > C , \quad F_{sL} = -\frac{H}{C}\frac{u_\infty}{2} , \quad F_{sR} = \frac{H}{C}\frac{u_\infty}{2} , \qquad t = 0$$

If a velocity disturbance $u(0, t) = u_0(t)$ occurs on the $x = 0$ boundary, the originating characteristic lines shown dashed correspond to

$$u_\infty < C , \quad F_R = \frac{H}{C}\left[u_0(t) - \frac{u_\infty}{2} \right] , \qquad x = 0$$

$$u_\infty = C , \quad F_{Rc} = \frac{H}{C} u_0(t) - \frac{H}{2} , \qquad x = 0 \tag{4.76}$$

$$u_\infty > C , \quad F_{sR} = \frac{H}{C}\left[u_0(t) - \frac{u_\infty}{2} \right] , \qquad x = 0$$

EXAMPLE 4.12: PROPAGATION AT VARIOUS STREAM SPEEDS
Water discharge from a gate is undergoing steady flow in a wave channel at velocity u_∞ and undisturbed depth H. A submerged flap at $x = 0$ begins small-amplitude cycling such that the local water velocity varies according to

$$u(0, t) = u_0(t) = u_\infty + a_0 \sin 2\pi ft$$

Determine the right traveling wave solution for subcritical, critical, and supercritical flows.

Equations (4.71), (4.72), and (4.73) with (4.76) yield

$$\eta(x, t) = \frac{H}{C} a_0 \sin 2\pi f\left(t - \frac{x}{C + u_\infty} \right) , \qquad u_\infty < C$$

$$\eta(x, t) = \frac{H}{C} a_0 \sin 2\pi f\left(t - \frac{x}{2C} \right) , \qquad u_\infty = C$$

and

$$\eta(x, t) = \frac{H}{C} a_0 \sin 2\pi f\left(t - \frac{x}{u_\infty + C} \right) , \qquad u_\infty > C$$

This section summarizes the basis for solving disturbance propagation problems in subcritical, critical, and supercritical flows.

4.12   Variable Channel Width

Consider the case of a channel in Fig. 4.1 with uniform depth H, $B = 0$, and variable width $D(x)$. It initially contains stationary liquid, $u_\infty = 0$, with no surface pressure variation. Small-amplitude wave behavior is described by Eqs. (4.9) and (4.10) in the forms

$$\frac{\partial \eta}{\partial t} + H \frac{\partial u}{\partial x} + \frac{H}{D} \frac{dD}{dx} u = 0 \qquad (4.77)$$

$$\frac{\partial u}{\partial t} + g \frac{\partial \eta}{\partial x} = 0 \qquad (4.78)$$

If u is eliminated,

$$\frac{\partial^2 \eta}{\partial t^2} - C^2 \frac{\partial^2 \eta}{\partial x^2} - C^2 \left(\frac{1}{D} \frac{dD}{dx} \right) \frac{\partial \eta}{\partial x} = 0 \qquad (4.79)$$

and if η is eliminated,

$$\frac{\partial^2 u}{\partial t^2} - C^2 \frac{\partial^2 u}{\partial x^2} - C^2 \frac{\partial}{\partial x} \left(\frac{1}{D} \frac{dD}{dx} u \right) = 0 \qquad (4.80)$$

When the channel width $D(x)$ is constant, Eqs. (4.79) and (4.80) reduce to Eq. (4.14) with $u_\infty = 0$, for which a solution is composed of right and left traveling disturbances. However, the general solution forms $F_L(t + x/C)$ and $F_R(t - x/C)$ do not satisfy Eqs. (4.79) and (4.80). Although a solution procedure for a general disturbance is described in Chapter 5, several special cases are discussed here for which analytical solutions can be formulated.

Consider a disturbance originating at $x = 0$ such that propagation is rightward at constant speed C. A function $F_R(t - x/C)$ with a space-dependent coefficient $f(x)$ yields a possible solution of the form

$$\phi(x, t) = f(x) F_R \left(t - \frac{x}{C} \right) \qquad (4.81)$$

Substitution into Eq. (4.79) gives

$$\frac{1}{C} \frac{F_R'}{F_R} = \frac{f'' + (D'/D)f'}{2f' + (D'/D)f} = \lambda$$

where λ is an arbitrary separation constant. The simplest solution, corresponding to $\lambda = 0$, is

$$F_R \left(t - \frac{x}{C} \right) = H_s \left(t - \frac{x}{C} \right)$$

where $H_s(t - x/C)$ is the Heaviside unit step, which is 1.0 for $t > x/C$, and 0 otherwise. Furthermore,

$$f(x) = K_1 + K_2 \int_0^x \frac{dx}{D(x)}$$

which yields

$$\eta(x, t) = \left[K_1 + K_2 \int_0^x \frac{dx}{D(x)} \right] H_s\left(t - \frac{x}{C} \right) \tag{4.82}$$

Substitution into Eqs. (4.77) and (4.78) and partial integration lead to

$$u(x, t) = \left[\frac{K_3}{D(x)} - \frac{K_2}{D(x)} gt \right] H_s\left(t - \frac{x}{C} \right) \tag{4.83}$$

Equation (4.82) shows that for a channel of variable width, an elevation disturbance sweeps through, varying with x, but remaining constant in time after it passes each location. Thus, $\lambda = 0$ corresponds to a step disturbance elevation at $x = 0$. Equation (4.83) shows that the velocity changes linearly with time at any x. The step entrance boundary condition is, therefore,

$$\eta(0, t) = N_0 = \text{constant}$$

for which Eq. (4.47) gives the initial entrance velocity as

$$u(0, 0) = \frac{CN_0}{H}, \qquad \lambda = 0$$

These conditions give $K_1 = N_0$ and $K_3 = CN_0 D(0)/H$. If the channel broadens such that $dD/dx > 0$ for all x; the distant boundary condition is given by

$$\eta(\infty, t) = 0, \quad \lambda = 0, \quad D'(x) > 0$$

for which

$$K_2 = - \frac{N_0}{\displaystyle\int_0^\infty \frac{dx}{D(x)}}$$

If the channel narrows with $dD/dx < 0$ for all x, closing at $x = L$, the corresponding boundary condition is

$$u(L, t) = 0, \quad \lambda = 0, \quad D'(x) < 0$$

and $K_2 = D(0)N_0/L$. It follows that the solution for a step elevation disturbance at the entrance for $dD/dx > 0$ is given by

$$\eta(x, t) = \left[1 - \frac{\int_0^x \dfrac{dx}{D}}{\int_0^\infty \dfrac{dx}{D}}\right] N_0 H_s\left(t - \frac{x}{C}\right)$$

$$u(x, t) = \left(1 + \frac{Ct}{D(0)\int_0^\infty \dfrac{dx}{D}}\right) \frac{D(0)}{D(x)} \frac{C}{H} N_0 H_s\left(t - \frac{x}{C}\right) \qquad (4.84)$$

$$\lambda = 0, \quad \frac{dD}{dx} > 0$$

and for $dD/dx < 0$ with $D(L) = 0$,

$$\eta(x, t) = \left(1 + \frac{D(0)}{L}\int_0^x \frac{dx}{D(x)}\right) N_0 H_s\left(t - \frac{x}{C}\right)$$

$$u(x, t) = \left(1 - \frac{Ct}{L}\right) \frac{D(0)}{D(x)} \frac{C}{H} N_0 H_s\left(t - \frac{x}{C}\right) \qquad (4.85)$$

$$\lambda = 0, \quad \frac{dD}{dx} < 0$$

Values of λ other than zero would yield other boundary conditions.

EXAMPLE 4.13: CHANNEL OF LINEARLY VARYING WIDTH

A channel of uniform depth H and width which varies according to

$$D(x) = D_0 \pm bx$$

initially contains stationary liquid. The constant b is positive. Determine the wave amplitude and velocity properties for an amplitude step to N_0 at the channel entrance.

It is easily shown that

$$\int_0^x \frac{dx}{D(x)} = \begin{cases} \ln\left(1 + \dfrac{bx}{D_0}\right)^{1/b}, & D(x) = D_0 + bx \\[2mm] \ln\left(1 - \dfrac{bx}{D_0}\right)^{-1/b}, & D(x) = D_0 - bx \end{cases}$$

Equations (4.84) therefore give

$$\eta(x, t) = N_0 H_s\left(t - \frac{x}{C}\right), \qquad D(x) = D_0 + bx$$

$$u(x, t) = \left(\frac{D_0}{D_0 + bx}\right)\frac{C}{H} N_0 H_s\left(t - \frac{x}{C}\right),$$

which shows that a step-wave amplitude at the entrance of a channel of linearly increasing width propagates as a uniform step, whereas the velocity diminshes with x.

The case for decreasing width corresponds to a wedge-shaped channel with $b = D_0/L$, where L is the channel length. Equation (4.85) yields for this case that,

$$\eta(x, t) = \left[1 + \ln\left(1 - \frac{x}{L}\right)^{-1}\right] N_0 H_s\left(t - \frac{x}{C}\right)$$

$$u(x, t) = \frac{1 - Ct/L}{1 - x/L} \frac{C}{H} N_0 H_s\left(t - \frac{x}{C}\right)$$

$$D(x) = D_0 - bx$$

The wave amplitude is seen to grow as its front moves through the narrowing wedge. If we follow the wave front by putting $x = Ct$, it can be seen that liquid velocity at the front remains constant during propagation. The amplitude remains constant at any x, but the velocity diminishes linearly with time after the front passes.

4.13 Variable Channel Depth

Next consider the case where a channel has uniform width D but continuously variable depth with floor elevation above the horizontal, given by $B = B(x)$. Equations (4.9) and (4.10) become, for initially stationary liquid,

$$\frac{\partial \eta}{\partial t} + (H - B) \frac{\partial u}{\partial x} = u \frac{dB}{dx} \tag{4.86}$$

$$\frac{\partial u}{\partial t} + g \frac{\partial \eta}{\partial x} = 0 \tag{4.87}$$

If we eliminate $u(x, t)$, we obtain

$$\frac{\partial^2 \eta}{\partial t^2} - g(H - B) \frac{\partial^2 \eta}{\partial x^2} + g \frac{dB}{dx} \frac{\partial \eta}{\partial x} = 0 \tag{4.88}$$

A variable fluid depth means that the wave propagation speed will vary with x according to

$$C(x) = \sqrt{g[H - B(x)]} \tag{4.89}$$

If a disturbance originates at $x = 0$, it will propagate into stationary liquid with the speed of Eq. (4.89), and its time to reach a given $x > 0$ will be $\int_0^x dx/C(x)$. Therefore, a solution of the form

$$\eta(x, t) = f(x) F_R\left(t - \int_0^x \frac{dx}{C(x)}\right) \tag{4.90}$$

is suggested. Substitution into Eq. (4.88) yields

$$\frac{F'_R}{F_R} = \frac{C(x)(d/dx)[(H-B)f']}{2(C(x)^2/g)f' - (dB/dx)f} = \lambda$$

The simplest case with $\lambda = 0$ yields

$$\eta(x, t) = \left(K_1 + K_2 \int_0^x \frac{dx}{C(x)^2} \right) H_s \left(t - \int_0^x \frac{dx}{C(x)} \right)$$

If we follow a procedure similar to that of Section 4.12, $u(x, t)$ is obtained by partial integration of Eqs. (4.86) and (4.87). A step change in elevation to N_0 at $x = 0$ yields

$$\frac{dB}{dx} < 0$$

$$\eta(x, t) = N_0 \left[1 - \frac{I(x)}{I(\infty)} \right] H_s[t - J(x)] \qquad (4.91)$$

$$u(x, t) = \frac{gN_0}{C^2(x)} \left(C(0) + \frac{t}{I(\infty)} \right) H_s[t - J(x)] \qquad (4.92)$$

$$I(x) = \int_0^x \frac{dx}{C^2(x)}, \qquad J(x) = \int_0^x \frac{dx}{C(x)} \qquad (4.93)$$

and

$$\frac{dB}{dx} > 0, \qquad B(L) = H$$

$$\eta(x, t) = N_0 \left[1 + C(0) \frac{I(x)}{J(L)} \right] H_s[t - J(x)] \qquad (4.94)$$

$$u(x, t) = \frac{gC(0)N_0}{C^2(x)} \left(1 - \frac{t}{J(L)} \right) H_s[t - J(x)] \qquad (4.95)$$

It is seen from Eqs. (4.91) and (4.94) that a step amplitude disturbance at $x = 0$ decreases if $dB/dx < 0$ or increases if $dB/dx > 0$, but remains constant at each x after it passes.

4.14 Wave Generation from Unsteady Surface Pressure

Consider a uniform channel of stationary liquid with initially undisturbed depth H. The surface pressure $P_\infty(x, t)$ begins to undergo time- and space-dependent variations, causing waves to be generated. Equations (4.9) and (4.10) provide

$$\frac{\partial \eta}{\partial t} + H \frac{\partial u}{\partial x} = 0 \tag{4.96}$$

and

$$\frac{\partial u}{\partial t} + g \frac{\partial \eta}{\partial x} = -\frac{g_0}{\rho} \frac{\partial P_\infty}{\partial x} \tag{4.97}$$

Let the channel be bounded at $x = 0$ so that $u(0, t) = 0$, which is equivalent to

$$\frac{\partial \eta(0, t)}{\partial x} = -\frac{g_0}{\rho g} \frac{\partial P_\infty}{\partial x}$$

from Eq. (4.97). Also assume that the channel is long so that the distant boundary condition is $\eta(\infty, t) = 0$. Suppose that a pressure disturbance front moves from $x = 0$ at some uniform speed u_f so that $P_\infty(x, t)$ is expressed as the general function

$$P_\infty(x, t) = P_0 f\left(t - \frac{x}{u_f}\right) H_s\left(t - \frac{x}{u_f}\right) \tag{4.98}$$

If u is eliminated from Eqs. (4.96) and (4.97), the full problem is

DE: $$\frac{\partial^2 \eta}{\partial t^2} - C^2 \frac{\partial^2 \eta}{\partial x^2} = \frac{C^2}{u_f^2} \frac{g_0 P_0}{\rho g} f''\left(t - \frac{x}{u_f}\right) H_s\left(t - \frac{x}{u_f}\right) \tag{4.99}$$

ICs: $$\eta(x, 0) = \frac{\partial \eta(x, 0)}{\partial t} = 0 \tag{4.100}$$

BCs: $$\frac{\partial \eta(0, t)}{\partial x} = \frac{1}{u_f} \frac{g_0 P_0}{\rho g} f'(t) H_s(t), \qquad \eta(\infty, t) = 0 \tag{4.101}$$

A solution for the special case where

$$f\left(t - \frac{x}{u_f}\right) = \sin 2\pi f_0\left(t - \frac{x}{u_f}\right)$$

can be obtained by use of Laplace transforms, and the result suggests that a solution for the general case is given by

$$\eta(x, t) = \frac{g_0 P_0}{\rho g} \left(\frac{C^2}{C^2 - u_f^2}\right) \left[\frac{u_f}{C} f\left(t - \frac{x}{C}\right) H_s\left(t - \frac{x}{C}\right)\right.$$

$$\left. - f\left(t - \frac{x}{u_f}\right) H_s\left(t - \frac{x}{u_f}\right)\right] \tag{4.102}$$

which can be verified by direct substitution. Equation (4.102) shows that for $C^2 > u_f^2$, a wave front of amplitude

$$\frac{g_0 P_0}{\rho g}\left(\frac{C^2}{C^2 - u_f^2}\right)\frac{u_f}{C} f\left(t - \frac{x}{C}\right)$$

precedes the pressure front, which may be positive or negative, depending
on $f(t - x/C)$. When the pressure front passes a given location, the second
term of Eq. (4.102) alters the wave character previously established. If
$u_f^2 > C^2$, the pressure front moves ahead of the wave front. The case where
$u_f = C$ is indeterminate from Eq. (4.102).

EXAMPLE 4.14: WAVES FROM A UNIFORM PRESSURE FRONT

A uniform pressure front of $P_0 = 0.05$ atm (0.74 psi) moves at a speed of u_f over a
water channel of 5 m (16.4 ft) undisturbed depth, for which the wave speed is
$C = 7$ m/s (23 ft/s). Describe the waves generated for cases where the pressure front
moves at speeds of $u_f = 4$ m/s (13 ft/s) and 10 m/s (32.8 ft/s).

Since the pressure front is a uniform step, we have from Eq. (4.98),

$$f\left(t - \frac{x}{u_f}\right) = 1.0$$

which also corresponds to $f(t-x/C) = 1.0$. The pressure front moves slower than the
wave speed for the case when $u_f = 4$ m/s (13 ft/s), for which Eq. (4.102) gives

$$\eta(x, t) = (0.435 \text{ m})H_s\left(t - \frac{x}{C}\right) - (0.762 \text{ m})H_s\left(t - \frac{x}{u_f}\right),$$

$$u_f = 4 \text{ m/s (13 ft/s)}, \quad C = 7 \text{ m/s (23 ft/s)}$$

It is seen that a uniform wave of 0.435 m (1.4 ft) positive amplitude first advances
into the channel, but drops to a depression of 0.435-0.762 = −0.327 m (−1.1 ft) when
the pressure front passes. When $u_f = 10$ m/s (32.8 ft/s), the pressure front moves
faster than the wave speed, and Eq. (4.108) gives

$$\eta(x, t) = (-0.707 \text{ m})H_s\left(t - \frac{x}{C}\right) + (0.494 \text{ m})H_s\left(t - \frac{x}{u_f}\right),$$

$$u_f = 10 \text{ m/s (32.8 ft/s)}, \quad C = 7 \text{ m/s (23 ft/s)}$$

It is seen that the surface first rises 0.494 m (1.62 ft) when the pressure front passes
and then drops to a depression of −0.707 + 0.494 = −0.213 m (0.7 ft) when the wave
front passes!

References

4.1 Lamb, H., *Hydrodynamics*, 6th ed., Dover, New York, 1945.

4.2 Milne-Thompson, L. M., *Theoretical Hydrodynamics*, 4th ed., Macmillan, New
York, 1960.

4.3 Hildebrand, F. B., *Advanced Calculus for Applications*, 2nd ed., Prentice-Hall,
Englewood Cliffs, N.J., 1976.

Problems

4.1   Equation (4.14) for the case of no initial flow becomes

$$\frac{\partial^2 \phi}{\partial t^2} - C^2 \frac{\partial^2 \phi}{\partial x^2} = 0$$

Show that if a solution of the form $\phi = f(\eta)$ is assumed with the similarity variable $\eta = tC/x$, the function $f(\eta)$ is given by

$$f(\eta) = A + \frac{B}{2} \ln\left(\frac{1+\eta}{1-\eta}\right)$$

where A and B are constants.

4.2   Obtain a similarity solution to Eq. (4.14) by assuming that $\phi = f(\eta)$, where η is the similarity variable $\eta = t(C - U_\infty)/x$.

Answer:

$$f(\eta) = A + B \frac{C - U_\infty}{2C} \ln \left| \frac{1+\eta}{(C - U_\infty)/(C + U_\infty) - \eta} \right|$$

4.3   Consider a uniform wave tank of length L and undisturbed liquid depth H, like the one shown in Fig. 4.4. The right wall is stationary, and the left wall suddenly undergoes a rightward step velocity $u(0, t) = u_0$ for a time interval $\tau = L/3C$, when it instantly reverses velocity to $u(0, t) = -u_0$ for a time interval $\tau < t < 2\tau$, stopping abruptly at its original position when $t = 2\tau$. Determine the resulting elevation η on both end walls.

Answer:

See Fig. P. 4.3.

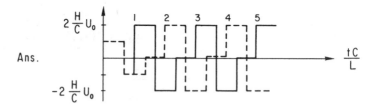

Figure P4.3

4.4   A uniform wave tank has length L and undisturbed liquid depth H. Suddenly the left wall moves rightward at constant velocity $u(0, t) = u_0$ for a time interval $0 \le t \le \tau$, then reverses its velocity to $-u_0$ for a time interval $\tau < t \le 2\tau$, coming to rest at its initial position. If

$\tau = 0.2L/C$, sketch the liquid profile which is propagating after $t = 2\tau$ and show that the velocity of the right end wall, u_r, which would be required to absorb the arriving wave with no reflected amplitude disturbance from elevation H is given by

$$0 \leq t \leq L/C, \qquad\qquad\qquad u_r = 0$$

$$L/C < t \leq L/C + \tau = 1.2(L/C), \qquad u_r = 2u_0$$

$$1.2(L/C) < t \leq L/C + 2\tau = 1.4(L/C), \qquad u_r = -2u_0$$

$$1.4(L/C) < t, \qquad\qquad\qquad u_r = 0$$

4.5 A horizontal, uniform rectangular tank is moving rightward at a constant velocity of 1.0 m/s (3.28 ft/s) and stops suddenly. The contained liquid is at uniform depth $H = 5$ m (16.4 ft) prior to stopping. If the tank horizontal length in the direction of motion is $L = 10$ m (32.8 ft) determine the liquid elevation at the left and right end walls as a function of time.

Answer:

Use of Table 4.2 and Fig. 4.8 yields the solution shown in Fig. P4.5.

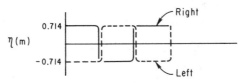

Figure P4.5

4.6 A stationary wave tank of length $L = 5$ m (16.4 ft) and undisturbed liquid depth $H = 1$ m (3.28 ft) has a disturbed surface at one instant of time (designated $t = 0$) in the form of a periodic wave with initial conditions given by

$$\eta(x, 0) = \eta_i(x_i) = \eta_0 \cos \frac{2n\pi x}{L}$$

$$u(x, 0) = u_i(x_i) = 0$$

where the amplitude is $\eta_0 = 0.1$ m (0.328 ft) and n is an integer.

(a) Determine the time-dependent elevations $\eta(0, t)$ and $\eta(L, t)$ at the two end walls, and in the section $(1, 2)$ of Fig. 4.8.

Answer:

Both left and right ends,

$$\eta = \eta_0 \cos 2\pi f_n t, \qquad f_n = 0.626\, n \text{ Hz}$$

and in section $(1, 2)$,

$$\eta_{(1,2)} = \eta_0 \cos 2\pi f_n t \cos \frac{2n\pi x}{L}$$

(**b**) Determine the frequency of oscillation for cases with $n = 1, 2$, and 3.

Answer:

n	$f_n = nc/L$ (Hz)
1	0.626
2	1.252
3	1.878

(**c**) Obtain a solution to Eq. (4.14) with $u_\infty = 0$ and $\phi = u$ by assuming a product solution of the form $u(x, t) = X(x)T(t)$. Obtain $\eta(x, t)$ by integrating Eqs. (4.12) and (4.13) with $u_\infty = 0$, and for full cycle waves at any instant verify that a solution is

$$\eta(x, t) = \eta_0 \cos \frac{2n\pi Ct}{L} \cos \frac{2n\pi x}{L}$$

which corresponds to the solution obtained for all sections of Fig. 4.8.

4.7 Consider a wave tank of length L and initially undisturbed liquid of depth H. Both tank end walls begin to move together with a periodic velocity

$$u(0, t) = u(L, t) = U_{max} \sin 2\pi f t$$

Use Table 4.1 and Fig. 4.8 to determine analytical expressions for the wave profile in regions 1, 2, 3, 4, $(1, 2)$, and $(3, 4)$. (This solution is patterned after Example 4.5.)

4.8 Consider a long-wave tank containing stationary water of undisturbed depth $H = 1$ m (3.28 ft). It is desirable to form a single sinusoidal wave of length $\lambda = 2$ m (6.56 ft) and amplitude $\eta_{max} = 0.1$ m (0.328 ft) by moving the left wall according to some specification.

(**a**) Show that the left wall should be moved according to

$$u(0, t) = -\frac{C}{H}\, \eta_{max}\, \sin \frac{2\pi C}{\lambda}\left(t - \frac{\lambda}{C}\right)$$

(b) Show that the energy required per unit width D of the wave tank for generating one sinusoidal wave is

$$\text{Energy}/D = \rho\, \frac{g}{g_0}\, \frac{\eta_{max}^2 \lambda}{2} = 98\ \text{J/m}\ (0.0283\ \text{B/ft})$$

4.9   The left wall velocity of a long-wave tank undergoes a ramp increase, followed by a ramp decrease according to

$$0 < t \le \tau, \qquad u(0, t) = at$$

$$\tau < t \le 2\tau, \qquad u(0, t) = a(2\tau - t)$$

$$2\tau < t, \qquad u(0, t) = 0$$

Show that the resulting liquid elevation on the wall is

$$0 < t \le \tau, \qquad \eta(0, t) = \frac{atH}{C}$$

$$\tau < t \le 2\tau, \qquad \eta(0, t) = \frac{a(2\tau - t)H}{C}$$

$$2\tau < t, \qquad \eta(0, t) = 0$$

4.10 The left wall of a long-wave tank undergoes sinusoidal motion according to $u(0, t) = U_{max} \sin 2\pi ft$. If $f = 1.0\ \text{Hz}$, $U_{max} = 0.2\ \text{m/s}$ (0.656 ft/s) and the undisturbed depth is $H = 1.163\ \text{m}$ (3.81 ft) show that the wave length propagating is 4 m (13.12 ft), and for comparison, verify that the wall back-and-forth motion amplitude is 0.03 m (0.0984 ft).

4.11 Show that the total energy in a sinusoidal small-amplitude wave is
(a) proportional to the wave length;
(b) proportional to the square of the wave amplitude. (Verify that the total energy of a single sinusoidal wave in a channel of width D is $(\rho g/g_0)\eta_{max}^2(\lambda/2)D$.

4.12 Consider an idealized wave of the form shown in Fig. P4.12 propagating rightward into stationary liquid with $H = 1\ \text{m}$ (3.28 ft) $\lambda = 4\ \text{m}$ (13.1 ft), $\eta_0 = 0.2\ \text{m}$ (0.656 ft) and a channel width $D = 1\ \text{m}$ (3.28 ft). If the liquid density is $1000\ \text{kg/m}^3$ (62.4 lbm/ft^3) show that the energy propagating with this wave is 784 J (0.74 B).

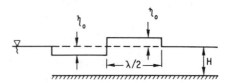

Figure P4.12

4.13 An incoming wave from the ocean has an amplitude of $\eta_0 = 2$ m
(6.56 ft) where the water has an undisturbed depth of $H = 60$ m
(197 ft). See Fig. P4.13. A reef lies a distance $L = 4$ m (13.12 ft)
below the undisturbed surface. Show that the wave amplitude transmitted toward the shore is 32/25 m (4.2 ft).

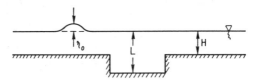

Figure P4.13

4.14 The incoming wave of 2 m (6.56 ft) amplitude in Problem 4.13 passes
over a ditch with $H = 4$ m (13.12 ft) and $L = 64$ m (210 ft). See Fig.
P4.14. Show that the transmited amplitude is 32/25 m (4.2 ft).

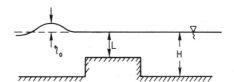

Figure P4.14

4.15 The sketch shows a long wave of amplitude η_0 arriving from deep
water at a two-step shelf from H_1 to H_2 to H_3 depth. Let $H_1 = 40$ m
(131 ft) and $H_3 = 10$ m (32.8 ft). See Fig. P4.15.
   (a) Show that the value of H_2 which causes the maximum transmitted
   wave amplitude is 20 m (65.6 ft) with a transmitted-to-oncoming
   amplitude ratio of 1.37.
   (b) Show that if the intermediate step H_2 was excluded, the transmitted-to-oncoming wave amplitude going directly from H_1 to H_3
   is 1.3.

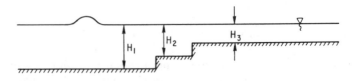

Figure P4.15

4.16 Refer to Fig. P4.16. Show that it a long wave of amplitude η_0 arrives at the series of steps shown, the amplitude transmitted at the last step N is

$$\frac{\eta_N}{\eta_0} = \prod_{n=1}^{N} \frac{2}{1 + \sqrt{H_n/H_{n-1}}}$$

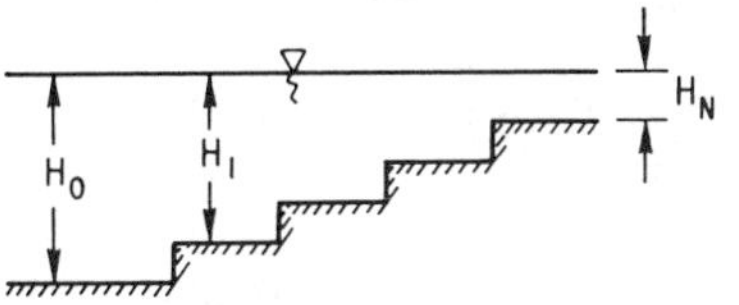

Figure P4.16

Also show that if all the steps are an equal fraction of the previous step with H_n/H_{n-1} = constant, then

$$\frac{\eta_N}{\eta_0} = \left(\frac{2}{1 + \sqrt{H_n/H_{n-1}}} \right)^N$$

(Note that η_N/η_0 becomes large as N increases.)

4.17 A wave tank like that of Fig. 4.13 has $D_I = 2$ m (6.56 ft), $D_{II} = 1$ m (3.28 ft), and $H_I = 2$ m (6.56 ft). Show that the value of H_{II} which will permit wave transmission into section II without amplitude change is 8 m (26.2 ft). (Two-dimensional effects would be important if good resolution were required in the vicinity of the transition.)

4.18 Liquid is flowing to the right in a long wave tank at velocity $u_\infty = 5$ m/s (16.4 ft/s). If the liquid depth is $H = 1$ m (3.28 ft) show that a periodic disturbance by a submerged flap of frequency $f_F = 1$ Hz creates traveling waves of wavelength 8.13 m (26.7 ft).

4.19 Consider a wave channel of uniform depth H. If the width varies exponentially according to $D = D_0 e^{ax}$, show that the equation describing propagation is

$$\frac{\partial^2 \eta}{\partial t^2} - C^2 \frac{\partial^2 \eta}{\partial x^2} - \frac{C^2}{a} \frac{\partial \eta}{\partial x} = 0$$

4.20 A step elevation disturbance η_0 enters a wave channel of uniform liquid depth H whose walls are shaped sinusoidally so that the width $D(x)$ varies as

$$D(x) = D_0 \left[1 + \epsilon \sin\left(\frac{2\pi x}{\lambda_w} \right) \right]$$

where $\epsilon < 1.0$ and λ_w is the wall wavelength. Verify that solutions of Eqs. (4.79) and (4.80) are given by

$$\eta(x, t) = \eta_0 H_s\left(t - \frac{x}{C}\right)$$

$$u(x, t) = \frac{C\eta_0}{H[1 + \epsilon \sin(2\pi x/\lambda_w)]} H_s\left(t - \frac{x}{C}\right)$$

(Note that η_0 propagates unchanged, but velocity increases and decreases according to the wall spacing.)

4.21 Consider a horizontal wave tank of undisturbed liquid with parallel sidewalls. Undisturbed liquid depth at the left end is H. The floor elevation varies according to $B(x)$. Consider a solution of Eq. (4.88) for the wave amplitude of the form $\eta(x, t) = f(x)F[t - g(x)]$. Show that for a floor variation

$$B(x) = H - \left\{[(H - B(L))^{3/4} - H^{3/4}]\frac{x}{L} + H^{3/4}\right\}^{4/3}$$

$\eta(x, t)$ is given by the general solution

$$\eta(x, t) = f_0\left[\int_0^x \frac{dx}{H - B(x)} + A\right]F\left(t - \int_0^x \frac{dx}{\sqrt{g(H - B(x))}}\right)$$

which is similar to the solutions of Eqs. (4.91) and (4.94). However, this solution is not restricted to a step.

4.22 Consider steady flow in a channel, described by Eqs. (4.6a) and (4.6b). Show that if the channel wall spacing $D(x) = $ constant and the floor elevation varies as $B(x)$, the velocity $u_\infty(x)$ can be obtained from a solution to the linear differential equation,

$$\frac{dB}{du_\infty} + \frac{1}{u_\infty}B = \frac{C_1}{u_\infty} - \frac{3}{2}\frac{u_\infty}{g}$$

where C_1 is a constant. If the conditions $B = 0$, $u_\infty = u_{\infty 0}$, and $H = H_0$ occur at $x = 0$, show that $H(x)$ and $u_\infty(x)$ for a known function $B(x)$ satisfy

$$B = \left(1 - \frac{u_{\infty 0}}{u_\infty}\right)H \quad \text{and} \quad H = H_0 + \frac{u_{\infty 0}^2 - u_\infty^2}{2g}$$

5 HYDROSTATIC WAVES AND LARGE-AMPLITUDE APPLICATIONS

Solutions for large-amplitude hydrostatic wave propagation in uniform and nonuniform channels are described in this chapter.

The equations developed in Chapter 4 are solved, without linearization for small-amplitude disturbances. Waves in uniform channels which propagate in one direction only are *simple waves*. The propagation of simultaneous right and left traveling disturbances results in *compound waves*. Analytical solutions are developed for large-amplitude simple-wave problems, such as the classical dam break. Compound-wave solutions are formulated with Riemann invariants.

The classical *method of characteristics* (MOC) is summarized in this chapter. A procedure is discussed for using the MOC to solve problems with eventual shock formation. A simplified solution also is described for cases where energy dissipation across a shock is small. A numerical procedure is outlined for using the MOC solution with various boundary conditions and nonuniform channels.

5.1 Uniform Channel, Simple-Wave Propagation

Large-amplitude waves in a uniform channel with constant surface pressure are described by Eqs. (4.1) and (4.2) with $B = 0$, and both D and P_∞ held constant, which yields

$$\frac{\partial y}{\partial t} + V \frac{\partial y}{\partial x} + y \frac{\partial V}{\partial x} = 0 \qquad (5.1)$$

233

$$\frac{\partial V}{\partial t} + V \frac{\partial V}{\partial x} + g \frac{\partial y}{\partial x} = 0 \tag{5.2}$$

It was shown in Section 4.7, Eq. (4.47), that for small-amplitude hydrostatic wave propagation in one direction, the elevation and velocity disturbances are related by

$$\eta(x, t) = \frac{H}{C} u(x, t)$$

This observation implies a possible starting point for a solution of large-amplitude wave propagation in one direction.

Since elevation and velocity are linearly related for small-amplitude waves, it is reasonable to assume that large-amplitude solutions to Eqs. (5.1) and (5.2) may exist for which the elevation y can be expressed as a general function of the velocity V,

$$y = y(V) \tag{5.3}$$

It follows that the derivatives of y are

$$\frac{\partial y}{\partial t} = \frac{dy}{dV} \frac{\partial V}{\partial t} , \qquad \frac{\partial y}{\partial x} = \frac{dy}{dV} \frac{\partial V}{\partial x}$$

Equations (5.1) and (5.2) become

$$\frac{dy}{dV} \frac{\partial V}{\partial t} = -\left(V \frac{dy}{dV} + y\right) \frac{\partial V}{\partial x} \tag{5.4}$$

and

$$\frac{\partial V}{\partial t} = -\left(V + g \frac{dy}{dV}\right) \frac{\partial V}{\partial x} \tag{5.5}$$

from which it follows that

$$\frac{dy}{dV} = \pm \sqrt{\frac{y}{g}} \tag{5.6}$$

Equation (5.6) can be integrated from the initial conditions y_i and V_i to give

$$V = V_i \pm 2(\sqrt{gy} - \sqrt{gy_i}) \tag{5.7}$$

Therefore, the assumption of Eq. (5.3) is valid if V and y are related by the form of Eq. (5.7).

The time and space dependence of V and y are obtained by substituting Eq. (5.6) into (5.4) or (5.5), either of which yields

$$\frac{\partial V}{\partial t} + (V \pm \sqrt{gy}) \frac{\partial V}{\partial x} = 0 \tag{5.8}$$

If V is expressed functionally as $V(x, t)$, its ordinary time derivative is given by

$$\frac{dV}{dt} = \frac{\partial V}{\partial t} + \frac{dx}{dt} \frac{\partial V}{\partial x} \tag{5.9}$$

Comparison of Eqs. (5.8) and (5.9) shows that for a path in the x, t plane on which $V =$ constant, the derivative dx/dt must be equal to $V \pm \sqrt{gy}$. However, if $V =$ constant, dx/dt actually becomes $(\partial x/\partial t)_V$, so

$$\left(\frac{\partial x}{\partial t} \right)_V = V \pm \sqrt{gy} \tag{5.10}$$

Equation (5.10) is integrated partially to obtain

$$x = (V \pm \sqrt{gy})t + f(V) \tag{5.11}$$

where the arbitrary function $f(V)$ must be determined. Since Eq. (5.7) shows that for a given value of V there is a unique value of y, Eq. (5.11) describes possible paths in the x, t plane on which both V and y are constant. It also is seen from Eq. (5.11) that constant values of V and y propagate on straight lines in the x, t plane with slopes $dt/dx = 1/(V \pm \sqrt{gy})$! This is true only for disturbance propagation in one direction, or *simple waves*. Moreover, Eq. (5.11) shows that the $+$ sign gives path lines sloping to the right for right traveling disturbances, and the $-$ sign corresponds to disturbances propagating to the left.

Other solution methods are required for problems where disturbances from opposite directions interfere with each other. However, Eqs. (5.7) and (5.11) provide the solution for a large class of interesting problems in which disturbances propagate in one direction only. Some of these problems are discussed in the following sections.

5.2 General Disturbance, Simple Waves

Consider the long uniform channel in Fig. 5.1 which initially contains stationary liquid of depth H. A vertical wall separates two regions of the liquid. If the wall moves to the right or left, it disturbs the liquid on both sides, and propagation in each region travels in one direction only. Leftward propagation occurs in the left region, and rightward propagation occurs in the right region. Suppose that the wall begins to move with a prescribed velocity

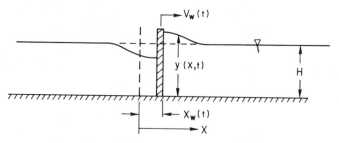

Figure 5.1 Wall Creating General Disturbance

$$V_w = V_w(t) \tag{5.12}$$

Its displacement is the velocity integral

$$x_w(t) = \int_0^t V_w \, dt \tag{5.13}$$

The function $f(V)$ in Eq. (5.11) is obtained wherever x, y, and t can be expressed in terms of liquid velocity V. A common procedure is to consider the liquid velocity V in contact with the wall such that $V = V_w$. Then Eq. (5.7) is used to express $y = y(V)$. These substitutions make it possible to rewrite Eq. (5.12) as $t = t(V_w) = t(V)$, from which $x_w(t)$ is expressed as $x(V)$ from Eq. (5.13), and $f(V)$ is obtained from Eq. (5.11). When $f(V)$ is determined, Eqs. (5.7) and (5.11) are combined to express either $V(x, t)$ or $y(x, t)$. This procedure is illustrated in the next section.

5.3 Accelerating Wall Withdrawal, Simple Waves

The long channel in Fig. 5.2 contains liquid of depth H. The bounding wall begins to move leftward with constant acceleration a_w so that

$$V_w = a_w t \tag{5.14}$$

The liquid velocity and elevation $V(x, t)$ and $y(x, t)$ are to be determined. The wall displacement is obtained from

$$x_w = \int_0^t V_w \, dt = \frac{1}{2} a_w t^2 \tag{5.15}$$

The function $f(V)$ is determined from Eq. (5.11) by expressing liquid velocity V at the moving wall. Since V and x are positive to the right, we have

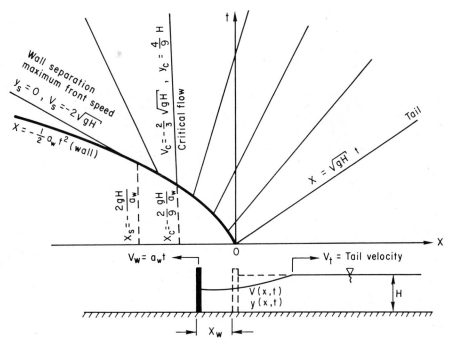

Figure 5.2 Accelerating Wall Withdrawal

$$V = -V_w = -a_w t$$

with the wall trajectory

$$x = -x_w = -\frac{1}{2} a_w t^2$$

which is drawn on the t, x diagram of Fig. 5.2. Parametric equations for x and t can be written in terms of V as

$$t = -\frac{V}{a_w}, \qquad x = -\frac{1}{2}\frac{V^2}{a_w} \qquad (5.16)$$

Furthermore, since wall motion causes disturbance propagation to the right, the $+$ signs are used in Eqs. (5.7) and (5.11) to yield

$$y = \frac{1}{g}\left(\sqrt{gH} + \frac{V}{2}\right)^2 \qquad (5.17)$$

and

$$x = (V + \sqrt{gy})t + f(V) = \left(\frac{3}{2}V + \sqrt{gH}\right)t + f(V) \qquad (5.18)$$

The function $f(V)$ is obtained by substituting for x and t from Eq. (5.16), which gives

$$f(V) = \frac{V^2}{a_w} + \frac{\sqrt{gH}}{a_w} V \tag{5.19}$$

Therefore, Eq. (5.18) gives the propagation lines in the t, x plane as

$$x = \left(\frac{3}{2} V + \sqrt{gH}\right) t + \frac{V^2}{a_w} + \frac{\sqrt{gH}}{a_w} V , \qquad V < 0 \tag{5.20}$$

which are drawn in Fig. 5.2, starting from the wall trajectory line. These propagation lines form an *expansion fan*. The value of y is expressed from Eq. (5.17). Note that one propagation line in Fig. 5.2 is vertical. This corresponds to

$$\left(\frac{\partial x}{\partial t}\right)_V = 0 \tag{5.21}$$

for which Eqs. (5.20) and (5.17) give the *critical flow* condition

Critical Flow

$$V = -(2/3)\sqrt{gH} , \qquad y = (4/9)H \tag{5.22}$$

where no disturbances propagate upstream. The location of critical flow for constant wall acceleration a_w is obtained from Eqs. (5.16) and (5.22) as

$$x = -\frac{1}{2} \frac{V^2}{a_w} = -\frac{1}{2} \frac{4}{9} \frac{gH}{a_w} = -\frac{2}{9} \frac{gH}{a_w} \tag{5.23}$$

Wall velocity eventually reaches a value where the liquid elevation y becomes zero, at which the liquid maximum front velocity from Eq. (5.17) is

Maximum Front Velocity

$$V_f = -2\sqrt{gH} \tag{5.24}$$

The corresponding location where this occurs is obtained from Eq. (5.16) as

Maximum Front Velocity Location

$$x_f = -\frac{2gH}{a_w} \tag{5.25}$$

The wall separates from the liquid after its velocity reaches $-2\sqrt{gH}$. The liquid front trajectory after separation from the wall is obtained from Eqs. (5.20) and (5.24) as

$$x_f = 2\left(\frac{gH}{a_w} - \sqrt{gH}\,t\right), \qquad t > \frac{2\sqrt{gH}}{a_w} \qquad (5.26)$$

The *tail* identifies that point to which the disturbance has propagated into the undisturbed liquid. Its location corresponds to $V = 0$ in Eqs. (5.20), for which

Tail Location

$$x_t = \sqrt{gH}\,t \qquad (5.27)$$

Equations (5.14) through (5.26) were employed to obtain Table 5.1, which gives a summary of the important relationships governing liquid velocity and elevation in a uniform channel with end wall acceleration away from the liquid region. Both dimensional and nondimensional forms are

TABLE 5.1   Uniform Liquid Channel, End Wall Acceleration Withdrawal

Liquid Velocity

$$V = \frac{1}{2}\left\{\left[\left(\sqrt{gH} + \frac{3}{2}a_w t\right)^2 - 4(a_w t\sqrt{gH} - a_w x)\right]^{1/2} - \left(\sqrt{gH} + \frac{3}{2}a_w t\right)\right\}$$

$$V^* = \frac{1}{2}\left\{\left[\left(1 + \frac{3}{2}t^*\right)^2 - 4(t^* - x^*)\right]^{1/2} - \left(1 + \frac{3}{2}t^*\right)\right\}$$

Elevation	*Point of Critical Flow*	
$y = \dfrac{1}{g}\left(\sqrt{gH} + \dfrac{V}{2}\right)^2$	$x_c = -\dfrac{2}{9}\dfrac{gH}{a_w}$	$x_c^* = -\dfrac{2}{9}$
$y^* = \left(1 + \dfrac{V^*}{2}\right)^2$	$t_c = \dfrac{2}{3}\dfrac{\sqrt{gH}}{a_w}$	$t_c^* = \dfrac{2}{3}$
Tail	$V_c = -\dfrac{2}{3}\sqrt{gH}$	$V_c^* = -\dfrac{2}{3}$
$V_t = 0 \qquad V_t^* = 0$		
$x_t = \sqrt{gH}\,t \qquad x_t^* = t^*$	$y_c = \dfrac{4}{9}H$	$y_c^* = \dfrac{4}{9}$
Wall	*Wall Separation*	
$x = -\dfrac{1}{2}a_w t^2 \qquad x^* = -\dfrac{1}{2}t^{*2}$	$x_s = -2\dfrac{gH}{a_w}$	$x_s^* = -2$
$V = -a_w t \qquad V^* = -t^*$	$t_s = 2\dfrac{\sqrt{gH}}{a_w}$	$t_s^* = 2$
Nondimensional Variables	$V_s = -2\sqrt{gH}$	$V_s^* = -2$
$x^* = \dfrac{xa_w}{gH} \qquad t^* = \dfrac{a_w t}{\sqrt{gH}}$	$y_s = 0$	$y_s^* = 0$
$V^* = \dfrac{V}{\sqrt{gH}} \qquad y^* = \dfrac{y}{H}$		

given. The nondimensionalized results are plotted in Fig. 5.3. The front continues to travel leftward at normalized velocity $V^* = V/\sqrt{gH} = -2$ after wall separation. The tail where liquid velocity $V^* = 0$ always travels rightward at constant propagation velocity so that its trajectory is given by $x^* = xa_w/gH = t^* = a_w t/\sqrt{gH}$.

EXAMPLE 5.1: WALL SEPARATION FROM LIQUID

Stationary liquid of depth $H = 2.0$ m is in a long uniform channel, bounded at the left end by a moveable wall. The wall begins to move leftward at an acceleration $a_w = 4.0$ m/s^2. Determine the times and locations where critical flow occurs, and where the wall separates from the liquid. Also determine the liquid elevation as a function of time at a location which coincides with the initial position of the wall.

Table 5.1 is used to determine the required quantities. Direct substitution gives properties at the point of critical flow as

$$x_c = -1.09 \text{ m}$$

$$V_c = -2.95 \text{ m/s}$$

$$y_c = 0.89 \text{ m}$$

and

$$t_c = 0.74 \text{ s}$$

Wall separation from the liquid corresponds to

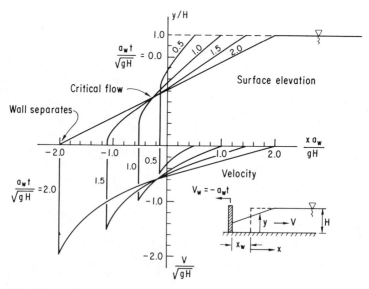

Figure 5.3 Wave Properties, Uniform Channel, Wall Acceleration away from Liquid

TABLE 5.2   Liquid Elevation at $x = 0$, Wall Acceleration Withdrawal from Liquid

$a_w t/\sqrt{gH}$	$\dfrac{y}{H}$	t, s	y, m
0.0	1.00	0.00	2.00
0.5	0.67	0.55	1.30
1.0	0.56	1.11	1.12
1.5	0.52	1.66	1.04
2.0	0.50	2.21	1.00

$$x_s = -9.8 \text{ m}$$

$$V_s = -8.85 \text{ m/s}$$

and

$$t_s = 2.21 \text{ s}$$

Liquid elevation at the initial location of the wall, $x = 0$, could be obtained either from Fig. 5.3 or Table 5.1 at given times. Figure 5.3 was used to obtain Table 5.2 for $y(0, t)$. A limiting process for V as $t \to \infty$ in Table 5.1 at $x = 0$ shows that the liquid velocity approaches

$$V \to -(2/3)\sqrt{gH}$$

for which $y = (4/9)H$. That is, for a wall withdrawing from liquid with finite acceleration, the critical flow state is closely approached but never fully achieved at $x = 0$. It is calculated in this example at $x_c = -1.09$ m and $t_c = 0.74$ s.

5.4 Wall Acceleration into Liquid, Simple Waves

A bounding wall at the left end of the stationary liquid region in Fig. 5.4 begins to move rightward at constant acceleration b so that its velocity and position are

$$V_w = bt, \qquad x_w = (1/2)bt^2 \tag{5.28}$$

The liquid velocity $V(x, t)$ and surface elevation $y(x, t)$ are to be determined.

The procedure of Sections 5.2 and 5.3 give the velocity and displacement of liquid in contact with the wall as

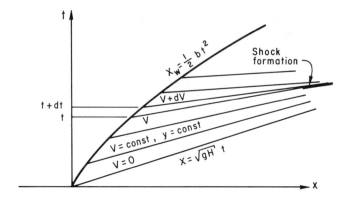

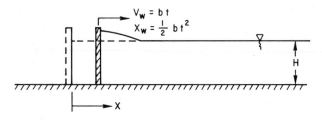

Figure 5.4 Wall Acceleration into Uniform Liquid Region

$$V = +V_w = bt \qquad \text{and} \qquad x = x_w = (1/2)bt^2$$

from which

$$t = \frac{V}{b} \qquad \text{and} \qquad x = \frac{1}{2}\frac{V^2}{b}$$

Wall motion causes disturbance propagation to the right so that the + signs of Eqs. (5.7) and (5.11) are used, giving

$$y = \frac{1}{g}\left(\sqrt{gH} + \frac{V}{2}\right)^2$$

and

$$x = (V + \sqrt{gy})t + f(V) = \left(\frac{3}{2}V + \sqrt{gH}\right)t + f(V)$$

If we substitute for x and t, we get

$$f(V) = -\frac{V^2}{b} - \frac{V}{b}\sqrt{gH}$$

and the propagation lines are described by

$$x = \left(\frac{3}{2} V + \sqrt{gH}\right)t - \frac{V^2}{b} - \frac{V}{b}\sqrt{gH} \qquad (5.29)$$

with

$$y = \frac{1}{g}\left(\sqrt{gH} + \frac{V}{2}\right)^2 \qquad (5.30)$$

The disturbance propagation speed $(\partial x/\partial t)_V$ is obtained by differentiating Eq. (5.29). This yields

$$\left(\frac{\partial x}{\partial t}\right)_V = \frac{3}{2} V + \sqrt{gH}$$

Since $V > 0$, the disturbance propagation speed relative to the earth increases with V. Thus, for wall acceleration *into* the liquid region, propagation lines generated at higher values of V will overtake those generated at lower V. This phenomenon leads to the formation of a hydrostatic shock, shown in the propagation diagram of Fig. 5.4.

Hydrostatic shock formation can be identified by considering two propagation lines, one corresponding to time t and velocity V, and the other corresponding to time $t + dt$ and velocity $V + dV$. These lines cross in Fig. 5.4 whenever they have a common x, t point. That is, one propagation line for $x(V, t)$ corresponds to Eq. (5.29). Another propagation line for $x(V + dV, t + dt)$ corresponds to

$$x(V + dV, t + dt) = \left[\frac{3}{2}(V + dV) + \sqrt{gH}\right](t + dt)$$
$$- \frac{(V + dV)^2}{b} - \frac{V + dV}{b}\sqrt{gH}$$

If $x(V, t)$ and $x(V + dV, t + dt)$ are equated, and second-order differentials are neglected, we find that

$$t_{\text{shock}} = \frac{1}{b}\left(\frac{4}{3} V + \frac{2}{3}\sqrt{gH}\right) \qquad (5.31)$$

The earliest appearance of a shock in this case corresponds to the smallest value of V, namely zero, for which

$$t_{\text{shock appears}} = \frac{2}{3}\frac{\sqrt{gH}}{b} \qquad (5.32)$$

The propagation line for $V = 0$ in Fig. 5.4 is given by $x = \sqrt{gH}\, t$, so shock formation begins at the location

$$x_{\text{shock appears}} = \sqrt{gH}\, t_{\text{shock appears}} = \frac{2}{3}\frac{gH}{b} \qquad (5.33)$$

The method just described for identifying a shock is based on the propagation diagram geometry. A rigorous procedure for shock identification corresponds to the condition where the partial derivatives $(\partial x/\partial V)_t$ and $(\partial^2 x/\partial V^2)_t$ are simultaneously zero. This procedure gives results identical to Eqs. (5.32) and (5.33) for walls with constant acceleration and is further discussed in Section 8.3.

Whenever a shock appears, it becomes a moving boundary condition from which disturbances propagate, and the problem no longer involves propagation in one direction only. Thus, the one-directional propagation solution described in Section 5.1 is not valid near a shock, and other solution methods must be used. Sections 5.10, 5.11, and 5.12 describe hydrostatic shock properties and a solution method for simultaneous propagation in both directions.

Equations (5.28) through (5.33) were employed to obtain Table 5.3, which gives a summary of the important relationships describing liquid velocity and elevation in a uniform channel with wall acceleration into the liquid region up to the time of shock formation. Dimensional and non-

TABLE 5.3   Uniform Channel, End Wall Acceleration into Liquid Region (Fig. 5.4)

Liquid Velocity

$$V = \frac{1}{2}\left\{\left[\left(\sqrt{gH} - \frac{3}{2}bt\right)^2 + 4(bt\sqrt{gH} - bx)\right]^{1/2} - \left(\sqrt{gH} - \frac{3}{2}bt\right)\right\}$$

$$V^* = \frac{1}{2}\left\{\left[\left(1 - \frac{3}{2}t^*\right)^2 - 4(t^* - x^*)\right]^{1/2} - \left(1 - \frac{3}{2}t^*\right)\right\}$$

Elevation

$$y = \frac{1}{g}\left(\sqrt{gH} + \frac{V}{2}\right)^2, \qquad y^* = \left(1 + \frac{V^*}{2}\right)^2$$

Shock Formation

$$x_{\text{shock appears}} = \frac{2}{3}gH/b \qquad t_{\text{shock appears}} = \frac{2}{3}\sqrt{gH}/b$$

$$x^*_{\text{shock appears}} = \frac{2}{3} \qquad t^*_{\text{shock appears}} = \frac{2}{3}$$

Wall

$$x_w = \frac{1}{2}bt^2 \qquad V_w = bt \qquad x^*_w = \frac{1}{2}t^{*2} \qquad V^*_w = t^*$$

Nondimensional Variables

$$x^* = \frac{bx}{gH} \qquad t^* = \frac{bt}{\sqrt{gH}} \qquad V^* = \frac{V}{\sqrt{gH}} \qquad y^* = \frac{y}{H}$$

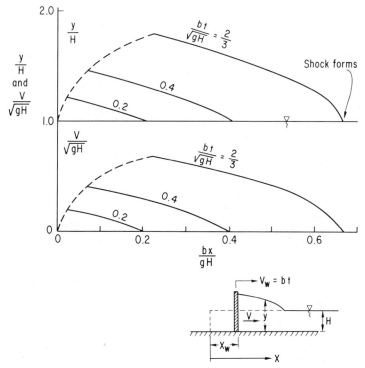

Figure 5.5   Wave Properties, Uniform Channel, Wall Acceleration into Liquid

dimensional forms are given. The nondimensional results are shown in Fig. 5.5.

EXAMPLE 5.2: SHOCK POSITION AND WATER RUN-UP
A wall begins to move rightward at acceleration $b = 2.0\,\text{m/s}^2$ (6.56 ft/s²) into a long channel of stationary liquid of depth $H = 1.0\,\text{m}$ (3.28 ft). Determine the time and location of shock formation and the liquid elevation on the wall, or *run-up*.
    The time of shock formation in Table 5.3 corresponds to

$$t_{\text{shock appears}} = \frac{2}{3}\frac{\sqrt{gH}}{b} = 1.04\,\text{s}$$

The location where shock formation begins is

$$x_{\text{shock appears}} = \frac{2}{3}\frac{gH}{b} = 3.27\,\text{m}\ (10.7\,\text{ft})$$

Liquid elevation on the wall at the time of shock formation is obtained from Fig. 5.5 as

$$y = (1.78)H = 1.78\,\text{m}\ (5.84\,\text{ft})$$

5.5   The Dam Break Problem

A classical problem involves the sudden disintegration or removal of the wall behind which a long channel of liquid exists. Figure 5.6 with disturbance propagation to the right helps describe this problem and its solution. If we put $y_i = H$, $V_i = 0$, and employ the $+$sign for rightward propagation, Eq. (5.7) becomes

$$V = 2(\sqrt{gy} - \sqrt{gH})\tag{5.34}$$

Movement of the leftward advancing front corresponds to $y = 0$, for which its velocity is

$$V = V_f = -2\sqrt{gH}\tag{5.35}$$

and its displacement is

$$x = x_f = -2\sqrt{gH}\,t\tag{5.36}$$

If we set $x = x_f$, $y = 0$, and $V = V_f$, it follows from Eq. (5.11) that $f(V) = 0$.

If we chose to follow the tail into undisturbed liquid at propagation velocity $V_t = \sqrt{gH}$, elevation $y = H$, and liquid velocity $V = 0$, we again would find that $f(V) = 0$. Equation (5.11) becomes, with the substitution of Eq. (5.34),

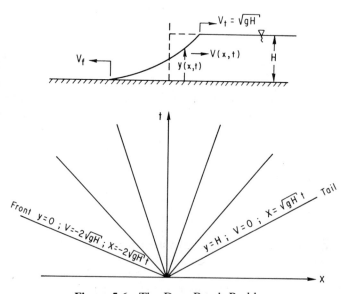

Figure 5.6   The Dam Break Problem

$$x = (V + \sqrt{gy})t = (\tfrac{3}{2}V + \sqrt{gH})t \tag{5.37}$$

which produces a family of propagation lines originating at $x = 0$ as shown in Fig. 5.6. These propagation lines form an expansion fan. Each value of fluid velocity between 0 and $-2\sqrt{gH}$ gives a different propagation line between the front and tail. The same result is obtained from Eq. (5.29) if the wall acceleration b increases without bound. Critical flow occurs at $x = 0$, with velocity $V = -(2/3)\sqrt{gH}$ and elevation $y = (4/9)H$, which is identical to results obtained for the accelerating wall withdrawal of Eq. (5.22).

Equations (5.34) and (5.37) can be written in the self-similar forms

$$\frac{V}{\sqrt{gH}} = 2\left(\frac{x}{3\sqrt{gH}\,t} - \frac{1}{3}\right) \tag{5.38}$$

and

$$\frac{y}{H} = \left(\frac{2}{3} + \frac{x}{3\sqrt{gH}\,t}\right)^2 \tag{5.39}$$

which are drawn in Fig. 5.7. The similarity variable $x/3\sqrt{gH}\,t$ involves both independent variables x and t, and implies that the same values of velocity and elevation which occur at one x, t set also occur at other x, t sets.

Suppose that a power plant was built on a very long lake behind a dam where it obtained its condenser cooling water. A sudden dam rupture would cause a rapid drop of the water surface to four ninths of its initial depth at the original dam location. It might be preferred to locate the power plant further upstream from the dam where the surface elevation would drop at a slower rate.

EXAMPLE 5.3: WATER FRONT VELOCITY AND LEVEL DROP
A disaster is postulated whereby the dam which helps contain one end of a long lake ruptures instantly. If the lake average depth is 20 m (65.6 ft), determine the front velocity and the elevation-time behavior 1.0 km (3280 ft) upstream from the dam.

The front velocity from Fig. 5.7 is

$$V = -2\sqrt{gH} = -28 \text{ m/s } (-92 \text{ ft/s})$$

Table 5.4 was constructed with the help of Fig. 5.7, using

$$\frac{x}{3\sqrt{gH}} = 23.8 \text{ s}$$

in order to determine the elevation-time behavior. The water elevation at 1.0 km (3280 ft) begins to drop at 71.4 s after dam rupture. It drops to half its 20-m (65.6 ft) depth in 2142 s, and its asymptotic elevation is 8.89 m (29 ft). This analysis does not apply after reflected waves return from the far end of the lake.

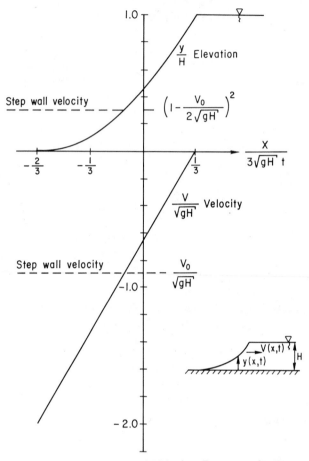

Figure 5.7 Elevation and Velocity, Rupture of a Dam

TABLE 5.4 Dam Rupture, Lake Elevation-Time Behavior

$\dfrac{y}{H}$	y, m (ft)		$\dfrac{x}{3\sqrt{gH}t}$	t, s
1.0	20	(66)	1.0	71.4 (start drop)
0.9	18	(59)	0.28	252
0.8	16	(52)	0.23	306
0.7	14	(46)	0.17	429
0.6	12	(39)	0.10	714
0.5	10	(33)	0.03	2142
4/9	8.9	(29)	0.0	∞

5.6   Wall Withdrawal at Step Velocity

The wall at the left end of a long channel of stationary liquid, depth H, suddenly begins to move leftward at constant velocity V_0, as shown in Fig. 5.8. The resulting liquid velocity and elevation are to be determined.

The tail corresponds to $V = 0$, $y = H$, and moves on the path $x = \sqrt{gH}\, t$, for which Eq. (5.11) gives $f(V) = 0$. Thus, Eqs. (5.7) and (5.11) yield propagation lines which begin at the origin and extend according to

$$x = (\tfrac{3}{2}V + \sqrt{gH})t$$

as in Eq. (5.37) for the dam break problem. If we move counterclockwise from the tail in Fig. 5.8, we encounter the propagation line

$$x = (-\tfrac{3}{2}V_0 + \sqrt{gH})t$$

at which the liquid velocity and wall velocity are equal. The shaded region corresponds to a uniform liquid velocity and elevation given by

$$V = -V_0 \qquad \text{and} \qquad y = \frac{1}{g}\left(\sqrt{gH} - \frac{V_0}{2}\right)^2 \qquad (5.40)$$

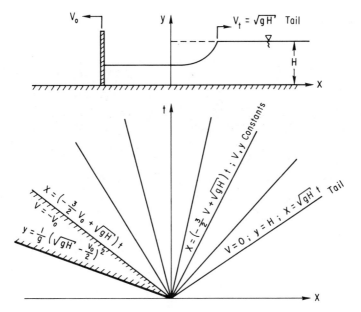

Figure 5.8   Wall Withdrawal at Step Velocity

Velocity and surface elevation profiles for the case of wall withdrawal at step velocity can be obtained from the dam break formulation of Fig. 5.7. If the wall velocity is V_0, the maximum leftward liquid velocity is $V = -V_0$, so the nondimensional velocity scale $V/\sqrt{gH}$ does not exceed $-V_0/\sqrt{gH}$, shown for arbitrary V_0 as a dashed horizontal line extending to the left, labeled *step wall velocity*. The corresponding liquid elevation profile also is shown as a dashed line, obtained from Eq. (5.40). The location of these example dashed lines depends on V_0

EXAMPLE 5.4: LIQUID ELEVATION, STEP WALL WITHDRAWAL
Determine liquid elevation profiles at several times for subcritical, critical, and supercritical step wall withdrawal velocities in a uniform wave tank of depth $H = 1.0$ m.

The critical velocity from Eq. (5.22) is

$$V = V_c = -\tfrac{2}{3}\sqrt{gH} = -2.087 \text{ m/s}$$

so we shall pick leftward step wall velocities of $V_0 = 1.0$, 2.087, and 4.0 m/s. The respective nondimensional liquid elevations on the wall are, from Eq. (5.40),

$$\frac{y}{H} = \left(1 - \frac{V_0}{2\sqrt{gH}}\right)^2 = 0.706, 0.444, \text{ and } 0.13$$

Figure 5.7 gives the corresponding similarity variable

$$\frac{x}{3\sqrt{gH}\,t} = 0.167, 0.0, \text{ and } -0.3$$

TABLE 5.5 Elevation Profiles at Several Times, Wall Withdrawal at Various Step Velocities ($H = 1.0$ m)

	V_0	y	$\dfrac{y}{H}$	$\dfrac{x}{3\sqrt{gH}\,t}$	x, m		
	m/s	m	(−)	(−)	$t = 1.0$ s	$t = 2.0$ s	$t = 3.0$ s
Subcritical	1.0	1.0	0.333	3.12	6.25	9.38
	1.0	0.9	0.283	2.66	5.32	7.97
	1.0	0.8	0.233	2.19	4.38	6.56
	1.0	0.706	0.167	1.57	3.14	4.70
Critical	2.087	0.6	0.10	0.94	1.88	2.82
	2.087	0.5	0.033	0.31	0.62	0.93
	2.087	0.444	0.0	0.0	0.0	0.0
Supercritical	4.0	0.3	−0.117	−1.10	−2.20	−3.30
	4.0	0.2	−0.217	−2.04	−4.08	−6.11
	4.0	0.13	−0.30	−2.82	−5.63	−8.45

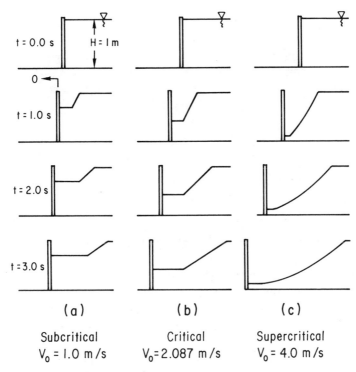

Figure 5.9   Liquid Elevation, Wall Withdrawal at (*a*) Subcritical ($V_0 = 1.0$ m/s), (*b*) Critical ($V_0 = 2.087$ m/s), (*c*) Supercritical ($V_0 = 4.0$ m/s) Velocities

Table 5.5 was obtained with the help of Fig. 5.7 for each V_0. The resulting liquid elevation profiles are shown in Fig. 5.9.

5.7   Large-Amplitude Wave Distortion

It can be seen from cases already discussed that a large-amplitude surface wave distorts as it propagates. The wave profile in Fig. 5.5 caused by wall acceleration into a liquid region becomes steeper as it advances, ultimately becoming vertical at the front where a hydrostatic shock forms. Distortive features of large-amplitude waves can be studied by comparing the time dependence at a downstream position with the originating disturbance. Consider a case in which the left wall velocity and displacement are periodic with

$$V_w = a_0 \sin \omega t \,, \qquad x_w = \frac{a_0}{\omega} \left(1 - \cos \omega t\right)$$

Liquid elevation on the wall is obtained from Eq. (5.7) as

$$y\left(\sqrt{H} + \frac{V_w}{2\sqrt{g}}\right)^2$$

The propagation diagram is given in Fig. 5.10, which shows sinusoidal wall motion and amplitude in terms of the normalized variables $t^* = \omega t$, $x^* = \omega x/a_0$, $V^* = V/\sqrt{gH}$, and $y^* = gy/a_0^2$ for the arbitrary case $\sqrt{gH}/a_0 = 4$. Equation (5.11) for propagation to the right becomes

$$x^* = \left(\frac{3}{2}V^* + \frac{\sqrt{gH}}{a_0}\right)t^* + f(V^*)$$

Since velocity, wave amplitude, and time are readily obtained on the moving wall, it is not necessary to solve explicitly for $f(V^*)$. Instead, the derivative

$$\left(\frac{\partial x^*}{\partial t^*}\right)_{V^*} = \frac{3}{2}V^* + \frac{\sqrt{gH}}{a_0} \tag{5.41}$$

is written, and simple wave characteristics are drawn as shown, for which V^* and y^* are constant on each. Amplitude-time graphs are obtained on the wall, or at any other fixed position, as shown for $x^* = 2$ and 10. The wave

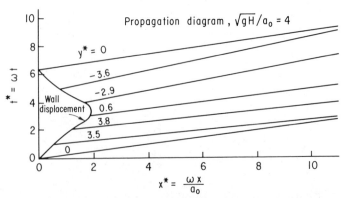

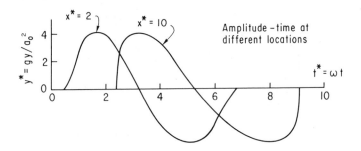

Figure 5.10 Amplitude-Time Distortion, Sinusoidal End Wall Motion

amplitude increases more rapidly and decreases more slowly as it propagates, whereas its period remains constant. Distortion effects can make it difficult to use measured amplitude-time behavior downstream and try to make conclusions about the disturbance source.

5.8   Compound Waves and the Method of Characteristics (MOC)

So far, we have considered large-amplitude disturbance propagation in one direction only. The waves are simple, which means that constant values of velocity V and elevation y propagate on straight characteristic lines in the t, x plane. However, there are many cases which may involve reflections, or more than one disturbance source, where large-amplitude disturbances from opposite directions intersect.

The *method of characteristics* (MOC) is a mathematical procedure for obtaining a general solution of wave propagation. The MOC solution describes left and right traveling characteristic lines in the t, x plane on which *functions* of V and y are constant, rather than the variables V and y. Whenever right and left traveling characteristic lines intersect, local values of V and y are uniquely determined.

The MOC solution described in this section for uniform channels reduces Eqs. (5.1) and (5.2) to four ordinary differential equations. This reduction is accomplished by first multiplying each term of Eq. (5.1) by an unknown function λ_1, each term of Eq. (5.2) by another unknown function λ_2, and then adding the two equations. This yields

$$\lambda_1 \frac{\partial y}{\partial t} + \lambda_2 \frac{\partial V}{\partial t} + (\lambda_1 V + \lambda_2 g) \frac{\partial y}{\partial x}$$
$$+ (\lambda_1 y + \lambda_2 V) \frac{\partial V}{\partial x} = 0 \qquad (5.42)$$

The yet unknown functions

$$y = y(x, t), \qquad V = V(x, t) \qquad (5.43)$$

are differentiated to give

$$\frac{dy}{dt} = \frac{\partial y}{\partial x} \frac{dx}{dt} + \frac{\partial y}{\partial t}, \qquad \frac{dV}{dt} = \frac{\partial V}{\partial x} \frac{dx}{dt} + \frac{\partial V}{\partial t} \qquad (5.44)$$

Thus, Eq. (5.42) becomes

$$\lambda_1 \frac{dy}{dt} + \lambda_2 \frac{dV}{dt} + \left[\lambda_1 \left(V - \frac{dx}{dt} \right) + \lambda_2 g \right] \frac{\partial y}{\partial x}$$
$$+ \left[\lambda_1 y + \lambda_2 \left(V - \frac{dx}{dt} \right) \right] \frac{\partial V}{\partial x} = 0 \qquad (5.45)$$

Equation (5.45) can be reduced to an ordinary differential equation in V and y if the coefficients of partial derivatives $\partial y/\partial x$ and $\partial V/\partial x$ are set equal to zero. This leads to the simple equations in matrix form

$$\begin{bmatrix} V - \dfrac{dx}{dt} & g \\ y & V - \dfrac{dx}{dt} \end{bmatrix} \begin{bmatrix} \lambda_1 \\ \lambda_2 \end{bmatrix} = \begin{bmatrix} 0 \\ 0 \end{bmatrix} \tag{5.46}$$

Since (5.46) gives two homogeneous equations, solutions for λ_1 and λ_2 require that the determinant of the coefficients be zero, or

$$\left(V - \frac{dx}{dt} \right)^2 = gy$$

from which

$$\frac{dx}{dt} = V \pm \sqrt{gy} \tag{5.47}$$

Equation (5.47) corresponds to paths or characteristic lines in the t, x plane on which Eq. (5.45) becomes an ordinary differential equation in y and V. The dx/dt of Eq. (5.47) is substituted into Eq. (5.46) to obtain

$$\frac{\lambda_2}{\lambda_1} = \pm \sqrt{\frac{y}{g}} \tag{5.48}$$

Since the coefficients of $\partial y/\partial x$ and $\partial V/\partial x$ were made zero by the conditions of Eq. (5.46), it follows that Eq. (5.45) can be written as

$$dy \pm \sqrt{\frac{y}{g}}\, dV = 0 \qquad \text{on} \quad \frac{dx}{dt} = V \pm \sqrt{gy} \tag{5.49}$$

Equation (5.49) expresses four ordinary differential equations which provide a full solution to Eqs. (5.1) and (5.2). The first two of these equations can be integrated from arbitrary initial values y_i and V_i to give

$$\mathscr{R}(V, y) = 2\sqrt{gy} + V = 2\sqrt{gy_i} + V_i = \text{constant} \quad \text{on the characteristic}$$
path, or $\mathscr{R}$ line

$$\frac{dx}{dt} = V + \sqrt{gy} \tag{5.50}$$

and

$$\mathscr{L}(V, y) = 2\sqrt{gy} - V = 2\sqrt{gy_i} - V_i = \text{constant} \quad \text{on the characteristic}$$
path, or $\mathscr{L}$ line

$$\frac{dx}{dt} = V - \sqrt{gy} \tag{5.51}$$

The functions $\mathcal{R}(V, y)$ and $\mathcal{L}(V, y)$, called *Riemann invariants*, remain constant on the characteristic path lines indicated, sometimes called $\mathcal{R}$ and $\mathcal{L}$ lines. This is a notable departure from the earlier formulation for simple waves in which V and y both were constant on straight characteristic lines.

Figure 5.11 shows a comparison of characteristic lines and quantities propagated for simple and compound waves. Constant values of V and y propagate on straight line characteristics for simple waves in Fig. 5.11(*a*). The Riemann invariants $\mathcal{L} = 2\sqrt{gy} - V$ and $\mathcal{R} = 2\sqrt{gy} + V$ are constant, respectively, on left and right traveling characteristics for compound waves in Fig. 5.11(*b*), but V and y can vary, causing the characteristic lines to be generally curved. Figure 5.11(*b*) also shows intersecting $\mathcal{L}$ and $\mathcal{R}$ lines. Since $\mathcal{L}$ and $\mathcal{R}$ remain constant on these lines, velocity V and surface elevation y are determined at the point of intersection from simultaneous solution of Eqs. (5.50) and (5.51), which gives

$$V = \frac{\mathcal{R} - \mathcal{L}}{2} \tag{5.52}$$

and

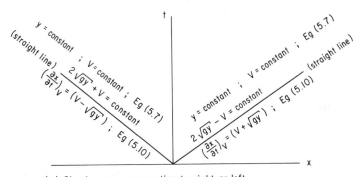

(a) Simple wave propagation to right or left

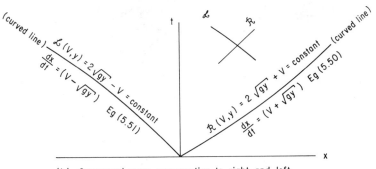

(b) Compound wave propagation to right and left

Figure 5.11 Characteristics for Simple and Compound Waves (a) Simple Wave Propagation to Right or Left (b) Compound Wave Propagation to Right and Left

$$\sqrt{gy} = \frac{\mathcal{R} + \mathcal{L}}{4} \tag{5.53}$$

It is more convenient to employ the variable $\sqrt{gy}$ rather than y alone because $\sqrt{gy}$ is required in determining the local slope of $\mathcal{L}$ and $\mathcal{R}$ lines from Eqs. (5.50) and (5.51). Equations (5.52) and (5.53) can be employed in (5.50) and (5.51) to give

$$\mathcal{R}(V, y) = \text{constant} \qquad \text{on} \qquad \frac{dx}{dt} = \frac{3\mathcal{R} - \mathcal{L}}{4} \tag{5.54}$$

and

$$\mathcal{L}(V, y) = \text{constant} \qquad \text{on} \qquad \frac{dx}{dt} = \frac{\mathcal{R} - 3\mathcal{L}}{4} \tag{5.55}$$

Initial and boundary conditions provide points of origination for $\mathcal{L}$ and $\mathcal{R}$ lines. If values of velocity and elevation at the initial time $t = 0$ are specified as V_i and y_i at all values of x_i, the values of $\mathcal{L}$ and $\mathcal{R}$ originating at $t = 0$, $x = x_i$, are obtained from Eqs. (5.52) and (5.53) as

$$\mathcal{R} = V_i + 2\sqrt{gy_i} \tag{5.56}$$

and

$$\mathcal{L} = -V_i + 2\sqrt{gy_i} \tag{5.57}$$

If the left boundary has a given velocity V_L, or a given elevation y_L, a known value of $\mathcal{L} = 2\sqrt{gy} - V$ arriving from the right creates a rightward propagating $\mathcal{R}$ which originates on the left boundary, and whose value according to either Eq. (5.52) or (5.53) is

$$\mathcal{R} = \begin{cases} 2V_L + \mathcal{L} & \text{if } V_L \text{ specified} \\ 4\sqrt{gy_L} - \mathcal{L} & \text{if } y_L \text{ specified} \end{cases} \tag{5.58}$$

Similarly, if the right boundary velocity is V_R, or the elevation is y_R, an arriving value of $\mathcal{R}$ creates a leftward propagating $\mathcal{L}$ which originates on the right boundary, whose value is obtained from either Eq. (5.52) or (5.53) as

$$\mathcal{L} = \begin{cases} \mathcal{R} - 2V_R & \text{if } V_R \text{ specified} \\ 4\sqrt{gy_R} - \mathcal{R} & \text{if } y_R \text{ specified} \end{cases} \tag{5.59}$$

Figure 5.12 gives a summary of relationships needed for a solution of hydrostatic wave propagation in uniform liquid channels using Riemann invariants.

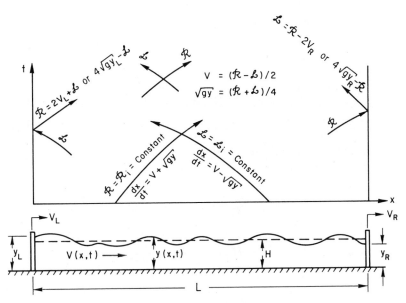

Values of $\mathcal{R}$ and $\mathcal{L}$ remain constant on characteristics

Initial conditions at $t = 0$, $x = x_i$, $V = V_i$, $\sqrt{gy} = \sqrt{gy_i}$

$$\mathcal{R} = \mathcal{R}_i = V_i + 2\sqrt{gy_i} = V + 2\sqrt{gy} = \text{constant} \qquad (5.50)$$

$$\mathcal{L} = \mathcal{L}_i = -V_i + 2\sqrt{gy_i} = -V + 2\sqrt{gy} = \text{constant} \qquad (5.51)$$

Boundary conditions
  Left wall, arriving $\mathcal{L}$ known

$$\begin{array}{ll} \mathcal{R} = 2V_L + \mathcal{L} & \text{if velocity } V_L \text{ specified} \\ \mathcal{R} = 4\sqrt{gy_L} - \mathcal{L} & \text{if elevation } y_L \text{ specified} \end{array} \qquad (5.58)$$

  Right wall, arriving $\mathcal{R}$ known

$$\begin{array}{ll} \mathcal{L} = \mathcal{R} - 2V_R & \text{if velocity } V_R \text{ specified} \\ \mathcal{L} = 4\sqrt{gy_R} - \mathcal{R} & \text{if elevation } y_R \text{ specified} \end{array} \qquad (5.59)$$

Slopes

$\mathcal{R}$ characteristics $\quad \dfrac{dx}{dt} = V + \sqrt{gy} = \dfrac{3\mathcal{R} - \mathcal{L}}{4} \quad$ on which $\mathcal{R} = \text{constant} \quad (5.54)$

$\mathcal{L}$ characteristics $\quad \dfrac{dx}{dt} = V - \sqrt{gy} = \dfrac{\mathcal{R} - 3\mathcal{L}}{4} \quad$ on which $\mathcal{L} = \text{constant} \quad (5.55)$

Local V and $\sqrt{gy}$ where $\mathcal{R}$ and $\mathcal{L}$ intersect

$$V = \frac{\mathcal{R} - \mathcal{L}}{2} \ (5.52), \quad \sqrt{gy} = \frac{\mathcal{R} + \mathcal{L}}{4} \ (5.53), \ \text{or } y = \frac{1}{g}\left(\frac{\mathcal{R} + \mathcal{L}}{4}\right)^2$$

Figure 5.12 Large-Amplitude Wave Solution, Uniform Channels.

5.9 Wall Withdrawal Acceleration by Riemann Invariants

The problem of wall withdrawal at constant acceleration from a stationary liquid region was solved in Section 5.3 for a long uniform channel by employing simple wave theory. The same problem is solved in this section with Riemann invariants for a uniform channel of length L with a stationary wall at the right end, as shown in Fig. 5.13. The solution for surface elevation $y(x, t)$ and velocity $V(x, t)$ should be the same as that for simple waves until the disturbance arrives at the stationary wall and is reflected. This analysis is generalized by employing the nondimensional quantities

$$
\begin{aligned}
&x^* = \frac{a}{gH}\, x \,, \quad V^* = \frac{V}{\sqrt{gH}} \,, \quad \mathscr{R}^* = \frac{\mathscr{R}}{\sqrt{gH}} = V^* + 2\sqrt{y^*} \\
&t^* = \frac{a}{\sqrt{gH}}\, t \,, \quad y^* = \frac{y}{H} \,, \quad \mathscr{L}^* = \frac{\mathscr{L}}{\sqrt{gH}} = -V^* + 2\sqrt{y^*}
\end{aligned}
\tag{5.60}
$$

Figure 5.12 was used to write the nondimensional solution equations in Table 5.6. The propagation diagram is shown in Fig. 5.14 with only a few characteristic lines, to provide a simple description of the solution process.

The fixed wall was arbitrarily placed at $x^* = aL/gH = 2$, and the moving wall trajectory is shown as $x_w^* = -t^{*2}/2$ from Table 5.6. Several values of fluid velocity on the wall, $V^* = -V_w^* = -t^*$, are shown on the wall trajectory at points from which $\mathscr{R}^*$ lines are drawn.

The $\mathscr{L}^*$ and $\mathscr{R}^*$ lines originating at $t^* = 0$ where $V^* = 0$ and $y^* = 1$ have the value $\mathscr{L}^* = R^* = 2$. Moreover, values of $\mathscr{L}^*$ originating at the right wall boundary in the interval $0 \le t^* \le 2$ also have the value $\mathscr{L}^* = 2$ according to the boundary conditions of Table 5.6. Values of R^*, originating at the wall trajectory wherever $\mathscr{L}^* = 2$ lines arrive, are expressed from Table 5.6 as

$$
\mathscr{R}^* = \mathscr{L}^* - 2t^* = 2(1 - t^*)
$$

The slopes of $\mathscr{L}^*$ and $\mathscr{R}^*$ lines originating at $t^* = 0$ are given by

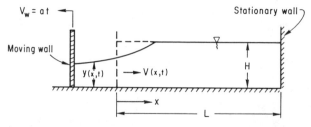

Figure 5.13   Accelerating Wall Withdrawal, Finite Channel Length

TABLE 5.6 Nondimensional Equations, Wall Withdrawal by Riemann Invariants

Initial Conditions at $t = 0$, $V_i = 0$, $y_i = H$

$$t^* = 0, \ V^* = 0, \ \sqrt{y^*} = 1$$

$$\mathscr{R}_i^* = V_i^* + 2\sqrt{y_i^*} = \text{constant} = 2$$

$$\mathscr{L}_i^* = -V_i^* + 2\sqrt{y_i^*} = \text{constant} = 2$$

Boundary Conditions

Left Wall: $\quad V^* = V_L^* = -V_w^* = -t^*, \quad x_w^* = -\dfrac{1}{2} t^{*2}$

$$\mathscr{R}^* = 2V_L^* + \mathscr{L}^* = \mathscr{L}^* - 2t^*$$

Right Wall: $\quad V^* = V_R^* = 0, \quad x^* = \dfrac{aL}{gH}$

$$\mathscr{L}^* = \mathscr{R}^* - 2V_R^* = \mathscr{R}^*$$

Slopes: $\quad \mathscr{R}^*$line $\quad \dfrac{dx^*}{dt^*} = V^* + \sqrt{y^*} = \dfrac{3\mathscr{R}^* - \mathscr{L}^*}{4}$

$\quad\quad\quad\quad \mathscr{L}^*$line $\quad \dfrac{dx^*}{dt^*} = V^* - \sqrt{y^*} = \dfrac{\mathscr{R}^* - 3\mathscr{L}^*}{4}$

Local Velocity: $\quad V^* = \dfrac{\mathscr{R}^* - \mathscr{L}^*}{2}$

Local Elevation: $\quad y^* = \dfrac{(\mathscr{R}^* + \mathscr{L}^*)^2}{16}$

$\mathscr{R}^*$ lines: $\quad \dfrac{dx^*}{dt^*} = \dfrac{3\mathscr{R}^* - \mathscr{L}^*}{4} = \dfrac{(3)(2) - 2}{4} = 1$

$\mathscr{L}^*$ lines: $\quad \dfrac{dx^*}{dt^*} = \dfrac{\mathscr{R}^* - 3\mathscr{L}^*}{4} = \dfrac{2 - (3)(2)}{4} = -1$

The $\mathscr{L}^*$ lines originating at the right wall between $0 \le t^* \le 2$ where $\mathscr{R}^* = \mathscr{L}^*$ also have slopes of $dx^*/dt^* = -1$.

The $\mathscr{R}^*$ lines originating at the moving wall trajectory with arriving $\mathscr{L}^* = 2$ lines have slopes

$$\frac{dx^*}{dt^*} = \frac{3\mathscr{R}^* - \mathscr{L}^*}{4} = \frac{3\mathscr{R}^* - 2}{4}$$

All $\mathscr{L}^*$ and $\mathscr{R}^*$ lines originating from the initial condition $t^* = 0$, from the

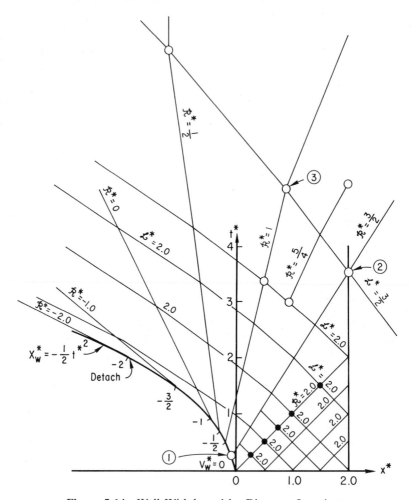

Figure 5.14 Wall Withdrawal by Riemann Invariants

moving wall trajectory, and from the fixed wall were lightly drawn straight at first with pencil. The $\mathscr{L}^*$ lines indeed are straight while they cross $\mathscr{R}^* = 2$ lines, but they change slope when crossing lines of $\mathscr{R}^* \neq 2$. Similarly, the $\mathscr{R}^*$ lines remain straight while crossing $\mathscr{L}^* = 2$ lines. However, $\mathscr{L}^* \neq 2$ for reflections from the right boundary, and whenever an $\mathscr{R}^*$ line crosses a line of $\mathscr{L}^* \neq 2$, the $\mathscr{R}^*$ slope changes. The heavy black dots show where the straight $\mathscr{L}^*$ lines begin to change slope and become curved. The open circles show where the straight $\mathscr{R}^*$ lines begin to curve. When an R^* line is curving, it is drawn from one $\mathscr{L}^*$ line to the next, at which a new slope is calculated. A similar procedure is used for drawing the $\mathscr{L}^*$ lines which are curving. Sample calculations at points labeled 1, 2, and 3 are summarized below:

POINT 1: ON LEFT WALL TRAJECTORY

$$\mathscr{L}^* = 2 \text{ arriving} , \qquad V^* = -V_w^* = -\frac{1}{4}$$

$$\mathscr{R}^* = 2V^* + \mathscr{L}^* = 2\left(-\frac{1}{4}\right) + 2 = \frac{3}{2}$$

$$\frac{dx^*}{dt^*} = \frac{3\mathscr{R}^* - \mathscr{L}^*}{4} = \frac{9/2 - 2}{4} = \frac{5}{8} \quad \text{for} \quad \mathscr{R}^* = \frac{3}{2} \text{ line}$$

$$\sqrt{y^*} = \frac{\mathscr{R}^* + \mathscr{L}^*}{4} = \frac{3/2 + 2}{4} = \frac{7}{8}$$

POINT 2: ON STATIONARY RIGHT WALL

$$\mathscr{R}^* = \frac{3}{2} \text{ arriving} , \qquad \mathscr{L}^* = \mathscr{R}^* = \frac{3}{2}$$

$$\frac{dx^*}{dt^*} = \frac{\mathscr{R}^* - 3\mathscr{L}^*}{4} = \frac{3/2 - (3)(3/2)}{4} = -\frac{3}{4}$$

$$\sqrt{y^*} = \frac{\mathscr{R}^* + \mathscr{L}^*}{4} = \frac{(3/2) + (3/2)}{4} = \frac{3}{4}$$

$$V^* = \frac{\mathscr{R}^* - \mathscr{L}^*}{2} = \frac{(3/2) - (3/2)}{2} = 0$$

POINT 3: ARBITRARY POINT

$$\mathscr{R}^* = 1 , \qquad \mathscr{L}^* = \frac{3}{2}$$

$$\frac{dx^*}{dt^*} = \frac{3\mathscr{R}^* - \mathscr{L}^*}{4} = \frac{3 - 3/2}{4} = \frac{3}{8} , \quad \mathscr{R}^* \text{ slope}$$

$$\frac{dx^*}{dt^*} = \frac{\mathscr{R}^* - 3\mathscr{L}^*}{4} = \frac{1 - 9/2}{4} = -\frac{7}{8} , \quad \mathscr{L}^* \text{ slope}$$

$$\sqrt{y^*} = \frac{\mathscr{R}^* + \mathscr{L}^*}{4} = \frac{1 + 3/2}{4} = \frac{5}{8}$$

$$V^* = \frac{\mathscr{R}^* - \mathscr{L}^*}{2} = \frac{1 - 3/2}{2} = -\frac{1}{4}$$

Points on the right boundary were used to sketch the liquid elevation of Fig. 5.15.

The calculational procedure is straightforward. However, it is possible to lose desired resolution as some characteristic lines fan out, leaving wide gaps between them. Although new characteristics can be introduced anywhere on

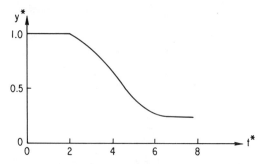

Figure 5.15 Liquid Elevation at Right Stationary Wall from Riemann Invariants

the propagation diagram to improve resolution, a generalized MOC is described in Section 5.14 which overcomes the resolution problem and is also convenient to use for nonuniform channels.

5.10 Hydrostatic Shocks

The formation of shocks was mentioned in Section 5.4 for end wall acceleration into a liquid channel. A hydrostatic shock is a region across which liquid velocity and elevation are discontinuous. When a shock forms in a hydrostatic wave propagation diagram, it becomes a moving boundary condition. The describing equations for hydrostatic waves developed in Chapter 4 are based on continuous liquid properties and do not allow for discontinuities. It is therefore necessary to relate discontinuous fluid properties across a shock in order to employ the equations for continuous flow on either side.

Figure 5.16 shows a discontinuity in elevation and velocity of liquid flowing in a uniform channel of width D. Liquid on the left with elevation H_d and velocity V_d has been disturbed, whereas liquid on the right is undisturbed with properties H_u and V_u. The shock discontinuity is moving rightward with speed S relative to the earth. A dotted CV straddles both sides of the shock, but is thin so that storage terms are negligible. Therefore, the mass conservation principle yields.

$$\dot{m}_{in} = \rho H_u D(S - V_u) = \dot{m}_{out} = \rho H_d D(S - V_d) \qquad (5.61)$$

The momentum principle is first written as

$$\dot{m}_{out}V_d - \dot{m}_{in}V_u = g_0(F_d - F_u) \qquad (5.62)$$

where the hydrostatic pressure forces are

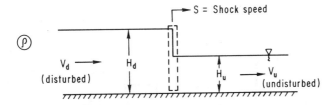

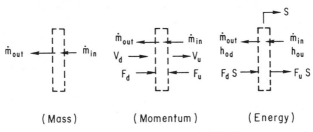

Figure 5.16 Hydrostatic Shock

$$F_d = \rho \, \frac{g}{g_0} \, \frac{H_d}{2} \, (DH_d) \quad \text{and} \quad F_u = \rho \, \frac{g}{g_0} \, \frac{H_u}{2} \, (DH_u) \tag{5.63}$$

Equation (5.62) becomes

$$(S - V_d)V_d H_d - (S - V_u)V_u H_u = \frac{g}{2} \, (H_d^2 - H_u^2) \tag{5.64}$$

Elimination of V_d from Eqs. (5.61) and (5.64) yields the shock speed S in the form

$$\frac{S - V_u}{\sqrt{gH_u}} = \sqrt{\frac{1}{2} \, \frac{H_d}{H_u} \left(1 + \frac{H_d}{H_u}\right)} \tag{5.65}$$

whereas if S is eliminated, the velocity difference is obtained as

$$\frac{V_d - V_u}{\sqrt{gH_u}} = \left(\frac{H_d}{H_u} - 1\right)\sqrt{\frac{1}{2} \, \frac{H_u}{H_d} \left(1 + \frac{H_d}{H_u}\right)} \tag{5.66}$$

Equations (5.65) and (5.66) are graphed in Fig. 5.17. It is seen that for stronger shocks with smaller H_u/H_d, the speed $S - V_u$, relative to the undisturbed liquid, and the velocity difference $V_d - V_u$ increase. Moreover, the limit of Eq. (5.65) as $H_d \rightarrow H_u$ yields $S - V_u = \sqrt{gH_u}$, which is the speed of a small-amplitude hydrostatic wave.

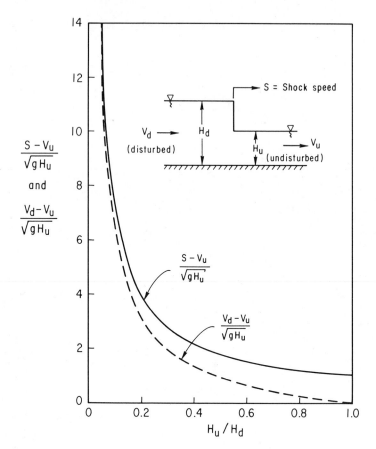

Figure 5.17   Hydrostatic Shock Properties

EXAMPLE 5.5: SHOCK PROPERTIES, SUDDEN WALL MOTION

The left end wall of a wave tank suddenly begins rightward motion at a constant velocity of 2 m/s (6.56 ft/s) into stationary liquid. If the initial depth is 1.0 m (3.28 ft) what amplitude propagates, and at what speed?

The stationary liquid is characterized by $V_u = 0$ and $H_u = 1.0$ m (3.28 ft). Disturbed liquid moves at the wall velocity $V_d = 2$ m/s (6.56 ft/s). If we enter Fig. 5.17 with

$$\frac{V_d - V_u}{\sqrt{gH_u}} = \frac{V_d}{\sqrt{gH_u}} = 0.64$$

we obtain $H_u/H_d = 0.56$ and $(S - V_u)/\sqrt{gH_u} = 1.6$, which gives a propagating amplitude of $H_d = 1.79$ m (5.9 ft) and a shock speed $S = 5$ m/s (16.4 ft/s).

5.11 Energy Dissipation across a Hydrostatic Shock

The energy diagram for a shock in Fig. 5.16 is employed to express energy conservation in the form

$$\dot{m}_{out} h_{0d} + F_u S - \dot{m}_{in} h_{0u} - F_d S = 0 \tag{5.67}$$

The stagnation enthalpy

$$h_0 = \frac{V^2}{2g_0} + \frac{gH}{g_0} + e + \frac{P}{\rho} \tag{5.68}$$

on either side of the shock can be expressed in terms of the average elevation $H = \bar{H} = H/2$ and the average hydrostatic pressure

$$P = \bar{P} = \left(\rho \, \frac{g}{g_0} \right) \frac{H}{2} \tag{5.69}$$

whereas V and e are uniform over the depth. Thus,

$$h_{od} = \frac{V_d^2}{2g_0} + \frac{gH_d}{g_0} + e_d \tag{5.70}$$

and

$$h_{0u} = \frac{V_u^2}{2g_0} + \frac{gH_u}{g_0} + e_u \tag{5.71}$$

Equations (5.61)–(5.63) for the mass flow rates and hydrostatic forces also are used in (5.67) to give

$$H_d(S - V_d)\left(\frac{V_d^2}{2g_0} + \frac{gH_d}{g_0} + e_d \right) - H_u(S - V_u)\left(\frac{V_u^2}{2g_0} + \frac{gH_u}{g_0} + e_u \right)$$
$$+ \frac{g}{2g_0}(H_u^2 - H_d^2)S = 0$$

The velocities V_u, V_d, and S can be eliminated with Eqs. (5.61) and (5.64) to obtain the internal energy increase across the shock as

$$e_d - e_u = \frac{gH_d}{2g_0}\left(1 - \frac{H_u}{H_d} \right)\left(\frac{1}{2}\frac{H_d}{H_u} + \frac{1}{2}\frac{H_u}{H_d} - 1 \right) \tag{5.72}$$

Equation (5.72) gives the available energy decrease or specific dissipation across the shock. It represents an entropy increase, which can be shown by writing the Gibbs equation for an incompressible liquid, $T \, ds = de$, which is integrated to give

$$s_d - s_u = \Delta s = \frac{e_d - e_u}{\bar{T}} \tag{5.73}$$

where $\bar{T}$ is the average temperature.

The specific energy required to form the shock is given by

$$\mathscr{E}_d - \mathscr{E}_u = \left(h_{0d} - \frac{P_d}{\rho}\right) - \left(h_{0u} - \frac{P_u}{\rho}\right) \tag{5.74}$$

The specific dissipation of Eq. (5.72) can be written as a fraction of the total energy difference across the shock. Equations (5.68)–(5.73) are employed with S and V_d from (5.65) and (5.66) to give

$$
\begin{aligned}
\mathscr{D}_s &= \frac{e_d - e_u}{\mathscr{E}_d - \mathscr{E}_u} = \frac{\bar{T}(s_d - s_u)}{\mathscr{E}_d - \mathscr{E}_u} \\[2mm]
&= \frac{\frac{1}{2}(H_d/H_u) + \frac{1}{2}(H_u/H_d) - 1}{H_d/H_u + V_u(H_d/H_u + 1)/\sqrt{(gH_d/2)(1 + H_d/H_u)}}
\end{aligned}
\tag{5.75}
$$

It is seen that when the elevations H_d and H_u are equal, $H_d/H_u = 1$, and $\mathscr{D}_s = 0$ as expected. If H_d/H_u is very large, $\mathscr{D}_s \to 1/2$. It follows that the maximum energy which is dissipated by a large hydrostatic shock never exceeds half the energy needed to form the shock.

EXAMPLE 5.6: HYDROSTATIC SHOCK ENERGY DISSIPATION
Determine the energy dissipation across the shock formed in Example 5.5 where the end wall begins sudden motion at 2 m/s into stationary liquid of 1.0 m depth.

It was found from Fig. 5.17 that $H_d = 1.79$ m. It follows from Eq. (5.75) with $V_u = 0$ that

$$\mathscr{D}_s = 0.097$$

Therefore, about 10 percent of the energy used to form this shock is dissipated.

5.12 The Method of Characteristics and Shock Formation

Shock formation from the acceleration of a wall into stationary liquid was discussed in Section 5.4. Figure 5.4 shows that a shock originates on the t, x diagram at the intersection of characteristic lines which are traveling in the same direction. When a shock forms, it becomes a line of discontinuous liquid elevation and velocity in the t, x plane. The MOC described in Section 5.8 applies to liquid regions upstream and downstream of the shock. The hydrostatic shock equations of Section 5.10 relate elevation and velocity values across the shock and provide the shock speed from which its path is obtained in the t, x plane.

The historical procedure described in this section is offered to show how the characteristic solution is blended with the hydrostatic shock relationships. It is not often used today because simpler approximate methods yield relatively accurate results. Some investigators have concluded that approximate shock solutions sacrifice less in accuracy than is sacrificed in complexity for a rigorous solution by the MOC with a moving shock boundary condition. The probable validity of this conclusion is demonstrated in Section 5.13, which gives a comparison of results obtained from the rigorous procedure and an approximate solution.

Consider the case where the left wall moves rightward at constant acceleration into the liquid. The wall velocity and displacement of Fig. 5.4 are described by $V_w = bt_w$ and $x_w = bt_w^2/2$, where t_w designates time on the wall trajectory. The large-amplitude wave solution described in Fig. 5.12 is nondimensionalized and summarized in Fig. 5.18. The nondimensional forms of Eqs. (5.65) and (5.66) for a moving shock also are included in Fig. 5.18.

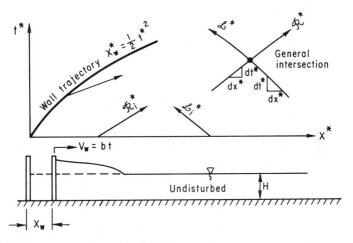

Initial conditions $t^* = 0$, $V^* = 0$, $y^* = 1$

$$\mathcal{R}_i^* = 2.0 \quad \text{on} \left(\frac{dt^*}{dx^*}\right)_i = 1.0 \;\; (5.50); \qquad \mathcal{L}_i^* = 2.0 \quad \text{on} \left(\frac{dt^*}{dx^*}\right)_i = -1 \;\; (5.51)$$

Wall trajectory

$$x_w^* = \frac{1}{2} t_w^{*2}, \qquad V_w^* = t_w^* \tag{5.28}$$

Left wall boundary condition, wall velocity specified

$$\mathcal{R}_w^* = 2V_w^* + \mathcal{L}^* = 2t_w^* + \mathcal{L}^* \;\; (5.58) \quad \text{on} \; \frac{dt^*}{dx^*} = \frac{4}{3\mathcal{R}^* - \mathcal{L}^*} \tag{5.54}$$

continued on next page

Figure 5.18 Method of Characteristics and Shock Formation, Wall Acceleration into Liquid.

Elevation y_d^* is eliminated from Eqs. (5.65) and (5.66) to obtain $\mathcal{L}^*$ and S^* as the functions of $\mathcal{R}^*$ shown in Fig. 5.19. When a known $\mathcal{R}^*$ line arrives at the shock, the reflected $\mathcal{L}^*$ line is obtained from the graph of Fig. 5.19. The shock speed S^* also is obtained from Fig. 5.19, which gives the shock slope dx_s^*/dt^*. The point of shock origination is obtained at the first location where two right traveling characteristics intersect. However, Eqs. (5.31) and (5.32) were employed to determine the time and location of the shock appearance for this case, which gave $x_s^* = t_s^* = 2/3$.

The procedure for constructing the propagation diagram is described in Table 5.7. The result of the construction described in Table 5.7 is shown in Fig. 5.20, for which Table 5.8 gives a tabulation of important quantities at each numbered point. Liquid elevation and velocity profiles at various times were constructed by obtaining $\mathcal{R}^*$ and $\mathcal{L}^*$ at various values of t^* for a range of x^* in Fig. 5.20, and then calculating V^* and y^* from Eqs. (5.52) and (5.53) of Fig. 5.18. These profiles are graphed as solid lines in Fig. 5.21. The dashed lines are based on the simplified approximate analysis discussed next.

Figure 5.18 (*continued*)

Right wall far away. No Boundary conditions required

General intersection of characteristics

$$\mathcal{R}^* \text{ slope } \frac{dt^*}{dx^*} = \frac{4}{3\mathcal{R}^* - \mathcal{L}^*} \;(5.54), \; V^* = \frac{\mathcal{R}^* - \mathcal{L}^*}{2} \tag{5.52}$$

$$\mathcal{L}^* \text{ slope } \frac{dt^*}{dx^*} = \frac{4}{\mathcal{R}^* - 3\mathcal{L}^*} \;(5.55), \; y^* = \frac{(\mathcal{R}^* + \mathcal{L}^*)^2}{16} \tag{5.53}$$

Moving shock speed S^* and disturbed velocity V_d^*

$$\frac{dx^*}{dt^*} = S^* = \sqrt{\frac{1}{2}\, y_d^*(1 + y_d^*)} = \sqrt{\frac{1}{2}\left(\frac{\mathcal{R}^* + \mathcal{L}^*}{4}\right)^2 \left[1 + \left(\frac{\mathcal{R}^* + \mathcal{L}^*}{4}\right)^2\right]} \tag{5.65}$$

$$V_d^* = (y_d^* - 1)\sqrt{\frac{1}{2}\frac{1}{y_d^*}(1 + y_d^*)}$$

$$= \left[\left(\frac{\mathcal{R}^* + \mathcal{L}^*}{4}\right)^2 - 1\right]\sqrt{\frac{1}{2}\left(\frac{4}{\mathcal{R}^* + \mathcal{L}^*}\right)^2 \left[1 + \left(\frac{\mathcal{R}^* + \mathcal{L}^*}{4}\right)^2\right]} \tag{5.66}$$

Nondimensional variables

$$x^* = \frac{bx}{gH} \quad V^* = \frac{V}{\sqrt{gH}} \quad \mathcal{R}^* = \frac{\mathcal{R}}{\sqrt{gH}} \quad V_d^* = \frac{V_d}{\sqrt{gH}} \quad V_w^* = \frac{V_w}{\sqrt{gH}}$$

$$t^* = \frac{bt}{\sqrt{gH}} \quad y^* = \frac{y}{H} \quad \mathcal{L}^* = \frac{\mathcal{L}}{\sqrt{gH}} \quad y_d^* = \frac{y_d}{H} \quad S^* = \frac{S}{\sqrt{gH}}$$

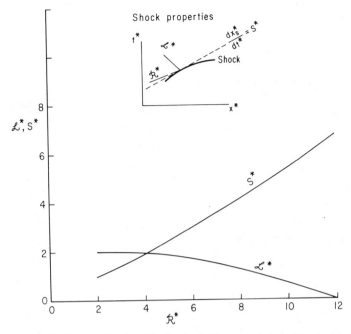

Figure 5.19 Shock Properties for Method of Characteristics, Shock Motion into Stationary Liquid

TABLE 5.7   Construction of Propagation Diagram with Shock (Fig. 5.18)

1. Plot wall trajectory $x_w^* = t_w^*/2$, and show point of shock origination

$$x_s^* = t_s^* = \frac{2}{3}$$

or determine it later from the first intersection of $\mathcal{R}^*$ lines.

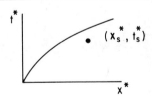

2. Lightly draw $\mathcal{L}_i^* = 2$ and $\mathcal{R}_i^* = 2$ lines, originating at the initial condition $t^* = 0$, with initial slopes $dt^*/dx^* = 1$ for $\mathcal{R}_i^*$ and $dt^*/dx^* = -1$ for $\mathcal{L}_i^*$, Eqs. (5.50) and (5.51).

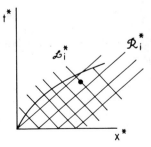

Table 5.7. (*continued*)

3. Lightly draw and label $\mathcal{R}_w^*$ lines originating at the moving wall. Values of $\mathcal{R}_w^*$ are obtained from $\mathcal{R}_w^* = 2t_w^* + \mathcal{L}^*$ (5.58), with $\mathcal{L}^* = 2$ (5.51) prior to shock formation, and slopes of these $\mathcal{R}_w^*$ lines are, from Eq. (5.54),

$$\frac{dt^*}{dx^*} = \frac{4}{(3\mathcal{R}^* - \mathcal{L}^*)}$$

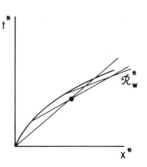

4. Where an $\mathcal{L}^*$ line crosses an $\mathcal{R}_w^*$ from the moving wall, although $\mathcal{L}^* = 2$ remains constant, its slope is determined from Eq. (5.55) as

$$\frac{dt^*}{dx^*} = \frac{4}{\mathcal{R}^* - 3\mathcal{L}^*}$$

However, as long as $\mathcal{R}^*$ lines cross $\mathcal{L}^* = 2$ lines, their slope is determined at the wall trajectory.

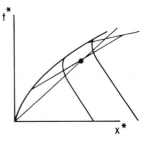

5. Extend the shock line a short distance with a slope dx_s^*/dt^* from Fig. 5.19 at $\mathcal{R}^*$. The $\mathcal{R}^*$ arriving at the shock yields new values of $\mathcal{L}^*$ and S^* from Fig. 5.19. The first $\mathcal{L}^*$ line originating at the shock is extended to other $\mathcal{R}^*$ lines. Slopes of $\mathcal{L}^*$ and $\mathcal{R}^*$ are computed from Eqs. (5.54) and (5.55) as

$$\frac{dt^*}{dx^*} = \frac{4}{(3\mathcal{R}^* - \mathcal{L}^*)}$$

for $\mathcal{R}^*$, or

$$\frac{dt^*}{dx^*} = \frac{4}{\mathcal{R}^* - 3\mathcal{L}^*}$$

for $\mathcal{L}^*$, until $\mathcal{L}^*$ arrives at the moving wall. Whenever $\mathcal{L}^*$ lines other than 2 reach the moving wall, the $\mathcal{R}^*$ coming from the moving wall should be computed from Eq. (5.58),

$$\mathcal{R}_w^* = 2t_w^* + \mathcal{L}^*$$

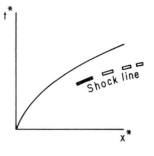

6. The process is continued in order to fill in the shocked region with characteristic lines. The undisturbed region remains at steady state.

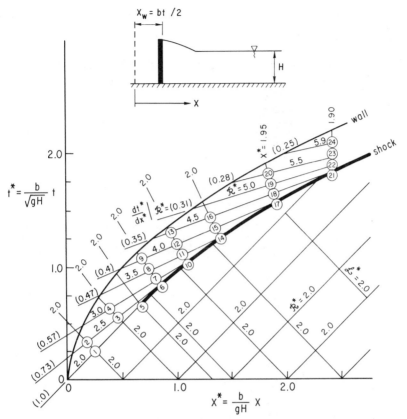

Figure 5.20   Method of Characteristics Construction with Moving Shock

5.13 Approximate MOC with Shock Formation

Although shock formation and propagation can be hand-calculated by methods in Section 5.12, the procedure is tedious. However, if the energy dissipation and entropy increase across a shock are small, the MOC can be used to obtain an approximate solution without the shock equations.

When the MOC is used without moving shock boundaries, *quasi-shocks* are estimated wherever adjacent characteristics in one direction intersect. Consider the case of accelerating wall motion and shock formation in section 5.12. The approximate solution does not require the shock equations listed in Fig. 5.18. Figures 5.18 and 5.20 show that all values of the $\mathcal{R}^*$ and $\mathcal{L}^*$ characteristics originating at $t^* = 0$ have the value 2.0, which is maintained until they arrive at a boundary. Each $\mathcal{R}^*$ line originating at the moving wall has a slope given by Eq. (5.54), Fig. 5.18, as

TABLE 5.8 **Computational Points (Fig. 5.20) Constant Wall Acceleration into Liquid**

Point ⓝ	$\mathscr{R}^*$	$\mathscr{L}^*$	$\dfrac{dt^*}{dx^*}\Big|_{\mathscr{R}^*}$	$\dfrac{dt^*}{dx^*}\Big|_{\mathscr{L}^*}$	$\dfrac{1}{S^*} = \dfrac{dt^*}{dx^*}\Big|_{shock}$
1	2.0	2.0	1.0	− 1.0	
2	2.5	2.0	0.73	− 1.14	
3	2.5	2.0	0.73	− 1.14	
4	3.0	2.0	0.57	− 1.33	
5	2.0	2.0		− 1.0	1.0
6	2.5	2.0		− 1.14	0.83
7	3.0	2.0	0.57	− 1.33	
8	3.5	2.0	0.47	− 1.6	
9	4.0	2.0	0.40	− 2.0	
10	3.0	2.0		− 1.33	0.71
11	4.0	2.0	0.47	− 2.0	
12	4.0	2.0	0.40	− 2.0	
13	4.5	2.0	0.35	− 2.7	
14	3.5	2.0		− 1.6	0.61
15	4.0	2.0	0.40	− 2.0	
16	4.5	2.0	0.35	− 2.7	
17	4.0	1.95		− 2.16	0.53
18	4.5	1.95	0.35	− 3.33	
19	5.0	1.95	0.31	− 5.71	
20	5.5	1.95	0.27	−20.0	
21	4.5	1.9		− 3.33	0.46
22	5.0	1.9	0.31	− 5.7	
23	5.5	1.9	0.27	−20.0	
24	5.9	1.9	0.25	−20.0	

$$x^* = \frac{b}{gH}\,x, \quad V^* = \frac{V}{\sqrt{gH}}, \quad \mathscr{R}^* = V^* + 2\sqrt{y^*}, \quad t^* = \frac{b}{\sqrt{gH}}\,t, \quad y^* = \frac{y}{H}, \quad \mathscr{L}^* = -V^* + 2\sqrt{y^*}$$

$$\frac{dt^*}{dx^*} = \frac{4}{3\mathscr{R}^* - \mathscr{L}^*}$$

where, in the absence of a shock, $\mathscr{L}^* = 2.0$. The $\mathscr{R}^*$ characteristics originating at the moving wall are obtained from Eq. (5.58) in Fig. 5.18 as

$$\mathscr{R}_w^* = 2t_w^* + \mathscr{L}^* = 2(t_w^* + 1)$$

It follows that $\mathscr{R}_w^*$ characteristics originating at the moving wall have the slope

$$\frac{dt^*}{dx^*} = \frac{1}{\frac{3}{2}t_w^* + 1} \tag{5.76}$$

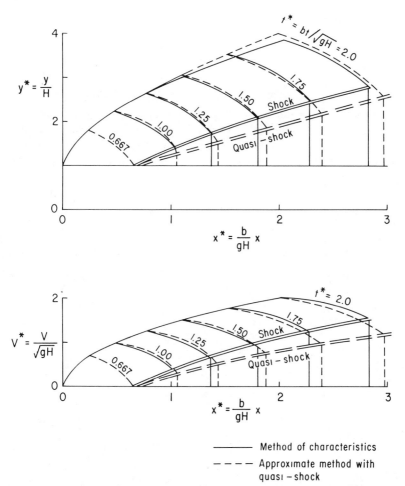

Figure 5.21   Elevation and Velocity Profiles, Wall Acceleration into Liquid Region

But t_w^* is the time at which a given $\mathcal{R}_w^*$ characteristic leaves the moving wall. Therefore, the $\mathcal{R}_w^*$ slope is constant as far as $\mathcal{R}_w^*$ extends. If we integrate Eq. (5.76) from t_w^* to t^* and x_w^* to x^*, and note from Eq. (5.28) in Fig. 5.18 that $x_w^* = (1/2)t_w^{*2}$, the equation of any $\mathcal{R}_w^*$ characteristic originating at the moving wall is

$$t^* = t_w^* + \frac{x^* - x_w^*}{\frac{3}{2}t_w^* + 1} = t_w^* + \frac{x^* - \frac{1}{2}t_w^{*2}}{\frac{3}{2}t_w^* + 1} \tag{5.77}$$

If a particular $\mathcal{R}_w^*$ line originates at t_w^*, and another originates at $t_w^* + dt_w^*$, the two will intersect if their trajectories have common x^*, t^* values. Use of

Eq. (5.77) yields the location of intersection, or the approximate shock location x_s^* as

$$x^* = x_s^* = \tfrac{2}{3} + \tfrac{4}{3}t_w^* + \tfrac{1}{2}t_w^{*2} \qquad (5.78)$$

Equations (5.77) and (5.78) describe a quasi-shock trajectory in the x^*, t^* plane without use of the moving normal shock equations. Note that for $t_w^* = 0$, the shock location is $x_s^* = t_s^* = 2/3$, which is the actual point of shock origination found in Fig. 5.20. The quasi-shock location is relatively close to the location found in Section 5.12, as seen from the comparison in Fig. 5.22.

The wave amplitude and velocity profiles between the moving wall and quasi-shock can be determined by drawing a horizontal line at t^* in Fig. 5.22. The horizontal line intersects the wall trajectory at

$$x^* = (1/2)t_w^{*2} \qquad (5.79)$$

which is the lower x^* limit of the profiles at t^* shown in fig. 5.21. Equation (5.77) is rewritten for $x^* = x_s^*$ and $t^* = t_s^*$ at the quasi-shock, and x_s^* is eliminated with Eq. (5.78). This yields the value of t_w^* on the wall from where the $\mathcal{R}_w^*$ line originates, which arrives at x_s^*, t_s^*. It follows that

$$t_w^* = (1/2)\left[(t_s^* + 14/9) + \sqrt{t_s^{*2} - (4/9)t_s^* + 52/81}\right] \qquad (5.80)$$

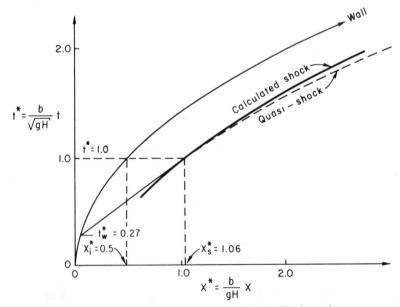

Figure 5.22 Calculated and Quasi-Shock Trajectories

The quasi-shock location x_s^* is obtained now from Eq. (5.78) for the upper spatial limit of the profile at time t^*. Since all $\mathcal{L}^*$ lines crossing the quasi-shock continue to have the value 2.0, which results directly from the approximation of negligible entropy change, the velocity and elevation profiles can be expressed from Eqs. (5.52), (5.28), and (5.58) with $\mathcal{L}^* = 2.0$ in Fig. 5.18 as $V^* = t_w^*$. Also, Eqs. (5.53) and (5.58) in Fig. 5.18 with $V^* = t_w^*$ yield

$$y^* = \left(1 + \frac{V^*}{2}\right)^2 \tag{5.81}$$

If Eq. (5.77) is solved for t_w^*, it follows that

TABLE 5.9 Velocity and Elevation Profiles with Quasi-Shock

t^*	(5.79) x_i^*	(5.80) t_w^*	(5.78) x_s^*	x_i^* to x_s^* x^*	(5.82) V^*	(5.81) y^*
0.67	0.22	0.0	0.67	0.22	0.67	1.78
				0.40	0.52	1.58
				0.67	0.00	1.00
1.25	0.78	0.49	1.44	0.78	1.25	2.64
				1.00	1.10	2.40
				1.20	0.93	2.14
				1.44	0.48	1.53
1.50	1.12	0.72	1.88	1.12	1.50	3.06
				1.20	1.46	2.98
				1.40	1.32	2.76
				1.60	1.16	2.50
				1.88	0.72	1.86
1.75	1.53	0.95	2.39	1.53	1.75	3.52
				1.80	1.59	3.23
				2.00	1.45	2.98
				2.20	1.27	2.67
				2.39	0.95	2.18
1.00	0.50	0.27	1.06	0.50	1.00	2.25
				0.60	0.93	2.15
				0.80	0.76	1.91
				1.06	0.25	1.26
2.00	2.00	1.19	2.96	2.00	2.00	4.00
				2.20	1.89	3.78
				2.40	1.77	3.56
				2.60	1.63	3.30
				2.80	1.45	2.97
				2.96	1.20	2.56

t^*, x^*, V^*, y^* defined in Fig. 5.18.

$$V^* = (1/2)\left[\sqrt{(1 - (3/2)t^*)^2 + 4(t^* - x^*)} - (1 - (3/2)t^*)\right] \quad (5.82)$$

Table 5.9 gives the computation of approximate velocity and elevation profiles.

The profiles obtained in Table 5.9 are graphed in Fig. 5.21 for comparison with the rigorous method involving moving shock calculations. Although the agreement becomes poorer as time increases, the quasi-shock procedure can be employed to obtain reasonable estimates of shock properties in subcritical flows. The rigorous elevation and velocity profiles behind the shock are almost indistinguishable from the approximate method.

The analysis in this section shows that the MOC gives approximate solutions if hydrostatic shocks form in subcritical flows. Shocks are identified by the intersection of characteristic lines which travel in the same direction. When two characteristic lines intersect, they terminate and do not propagate through the shock. The MOC does not give an approximate solution if shocks occur in a stream which flows near critical or supercritical velocity, because it does not permit the quasi-shock to propagate upstream. An auxiliary equation from the mass conservation principle can be used to correct this deficiency, and the procedure is discussed in Chapter 8 for large-amplitude pressure waves in compressible flows.

A hybrid procedure for analyzing problems with shock formation has been developed [5.3]. It involves a finite difference formulation of the governing partial differential equations with the MOC employed at the boundaries.

5.14 The MOC and Nonuniform Channels

Equations (4.1) and (4.2) for large-amplitude hydrostatic wave propagation in a nonuniform channel are rewritten here as

$$\frac{\partial \bar{y}}{\partial t} + V \frac{\partial \bar{y}}{\partial x} + \bar{y} \frac{\partial V}{\partial x} = F(x, t) \quad (5.83)$$

and

$$\frac{\partial V}{\partial t} + V \frac{\partial V}{\partial x} + g \frac{\partial \bar{y}}{\partial x} = G(x, t) \quad (5.84)$$

where

$$F(x, t) = -\frac{\bar{y}}{D} V \frac{dD}{dx}$$

$$G(x, t) = -\left(\frac{g_0}{\rho} \frac{\partial P_\infty}{\partial x} + g \frac{dB}{dx}\right) \quad (5.85)$$

and

$$\bar{y} = \bar{y}(x, t) = y(x, t) - B(x) \quad (5.86)$$

The channel width $D(x)$ and bottom elevation $B(x)$ are functions of x only, whereas the imposed surface pressure $P_\infty(x, t)$ may vary both in space and time. If the channel bottom is flat, $B = 0$ and $\bar{y} = y$. The procedure of Section 5.8 is applied to Eqs. (5.83) and (5.84), casting them into the forms

$$d\bar{y} + \sqrt{\frac{\bar{y}}{g}}\, dV = \left(F + \sqrt{\frac{\bar{y}}{g}}\, G\right) dt \quad \text{on} \quad \frac{dx}{dt} = V + C \qquad (5.87)$$

$$d\bar{y} - \sqrt{\frac{\bar{y}}{g}}\, dV = \left(F - \sqrt{\frac{\bar{y}}{g}}\, G\right) dt \quad \text{on} \quad \frac{dx}{dt} = V - C \qquad (5.88)$$

where

$$C = \sqrt{g\bar{y}} \qquad (5.89)$$

Equations (5.87) and (5.88) are suited for stepwise calculations on the uniform t, x grid shown in Fig. 5.23. Current values of $\bar{y}$ and V are presumed to be known at each mesh point in subcritical flow, shown as black dots, in Fig. 5.23. Points R and L designate the origins of right and left traveling characteristics which intersect mesh point 4 at x, $t + \Delta t$ where $\bar{y}$ and V are to be calculated. Since $\bar{y}_1, V_1, C_1, \bar{y}_2, V_2, C_2$, and $\bar{y}_3, V_3, C_3$ are known at points 1, 2, and 3, linear interpolation between 1 and 2, and 2 and 3 yields

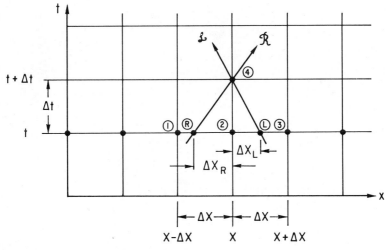

Figure 5.23 Method of Characteristics Computational Grid, Procedure for Subcritical Flows

SUBCRITICAL FLOWS

$$
\left.
\begin{aligned}
V_R &= V_2 + (V_1 - V_2)\frac{\Delta x_R}{\Delta x} \\[2mm]
\bar{y}_R &= \bar{y}_2 + (\bar{y}_1 - \bar{y}_2)\frac{\Delta x_R}{\Delta x} \\[2mm]
C_R &= C_2 + (C_1 - C_2)\frac{\Delta x_R}{\Delta x} \\[2mm]
f_R &= f_2 + (f_1 - f_2)\frac{\Delta x_R}{\Delta x}, \qquad f = F \;\; \text{or} \;\; G, \; \text{Eq. (5.85)}
\end{aligned}
\right\}
\tag{5.90}
$$

$$
\left.
\begin{aligned}
V_L &= V_2 + (V_3 - V_2)\frac{\Delta x_L}{\Delta x} \\[2mm]
\bar{y}_L &= \bar{y}_2 + (\bar{y}_3 - \bar{y}_2)\frac{\Delta x_L}{\Delta x} \\[2mm]
C_L &= C_2 + (C_3 - C_2)\frac{\Delta x_L}{\Delta x} \\[2mm]
f_L &= f_2 + (f_3 - f_2)\frac{\Delta x_L}{\Delta x}, \qquad f = F \;\; \text{or} \;\; G, \; \text{Eq. (5.85)}
\end{aligned}
\right\}
\tag{5.91}
$$

The geometry of Fig. 5.23 and the paths designated in Eqs. (5.87) and (5.88) yield

$$
\frac{\Delta x_R}{\Delta t} \approx \left.\frac{dx}{dt}\right|_R = V_R + C_R
$$

$$
\frac{\Delta x_L}{\Delta t} \approx -\left.\frac{dx}{dt}\right|_L = -V_L + C_L
$$

Equations (5.90) and (5.91) are used to obtain

$$
\frac{\Delta x_R}{\Delta x} = \frac{V_2 + C_2}{\Delta x/\Delta t + V_2 + C_2 - (V_1 + C_1)}
\tag{5.92}
$$

and

$$
\frac{\Delta x_L}{\Delta x} = \frac{-V_2 + C_2}{-V_2 + C_2 + (V_3 - C_3) + \Delta x/\Delta t}
\tag{5.93}
$$

The finite difference forms of Eq. (5.87) between points R and 4 and Eq. (5.88) between L and 4 are

$$
\bar{y}_4 - \bar{y}_R + \sqrt{\frac{\bar{y}_R}{g}}\,(V_4 - V_R) = \left(F_R + \sqrt{\frac{\bar{y}_R}{g}}\,G_R\right)\Delta t
\tag{5.94}
$$

and

$$\bar{y}_4 - \bar{y}_L - \sqrt{\frac{\bar{y}_L}{g}}\,(V_4 - V_L) = \left(F_L - \sqrt{\frac{\bar{y}_L}{g}}\,G_L\right)\Delta t \qquad (5.95)$$

Equations (5.94) and (5.95) are solved for V_4 and y_4 for a general mesh point not touching a boundary, which gives

$$V_4 =$$

$$\frac{\bar{y}_R - \bar{y}_L + V_R\sqrt{\bar{y}_R/g} + V_L\sqrt{\bar{y}_L/g} + (F_R - F_L + \sqrt{\bar{y}_R/g}\,G_R + \sqrt{\bar{y}_L/g}\,G_L)\Delta t}{\sqrt{\bar{y}_R/g} + \sqrt{\bar{y}_L/g}}$$

$$(5.96)$$

and

$$\bar{y}_4 = \sqrt{\frac{\bar{y}_L\bar{y}_R}{g^2}}$$

$$\times \left[\frac{\sqrt{g\bar{y}_L} + \sqrt{g\bar{y}_R} + V_R - V_L + (\sqrt{g/\bar{y}_R}\,F_R + \sqrt{g/\bar{y}_L}\,F_L + G_R - G_L)\Delta t}{\sqrt{\bar{y}_R g} + \sqrt{\bar{y}_L/g}}\right]$$

$$(5.97)$$

Equations (5.85)–(5.93) and (5.96) and (5.97) enable the computation of V_4 and y_4 in terms of properties at points $1, 2,$ and 3. Points R and L must lie within a distance Δx from point 2, that is, $0 \le \Delta x_R/\Delta x \le 1$ and $0 \le \Delta x_L/\Delta x \le 1$. Thus, Eqs. (5.92) and (5.93) yield the conditions

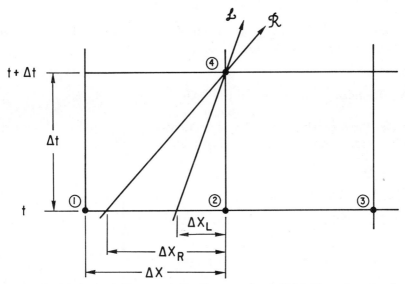

Figure 5.24 Method of Characteristics Computational Grid, Procedure for Supercritical Flows

$$\frac{\Delta x}{\Delta t} \geq V_1 + C_1 \quad \text{or} \quad \frac{\Delta x}{\Delta t} \geq C_3 - V_3 \tag{5.98}$$

Equations (5.98) are one form of the Courant stability criterion, employed in many numerical solutions of unsteady fluid flow. The larger $\Delta x / \Delta t$ of Eq. (5.98) should be used.

If supercritical rightward flow occurs where $V > C$, the computational procedure for V_4 and y_4 must correspond to the characteristic lines of Fig. 5.24. Here, the $\mathcal{R}$ characteristic still lies to the left of point 2 at time t, but so does the $\mathcal{L}$ characteristic. The interpolation formulas for V_L, y_L, and C_L become

SUPERCRITICAL RIGHTWARD FLOW

$$\begin{aligned}
V_L &= V_2 + (V_1 - V_2)\frac{\Delta x_L}{\Delta x} \\[2mm]
\bar{y}_L &= \bar{y}_2 + (\bar{y}_1 - \bar{y}_2)\frac{\Delta x_L}{\Delta x} \\[2mm]
C_L &= C_2 + (C_1 - C_2)\frac{\Delta x_L}{\Delta x} \\[2mm]
f_L &= f_2 + (f_1 - f_2)\frac{\Delta x_L}{\Delta x}, \qquad f = F \quad \text{or} \quad G, \text{ Eq. (5.85)}
\end{aligned} \right\} \tag{5.99}$$

with

$$\frac{\Delta x_L}{\Delta x} = \frac{V_2 - C_2}{V_2 - C_2 - (V_1 - C_1) + \Delta x/\Delta t} \tag{5.100}$$

The computation of V_4 and $\bar{y}_4$ from Eqs. (5.96) and (5.97) is the same for subcritical or supercritical flows.

5.15   Boundary Conditions in the MOC

Consider the case where boundary conditions are known at fixed values of x in Fig. 5.25. If the velocity $V = V_4$ or elevation $\bar{y} = \bar{y}_4$ are known as a

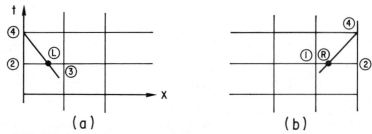

Figure 5.25   Boundary Conditions at Fixed Locations in the Method of Characteristics

function of time at the left fixed boundary, Eq. (5.95) and Fig. 5.25(*a*) yield either

LEFT

$$\bar{y}_4 = \bar{y}_L + \sqrt{\frac{\bar{y}_L}{g}} \, (V_4 - V_L) + \left(F_L - \sqrt{\frac{\bar{y}_L}{g}} \, G_L \right) \Delta t$$

$$V_4 \quad \text{specified}$$

or   (5.101)

$$V_4 = V_L + \sqrt{\frac{g}{\bar{y}_L}} \, (\bar{y}_4 - \bar{y}_L) + \left(G_L - \sqrt{\frac{g}{\bar{y}_L}} \, F_L \right) \Delta t$$

$$\bar{y}_4 \quad \text{specified}$$

Similarly, if $V = V_4$ or $\bar{y} = \bar{y}_4$ are known at the right fixed boundary, Eq. (5.94) and Fig. 5.25(*b*) yield

RIGHT

$$\bar{y}_4 = \bar{y}_R - \sqrt{\frac{\bar{y}_R}{g}} \, (V_4 - V_R) + \left(F_R + \sqrt{\frac{\bar{y}_R}{g}} \, G_R \right) \Delta t$$

$$V_4 \quad \text{specified}$$

or   (5.102)

$$V_4 = V_R - \sqrt{\frac{g}{\bar{y}_R}} \, (\bar{y}_4 - \bar{y}_R) + \left(G_R + \sqrt{\frac{g}{\bar{y}_R}} \, F_R \right) \Delta t$$

$$\bar{y}_4 \quad \text{specified}$$

If a boundary is moving with a known displacement function of time, its motion often is confined to the computational mesh by a series of steps as shown in Fig. 5.26. This method simplifies the computational procedure

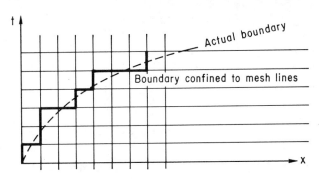

Figure 5.26 Simple Simulation of Moving Boundary

considerably from what would be required to follow the moving boundary continuously. Suppose that the left wall velocity V_w is known as a function of time. The boundary value of velocity at each corner shown on the mesh corresponds to that value of V_w at the time t where the corner exists. Movement of the right boundary can be treated in a similar way.

5.16   A General Computational Method

A general computational mesh is shown in Fig. 5.27 with equally spaced mesh lines at $x_1, x_2, \ldots, x_i, \ldots, x_{max}$ and $t_1, t_2, \ldots, t_j, \ldots, t_{max}$. Properties are prescribed at the initial time t_1, and boundary conditions at x_1 and x_{max} are specified. Channel properties $B(x)$ and $D(x)$ are either tabulated or expressed as equations in terms of x. Computations are carried out by advancing time one increment, sweeping all mesh points from x_1 to x_{max} and repeating the process until the maximum time t_{max} is reached. All fluid properties and channel dimensions are known at the initial time t_1 and corresponding mesh points at $x_1, x_2, \ldots, x_i, \ldots, x_{max}$. The first sweep is made in order to calculate properties at time t_2 and $x_1, x_2, \ldots$ The left boundary condition, Eq. (5.101), is employed first in the sweep at mesh point x_1, t_2. However, before calculating either $\bar{y}_4 = \bar{y}_4(x_1, t_2)$ or $V_4 = V_4(x_1, t_2)$ from Eq. (5.101), it is first necessary to obtain $\bar{y}_L$, V_L, C_L, F_L, and G_L from either Eqs. (5.91) or (5.99) with $\Delta x_L/\Delta x$ calculated from Eq. (5.93) or (5.100), depending on whether the flow is subcritical or supercritical. The sweep then proceeds to mesh points $x_2, t_2, x_3, t_2, \ldots$ until the right boundary is reached.

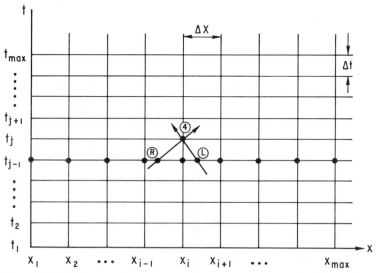

Figure 5.27   General Computational Scheme, Method of Characteristics

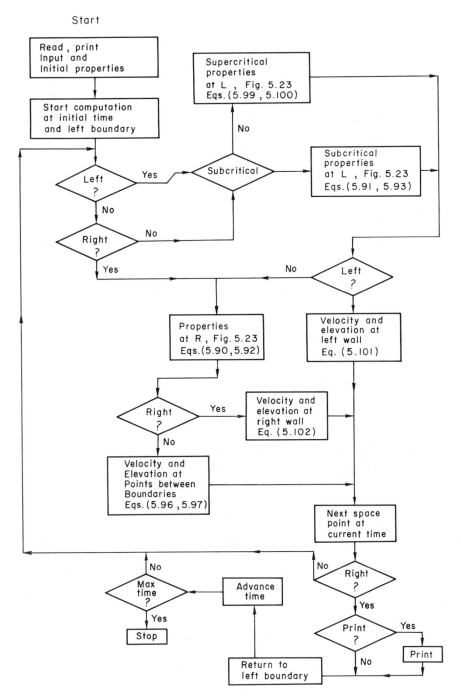

Figure 5.28 Flowchart, Large-Amplitude Hydrostatic Waves in a Uniform Channel

Computations at a general mesh point x_i, t_j first yield properties at L and R in Figs. 5.23 or 5.27 from Eqs. (5.90) and (5.92), (5.91) and (5.93) if subcritical flow, or (5.99) and (5.100) if supercritical flow, after which properties at 4 are obtained from (5.96) and (5.97). The right boundary is calculated with Eqs. (5.90), (5.92), and (5.102). This computational procedure is shown in the flowchart of Fig. 5.28.

Whether the local flow is subcritical or supercritical is determined by noting whether the local velocity V exceeds the local wave speed C; that is,

$$\text{Subcritical flow if}\quad V < C$$

$$\text{Supercritical flow if}\quad V > C$$

The computational flowchart of Fig. 5.28 was programmed and solved on a personal computer. Two cases involved liquid extraction from the left end of two horizontal channels of length L with $B(x) = 0$, shown in Figs. 5.29 and 5.30. One channel is rectangular with $D = \text{constant}$. The other is an annular tank (see Problem 5.14), with the left wall at $x = r_{\text{left}}$ and the right wall at $x = r_{\text{right}}$ with total radial length L. The initial liquid depth is H, and the non-dimensional variables employed were

$$\text{elevation}\ \frac{y}{H}, \qquad \text{distance}\ \frac{x}{H}$$

$$\text{velocity}\ \frac{V}{\sqrt{gH}}, \qquad \text{time}\ t\sqrt{\frac{g}{H}}$$

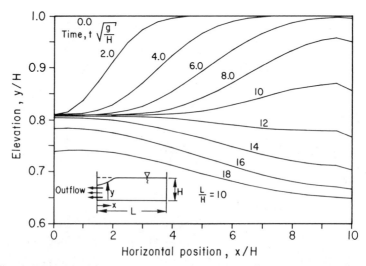

Figure 5.29 Liquid Surface Elevation, Outflow from Rectangular Tank

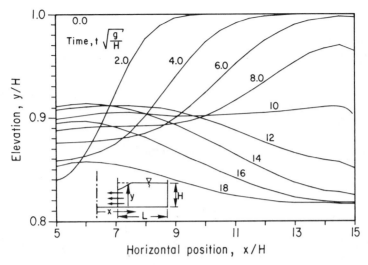

Figure 5.30   Liquid Surface Elevation, Outflow from Annular Tank

The left boundary was one of liquid extraction at velocity $V/\sqrt{gH} = -0.2$. Elevation profiles shown in Figs. 5.29 and 5.30 were obtained for 20 equal horizontal space divisions with a time increment $\Delta t\sqrt{g/H} = 0.1$.

References

5.1   Lamb, H., *Hydrodynamics*, 6th ed., Dover, New York, 1945.

5.2   Streeter, V. L., and E. B. Wylie, *Fluid Mechanics*, 8th ed., McGraw-Hill, New York, 1985.

5.3   Shin, Y. W., and A. H. Wiedermann, "A Hybrid Numerical Method for Homogeneous Two-Phase Flow in One Space Dimension," *ASME J. Pressure Vessel Technology*, **103**(1), (1981).

5.4   Rudinger, G., *Nonsteady Duct Flow: Wave-Diagram Analysis*, Dover, New York, 1959.

Problems

5.1   Equations (5.1) and (5.2) can be solved for certain cases with a similarity variable containing both independent variables x and t. A nondimensionalization of the governing equations can help to determine appropriate similarity variables in some cases. Introduce the nondimensional variables $y^* = y/y_R$, $V^* = V/V_R$, $x^* = x/x_R$, and $t^* = t/t_R$ into Eqs. (5.1) and (5.2), and show that the nondimensional parameter group $V_R t_R/x_R$ results. This group contains t_R/x_R, which suggests a similarity variable of the form $z = bt/x$, where b is a constant. Let $y(x, t) = f_1(z)$ and $V(x, t) = f_2(z)$, and show that (5.1) and (5.2) become

$$f_1' - \frac{z}{b} f_2 f_1' - \frac{z}{b} f_1 f_2' = 0$$

$$f_2' - \frac{z}{b} f_2 f_2' - g \frac{z}{b} f_1' = 0$$

where primes indicate differentiation with respect to z. Combine to verify that

$$\frac{f_1'}{f_2'} = \frac{df_1}{df_2} = \pm \sqrt{\frac{f_1}{g}}$$

Integrate to obtain f_1 as a function of f_2 and an integration constant B. Then use the results to show that

$$y = f_1 = \left(\frac{1}{3\sqrt{g}} \frac{x}{t} + \frac{B}{3} \right)^2, \qquad V = \frac{2}{3} \frac{x}{t} - \frac{B\sqrt{g}}{3}$$

Show that for the dam break problem with $y = H$ and $V = 0$, x/t can be evaluated to obtain B. Compare with Eqs. (5.38) and (5.39).

5.2   What volume flow rate of liquid is required to make the horizontal channel in Fig. P5.2 with a rectangular cross section $B = 1$ m (3.28 ft),

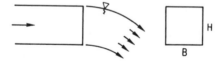

Figure P5.2

$H = 2$ m (6.56 ft), remains full? (Hint: It would run partially full if an elevation disturbance could propagate into the oncoming liquid.

Answer:

6.26 m³/s (221 ft³/s)

5.3   Consider the triangular channel in Fig. P5.3 with known angle θ. Follow the analysis leading to Eqs. (4.1) and (4.2) to show that long waves in a channel of triangular cross-section are described by

$$\frac{\partial y}{\partial t} + V \frac{\partial y}{\partial x} + \frac{1}{2} y \frac{\partial V}{\partial x} = 0$$

$$\frac{\partial V}{\partial t} + V \frac{\partial V}{\partial x} + g \frac{\partial y}{\partial x} = 0$$

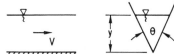

Figure P5.3

Also, follow the solution of Section 5.1 to show that for wave propagation only in one direction, a solution is

$$V = V_i \pm 2(\sqrt{2gy} - \sqrt{2gH})$$

$$x = \left(V \pm \sqrt{\frac{gy}{2}}\right)t + f(V)$$

Finally, show that the long-wave propagation speed into stationary liquid in the triangular channel is given by $\sqrt{gy/2}$. How does this wave speed compare with that in a rectangular channel?

5.4   The Big Thompson Canyon near Estes Park, Colorado, was the scene of disaster during the late 1970s when heavy rains caused flash flooding. Some who escaped reported a water wall 9 m (29.5 ft) high cascading down the canyon. Assume that the canyon is approximated by a triangular channel. Use results from Problem 5.3 to estimate the velocity of water that would have to occur to cause a 9 m (29.5 ft) high water wall. (Hint: The water velocity would have to equal or exceed the front velocity resulting from a dam break)

Answer:

26.6 m/s (87.2 ft/s)

5.5   Water is to be extracted from a rectangular wave tank of width $D = 1$ m (3.28 ft) and undisturbed liquid elevation $H = 1$ m (3.28 ft) from a hole in the bottom which is midway between the two distant ends. Estimate the maximum volume rate which a pump can extract.

Answer:

1.86 m³/s (65.6 ft³/s)

5.6   Determine the volume rate which can be extracted from the wave tank of Problem 5.5 with $H = D = 1$ m (3.28 ft) through a circular hole in the bottom of area $A = 0.4$ m² (4.3 ft²) at an ideal gravity draining rate $\dot{V} = A\sqrt{2gy}$, where y is the liquid elevation above the hole. Show that two solutions exist corresponding to 0.03 m³/s (1.06 ft³/s) with $y = 0.029$ m (0.095 ft) and 1.49 m³/s (52.6 ft³/s) with $y = 0.68$ m (2.23 ft). Explain why the smaller volume rate cannot occur.

5.7   A lake is in the form of a long rectangular channel of width $D = 30$ m (98.4 ft) and undisturbed water depth $H = 20$ m (65.6 ft). A dam at the end of the lake suddenly disintegrates. Determine the volume discharge rate into the valley from the lake.

Answer:

2489 m^3/s (87, 831 ft^3/s)

5.8   The left wall of a long rectangular wave tank moves rightward into undisturbed liquid of depth H. The wall moves according to

$$V_w = V_\infty(1 - e^{-t/\tau})$$

where V_∞ is the asymptotic velocity and τ is the time constant,
(a) Show that the function $f(V)$ in Eq. (5.11) is

$$f(V) = -V_\infty \tau \left[\frac{V}{V_\infty} + \ln\left(1 - \frac{V}{V_\infty}\right) \right] + \left(\frac{3}{2} V + \sqrt{gH} \right) \tau \ln\left(1 - \frac{V}{V_\infty} \right)$$

(b) One test for determining the formation time and location of a shock, if one forms, is to set the first and second derivatives of velocity with respect to x at a fixed time equal to infinity. An infinite first derivative corresponds to a vertical tangent to the velocity profile on a V, x plot, and an infinite second derivative corresponds to a point of inflection in the velocity profile at the condition of so-called wave breaking. These test conditions also can be inverted to give the alternate conditions

$$\left(\frac{\partial x}{\partial V} \right)_t = 0 \quad \text{and} \quad \left(\frac{\partial^2 x}{\partial V^2} \right)_t = 0$$

If the second derivative condition fails to give meaningful results, the shock occurs at the frontmost part of the propagating disturbance, obtained by setting $V = 0$ in the first derivative condition. Show that the second derivative would give a liquid velocity greater than V_∞ and, therefore, is nonsense. Then show that the first derivative with $V = 0$ gives the time and location of shock formation as

$$t_{shock} = \frac{2}{3} \sqrt{gH} \frac{\tau}{V_\infty}, \qquad x_{shock} = \frac{2}{3} gH \frac{\tau}{V_\infty}$$

5.9   A wave tank of length $L = 9.8$ m (32.1 ft) has an undisturbed liquid depth of $H = 1$ m (3.28 ft). Show that the maximum constant acceleration the left wall can undergo into the liquid which will not cause a hydrostatic shock to form until after the disturbance arrives at the right end is $2/3 \, \text{m/s}^2$ $(2.19 \, \text{ft/s}^2)$. Also show that when the disturbance arrives at the right end, the left wall will have moved $3.27 \, \text{m}$ $(10.7 \, \text{ft})$.

5.10 The left wall of a long rectangular wave tank undergoes one cycle of velocity according to

$$V(0, t) = V_0 \sin 2\pi ft$$

and then returns to its starting position. Consider x, V, and t at the moving wall and show that the function $f(V)$ is given by

$$f(V) = \frac{V_0}{2\pi f}\left[1 - \sqrt{1 - \left(\frac{V}{V_0}\right)^2} - \left(\frac{3}{2}\frac{V}{V_0} + \frac{\sqrt{gH}}{V_0}\right)\sin^{-1}\frac{V}{V_0}\right]$$

5.11 The undisturbed water level behind a wall in Fig. P5.11 is H and the

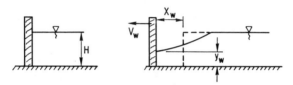

Figure P5.11

wall mass is M. The wall is suddenly released and is free to slide leftward on a frictionless track, remaining upright, and driven by hydrostatic pressure on its right side. The hydrostatic force against the wall of width D into the page is given by

$$F = \rho \frac{g}{g_0} \frac{D y_w^2}{2}$$

(a) Verify that the dynamic equation for the wall is

$$\frac{dV_w}{dt} = \frac{d^2 x_w}{dt^2} = \frac{\rho g D}{2M} y_w^2$$

(b) Use the hydrostatic wave solution of Eqs. (5.7) and (5.11) with the wall dynamic equation to obtain y_w, V_w, and x_w as functions of time. One possible procedure is to eliminate y_w, solve for V_w, and integrate for x_w. Another approach is to eliminate V_w and express the wall dynamic equation as a differential equation in y_w, integrate to obtain y_w, then plug into the general solution for V, giving V_w, and integrate to obtain x_w. Verify the following:

$$y_w = H(1 + at)^{-2/3}$$

$$V_w = 2\sqrt{gH}\,[1 - (1 + at)^{-1/3}]$$

$$x_w = 2\sqrt{gH}\,t - \frac{4M}{\rho DH}[(1 + at)^{2/3} - 1]$$

$$a = \frac{3\rho DH\sqrt{gH}}{4M}$$

(c) Show that the arbitrary function $f(V)$ is

$$f(V) = \frac{2MV}{\rho DH\sqrt{gH}}$$

Here are other steps which will give additional insight to this problem.

(d) Sketch a propagation diagram in the t, x plane, showing the wall trajectory and various characteristic lines with associated V and y values.

(e) Show at what position critical flow would be reached; that is, where does the liquid velocity become equal to the wave propagation speed? This condition should correspond to a vertical characteristic on the t, x plane.

(f) Indicate at what time and location the wall will stop accelerating. Or will it ever?

(g) Show that your results can be used to predict the limiting case where wall mass M becomes zero, and you have sudden disintegration of a dam.

5.12 Consider the propagation of axisymmetric cylindrical waves in a large expanse of liquid resting on a horizontal floor in the x, z plane with undisturbed elevation H and uniform top pressure P_∞. Disturbances propagate outward in the radial coordinate r. Employ a stationary cylindrical shell control volume of inner radius r, thickness dr, and height y to the liquid free surface to show that the mass conservation and momentum equations can be written in terms of elevation y and radial velocity V_r as

$$\frac{\partial y}{\partial t} + V_r\frac{\partial y}{\partial r} + y\frac{\partial V_r}{\partial r} + \frac{y}{r}V_r = 0$$

$$\frac{\partial V_r}{\partial t} + V_r\frac{\partial V_r}{\partial r} + g\frac{\partial y}{\partial r} = 0$$

5.13 Apply the method of characteristics described in Section 5.8 to the equations obtained in Problem 5.12 for cylindrical hydrostatic waves, and show that the characteristic formulation corresponding to Eq. (5.49) is given by

$$\frac{dy}{dt} \pm \sqrt{\frac{y}{g}} \frac{dV_r}{dt} + \frac{y}{r} V_r = 0 \qquad \text{on paths } \frac{dr}{dt} = V_r \pm \sqrt{gy}$$

5.14 Consider hydrostatic waves with circular symmetry about the y axis, where the horizontal radial coordinate is $r = x$. Show that the equations derived in Problem 5.12 could be obtained from Eqs. (5.83)–(5.86) for a horizontal pie-slice section with $D(x) = (\text{constant})x$, $V = V_r$, $x = r$, and $y = y$, giving

$$\frac{\partial y}{\partial t} + V \frac{\partial y}{\partial x} + y \frac{\partial V}{\partial x} + \frac{n}{x} yV = 0, \qquad n = \begin{cases} 1, & \text{axisymmetric} \\ 0, & \text{rectangular} \end{cases}$$

$$\frac{\partial V}{\partial t} + V \frac{\partial V}{\partial x} + g \frac{\partial y}{\partial x} = 0$$

Follow the MOC procedure of Section 5.14 to obtain the formulation

$$\frac{dy}{dt} \pm \sqrt{\frac{y}{g}} \frac{dV}{dt} + n \frac{y}{x} V = 0 \qquad \text{on paths } \frac{dx}{dt} = V \pm \sqrt{gy}$$

Also show that for the computational grid of Fig. 5.23, the full characteristic solution in finite difference form for any interior mesh point 4 is given as follows:

$$V_4 = \frac{(a_L - a_R)\Delta t + g(y_R - y_L) + V_R \sqrt{gy_R} + V_L \sqrt{gy_L}}{\sqrt{gy_L} + \sqrt{gy_R}}$$

$$y_4 = \frac{y_R \sqrt{gy_L} + y_L \sqrt{gy_R} + (V_R - V_L)\sqrt{y_R y_L} - (b_{LR} - b_{RL})\Delta t}{\sqrt{gy_L} + \sqrt{gy_R}}$$

$$a_i = \frac{gy_i V_i}{x_i} \quad (i = R, L), \qquad b_{LR} = \frac{a_L \sqrt{gy_R}}{g}, \qquad b_{RL} = \frac{a_R \sqrt{gy_L}}{g}$$

with

$$f_i = f_1 + (f_2 - f_1) \frac{\Delta x_R}{\Delta x}, \qquad f = \sqrt{y_R}, V_R, x_R$$

$$\frac{\Delta x_R}{\Delta x} = -\frac{V_1 + \sqrt{gy_1}}{\sqrt{gy_1} - \sqrt{gy_2} + V_1 - V_2 - \Delta x/\Delta t}$$

and

$$g_i = g_2 + (g_3 - g_2) \frac{\Delta x_L}{\Delta x}, \qquad g = \sqrt{y_L}, V_L, x_L$$

$$\frac{\Delta x_L}{\Delta x} = -\frac{V_2 - \sqrt{gy_2}}{\sqrt{gy_3} - \sqrt{gy_2} + V_2 - V_3 + \Delta x/\Delta t}$$

5.15 A hydraulic bore (hydrostatic shock) propagates at a speed of $S = 12\,\text{m/s}$ (39.4 ft/s) in a horizontal channel of rectangular cross section which contains stationary undisturbed liquid of depth $H_u = 10\,\text{m}$ (32.8 ft). Show that the bore amplitude and velocity of liquid in the bore are 12.9 m (42.3 ft) and 2.7 m/s (8.86 ft/s), respectively.

5.16 A hydrostatic shock of elevation 2 m (6.56 ft) above the stationary undisturbed water of depth $H_u = 10\,\text{m}$ (32.8 ft) arrives at a stationary vertical wall. Show that the oncoming shock speed is 11.4 m/s (37.4 ft/s); the reflected amplitude is 1.95 m (6.4 ft) above the oncoming amplitude [(3.95 m (12.96 ft) above the initial 10-m (32.8-ft) depth)]; and that the reflected shock speed is 9.5 m/s (31.2 ft/s).

5.17 A partition separates two stationary liquid regions in a rectangular wave channel which extends long distances in both directions (See Fig. P5.17). The initial elevation $H_R = 1\,\text{m}$ (3.28 ft) is on the right.

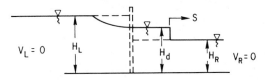

Figure P5.17

Sudden removal of the partition must cause a hydrostatic shock of $H_d = 2\,\text{m}$ (6.56 ft) propagating to the right. What initial elevation of liquid on the left, H_L, is required? (Hint: Long-wave propagation will occur to the left.)

Answer:

3.4 m (11.2 ft)

5.18 Determine the force per unit width which must be applied to the end wall of a wave tank to have it move into water of elevation $H = 1\,\text{m}$ (3.28 ft) at constant velocity $V(0, t) = 2\,\text{m/s}$ (6.56 ft/s).

Answer:

15.6 kN/m (1069 lbf/ft). Compare with the undisturbed liquid force on the wall.

5.19 A long horizontal wave tank of width $D = 1\,\text{m}$ (3.28 ft) has an undisturbed water depth of $H_u = 1\,\text{m}$ (3.28 ft). The left wall suddenly begins moving rightward with constant velocity $V_d = 2\,\text{m/s}$ (6.56 ft/s), as in Problem 5.18.

(a) Verify that the hydrostatic shock has an elevation of $H_d = 1.79$ m (5.87 ft), and the external force required to keep the wall moving is $F = 15.6$ kN (3507 lbf).

(b) Show that the power input to the liquid from the force in part (a) is 31.2 kW (29.6 B/s).

(c) Show that the energy (kinetic plus gravitational potential energy) per unit mass in the shocked liquid, above the potential energy per unit mass in the undisturbed liquid, is

$$\frac{\Delta E}{\Delta M} = \frac{V_d^2}{2g_0} + \frac{gH_d}{2g_0}\left(1 - \frac{H_u^2}{H_d^2}\right)$$

(d) Show that the energy rate passing any fixed location (available power) in the shocked region is

$$\mathscr{P}_{available} = \rho DH_dV_d\frac{\Delta E}{\Delta M}$$

(e) Show that the available power is less than the power input by an amount

$$\mathscr{P}_{loss} = \mathscr{P}_{in} - \mathscr{P}_{available} = 2.65 \text{ kW } (2.5 \text{ B/s})$$

(f) Show that the loss of available power could have been obtained from

$$\mathscr{P}_{loss} = \mathscr{D}_s\mathscr{P}_{in}$$

where $\mathscr{D}_s$ is obtained from Eq. (5.75).

5.20 Follow the procedure of Section 5.10 to show that hydrostatic shock speed, liquid velocity, and elevation in a triangular channel are governed by

$$\frac{S - V_u}{\sqrt{gH_u}} = \sqrt{\frac{1}{3}\frac{(H_d/H_u)^3 - 1}{1 - (H_u/H_d)^2}}$$

$$\frac{V_d - V_u}{\sqrt{gH_u}} = \sqrt{\frac{1}{3}\left[\left(\frac{H_d}{H_u}\right)^3 - 1\right]\left[1 - \left(\frac{H_u}{H_d}\right)^2\right]}$$

Also show that in the limit when $H_d/H_u \to 1$, the disturbance propagation speed reduces to that obtained in Problem 5.3 for a triangular channel.

6 UNSTEADY THERMOFLUID SYSTEMS AND NORMALIZATION

This chapter is designed to show how rigorous problem formulations can be simplified to provide quick estimates of unsteady thermofluid system behavior.

The differential equations are formulated for systems of one-dimensional and multidimensional flows and for spatially uniform systems. A normalization procedure is presented for the purpose of revising the equations in terms of non-dimensional variables, which produce nondimensional model coefficients. Magnitude estimates of the model coefficients display terms which are important or negligible in the differential equations. Simplification by dropping negligible terms often yields a formulation which can be solved by pencil-and-paper methods, or otherwise interpreted.

Methods discussed in this chapter also can be employed in the design and interpretation of scale model tests, or in the correlation of test data to predict the behavior response of full-size thermofluid systems. Some non-dimensional terms obtained in this chapter could result from the Buckingham Pi Theorem; however, the Pi Theorem does not indicate which effects are negligible, whereas methods presented here overcome that difficulty.

6.1 One-Dimensional Flows

Figure 6.1(a) helps to describe a general model for the analysis of one-dimensional flow in a pipe or other flow passage. Length in the flow direction is designated by z, elevation by y. The flow area $A(z, t)$ can vary continuously in the direction of flow, and also in time, for application to

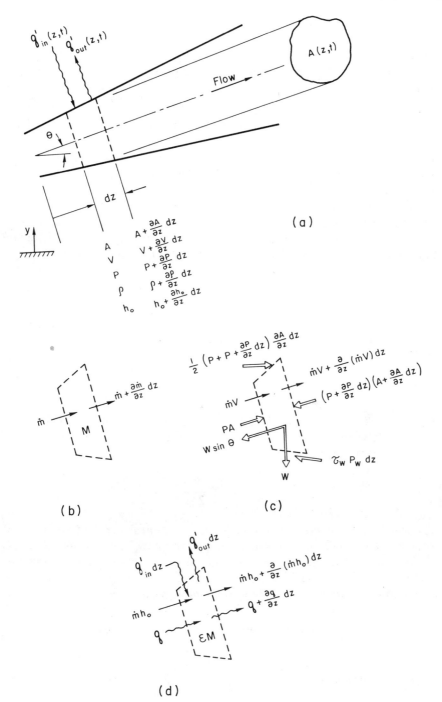

Figure 6.1 One-Dimensional Flow Model. (*a*) General Model; (*b*) Mass Conservation; (*c*) Momentum; (*d*) Energy Conservation

flexible tubes. Heat transfer to or from the passage walls is included, and fluid heat conduction occurs in the z direction. The fluid is arbitrary, compressible, or incompressible, and wall friction effects are included.

The mass conservation principle is written for the stationary CV in Fig. 6.1(b) as

$$\frac{\partial \dot{m}}{\partial z} dz + \frac{\partial M}{\partial t} = 0 \tag{6.1}$$

The partial derivative $\partial \dot{m}/\partial z = (\partial \dot{m}/\partial z)_t$ is taken at a given instant of time, and $\partial M/\partial t = (\partial M/\partial t)_z$ is taken at a given location. The subscripts usually are omitted in analyses.

The momentum principle is formulated from Fig. 6.1(c). Active forces include the wall shear force, a force due to the fluid weight, and pressure forces on both surfaces normal to z, plus an average pressure force acting on the projected area $(\partial A/\partial z) dz$. Thus, the total force is

$$F = -\left(A \frac{\partial P}{\partial z} dz + \tau_w P_w dz + W \sin \theta \right) \tag{6.2}$$

The wall shear stress τ_w historically was expressed for steady flow in terms of the Fanning, or *small*, friction factor f_F as

$$\tau_{w,\,\text{steady}} = f_F \frac{V^2}{2g_0} \rho \tag{6.3}$$

However, the *large* friction factor, $f = 4f_F$, generally is used for pipe flow analysis. It would be appropriate to use an unsteady friction factor f_u for time-dependent problems, that is,

$$\tau_w = \frac{f_u}{4} \frac{V^2 \rho}{2g_0} \tag{6.4}$$

The classical steady flow friction factor f is given in Fig. B.1 of Appendix B in terms of the pipe relative roughness and the flow Reynolds number. It is often permissible to use the steady value of f in place of f_u in unsteady flow problems (see Example 2.2).

The rate of creation (ROC) of momentum in the CV is

$$\text{ROC (momentum)} = \frac{\partial}{\partial z} (\dot{m}V) dz + \frac{\partial}{\partial t} (MV) \tag{6.5}$$

Equation (6.5) is expanded, and Eq. (6.1) is employed to write the momentum principle as

$$\dot{m} \frac{\partial V}{\partial t} dz + M \frac{\partial V}{\partial t} = g_0 F \tag{6.6}$$

Energy storage and transfer terms are shown in Fig. 6.1(d), which include heating or cooling at the walls, heat conduction in the flow direction, and convected energy by mass flow. Since the flow area may change with time, a compressive power term also is shown. Energy added or extracted by a pump, fan, or other machine, would be included as a boundary condition in a flow passage and does not appear in the differential CV of Fig. 6.1(d). The energy conservation principle is written as

$$\frac{\partial}{\partial z}(\dot{m}h_0)\,dz + \frac{\partial q}{\partial z}\,dz + (q'_{out} - q'_{in})\,dz + \frac{\partial}{\partial t}(\mathscr{E}M) = 0 \qquad (6.7)$$

The product terms $\dot{m}h_0$ and $\mathscr{E}M$ are expanded and combined with both Eq. (6.1) and the stagnation enthalpy $h_0 = \mathscr{E} + P/\rho$ to give

$$\dot{m}\frac{\partial h_0}{\partial z}\,dz + \frac{P}{\rho}\frac{\partial \dot{m}}{\partial z}\,dz + \frac{\partial q}{\partial z}\,dz + (q'_{out} - q'_{in})\,dz + M\frac{\partial \mathscr{E}}{\partial t} = 0 \quad (6.8)$$

Equations (6.1), (6.6), and (6.8) next are written with the substitutions

$$\dot{m} = \rho A V, \qquad \mathscr{E} = \frac{V^2}{2g_0} + \frac{g}{g_0}y + e = h'_0 - \frac{P}{\rho}$$

$$M = \rho A\,dz, \quad q = -\kappa A\frac{\partial T}{\partial z}, \quad \mathscr{V} = A\,dz \qquad (6.9)$$

$$W = M\frac{g}{g_0}, \qquad \frac{g_0\tau_w P_w}{\rho A} = \frac{fV|V|}{2D_h}$$

where D_h is the hydraulic diameter $4A/P_w$, and V^2 is written as $|V|V$ so that the wall friction force always opposes the direction of flow. This results in

Mass Conservation

$$\frac{\partial \rho}{\partial t} + V\frac{\partial \rho}{\partial z} + \rho\frac{\partial V}{\partial z} + \frac{\rho}{A}\left(\frac{\partial A}{\partial t} + V\frac{\partial A}{\partial z}\right) = 0 \qquad (6.10)$$

Momentum Creation

$$\frac{\partial V}{\partial t} + V\frac{\partial V}{\partial z} + \frac{g_0}{\rho}\frac{\partial P}{\partial z} + \frac{f}{D_h}\frac{V|V|}{2} + g\sin\theta = 0 \qquad (6.11)$$

Energy Conservation

$$\frac{\partial h_0}{\partial t} + V\frac{\partial h_0}{\partial z} - \frac{1}{\rho}\frac{\partial P}{\partial t} - \frac{\kappa}{\rho}\left(\frac{\partial^2 T}{\partial z^2} + \frac{1}{A}\frac{\partial A}{\partial z}\frac{\partial T}{\partial z}\right) + \frac{q'_{out} - q'_{in}}{\rho A} = 0$$
$$(6.12)$$

Equations (1.69) and (1.70) are employed in (6.12) to eliminate the stagnation enthalpy. Further elimination of static enthalpy with Eq. (2.118), and subsequent combination with (6.11) and (6.10) give mass and energy consevation in the alternate forms,

Mass Conservation

$$\frac{\partial P}{\partial t} + V \frac{\partial P}{\partial z} + \frac{\rho C^2}{g_0} \frac{\partial V}{\partial z} + \frac{\rho C^2}{A g_0} \left(\frac{\partial A}{\partial t} + V \frac{\partial A}{\partial z} \right) = \frac{\beta C^2}{g_0 c_p} \mathscr{F}_1 \qquad (6.13)$$

Energy Conservation

$$\frac{g_0}{C^2} \left(\frac{\partial P}{\partial t} + V \frac{\partial P}{\partial z} \right) - \left(\frac{\partial \rho}{\partial t} + V \frac{\partial \rho}{\partial z} \right) = \frac{\beta}{c_p} \mathscr{F}_1 \qquad (6.14)$$

Further combination of Eq. (6.14) with (2.119) or (2.116) yields two other useful forms of energy conservation, namely

$$T \left(\frac{\partial s}{\partial t} + V \frac{\partial s}{\partial z} \right) = \frac{1}{\rho} \mathscr{F}_1 \qquad (6.15)$$

or

$$\frac{\partial T}{\partial t} + V \frac{\partial T}{\partial z} - \frac{\beta T}{\rho c_p} \left(\frac{\partial P}{\partial t} + V \frac{\partial P}{\partial z} \right) = \frac{1}{\rho c_p} \mathscr{F}_1 \qquad (6.16)$$

where

$$\mathscr{F}_1 = \kappa \left(\frac{\partial^2 T}{\partial z^2} + \frac{1}{A} \frac{\partial A}{\partial z} \frac{\partial T}{\partial z} \right) + \frac{f}{D_h} \rho \frac{V^3}{2 g_0} + \frac{q'_{in} - q'_{out}}{A} \qquad (6.17)$$

Equations, (6.13), (6.11) and (6.14) are useful in formulating problems which involve the propagation of pressure waves. If the flow passage is rigid, $\partial A / \partial t = 0$, although A may vary with distance z. The function $\mathscr{F}_1$ includes the heating effect of conduction in the fluid, wall friction, and wall heating on the fluid density, which usually is negligible in liquid flows. Moreover, $\mathscr{F}_1$ is also an energy dissipation term because it plays a major role in the entropy production of Eq. (6.15).

Equation (6.16) is useful in determining the unsteady temperature in a passage with heated or cooled fluid flow. It can be used when it is necessary to determine the inside wall temperature and resulting thermal stress which occurs if hotter or cooler fluid enters a pipe at some initial temperature. When pressure effects, axial heat conduction in the fluid, and wall friction terms are negligible, Eq. (6.16) reduces to Eq. (3.62).

6.2 Multidimensional Flows, Mass Conservation

Multidimensional flows are characterized by property variations in two or three space dimensions. The formulations summarized in this chapter are based on three-dimensional x, y, z geometry. Two-dimensional flows in the x, y plane are readily described by specifying no property variation in the z direction.

It is convenient here to adopt the classical approach which employs a Lagrangian reference frame. That is, we follow a discrete element of fluid which moves and deforms relative to a stationary reference frame fixed to the earth.

Figure 6.2 shows a rectangular element of fluid with dimensions δx, δy, δz at time t. The volume $\mathcal{V}$ is $\delta x \, \delta y \, \delta z$, and the contained mass M is $\rho \mathcal{V} = \rho \, \delta x \, \delta y \, \delta z$. Since no mass crosses its boundaries, the mass conservation law is

$$\frac{dM}{dt} = \rho \, \frac{d\mathcal{V}}{dt} + \mathcal{V} \, \frac{d\rho}{dt} = 0 \tag{6.18}$$

If the density is expressed functionally as $\rho(x, y, z, t)$, its time derivative is

$$\frac{d\rho}{dt} = \frac{\partial \rho}{\partial t} + \mathbf{V} \cdot \nabla \rho \tag{6.19}$$

Moreover, the volume derivative $d\mathcal{V}/dt$ includes the time derivatives $d\,\delta x/dt$, $d\,\delta y/dt$, and $d\,\delta z/dt$, each of which represents a stretching or contracting rate. That is, the length δx has collinear velocity components u at the left end and $u + (\partial u/\partial x)\,\delta x$ at the right end so that $d\,\delta x = (\partial u/\partial x)\,\delta x\,dt$. Similar analyses apply to lengths y and z so that

$$\frac{d\,\delta x}{dt} = \frac{\partial u}{\partial x}\,\delta x \,, \quad \frac{d\,\delta y}{dt} = \frac{\partial v}{\partial y}\,\delta y \,, \quad \frac{d\,\delta z}{dt} = \frac{\partial w}{\partial z}\,\delta z$$

It follows from Eq. (6.18) that multidimensional mass conservation can be expressed by

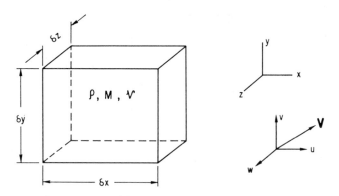

Figure 6.2 Fluid Element for Mass Conservation

$$\frac{d\rho}{dt} = \frac{\partial \rho}{\partial t} + \mathbf{V} \cdot \nabla \rho = -\rho \nabla \cdot \mathbf{V} \tag{6.20}$$

If the flow is incompressible, $\rho =$ constant and

$$\nabla \cdot \mathbf{V} = 0 \tag{6.21}$$

6.3 Multidimensional Flows, Momentum Creation

A large class of multidimensional problems can be treated with inviscid flow theory. Therefore, the momentum equation is formulated with the help of Fig. 6.3, where only pressure and body forces act on a rectangular fluid increment. If pressure P occurs at the center, the surface force component in the x direction is $-(\partial P/\partial x)\,\delta x\,\delta y\,\delta z$. Similar force components for the y and z directions yield the total surface force

$$\mathbf{F}_s = -\nabla P\,\delta x\,\delta y\,\delta z \tag{6.22}$$

The body force is given by

$$\mathbf{F}_b = -\frac{M}{g_0}\mathbf{b} \tag{6.23}$$

where the body force acceleration vector is

$$\mathbf{b} = b_x \mathbf{n}_x + b_y \mathbf{n}_y + b_z \mathbf{n}_z \tag{6.24}$$

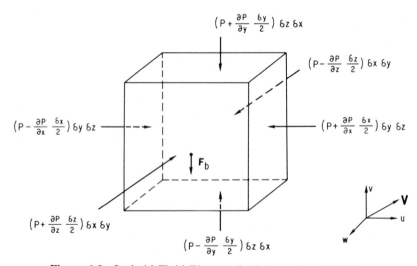

Figure 6.3 Inviscid Fluid Element for Momentum Principle

The most general body force acceleration is a function of time $\mathbf{b}(t)$, as in the case of an accelerating container. However, $\mathbf{b}$ is constant for reference frames which are not accelerating relative to the earth. If the body force is due to gravity, $\mathbf{b} = g\mathbf{n}_y$. The momentum principle yields

$$\mathbf{F} = \mathbf{F}_s + \mathbf{F}_b = \frac{M}{g_0} \frac{d\mathbf{V}}{dt} \tag{6.25}$$

If we write

$$M = \rho \, \delta x \, \delta y \, \delta z \tag{6.26}$$

and

$$\frac{d\mathbf{V}}{dt} = \frac{\partial \mathbf{V}}{\partial t} + \mathbf{V} \cdot \nabla \mathbf{V} \tag{6.27}$$

Eq. (6.25) becomes, for inviscid flow,

$$\frac{\partial \mathbf{V}}{\partial t} + \mathbf{V} \cdot \nabla \mathbf{V} + \frac{g_0}{\rho} \nabla P + \mathbf{b} = 0 \tag{6.28}$$

The fluid increment of Fig. 6.3 also can be treated as a deformable body under the action of viscous forces in addition to pressure and body forces. The viscous force components are rigorously formulated in texts on boundary layer and viscous flow theory, e.g., references [6.3], [6.4], and [6.5]. When these forces are employed in Eq. (6.25), the momentum formulation becomes

$$\frac{\partial \mathbf{V}}{\partial t} + \mathbf{V} \cdot \nabla \mathbf{V} + \frac{g_0}{\rho} \nabla P + \mathbf{b} = \frac{g_0}{\rho} \nabla \cdot \Gamma = \frac{g_0}{\rho} \left[\mu \nabla^2 \mathbf{V} + \left(\zeta + \frac{1}{3} \mu \right) \nabla(\nabla \cdot \mathbf{V}) \right] \tag{6.29}$$

where the shear stress dyadic in rectangular coordinates is defined as in Eq. (1.41), by

$$\begin{aligned}
\Gamma = \sum_j \sum_i \mathbf{n}_i \tau_{ij} \mathbf{n}_j , \quad & i, j = x, y, z \\
= (\mathbf{n}_x \tau_{xx} + \mathbf{n}_y \tau_{yx} + \mathbf{n}_z \tau_{zx}) \mathbf{n}_x & \\
+ (\mathbf{n}_x \tau_{xy} + \mathbf{n}_y \tau_{yy} + \mathbf{n}_z \tau_{zy}) \mathbf{n}_y & \\
+ (\mathbf{n}_x \tau_{xz} + \mathbf{n}_y \tau_{yz} + \mathbf{n}_z \tau_{zz}) \mathbf{n}_z &
\end{aligned} \tag{6.30}$$

The corresponding shear stress components are given by

$$\tau_{xx} = \mu\left[\left(2 + \frac{\zeta}{\mu}\right)\frac{\partial u}{\partial x} - \frac{2}{3}\nabla\cdot\mathbf{V}\right]$$

$$\tau_{yy} = \mu\left[\left(2 + \frac{\zeta}{\mu}\right)\frac{\partial v}{\partial y} - \frac{2}{3}\nabla\cdot\mathbf{V}\right]$$

$$\tau_{zz} = \mu\left[\left(2 + \frac{\zeta}{\mu}\right)\frac{\partial w}{\partial z} - \frac{2}{3}\nabla\cdot\mathbf{V}\right]$$

$$\tau_{xy} = \tau_{yx} = (\mu + \zeta)\left(\frac{\partial u}{\partial y} + \frac{\partial v}{\partial x}\right)$$

$$\tau_{yz} = \tau_{zy} = (\mu + \zeta)\left(\frac{\partial v}{\partial z} + \frac{\partial w}{\partial y}\right)$$

$$\tau_{zx} = \tau_{xz} = (\mu + \zeta)\left(\frac{\partial w}{\partial x} + \frac{\partial u}{\partial z}\right) \tag{6.31}$$

The operation $\nabla\cdot\Gamma$ means the divergence of each term in parentheses of Eq. (6.30). It is seen that the last term of Eq. (6.29) in brackets vanishes for incompressible flow with $\nabla\cdot\mathbf{V} = 0$. Both the dynamic viscosity μ and the so-called second viscosity ζ are appropriate in formulating shear stresses which vanish in either uniform translational or uniform rotational (solid body rotational) flows.

6.4 Multidimensional Flows, Energy Conservation

The energy principle for multidimensional, inviscid flow is formulated for the moving fluid increment of Fig. 6.4. Energy flows across the boundary are due to heat conduction and mechanical power from action of the surface forces. Thus, energy conservation is written for the incremental mass M in Fig. 6.4(a) as

$$q_{\text{out}} - q_{\text{in}} + \mathscr{P}_{\text{out}} - \mathscr{P}_{\text{in}} + M\frac{d}{dt}\mathscr{E} = 0 \tag{6.32}$$

The heat conduction terms are shown in Fig. 6.4(b), for which

$$q_x = -\kappa\frac{\partial T}{\partial x}\,\delta y\,\delta z$$

$$q_y = -\kappa\frac{\partial T}{\partial y}\,\delta z\,\delta x \tag{6.33}$$

$$q_z = -\kappa\frac{\partial T}{\partial z}\,\delta x\,\delta y$$

Therefore, the summation of heat conduction terms in Eq. (6.32) becomes

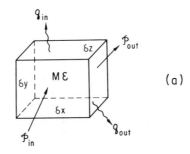

(a)

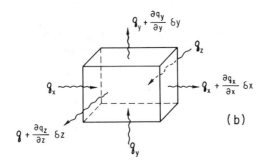

(b)

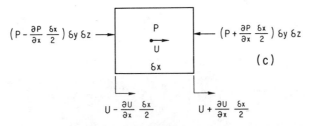

Figure 6.4 Diagram for Energy Conservation, Inviscid Flow. (*a*) Energy Terms; (*b*) Heat Transfer Terms. (*c*) Mechanical Power Terms, *x* Directions

$$q_{\text{out}} - q_{\text{in}} = -\nabla \cdot (\kappa \nabla T)\, \delta x\, \delta y\, \delta z \tag{6.34}$$

Mechanical power terms are expressed with the help of pressure components shown in Fig. 6.3. The associated pressure forces acting in the *x* direction are shown in Fig. 6.4(*c*), with the corresponding velocity components. The mechanical power for motion in the *x* direction is

$$(\mathscr{P}_{\text{out}} - \mathscr{P}_{\text{in}})_x = \frac{\partial}{\partial x}(Pu)\, \delta x\, \delta y\, \delta z$$

Similar expressions for mechanical power in the y and z directions are

$$(\mathscr{P}_{\text{out}} - \mathscr{P}_{\text{in}})_y = \frac{\partial}{\partial y}(Pv)\,\delta x\,\delta y\,\delta z$$

and

$$(\mathscr{P}_{\text{out}} - \mathscr{P}_{\text{in}})_z = \frac{\partial}{\partial z}(Pw)\,\delta x\,\delta y\,\delta z$$

The total mechanical power term for inviscid flow is the sum, given by

$$\mathscr{P}_{\text{out}} - \mathscr{P}_{\text{in}} = \nabla\cdot(P\mathbf{V})\,\delta x\,\delta y\,\delta z = (P\nabla\cdot\mathbf{V} + \mathbf{V}\cdot\nabla P)\,\delta x\,\delta y\,\delta z \quad (6.35)$$

If we consider the general body force acceleration vector of Eq. (6.24), the total stored energy per unit mass is given by

$$\mathscr{E} = \frac{\mathbf{V}\cdot\mathbf{V}}{2g_0} + \frac{1}{g_0}\mathbf{b}\cdot\mathbf{r} + e \quad (6.36)$$

where r is the position vector

$$\mathbf{r} = x\mathbf{n}_x + y\mathbf{n}_y + z\mathbf{n}_z \quad (6.37)$$

If we write $d\mathbf{r}/dt = \mathbf{V}$ and $d\mathbf{b}/dt = \dot{\mathbf{b}}$, Eq. (6.36) yields

$$\frac{d\mathscr{E}}{dt} = \frac{1}{g_0}\mathbf{V}\cdot\frac{d\mathbf{V}}{dt} + \frac{1}{g_0}\mathbf{b}\cdot\mathbf{V} + \frac{1}{g_0}\dot{\mathbf{b}}\cdot\mathbf{r} + \frac{de}{dt}$$

Combining with Eqs. (6.32), (6.34), (6.35), and $M = \rho\,\delta x\,\delta y\,\delta z$ yields

$$-\nabla\cdot(\kappa\nabla T) + (P\nabla\cdot\mathbf{V} + \mathbf{V}\cdot\nabla P) + \frac{\rho}{g_0}\left(\mathbf{V}\cdot\frac{d\mathbf{V}}{dt} + \mathbf{b}\cdot\mathbf{V} + \dot{\mathbf{b}}\cdot\mathbf{r}\right) + \rho\frac{de}{dt} = 0$$

$$(6.38)$$

Since this analysis is for inviscid flow, we may employ Eq. (6.28) with

$$\frac{d\mathbf{V}}{dt} = \frac{\partial\mathbf{V}}{\partial t} + \mathbf{V}\cdot\nabla\mathbf{V}$$

to obtain from Eq. (6.38),

$$-\nabla\cdot(\kappa\nabla T) + P\nabla\cdot\mathbf{V} + \frac{\rho}{g_0}\dot{\mathbf{b}}\cdot\mathbf{r} + \rho\frac{de}{dt} = 0$$

If we differentiate $e = h - P/\rho$ with respect to time and employ Eq. (6.20), the energy equation for inviscid fluid becomes

$$\frac{\partial h}{\partial t} + \mathbf{V} \cdot \nabla h - \frac{1}{\rho}\left(\frac{\partial P}{\partial t} + \mathbf{V} \cdot \nabla P\right) + \frac{1}{g_0}\, \dot{\mathbf{b}} \cdot \mathbf{r} = \frac{1}{\rho}\, \nabla \cdot (\kappa \nabla T) \qquad (6.39)$$

The static enthalpy of Eq. (6.39) can be eliminated with (2.118), and further combination with (6.29), (6.20), (2.119), and (2.116) yields the alternative forms,

$$T\left(\frac{\partial s}{\partial t} + \mathbf{V} \cdot \nabla s\right) = \frac{1}{\rho}\, \nabla \cdot (\kappa \nabla T) \qquad (6.40)$$

and

$$\frac{\partial T}{\partial t} + \mathbf{V} \cdot \nabla T = \frac{\beta T}{\rho c_p}\left(\frac{\partial P}{\partial t} + \mathbf{V} \cdot \nabla P\right) + \frac{1}{\rho c_p}\, \nabla \cdot (\kappa \nabla T) - \frac{1}{g_0 c_p}\, \dot{\mathbf{b}} \cdot \mathbf{r} \qquad (6.41)$$

If the viscous force components were employed, the heat conduction term $\nabla \cdot (\kappa \nabla T)$ in Eqs. (6.38)–(6.41) would be replaced according to

$$\nabla \cdot (\kappa \nabla T) \rightarrow \nabla \cdot (\kappa \nabla T) + \mathbf{\Gamma} \cdot \nabla \cdot \mathbf{V} \qquad (6.42)$$

where $\mathbf{\Gamma}$ is given by Eq. (6.30). The product $\mathbf{\Gamma} \cdot \nabla \cdot \mathbf{V}$ means that the three parenthetical coefficients of $\mathbf{n}_x$, $\mathbf{n}_y$, $\mathbf{n}_z$ in Eq. (6.30) are first dotted with the operator ∇, and then the resulting terms of $\mathbf{\Gamma} \cdot \nabla$ are dotted with $\mathbf{V}$. It is seen from Eq. (6.40) that (6.42) is a dissipative term which causes entropy production.

Equation (6.20) can be written in terms of pressure rather than density derivatives by a procedure similar to that employed to obtain Eq. (6.13) for one dimensional flow, which gives

Mass Conservation

$$\frac{\partial P}{\partial t} + \mathbf{V} \cdot \nabla P + \frac{\rho C^2}{g_0}\, \mathbf{\nabla} \cdot \mathbf{V}$$

$$= \frac{\beta C^2}{g_0 c_p}[\mathbf{\nabla} \cdot (\kappa \nabla T) + \mathbf{\Gamma} \cdot \mathbf{\nabla} \cdot V] \qquad (6.42a)$$

An intermediate form of the energy equation, obtained in the course of deriving Eqs. (6.40) and (6.41) with both heat conduction and viscous terms, is given by

$$\frac{g_0}{C^2}\left(\frac{\partial P}{\partial t} + \mathbf{V} \cdot \nabla P\right) - \frac{\partial \rho}{\partial t} - V \cdot \nabla \rho$$

$$= \frac{\beta}{c_p}[\mathbf{\nabla} \cdot (\kappa \nabla T) + \mathbf{\Gamma} \cdot \mathbf{\nabla} \cdot V] \qquad (6.42b)$$

6.5   Introduction to Normalization

The previous sections of this chapter have provided the equations which describe one- and multidimensional flow systems. When these equations are properly normalized, application to problems with known parameters and disturbances usually results in a simpler, reduced formulation. Sometimes analysts are tempted to assume that certain terms can be neglected from an equation, and an important effect may be lost from the solution. The normalization procedure described here is designed to provide a rational basis for simplifying the governing equations without neglecting important phenomena. The method is introduced next by application to a simple unsteady flow system.

Consider a straight, downward-pointing circular pipe of diameter D, submerged in a liquid pool as shown in Fig. 6.5. Both the submerged depth and the initial liquid column length are L. Pressure initially on both the inside and outside liquid surfaces is P_∞, and the discharge end pressure is hydrostatic, that is, $P_e = P_\infty + \rho g L / g_0$. The column surface pressure suddenly increases to a constant P_0, and it is necessary to determine the time required to expel the liquid column.

The equation of motion and initial conditions for this problem were given in Section 1.10 and are rewritten here for a step pressure with $P(t) = P_0$ as

DE: $$(L - y) \frac{d^2 y}{dt^2} + gy + \left(\frac{f_u P_w}{8A} \right)(L - y)\left(\frac{dy}{dt} \right)^2 = \frac{g_0}{\rho}(P_0 - P_\infty)$$

$$(6.43)$$

IC: $$t = 0, \quad y = 0, \quad \frac{dy}{dt} = 0 \qquad (6.44)$$

The procedure for normalizing and simplifying is designed for equations which contain derivatives no higher than first order. Therefore, the term dy/dt is replaced by the water column velocity V; that is,

$$\frac{dy}{dt} = V \qquad (6.45)$$

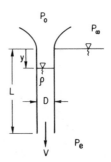

Figure 6.5   Expulsion of Liquid Column

If we write the wetted perimeter P_w and area A as πD and $\pi D^2/4$, respectively, the dynamic equation and initial conditions become

$$(L - y) \frac{dV}{dt} + gy + \left(\frac{f_u}{2D}\right)(L - y)V^2 = \frac{g_0}{\rho}(P_0 - P_\infty) \qquad (6.46)$$

$$t = 0, \quad y = 0, \quad V = 0 \qquad (6.47)$$

Equations (6.43) and (6.44) have been replaced by (6.45)–(6.47). The dependent variables are y and V, and the independent variable is t. It is necessary to estimate the maximum change of each variable before it is normalized.

The displacement y changes from 0 to L during expulsion, so the maximum change of y is

$$y_{\max} - 0 = \Delta y = L \qquad (6.48)$$

The maximum change of velocity V is not obvious. However, a simple estimate can be obtained by applying a force $(P_0 - P_\infty)A$ to an average liquid mass of half the initial length L (that is, $M = \rho AL/2$) and calculating its motion. Newton's law gives

$$P_0 - P_\infty = \frac{\rho L}{2g_0} \frac{dV}{dt}$$

If we write

$$\frac{dV}{dt} = \frac{dV}{dy} \frac{dy}{dt} = V \frac{dV}{dy}$$

a solution for V is

$$V = 2\sqrt{\frac{g_0}{\rho L}(P_0 - P_\infty)y}$$

Thus, as y goes from 0 to L, the change in velocity is

$$V_{\max} - 0 = \Delta V = 2\sqrt{\frac{g_0}{\rho}(P_0 - P_\infty)} \qquad (6.49)$$

If the liquid surface travels at an average velocity of $\Delta V/2$, the expulsion time, or system response time, Δt is

$$\Delta t = \frac{L}{\Delta V/2} = L\sqrt{\frac{\rho}{g_0(P_0 - P_\infty)}} \qquad (6.50)$$

Now we may write normalized variables as

$$y^* = \frac{y}{\Delta y} = \frac{y}{L} \qquad (6.51)$$

$$V^* = \frac{V}{\Delta V} = \frac{V}{2\sqrt{g_0(P_0 - P_\infty)/\rho}} \qquad (6.52)$$

and

$$t^* = \frac{t}{\Delta t} = \frac{t}{L}\sqrt{\frac{g_0(P_0 - P_\infty)}{\rho}} \qquad (6.53)$$

It follows that the approximate magnitudes of y^*, V^*, and t^* lie between 0 and 1.0. That is, the relative magnitude of each normalized variable is $O(1)$, which means *order of magnitude of* 1.0. Moreover, the derivatives dy^*/dt^* and dV^*/dt^* also are $O(1)$. This is easily seen by dividing the changes in y^* and V^* by the change in t^*. That is, since $\Delta V^* = 1$ and $\Delta t^* = 1$,

$$\frac{dV^*}{dt^*} \approx \frac{\Delta V^*}{\Delta t^*} = \frac{1}{1} = 1$$

If we substitute for y, V, and t from Eqs. (6.51)–(6.53), after rearrangement, the full problem becomes

$$\frac{dy^*}{dt^*} = 2V^* \qquad (6.54)$$

$$(1 - y^*)\frac{dV^*}{dt^*} + \left[\frac{\rho g L}{2g_0(P_0 - P_\infty)}\right]y^* + \left[\frac{f_u L}{D}\right](1 - y^*)V^{*2} = \frac{1}{2} \qquad (6.55)$$

$$t^* = 0, \quad y^* = 0, \quad V^* = 0 \qquad (6.56)$$

Since the normalized variables and first derivatives are $O(1)$, the magnitudes of the coefficients in brackets determine whether the y^* or $(1 - y^*)V^{*2}$ terms can be neglected.

TABLE 6.1   Parameters, Example Normalization

$\rho = 1000 \text{ kg/m}^3$
$L = 10 \text{ m}$
$D = 1.0 \text{ m}$
$P_0 - P_\infty = 1.0 \text{ MPa}$
$f_u = 0.01$ (approximate steady value)

Consider a case with the parameters given in Table 6.1. The unsteady friction factor f_u depends on development of a boundary layer when liquid motion starts. Therefore, the maximum value of f_u would not exceed the steady-state value. It follows that

$$\frac{\rho g L}{2 g_0 (P_0 - P_\infty)} = 0.05$$

and

$$\frac{f_u L}{D} = 0.1$$

Since both of the coefficients are small (that is, their sizes are one order of magnitude less than the other terms), the equations of motion can be simplified to

Nondimensional	Dimensional

$$\frac{dy^*}{dt^*} = 2V^* \qquad\qquad \frac{dy}{dt} = V$$

$$(1 - y^*)\frac{dV^*}{dt^*} = \frac{1}{2} \qquad (L - y)\frac{dV}{dt} = \frac{g_0}{\rho}(P_0 - P_\infty)$$

If we recombine the two first-order differential equations by eliminating V, we obtain

$$(L - y)\frac{d^2 y}{dt^2} = \frac{g_0}{\rho}(P_0 - P_\infty) \tag{6.57}$$

with initial conditions

$$t = 0, \quad y = 0, \quad \frac{dy}{dt} = 0 \tag{6.58}$$

Equation (6.57) is considerably reduced from the form of Eq. (6.43).

The systematic normalization procedure showed that in this case, the gravitational and frictional terms are of minor importance and can be neglected from the analysis for the parameters of Table 6.1. The solution to Eqs. (6.57) and (6.58) is given by Eq. (1.51). Success of the normalization procedure depends on selection of the appropriate response time and reference parameters. This selection only requires estimates which are within an order of magnitude of actual behavior, and usually can be obtained by inspection or from simple approximations.

6.6   The General Normalization Procedure

The describing equations with initial and boundary conditions provide a starting point for the solution of many problems in unsteady thermofluid mechanics. Most problems are formulated as ordinary differential equations or as several partial differential equations. The normalization procedure discussed in Section 6.5 is designed for differential equations of first order. Second- or higher-order differential equations should be reduced to a system of first-order differential equations before normalization so that both the variables and first derivatives are readily normalized to $O(1)$.

Time t is normalized with respect to the estimated response time Δt or other reference time such that

$$t^* = \frac{t}{\Delta t} \tag{6.59}$$

Since time increases from $t = 0$ to $t = \Delta t$ during the system response, it follows that t^* is $O(1)$.

Length variables are normalized with respect to the size of the region or displacement in the system being studied. If the analysis involves the expected displacement L_r of fluid particles or an object, then Cartesian coordinates x, y, z are normalized as

$$x^* = \frac{x}{L_r}, \quad y^* = \frac{y}{L_r}, \quad z^* = \frac{z}{L_r} \tag{6.60}$$

of which at least one is $O(1)$.

A general dependent variable such as pressure, velocity, or temperature is designated here by ϕ. If it appears in a derivative, it is normalized with respect to its estimated change

$$\Delta \phi = \phi_r - \phi_i \tag{6.61}$$

where ϕ_i is its initial value and ϕ_r is its expected maximum departure from ϕ_i. Then

$$\phi^* = \frac{\phi - \phi_i}{\phi_r - \phi_i} = \frac{\phi - \phi_i}{\Delta \phi} \tag{6.62}$$

It follows that the magnitude of ϕ^* and both its time and first spatial derivatives are $O(1)$. All other variables not appearing in derivatives are designated here by ψ and are normalized with respect to an appropriate reference value ψ_r, which could be anywhere between the initial and maximum values of ψ. Thus,

$$\psi^* = \frac{\psi}{\psi_r} \tag{6.63}$$

which also is $O(1)$.

When the normalized variables are substituted into the describing equations, nondimensional groups $\pi_1, \pi_2, \ldots, \pi_n$ appear, which are called *model coefficients*. The relative magnitude of each model coefficient shows which terms are important and which can be neglected from further analysis.

The describing equations for various systems discussed earlier in this chapter are normalized in the following sections.

6.7 Normalization of Unsteady Flow Systems

The differential equations for one-dimensional flows in Section 6.1 and multidimensional flows in Sections 6.2–6.4 contain pressure, velocity, and

TABLE 6.2 Normalized Variables, One- and Multidimensional Flows

$P^* = \dfrac{P - P_i}{\Delta P}$	$\mathbf{b}^* = \dfrac{\mathbf{b}}{b_r}$
$\mathbf{V}^* = \dfrac{\mathbf{V} - \mathbf{V}_i}{\Delta V}$	$c_p^* = \dfrac{c_p}{c_{pr}}$
$\rho^* = \dfrac{\rho - \rho_i}{\Delta \rho}$	$\beta^* = \dfrac{\beta}{\beta_r}$
$T^* = \dfrac{T - T_i}{\Delta T}$	$\kappa^* = \dfrac{\kappa}{\kappa_r}$
$t^* = \dfrac{t}{\Delta t}$	$\Gamma^* = \dfrac{\Gamma L_r}{\mu_r \Delta V}$
$x^* = \dfrac{x}{L_r}$	$q^* = \dfrac{q}{q_r}$
$y^* = \dfrac{y}{L_r}$	$D^* = \dfrac{D}{D_r}$
$z^* = \dfrac{z}{L_r}$	$A^* = \dfrac{A}{A_r}$
$C^* = \dfrac{C}{C_r}$	$\mu^* = \dfrac{\mu}{\mu_r}$

$$\nabla^* = L_r \nabla = \mathbf{n}_x \frac{\partial}{\partial x^*} + \mathbf{n}_y \frac{\partial}{\partial y^*} + \mathbf{n}_z \frac{\partial}{\partial z^*}$$

$$\nabla^{*2} = \nabla^* \cdot (\nabla^*) = L_r^2 \nabla^2$$

Table 6.3 Model Coefficients, One- and Multidimensional Flows

$$\pi_1 = \frac{g_0\,\Delta P}{\rho_r C_r^2} \qquad\qquad \pi_{10} = \frac{\beta_r\,\Delta P}{\rho_r c_{pr}}$$

$$\pi_2 = \frac{\Delta t\,\Delta V}{L_r} \qquad\qquad \pi_{11} = \frac{\kappa_r\,\Delta t\,\Delta T}{\rho_r T_r c_{pr} L_r^2}$$

$$\pi_2^0 = \frac{V_i\,\Delta t}{L_r} \qquad\qquad \pi_{12} = \frac{\nu_r\,\Delta t(\Delta V)^2}{g_0 L_r^2 c_{pr} T_r}$$

$$\pi_3 = \frac{\Delta\rho}{\rho_r} \qquad\qquad \pi_{13} = \frac{\beta_r q_r'\,\Delta t}{\rho_r A_r c_{pr}}$$

$$\pi_4 = \frac{\kappa_r \beta_r\,\Delta T\,\Delta t}{\rho_r L_r^2 c_{pr}} \qquad\qquad \pi_{14} = \frac{f_u L_r}{D_r}$$

$$\pi_5 = \frac{\nu_r \beta_r\,\Delta t(\Delta V)^2}{g_0 L_r^2 c_{pr}} \qquad\qquad \pi_{15} = \frac{\beta_r L_r^2}{2 g_0 c_{pr}(\Delta t)^2}$$

$$\pi_6 = \frac{g_0\,\Delta P(\Delta t)^2}{\rho_r L_r^2} \qquad\qquad \pi_{16} = \frac{g(\Delta t)^2}{L_r}\sin\theta$$

$$\pi_7 = \frac{b_r(\Delta t)^2}{L_r} \qquad\qquad \pi_{17} = \frac{q_r'\,\Delta t}{\rho_r T_r A_r c_{pr}}$$

$$\pi_8 = \frac{\nu_r\,\Delta V(\Delta t)^2}{L_r^3} \qquad\qquad \pi_{18} = \frac{L_r^2}{g_0 T_r c_{pr}(\Delta t)^2}$$

$$\pi_9 = \frac{\Delta T}{T_r}$$

density as dependent variables. Other properties can be expressed as functions of these dependent variables, controlled externally, or otherwise specified. The independent variables are time and the space coordinates. Normalized values of all the variables in one- and multidimensional flows are listed in Table 6.2.

The normalized variables of Table 6.2 were substituted into (6.13), (6.11), (6.14), and (6.16) respectively, for mass, momentum, and energy conservation in one-dimensional flows. The dimensional and nondimensional forms are written in Eqs. (6.64) to (6.68) to readily identify corresponding terms. The model coefficients $\pi_1, \pi_2, \ldots, \pi_n$ are listed in Table 6.3.

ONE-DIMENSIONAL FLOW EQUATIONS

MASS

Nondimensional

$$\pi_1 \frac{1}{C^{*2}} \frac{\partial P^*}{\partial t^*}$$

$$+ \pi_1 \frac{1}{C^{*2}} (\pi_2 V^* + \pi_2^0) \frac{\partial P^*}{\partial z^*}$$

$$+ \pi_2(\pi_3 \rho^* + 1) \frac{\partial V^*}{\partial z^*}$$

$$+ (\pi_3 \rho^* + 1) \frac{1}{A^*} \frac{\partial A^*}{\partial t^*}$$

$$+ (\pi_3 \rho^* + 1) \frac{1}{A^*} (\pi_2 V^*$$

$$+ \pi_2^0) \frac{\partial A^*}{\partial z^*}$$

$$= \frac{\beta^*}{c_p^*} \mathscr{F}^* \qquad (6.64\text{a})$$

Dimensional

$$\frac{1}{C^2} \frac{\partial P}{\partial t}$$

$$+ \frac{1}{C^2} V \frac{\partial P}{\partial z}$$

$$+ \frac{\rho}{g_0} \frac{\partial V}{\partial z}$$

$$+ \frac{\rho}{g_0} \frac{1}{A} \frac{\partial A}{\partial t}$$

$$+ \frac{\rho}{g_0} \frac{1}{A} V \frac{\partial A}{\partial z}$$

$$= \frac{\beta}{g_0 c_p} \mathscr{F} \qquad (6.64\text{b})$$

MOMENTUM

Nondimensional

$$\pi_2(\pi_3 \rho^* + 1) \frac{\partial V^*}{\partial t^*}$$

$$+ \pi_2(\pi_3 \rho^* + 1)$$

$$\times (\pi_2 V^* + \pi_2^0) \frac{\partial V^*}{\partial z^*}$$

$$+ \pi_6 \frac{\partial P^*}{\partial z^*}$$

$$+ \frac{1}{2} \pi_{14}(\pi_3 \rho^* + 1)$$

$$\times (\pi_2 V^* + \pi_2^0)^2$$

$$+ \pi_{16}(\pi_3 \rho^* + 1) = 0$$

$$(6.65\text{a})$$

Dimensional

$$\frac{\partial V}{\partial t}$$

$$+ V \frac{\partial V}{\partial z}$$

$$+ \frac{g_0}{\rho} \frac{\partial P}{\partial z}$$

$$+ \frac{f}{4} \frac{P_w}{A} \frac{|V|}{2} V$$

$$+ g \sin \theta = 0 \qquad (6.65\text{b})$$

ENERGY

Nondimensional	*Dimensional*

$$\pi_1 \frac{1}{C^{*2}} \frac{\partial P^*}{\partial t^*}$$

$$+ \pi_1 \frac{1}{C^{*2}} (\pi_2 V^* + \pi_2^0) \frac{\partial P^*}{\partial z^*}$$

$$- \pi_3 \frac{\partial \rho^*}{\partial t^*}$$

$$- \pi_3 (\pi_2 V^* + \pi_2^0) \frac{\partial \rho^*}{\partial z^*}$$

$$= \frac{\beta^*}{c_p^*} \mathcal{F}^* \qquad (6.66a)$$

$$\frac{g_0}{C^2} \frac{\partial P}{\partial t}$$

$$+ \frac{g_0}{C^2} V \frac{\partial P}{\partial z}$$

$$- \frac{\partial \rho}{\partial t}$$

$$- V \frac{\partial \rho}{\partial z}$$

$$= \frac{\beta}{c_p} \mathcal{F} \qquad (6.66b)$$

$$\pi_9 (\pi_3 \rho^* + 1) c_p^* \frac{\partial T^*}{\partial t^*}$$

$$+ \pi_9 (\pi_3 \rho^* + 1) c_p^* (\pi_2 V^*$$

$$+ \pi_2^0) \frac{\partial T^*}{\partial z^*}$$

$$= \pi_{10} \beta^* (\pi_9 T^* + 1) \frac{\partial P^*}{\partial t^*}$$

$$+ \pi_{10} \beta^* (\pi_9 T^* + 1)$$

$$\times (\pi_2 V^* + \pi_2^0) \frac{\partial P^*}{\partial z^*}$$

$$+ \mathcal{F}^{**} \qquad (6.67a)$$

$$\rho c_p \frac{\partial T}{\partial t}$$

$$+ \rho c_p V \frac{\partial T}{\partial z}$$

$$= \beta T \frac{\partial P}{\partial t}$$

$$+ \beta T V \frac{\partial P}{\partial z}$$

$$+ \mathcal{F} \qquad (6.67b)$$

$$\left\{ \begin{matrix} \mathcal{F}^* \\ \mathcal{F}^{**} \end{matrix} \right\} = \left\{ \begin{matrix} \pi_4 \\ \pi_{11} \end{matrix} \right\} \frac{1}{A^*}$$

$$\times \frac{\partial}{\partial z^*} \left(\kappa^* A^* \frac{\partial T^*}{\partial z^*} \right)$$

$$+ \left\{ \begin{matrix} \pi_{13} \\ \pi_{17} \end{matrix} \right\} \frac{q'^*_{in} - q'^*_{out}}{A^*}$$

$$+ \left\{ \begin{matrix} \pi_{14} \pi_{15} \\ \pi_{14} \pi_{18} \end{matrix} \right\}$$

$$\times \frac{(\pi_3 \rho^* + 1)(\pi_2 V^* + \pi_2^0)^3}{2D^*}$$

$$(6.68a)$$

$$\mathcal{F} = \frac{1}{A} \frac{\partial}{\partial z} \left(\kappa A \frac{\partial T}{\partial z} \right)$$

$$+ \frac{q'_{in} - q'_{out}}{A}$$

$$+ \frac{f}{4} \frac{P_w}{A} \frac{\rho V^3}{2g_0} \qquad (6.68b)$$

The normalized variables of Table 6.2 also were employed in Eqs. (6.42a), (6.29), (6.42b), and (6.41) respectively, with the dissipative term of (6.42) to obtain the nondimensional equations for multidimensional flows and constant body force acceleration **b**. Both dimensional and nondimensional forms are written in Eqs. (6.69) to (6.72), where the model coefficients are listed in Table 6.3.

When a system is to be analyzed, the numerical values of the model coefficients are calculated in order to determine which terms dominate and which terms can be neglected from the equations. However, success of this process depends on system space interval and response time estimates which should be accurate to within an order of magnitude. This is not a serious restriction of the process, as will be evident from examples given later in this chapter.

MULTIDIMENSIONAL FLOW EQUATIONS

MASS

Nondimensional	*Dimensional*

$$\pi_1 \frac{1}{C^{*2}} \frac{\partial P^*}{\partial t^*}$$

$$+ \pi_1 \frac{1}{C^{*2}} (\pi_2 \mathbf{V}^* + \boldsymbol{\pi}_2^0) \cdot \nabla^* P^*$$

$$+ \pi_2(\pi_3 \rho^* + 1)\nabla^* \cdot \mathbf{V}^*$$

$$= \frac{\beta^*}{c_p^*} [\pi_4 \nabla^* \cdot (\kappa^* \nabla^* T^*)]$$

$$+ \frac{\beta^*}{c_p^*} (\pi_5 \Gamma^* \cdot \nabla^* \cdot \mathbf{V}^*) \quad (6.69a)$$

$$\frac{1}{C^2} \frac{\partial P}{\partial t}$$

$$+ \frac{1}{C^2} \mathbf{V} \cdot \nabla P$$

$$+ \frac{\rho}{g_0} \nabla \cdot \mathbf{V}$$

$$= \frac{\beta}{g_0 c_p} \nabla \cdot (\kappa \nabla T)$$

$$+ \frac{\beta}{g_0 c_p} \Gamma \cdot \nabla \cdot \mathbf{V} \quad (6.69b)$$

MOMENTUM

Nondimensional	*Dimensional*

$$\pi_2(\pi_3 \rho^* + 1) \frac{\partial \mathbf{V}^*}{\partial t^*}$$

$$+ \pi_2(\pi_3 \rho^* + 1)(\pi_2 \mathbf{V}^* + \boldsymbol{\pi}_2^0)$$

$$\cdot \nabla^* \mathbf{V}^*$$

$$+ \pi_6 \nabla^* P^* + \pi_7(\pi_3 \rho^* + 1)\mathbf{b}^*$$

$$= \pi_8 \nabla^* \cdot \Gamma^* \quad (6.70a)$$

$$\frac{\partial \mathbf{V}}{\partial t}$$

$$+ \mathbf{V} \cdot \nabla \mathbf{V} + \frac{g_0}{\rho} \nabla P + \mathbf{b}$$

$$= \frac{g_0}{\rho} \nabla \cdot \Gamma \quad (6.70b)$$

ENERGY

<table>
<tr><td align="center">Nondimensional</td><td align="center">Dimensional</td></tr>
</table>

$$\pi_1 \frac{1}{C^{*2}} \frac{\partial P^*}{\partial t^*} \qquad\qquad \frac{g_0}{C^2} \frac{\partial P}{\partial t}$$

$$+ \pi_1 \frac{1}{C^{*2}} (\pi_2 \mathbf{V}^* + \boldsymbol{\pi}_2^0) \cdot \nabla^* P^* \qquad\qquad + \frac{g_0}{C^2} \mathbf{V} \cdot \nabla P$$

$$- \pi_3 \frac{\partial \rho^*}{\partial t^*} - \pi_3 (\pi_2 \mathbf{V}^* + \boldsymbol{\pi}_2^0) \cdot \nabla^* \rho^* \qquad\qquad - \frac{\partial \rho}{\partial t} - \mathbf{V} \cdot \nabla \rho$$

$$= \frac{\beta^*}{c_p^*} [\pi_4 \nabla^* \cdot (\kappa^* \nabla^* T^*)] \qquad\qquad = \frac{\beta}{c_p} \nabla \cdot (\kappa \nabla T)$$

$$+ \frac{\beta^*}{c_p^*} (\pi_5 \Gamma^* \cdot \nabla^* \cdot \mathbf{V}^*) \quad (6.71a) \qquad\qquad + \frac{\beta}{c_p} \Gamma \cdot \nabla \cdot \mathbf{V} \qquad\qquad (6.71b)$$

$$\pi_9 (\pi_3 \rho^* + 1) c_p^* \frac{\partial T^*}{\partial t^*} \qquad\qquad \rho c_p \frac{\partial T}{\partial t}$$

$$+ \pi_9 (\pi_3 \rho^* + 1) \qquad\qquad + \rho c_p \mathbf{V} \cdot \nabla T$$

$$\times c_p^* (\pi_2 \mathbf{V}^* + \boldsymbol{\pi}_2^0) \cdot \nabla^* T^* \qquad\qquad = \beta T \frac{\partial P}{\partial t}$$

$$= \pi_{10} \beta^* (\pi_9 T^* + 1) \frac{\partial P^*}{\partial t^*} \qquad\qquad + \beta T \mathbf{V} \cdot \nabla P$$

$$+ \pi_{10} \beta^* (\pi_9 T^* + 1) \qquad\qquad + \nabla \cdot (\kappa \nabla T)$$

$$\times (\pi_2 \mathbf{V}^* + \boldsymbol{\pi}_2^0) \cdot \nabla^* P^* \qquad\qquad + \Gamma \cdot \nabla \cdot \mathbf{V} \qquad\qquad (6.72b)$$

$$+ \pi_{11} \nabla^* \cdot (\kappa^* \nabla^* T^*)$$

$$+ \pi_{12} \Gamma^* \cdot \nabla^* \cdot \mathbf{V}^* \qquad (6.72a)$$

6.8 Selection of Space Intervals and Response Time

The choice of a space interval in one- and multidimensional flows, character-ized by a reference length L_r, is determined by the phenomenon to be analyzed. We could choose, for example, the pipe length in a one-dimen-sional flow or the average distance through the region which is to be studied in a multidimensional flow.

The response time should correspond to how fast a system state responds to a given disturbance at the system boundary. System response may be dominated by either *bulk flow* or *propagative flow*, as discussed in Chapter 1. Let the propagation speed be V_p, which designates the speed of sound, moving shocks, or water waves, or the diffusive speed of viscous shear or heat. Recall from Eq. (1.1) that if L_r is the space dimension through which a disturbance propagates, the *propagation time* is defined by

$$t_p = \frac{L_r}{V_p} \tag{6.73}$$

The *disturbance time* t_d corresponds to the time interval over which a disturbance causes transient property changes to occur at the system boundary. Suppose that fluid flow in a rigid pipe of length $L_r = L$ is to be terminated by a valve with closure time t_v, and the pipe pressure during closure is to be determined. The disturbance time is $t_d = t_v$, and the propagation time t_p is approximately L/C, where C is the acoustic speed in the fluid. If the valve closure time is long compared with the propagation time (that is, $t_v \gg t_p$), the effect of individual pressure waves would be small perturbations on the increasing pressure. Hence, if the overall pressure rise was required, bulk flow would be appropriate and response time Δt would correspond to valve closure time t_v. However, if it was desirable to study pressure wave propagation in the pipe, or if t_v was within an order of magnitude of t_p, that is, if $t_v = O(t_p)$, propagation effects would be important, and t_p would be an appropriate response time. The criterion for response time is summarized by

$$\Delta t = \begin{cases} t_d, & t_d \gg \dfrac{L}{V_p} \\[2ex] \dfrac{L}{V_p}, & t_d = O\!\left(\dfrac{L}{V_p}\right) \end{cases} \tag{6.74}$$

Occasionally the appropriate response time is not easily determined. Whenever two or more possible choices exist, a simple rule is that the *shorter* response time should be chosen. This choice will ensure that both the fine detail and coarse structure of the phenomenon to be studied will be preserved.

EXAMPLE 6.1: REFERENCE LENGTH AND RESPONSE TIME FOR BUBBLE STUDY

A study is required to determine the shape and motion characteristics of stable gas bubbles rising through 25 cm (9.8 in) of room temperature water. Suppose that no information is available on the stable bubble size or rise velocity. Determine an approximate reference length L_r and response time Δt.

Since individual bubble behavior is to be studied, it is reasonable to choose the bubble diameter as a reference length, which is not known. Crude estimates of bubble size and rise velocity can be obtained by considering the idealized spherical bubble of radius R in Fig. 6.6, rising at velocity V_∞ in stationary liquid of density ρ. If the bubble was a solid sphere rising at velocity V_∞, the dynamic pressure at the top would be $\rho V_\infty^2/2g_0$. If this pressure was exerted throughout the bubble gas, it would be the approximate pressure difference across the gas-liquid interface caused by

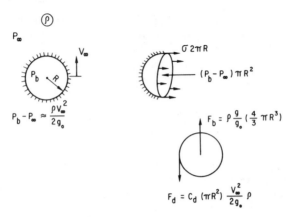

Figure 6.6 Approximate Bubble Size and Rise Velocity

surface tension. A force balance for a hemispherical section of the bubble requires that the exerted pressure force $(\rho V_\infty^2/2g_0)\pi R^2$ be equal to the surface tension force $2\pi R\sigma$ in order to prevent a stable bubble from being pulled apart. This leads to an approximate bubble radius

$$R \approx \frac{4\sigma g_0}{\rho V_\infty^2}$$

However, V_∞ is not yet known. If we equate the buoyant force $(\rho g/g_0)(4\pi R^3/3)$ to the drag force $\rho C_D \pi R^2 V_\infty^2/2g_0$ on a rising sphere we obtain the approximate rise velocity

$$V_\infty \approx \sqrt{\frac{8}{3}\frac{gR}{C_d}}$$

If the last two equations are solved for R and V_∞ with a drag coefficient of about 1.0, we find that

$$R \approx 1.22\sqrt{\frac{g_0\sigma}{g\rho}} \quad \text{and} \quad V = 1.8\left(\frac{gg_0\sigma}{\rho}\right)^{1/4}$$

It is emphasized that the values of R and V_∞ from approximate considerations need only be accurate to within an order of magnitude to serve a useful purpose in later evaluations of the model coefficients. A rigorous analysis by Peebles and Garber [6.15] yields

$$R = 0.89\sqrt{\frac{g_0\sigma}{g\rho}} \quad \text{and} \quad V_\infty = 1.53\left(\frac{gg_0\sigma}{\rho}\right)^{1/4}$$

for bubbles near the stable size limit, which are close to the crude estimates just presented.

If we employ $\sigma = 70$ dynes/cm (0.005 lbf/ft) and $\rho = 1.0$ g/cm^3 (62.4 lbm/ft^3), we obtain $R \approx 0.33$ cm (0.13 in) and $V_\infty \approx 25$ cm/s (9.8 in/s) from the crude estimates. It follows that the average bubble diameter or reference length is about $D = 2R \approx 0.67$ cm (0.26 in).

An estimate of the response time is still needed. One possibility is the time required for a bubble to rise through the entire 25 cm (9.8 in) of water depth at velocity V_∞. However, in order to study the unsteady bubble shape and motion as it rises, a shorter response time would be better—but how short? A response time equal to the time required for a bubble to rise a length equal to its own diameter would be more consistent with the behavior to be studied. Therefore,

$$\Delta t \approx \frac{D}{V_\infty} = \frac{2R}{V_\infty} = \left(\frac{g_0 \sigma}{\rho g^3}\right)^{1/4} \approx 0.015 \text{ s}$$

This approximate analysis employs simple, basic ideas to estimate a reference length and response time. Also, V_∞ would be a good reference velocity, if required.

6.9 System Disturbances

When the space interval and response time are selected for a given phenomenon or system, appropriate values for disturbed system properties must be determined. An originating disturbance $\Delta\phi$ represents the known or specified disturbance value of a property. Usually when a disturbance is imposed, other property changes are uniquely determined. Consider the sudden stoppage of liquid whose initial velocity is $V_i = V$. The velocity is changed to zero, $V_r = 0$, or disturbed an amount

$$\Delta V = V_r - V_i = 0 - V = -V \qquad (\text{but use } V)$$

Even though the change of a disturbed property can be negative, the absolute value is used in this procedure. Sudden stoppage implies that the disturbance propagates at approximately sonic speed. The corresponding pressure disturbance ΔP is estimated from waterhammer theory. However, if the liquid flow was stopped slowly without significant propagation effects, ΔP would be the dynamic head.

Table 6.3 contains model coefficients which require known or estimated values of all property disturbances associated with the originating disturbance. It is important that all disturbed property values $\Delta\phi$ be consistent with the known or prescribed originating disturbance. Precision is not required in determining the $\Delta\phi$ values, but estimates should be accurate to within an order of magnitude. Usually, approximate analyses can be employed to obtain simple algebraic expressions for the $\Delta\phi$. The following paragraphs include discussions of several originating disturbances and estimates of associated property disturbances.

Bulk Flows Bulk fluid flows are characterized by negligible propagation effects. If fluid velocity is disturbed from V_i to V_r, giving a disturbance $\Delta V = V_r - V_i$, the corresponding pressure disturbance ΔP can be approximated from the differential change in dynamic pressure $\Delta P = \Delta(\rho V^2/2g_0)$ by writing

$$\Delta P \approx \frac{\rho V \, \Delta V}{g_0} \tag{6.75}$$

Whether a velocity change has disturbed pressure, or vice versa, Eq. (6.75) relates ΔP and ΔV for bulk fluid flows. When a disturbance pressure ΔP is known, corresponding density and temperature disturbances, $\Delta \rho$ and ΔT, can be approximated from the thermodynamic property differentials of Eqs. (2.116)–(2.119) by employing the dT and ds formulas with $T \, ds = dQ$ as

$$\Delta \rho \approx \frac{g_0}{C_r^2} \Delta P - \frac{\beta_r \rho_r}{c_{pr}} \left\{ \begin{matrix} T_0 \, \Delta s \\ \Delta Q \end{matrix} \right\} \tag{6.76}$$

and

$$\Delta T \approx \left(\frac{g_0}{\rho_r \beta_r C_r^2} + \frac{\beta_r T_0}{\rho_r c_{pr}} \right) \Delta P - \frac{1}{\beta_r \rho_r} \Delta \rho$$

$$= \left(\frac{\beta_r T_0}{\rho_r c_{pr}} \right) \Delta P + \frac{1}{c_{pr}} \left\{ \begin{matrix} T_0 \, \Delta s \\ \Delta Q \end{matrix} \right\} \tag{6.77}$$

Propagation Flows, Acoustic Rapid pressure or velocity disturbances which generate acoustic waves are related approximately by the water-hammer equation

$$\Delta P = \frac{\rho_r C_r}{g_0} \Delta V \tag{6.78}$$

for which the density disturbance is estimated from Eq. (6.76) at constant entropy, giving

$$\Delta \rho = \frac{g_0}{C_r^2} \Delta P \tag{6.79}$$

Also, the temperature disturbance can be estimated from Eqs. (6.77) and (6.79) as

$$\Delta T = \frac{\beta_r T_0}{\rho_r c_{pr}} \Delta P \tag{6.80}$$

Propagation Flows, Moving Shocks Mass and momentum equations across a moving normal shock are discussed in Chapter 8, through which the undisturbed properties P_i, V_i, and ρ_i change to P_r, V_r, and ρ_r. These can be written as

$$\rho_r(S - V_r) = \rho_i S \tag{6.81}$$

and

$$\rho_i S V_r = g_0(P_r - P_i) \tag{6.82}$$

where S is the shock velocity. Equations (6.81) and (6.82) are for an arbitrary fluid. Proper determination of S requires energy conservation also, but for estimates $S \approx C$ is adequate. Therefore, pressure, density, and temperature disturbances are estimated from Eqs. (6.78)–(6.80).

EXAMPLE 6.2: PROPERTY DISTURBANCES FROM ABRUPT PISTON MOTION
A piston suddenly begins moving at 500 m/s in a long cylinder containing standard air at 101 kPa, 20°C, 1.2 kg/m^3 density, 343 m/s sound speed, and a constant pressure specific heat of 1004 J/kg-K. Estimate the pressure, temperature, and density disturbances in the air adjacent to the piston.
    The piston velocity has been disturbed by an amount

$$\Delta V = 500 \text{ m/s}$$

from an initially stationary condition. Since the pipe is long and the piston moves suddenly, propagation effects should dominate the property disturbances. Therefore, Eqs. (6.78)–(6.80) yield

$$\Delta P = \frac{\rho_r C_r}{g_0} \Delta V = 206 \text{ kPa}$$

$$\Delta \rho = \frac{g_0}{C_r^2} \Delta P = 1.75 \text{ kg/m}^3$$

Since $\beta = 1/T$ for a perfect gas is

$$\Delta T = \frac{\beta_r T_0}{\rho_r c_{pr}} \Delta P = 171 \text{ K}$$

If a sudden valve opening should cause fluid discharge into a pipe of flow area A, it is convenient to relate the disturbed state properties to the valve charging rate $\dot{m}_r$. Equation (6.81) can be written as

$$\Delta \rho = \rho_r - \rho_i = \frac{\rho_r V_r}{S} = \frac{\dot{m}_r/A}{S} \approx \frac{\dot{m}_r}{A C_r} \tag{6.83}$$

Moreover, Eqs. (6.78) and (6.80) with

$$\Delta V = V_r \tag{6.84}$$

yield

$$\Delta P \approx \frac{\dot{m}_r C_r}{g_0 A} \tag{6.85}$$

and

$$\Delta T \approx \frac{\beta_r T_0}{\rho_r g_0} \frac{\dot{m}_r}{A} \frac{C_r}{c_{pr}} \tag{6.86}$$

Propagation Flows, Progressive Surface Waves: Periodic surface waves, discussed in Chapter 9, have elevations of the form

$$\eta = \eta_0 \sin \frac{x - Ct}{\lambda} \tag{6.87}$$

and propagate at velocity

$$C = \sqrt{g\lambda \tanh \frac{H}{\lambda}} \quad \text{or} \quad C \approx \sqrt{g\lambda} \quad \text{for } H \to \infty \tag{6.88}$$

where η_0 is the amplitude above the undisturbed surface elevation, λ is the wavelength, and H is the undisturbed liquid depth. The maximum velocity disturbance of a surface liquid particle is obtained by differentiating Eq. (6.87), which leads to

$$\Delta V \approx \eta_0 \sqrt{\frac{g}{\lambda}} \tag{6.89}$$

The maximum pressure disturbance below a wave of this amplitude is approximated by the hydrostatic value

$$\Delta P = \rho \frac{g}{g_0} \eta_0 \tag{6.90}$$

Numerous other phenomena and associated property disturbances exist in thermofluid systems, which are not included here. However, the cases examined in this section show how a known property disturbance can be employed with basic equations of various phenomena to estimate consistent disturbances of other properties.

6.10   Simplification of Equations

Equations for one- and multidimensional flows were normalized in Section 6.7, which resulted in nondimensional model coefficients, listed in Table 6.3.

Sections 6.8 and 6.9 give examples of how terms in the model coefficients can be evaluated. These coefficients are calculated from fluid and system

TABLE 6.4   Procedure for Simplification of Equations

1. Determine the boundaries of the system to be analyzed.
2. Characterize the phenomena to be studied.
3. Estimate the system space interval L_r and response time Δt.
4. Estimate the property changes $\Delta \phi$ from a known disturbance.
5. Calculate all the appropriate model coefficients, $\pi_1, \pi_2, \ldots, \pi_n$, based on known or estimated parameters.
6. Neglect all terms with relatively small π_n values, because they are associated with negligible physical effects.
7. Write the equations without the negligible terms. These are the simplified equations.

parameters, estimated response times, and property disturbances. The relative numerical size of the model coefficients displays terms which are significant in the equations and terms which can be neglected from analysis. A procedure for simplifying the equations is given in Table 6.4.

EXAMPLE 6.3: TRANSIENT PIPE FLOW PROPERTIES

A uniform pipe of length $L = 100$ m (328 ft) and diameter $D = 0.1$ m (0.33 ft) is attached to a large tank of compressed air at pressure $P_0 = 10.1$ MPa (1470 psia), temperature $T_0 = 20°C = 293°K$ (527°R) and density $\rho_0 = 120$ kg/m^3 (7.48 lbm/ft^3). The pipe, which contains straight segments between elbows, is normally closed. However, a valve is installed at the discharge end for occasional pressure reduction in the tank. It is necessary to predict unsteady flow properties in the pipe when discharge occurs in order to determine time-dependent forces on each segment. This will make it possible to specify adequate pipe-supporting structures. Determine the appropriate transient flow equations for valve opening times of $\tau_1 = 0.1$ s and $\tau_2 = 3.0$ s. Ambient pressure is $P_\infty = 101$ kPa (14.7 psia).

The steps of Table 6.4 are summarized next for this example.

1. The system boundaries include the pipe, its attachment to the vessel with its entrance properties, and the valve.

2. The phenomenon to be studied involves one-dimensional flow.

3. The space interval of interest is the pipe length $L_r = L = 100$ m (328 ft). Propagation time in the pipe is $t_p = L/C$. Sound speed for the given air condition is

$$C = \sqrt{kg_0 R T_0} = 343 \text{ m/s } (1125 \text{ ft/s})$$

Therefore,

$$t_p = 0.29 \text{ s}$$

Hence, for a 0.1-s valve, the flow is dominated by propagation effects, whereas for a 3.0-s valve bulk flow dominates. Therefore, the response times for both cases are

$$\Delta t_p = t_p = 0.29\,\text{s} = O(\tau_1)$$

and

$$\Delta t_b = \tau_2 = 3\,\text{s} \gg t_p$$

4. The imposed disturbance is approximately the reduction of discharge pressure from P_0 to the discharge value. Equation (2.59) yields the discharge pressure as the critical value

$$P_r = P_{\text{discharge}} \approx P_c = \left(\frac{2}{k+1}\right)^{k/(k-1)} P_0$$

$$= 5252\,\text{kPa} \ (764\,\text{psia}) > P_\infty = 101\,\text{kPa} \ (14.7\,\text{psia})$$

The corresponding density is, based on isentropic state changes,

$$\rho_r = \rho_{\text{discharge}} = \rho_c = \left(\frac{P_{\text{discharge}}}{P_0}\right)^{1/k} \rho_0$$

$$= 76\,\text{kg/m}^3 \ (4.74\,\text{lbm/ft}^3)$$

Therefore, with $P_i = P_0$ $\rho_i = \rho_0$, $P_r = P_c$, and $\rho_r = \rho_c$, Eq. (6.61) yields

$$\Delta P = P_0 - P_c = 10.1 - 5.25 = 4.9\,\text{MPa} \ (713\,\text{psia})$$

$$\Delta\rho = \rho_0 - \rho_c = 120 - 76 = 44\,\text{kg/m}^3 \ (2.74\,\text{lbm/ft}^3)$$

The temperature disturbance is obtained from Eq. (2.116) as

$$\Delta T = \left(\frac{g_0}{\rho\beta C^2} + \frac{\beta T}{\rho c_p}\right)_0 \Delta P - \frac{1}{\beta_0 \rho_0} \Delta\rho$$

$$= T_0\left(\frac{\Delta P}{P_0} - \frac{\Delta\rho}{\rho_0}\right) = 35\,\text{K} \ (63\,\text{F}°)$$

Moreover,

$$\Delta V = C_0 \approx 343\,\text{m/s} \ (1125\,\text{ft/s})$$

Other parameters for the compressed air are

$$\kappa_r = 0.026\,\text{W/m} - \text{K} \ (0.015\,\text{B/h-ft-°F}) \qquad b_r = g = 9.8\,\text{m/s}^2 \ (32.2\,\text{ft/s}^2)$$

$$c_{pr} = 1.0\,\text{J/g-K} \ (0.2396\,\text{B/lbm-°F}) \qquad f_u = 0.01$$

$$\nu_r = 0.19\,\text{cm}^2/\text{s} \ (0.029\,\text{in}^2/\text{s}) \qquad \theta = \pi/2$$

5. The following model coefficients of Table 6.3 were obtained for the two valve opening times:

π_n	0.1 s	3.0 s	π_n	0.1 s	3.0 s
π_1	0.34	0.34	π_9	0.11	0.11
π_2	1.0	10.3	π_{10}	0.14	0.14
π_2^0	0.0	0.0	π_{11}	7×10^{-13}	7×10^{-13}
π_3	0.37	0.37	π_{12}	2.2×10^{-10}	2.2×10^{-9}
π_4	4.2×10^{-12}	4.2×10^{-11}	π_{13}	0.0	0.0
π_5	2.2×10^{-10}	2.2×10^{-9}	π_{14}	10.0	10.0
π_6	0.34	34.0	π_{15}	0.2	0.002
π_7	0.008	0.8	π_{16}	0.008	0.8
π_8	5.6×10^{-10}	5.6×10^{-8}	π_{17}	0.0	0.0
			π_{18}	0.4	0.004

6. The dominant model coefficients for a valve opening time of 0.1 s are π_1, π_2, π_3, π_6, π_9, π_{10}, π_{14}, π_{15}, and π_{18}. The dominant coefficients for a valve opening time of 3.0 s are π_2, π_6, and π_{14}.

7. Equations (6.64)–(6.68) and the dominant model coefficients just obtained yield the following equations:

VALVE $\tau = 0.1$ s

$$\frac{\partial P}{\partial t} + V \frac{\partial V}{\partial x} + \frac{\rho C^2}{g_0} \frac{\partial V}{\partial x} = \frac{\beta C^2}{g_0 c_p} \frac{f}{4} \frac{P_w}{A} \frac{\rho V^3}{2 g_0}$$

$$\rho \left(\frac{\partial V}{\partial t} + V \frac{\partial V}{\partial x} \right) + g_0 \frac{\partial P}{\partial x} + \frac{f}{4} \frac{P_w}{A} \frac{V^2}{2} \rho = 0$$

$$\frac{g_0}{C^2} \left(\frac{\partial P}{\partial t} + V \frac{\partial P}{\partial x} \right) - \left(\frac{\partial \rho}{\partial t} + V \frac{\partial \rho}{\partial x} \right) = \frac{\beta}{c_p} \frac{f}{4} \frac{P_w}{A} \frac{\rho V^3}{2 g_0}$$

VALVE $\tau = 3.0$ s

$$\frac{\partial V}{\partial x} = 0$$

$$\frac{\partial V}{\partial t} + V \frac{\partial V}{\partial x} + \frac{g_0}{\rho} \frac{\partial P}{\partial x} + \frac{f}{4} \frac{P_w}{A} \frac{V^2}{2} = 0$$

We see for a valve opening time of 0.1 s that the unsteady flow equations are required. However, for a valve opening time of 3.0 s, $\partial V / \partial x \approx 0$ and flow in the pipe can be approximated by a solution to

$$\frac{\partial V}{\partial t} + \frac{g_0}{\rho} \frac{\partial P}{\partial x} + \frac{f}{4} \frac{P_w}{A} \frac{V^2}{2} = 0$$

This differential equation could first be integrated with respect to x for the pipe length L, then integrated with respect to time in order to obtain a solution.

EXAMPLE 6.4: SUBMERGED, OSCILLATING CYLINDER

A long cylinder of radius R is submerged in water. It begins horizontal oscillation of amplitude $x_0 = 0.1$ m (0.33 ft) at a frequency of $f_1 = 10$ Hz in one case and $f_2 = 1000$ Hz in another case, with a displacement

$$x = x_0 \sin 2\pi ft$$

The water boundary is an average distance $r = 3$ m (9.8 ft) from the cylinder, as shown in Fig. 6.7. Determine the appropriate equations to be used in determining the liquid motion and pressure field for both frequencies. Sound speed in water is about 1500 m/s (4920 ft/s), its density is 1000 kg/m^3 (62.4 lbm/ft^3), $\beta = 1.8 \times 10^{-4}$/K (0.00032/F), $\kappa = 0.7$ W/m-K (0.4 B/h-ft-°F), $c_p = 4.17$ J/g-K (1.0 B/lbm-°F) and $T = 293$ K (527°R). The following steps are in accordance with the procedure of Table 6.4:

1. The system boundaries are the cylinder and containing vessel walls and the water free surface.

2. The liquid motion is multidimensional.

3. The space interval of interest is the characteristic liquid dimension r, such that

$$L_r = r = 3 \text{ m}$$

The fluid response time is determined by comparing the propagation time

$$t_p = \frac{r}{C} = \frac{3 \text{ m}}{1500 \text{ m/s}} = 0.002 \text{ s} = 2 \text{ ms}$$

with the disturbance times, or periods of oscillation,

$$t_\infty = \begin{cases} \dfrac{1}{f_1} = \dfrac{1}{10} \text{ s} = 100 \text{ ms} \\[2mm] \dfrac{1}{f_2} = \dfrac{1}{1000} \text{ s} = 1 \text{ ms} \end{cases}$$

Since $t_\infty \gg t_p$ for the 10-Hz oscillation, this problem is one of bulk flow. However, for the 1000-Hz oscillation, t_∞ and t_p are comparable, which indicates that propagation effects dominate the water response. Thus,

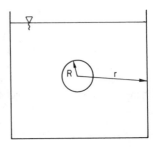

Figure 6.7 Submerged Oscillating Cylinder

$$\Delta t = \begin{cases} t_{\infty} = \dfrac{1}{f_1} = 100 \text{ ms} , & f_1 = 10 \text{ Hz} \\ \dfrac{r}{C} = 2 \text{ ms} , & f_2 = 1000 \text{ Hz} \end{cases}$$

4. It is assumed that liquid remains in contact with the cylinder so that fluid velocity is disturbed according to

$$\frac{dx}{dt} = \Delta V \cos 2\pi f t$$

where

$$\Delta V = 2\pi f x_0$$

The corresponding bulk flow pressure disturbance can be estimated from the Bernoulli equation as

$$\Delta P = \frac{(\Delta V)^2}{2g_0} \rho = \frac{(2\pi f_1 x_0)^2}{2g_0} \rho , \qquad f_1 = 10 \text{ Hz}$$

and the propagation pressure disturbance would be

$$\Delta P = \frac{\rho C \, \Delta V}{g_0} = \frac{\rho C}{g_0} 2\pi f_2 x_0 , \qquad f_2 = 1000 \text{ Hz}$$

Water with relatively constant density and Eq. (2.116) can be used to give the temperature disturbance as $\Delta T \approx 0$.

5. The model coefficients π_1 through π_{12} in Table 6.3 are associated with multidimensional flows. The following values were obtained for the two cases:

π_n	10 Hz	1000 Hz	π_n	10 Hz	1000 Hz
π_1	8.8×10^{-6}	0.4	π_7	0.033	0
π_2	0.21	0.4	π_8	0	0
π_3	0	0	π_9	0	0
π_4	0	0	π_{10}	0	0
π_5	0	0	π_{11}	0	0
π_6	0.022	0.4	π_{12}	0	0

6. It is seen that for $f_1 = 10$ Hz, the largest model coefficient is $\pi_2 = 0.21$ and the next one is $\pi_6 = 0.022$, an order of magnitude smaller. If we neglect terms with small π_n values, it follows from Eqs. (6.69)–(6.72) that for the $f = 10$ Hz case, the nondimensional equations for water become

$$\nabla^* \cdot \mathbf{V}^* = 0 \qquad \text{and} \qquad \frac{\partial \mathbf{V}^*}{\partial t^*} = 0$$

for which the dimensional forms are

$$\nabla \cdot \mathbf{V} = 0 \qquad \text{and} \qquad \frac{\partial \mathbf{V}}{\partial t} = 0$$

This result shows that for the slow 10-Hz oscillation, the fluid motion can be solved as a steady flow problem at every instant of time. This determination probably could save many hours of analysis and perhaps unnecessary numerical procedures.

The case with $f_2 = 1000\,\text{Hz}$ is seen to have a different set of dominant model coefficients. These with Eqs. (6.69)–(6.72) yield

$$0.4\left(\frac{\partial P^*}{\partial t^*} + 0.4\mathbf{V}^* \cdot \nabla^* P^* + \nabla^* \cdot \mathbf{V}^*\right) = 0$$

$$0.4\left(\frac{\partial V^*}{\partial t^*} + 0.4\mathbf{V}^* \cdot \nabla^* \mathbf{V}^* + \nabla^* P^*\right) = 0$$

for which the dimensional forms become

$$\frac{\partial P}{\partial t} + \mathbf{V} \cdot \nabla P + \frac{\rho C^2}{g_0} \nabla \cdot \mathbf{V} = 0$$

and

$$\frac{\partial V}{\partial t} + \mathbf{V} \cdot \nabla \mathbf{V} + \frac{g_0}{\rho} \nabla P = 0$$

Note that the convective terms $\mathbf{V} \cdot \nabla P$ and $\mathbf{V} \cdot \nabla \mathbf{V}$ are important in this high-frequency case.

It would be natural to discuss procedures for scale model design and interpretation of one- and multidimensional flows at this point. However, we will first consider the describing equations, normalization, and model coefficients of spatially uniform systems, because of the frequent need to obtain scale model laws for unsteady thermodynamic processes. Then in Section 6.13 we will discuss scale model design and interpretation for all systems considered in this chapter.

6.11   Spatially Uniform Systems

The three-dimensional region of Fig. 6.8 contains fluid whose quasi-equilibrium properties are spatially uniform, although changing with time. Mass and energy flow across the boundary and its volume can be time-dependent. Kinetic and potential energy components and spatial variation of properties are negligible. Therefore, the momentum law is not required for a determination of the unsteady state.

The mass and energy conservation principles are written for the model of Fig. 6.8 as

Mass:
$$\dot{m}_{\text{out}} - \dot{m}_{\text{in}} + \frac{dM}{dt} = 0 \qquad (6.91)$$

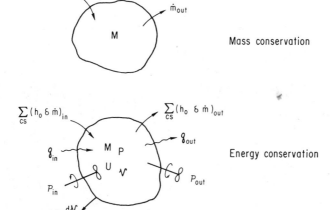

Figure 6.8   Spatially Uniform System

Energy:
$$\mathcal{P}_{\text{out}} - \mathcal{P}_{\text{in}} + q_{\text{out}} - q_{\text{in}} + P\,\frac{d\mathcal{V}}{dt}$$

$$+ \sum_{cs} (h_0\,\delta\dot{m})_{\text{out}} - \sum_{cs} (h_0\,\delta\dot{m})_{\text{in}} + \frac{dU}{dt} = 0 \tag{6.92}$$

A state equation of the form

$$e = \frac{U}{M} = e(P,\rho) = e\left(P,\frac{M}{\mathcal{V}}\right) \tag{6.93}$$

is appropriate for relating U, M, P, and $\mathcal{V}$. Equation (6.93) is differentiated and combined with (6.91), (6.92), and several thermodynamic relationships from Eqs. (2.116)–(2.119) to give

$$\frac{dU}{dt} = \left(\frac{U}{M} + \frac{P\mathcal{V}}{M} - \frac{c_p}{\beta}\right)\frac{dM}{dt} + \left(\frac{g_0 c_p\,\mathcal{V}}{\beta C^2}\right)\frac{dP}{dt} + \left(\frac{c_p}{\beta}\,\frac{M}{\mathcal{V}} - P\right)\frac{d\mathcal{V}}{dt} \tag{6.94}$$

Equations (6.91), (6.92), and (6.94) describe the behavior of spatially uniform systems containing fluids with known state equations.

6.12   Normalization of Spatially Uniform Systems

A normalization procedure similar to that used for one- and multidimensional flows is applied to Eqs. (6.91), (6.92), and (6.94). If a property

TABLE 6.5 Normalized Variables, Spatially Uniform Systems

$$M^* = \frac{M - M_i}{\Delta M} \qquad\qquad P^* = \frac{P - P_i}{\Delta P}$$

$$U^* = \frac{U - U_i}{\Delta U} \qquad\qquad c_p^* = \frac{c_p}{c_{pr}}$$

$$\mathcal{V}^* = \frac{\mathcal{V} - \mathcal{V}_i}{\Delta \mathcal{V}} \qquad\qquad \beta^* = \frac{\beta}{\beta_r}$$

$$\dot{m}^* = \frac{\dot{m}}{\dot{m}_r} \qquad\qquad C^* = \frac{C}{C_r}$$

$$h^* = \frac{h}{h_r} \qquad\qquad t^* = \frac{t}{\Delta t}$$

$$q^* = \frac{q}{q_r}$$

Note: If mass M, energy U, pressure P, or volume $\mathcal{V}$ are constant, replace the denominators ΔM with M_r, $\Delta \mathcal{V}$ with $\mathcal{V}_r$, ΔU with U_r, and ΔP with P_r where subscript r is an appropriate reference value, such as the initial value.

TABLE 6.6 Model Coefficients, Spatially Uniform Systems

$$\pi_{19} = \frac{\dot{m}_r \Delta t}{\Delta M} \qquad\qquad \pi_{26} = \frac{P_i \mathcal{V}_i}{\Delta U}$$

$$\pi_{20} = \frac{P_i \Delta \mathcal{V}}{\Delta U} \qquad\qquad \pi_{27} = \frac{\Delta P \mathcal{V}_i}{\Delta U}$$

$$\pi_{21} = \frac{\Delta P \Delta \mathcal{V}}{\Delta U} \qquad\qquad \pi_{28} = \frac{c_{pr}}{\beta_r} \frac{\Delta M}{\Delta U}$$

$$\pi_{22} = \frac{\dot{m}_r h_{0r} \Delta t}{\Delta U} \qquad\qquad \pi_{29} = \frac{g_0 c_{pr} \Delta P \mathcal{V}_i}{\Delta U \beta_r C_r^2}$$

$$\pi_{23} = \frac{q_r \Delta t}{\Delta U} \qquad\qquad \pi_{30} = \frac{g_0 c_{pr} \Delta P \Delta \mathcal{V}}{\Delta U \beta_r C_r^2}$$

$$\pi_{24} = \frac{U_i}{\Delta U} \qquad\qquad \pi_{31} = \frac{\mathcal{V}_i}{\Delta \mathcal{V}}$$

$$\pi_{25} = \frac{M_i}{\Delta M}$$

Note: ΔM, ΔU, or $\Delta \mathcal{V}$ should be considered equal to their initial values M_i, U_i, and $\mathcal{V}_i$ in systems where M, U, or $\mathcal{V}$ are constants.

change is going to occur, it should be estimated, for example, as ΔM, $\Delta \mathcal{V}$, ΔP, or ΔU. If a property remains constant, its time derivative is deleted from the equations. The normalized variables for spatially uniform systems are given in Table 6.5. The dimensional and nondimensional equations for spatially uniform systems are listed in Eqs. (6.95)–(6.97) and the model coefficients are given in Table 6.6. Tables 6.5 and 6.6 and Eqs. (6.95)–(6.97) can be employed to simplify the spatially uniform system equations by following the same procedure given in Table 6.4 for one- and multidimensional flows. Simplifications usually are obvious, but the greater importance of Tables 6.5 and 6.6 and Eqs. (6.95)–(6.97) comes in the design of small-scale models to be used in predicting full-size system behavior.

SPATIALLY UNIFORM SYSTEM EQUATIONS

MASS

$$\quad\quad\quad Nondimensional \quad\quad\quad\quad\quad\quad\quad\quad\quad\quad\quad\quad Dimensional$$

$$\pi_{19}(\dot{m}_{\text{out}}^* - \dot{m}_{\text{in}}^*) + \frac{dM^*}{dt^*} = 0 \quad (6.95\text{a}) \quad\quad \dot{m}_{\text{out}} - \dot{m}_{\text{in}} + \frac{dM}{dt} = 0 \quad (6.95\text{b})$$

ENERGY

$$(\pi_{20} + \pi_{21}P^*) \frac{d\mathcal{V}^*}{dt^*} + \frac{dU^*}{dt^*}$$
$$+ \pi_{22}[(\dot{m}^* h_0^*)_{\text{out}} - (\dot{m}^* h_0^*)_{\text{in}}]$$
$$+ \pi_{23}(q_{\text{out}}^* - q_{\text{in}}^*) = 0 \quad (6.96\text{a})$$

$$P\frac{d\mathcal{V}}{dt} + \frac{dU}{dt}$$
$$+ (\dot{m} h_0)_{\text{out}} - (\dot{m} h_0)_{\text{in}}$$
$$+ q_{\text{out}} - q_{\text{in}} = 0 \quad (6.96\text{b})$$

STATE

$$dU^* = (\pi_{29} + \pi_{30} \mathcal{V}^*) \frac{c_p^*}{\beta^* C^{*2}} dP^*$$

$$+ \left[\frac{\pi_{24} + U^*}{\pi_{25} + M^*} + \pi_{28} \frac{c_p^*}{\beta^*} \right] dM^*$$

$$+ \left[\frac{\pi_{26} + \pi_{20} \mathcal{V}^* + \pi_{27}P^* + \pi_{21}P^* \mathcal{V}^*}{\pi_{25} + M^*} \right]$$

$$\times \, dM^*$$

$$+ \left[\pi_{28} \frac{c_p^*}{\beta^*} \frac{\pi_{25} + M^*}{\pi_{31} + \mathcal{V}^*} \right.$$

$$\left. - (\pi_{20} + \pi_{21}P^*) \right] d\mathcal{V}^* \quad (6.97\text{a})$$

$$dU = \frac{g_0 c_p \mathcal{V}}{\beta C^2} dP$$

$$+ \left[\frac{U}{M} - \frac{c_p}{\beta} \right] dM$$

$$+ \left[\frac{P\mathcal{V}}{M} \right] dM$$

$$+ \left(\frac{c_p}{\beta} \frac{M}{\mathcal{V}} - P \right) d\mathcal{V} \quad (6.97\text{b})$$

Note: If M, $\mathcal{V}$, U, or P are constant, the corresponding derivatives or differential terms in Eqs. (6.95)–(6.97) are zero. This will diminish the number of required model coefficients in Table 6.6.

6.13   Scale Model Design and Interpretation

The normalization process and development of model coefficients presented in this chapter can also be employed to help design scale model tests for the purpose of studying full-scale systems. Once the describing equations are nondimensionalized with the normalized variables of Tables 6.2 and 6.5, any two systems with identical numerical values of the resulting model coefficients will have identical solutions of the normalized variables if the corresponding boundary and initial values are the same.

Consider the case of liquid sloshing in a rigid container. If the liquid is incompressible and of low viscosity, the mass conservation and momentum principles of Eqs. (6.69) and (6.70) reduce to

$$\nabla^* \cdot \mathbf{V}^* = 0$$

and

$$\pi_2\left(\frac{\partial \mathbf{V}^*}{\partial t^*} + \pi_2 V^* \cdot \nabla^* \mathbf{V}^*\right) + \pi_6 \nabla^* P^* + \pi_7 \mathbf{b}^* = 0$$

If the coefficients π_2, π_6, and π_7 all are evaluated for the full-size system and found to be of comparable size, then equal values of these coefficients are required for a scale model. The mathematical solutions for $\mathbf{V}^*$ and P^* would then be identical for both systems if the normalized boundary and initial conditions were the same. Therefore, results from a scale model would yield $\mathbf{V}^*$ and P^*, which also are the same in full size. It follows that if subscripts f and m designate the full-size and scale model systems, then

$$\mathbf{V}^* = \frac{\mathbf{V}_f}{\Delta V_f} = \frac{\mathbf{V}_m}{\Delta V_m}$$

and

$$P^* = \frac{P_f}{\Delta P_f} = \frac{P_m}{\Delta P_m}$$

The full-size variables can be obtained from values measured in small scale by employing the scale-up formulas

$$\mathbf{V}_f = \mathbf{V}_m \frac{\Delta V_f}{\Delta V_m} \tag{6.98}$$

$$P_f = P_m \frac{\Delta P_f}{\Delta P_m} \qquad (6.99)$$

The time scales also imply that

$$t^* = \frac{t_f}{\Delta t_f} = \frac{t_m}{\Delta t_m}$$

from which

$$t_f = t_m \frac{\Delta t_f}{\Delta t_m} \qquad (6.100)$$

Although geometric similarity is not necessary in spatially uniform systems, it generally is a required condition for full-size and scale model systems in one- and multidimensional flows. Therefore since

$$x^*, y^*, \text{ or } z^* = \frac{x_f, y_f, \text{ or } z_f}{L_{rf}} = \frac{x_m, y_m, \text{ or } z_m}{L_{rm}}$$

we have

$$x_f, y_f, \text{ or } z_f = (x_m, y_m, \text{ or } z_m) \frac{L_{rf}}{L_{rm}} \qquad (6.101)$$

The scale-modeling procedure is summarized in Table 6.7.

Scale model laws are developed in the next few sections for several fundamental problems in unsteady thermofluid mechanics. Although there is

TABLE 6.7 Scale Model Design and Interpretation

1. Determine the boundaries of the system to be modeled.
2. Characterize the phenomena to be studied according to one- or multidimensional flow or spatially uniform systems.
3. Estimate the full-size system space interval L and response time Δt.
4. Estimate the full-size property changes $\Delta\phi$ from a known disturbance. Also, list other reference properties which do not appear in derivatives of the describing equations.
5. Express all model coefficients algebraically in terms of the given parameter symbols. Often symbols will cancel from a given π_n, leaving a concise nondimensional group for each model coefficient. Then obtain numerical values for all the π's based on the full-size system.
6. Neglect all relatively small π values and verify that the neglected π's do not become too large to neglect in the scale model.
7. Design the scale model experiment, based on the significant π's, and obtain data.
8. Predict the full-size property and time behavior from scale model data and the normalized variables, which are equal in both systems.

an almost unlimited number of specific applications, the selected developments help to show how the model coefficients are obtained in one- and
multidimensional flow and in spatially uniform systems.

6.14 Liquid Sloshing During an Earthquake

Suppose that a small-scale experiment is to be performed to determine if the
motion of water during an earthquake has amplitudes large enough to slosh
over the container sides. The container is a cylinder of diameter D and
height $2H$, where H is the undisturbed water depth. The ground acceleration vector $\mathbf{G}(t)$, which corresponds to $\mathbf{b}$ in Eq. (6.70), was obtained from
data recorded during an actual earthquake. The scale model tank will
contain water and is attached to a shaker table which is programmed to
provide the appropriately scaled earthquake. The steps of Table 6.7 are
outlined next.

1. Boundaries are defined by the cylindrical container, diameter D, and
the free surface of the liquid.

2. The phenomenon is multidimensional flow.

3. A suitable space interval L_r for sloshing waves across the container is
the diameter D. A hydrostatic wave has speed $V_p = \sqrt{gH}$, Section A.2 of
Appendix A, which is an approximate propagation speed for the wave
response time, $\Delta t = L_r/V_p = D/\sqrt{gH}$.

4. The full-size maximum horizontal ground acceleration never exceeded
that of gravity. If a sideways acceleration of g occurred, pressure on one side
of the container would increase an amount $\Delta P = \rho g D/g_0$, for which the
elevation would increase $\Delta H = \Delta P/(\rho g/g_0) = D$. A reference change in
velocity is $\Delta V = \Delta H/\Delta t = \sqrt{gH}$.

5. The water is considered incompressible with negligible viscosity and
thermal conductivity. A reference acceleration is $b_r = g$. Sloshing action is
determined by the mass conservation and momentum principles without
need for the energy equation. Equations (6.69) and (6.70) show that
without the need to model energy, only model coefficients π_1 through π_8 are
required. The resulting model coefficients of Table 6.3 are

$$\pi_1 \approx 0 \qquad \pi_5 = 0 \qquad \pi_9 = 0$$
$$\pi_2 = 1 \qquad \pi_6 = \frac{D}{H} \qquad \pi_{10} = 0$$
$$\pi_3 = 0 \qquad \pi_7 = \frac{D}{H} \qquad \pi_{11} = 0$$
$$\pi_4 = 0 \qquad \pi_8 = 0 \qquad \pi_{12} = 0$$

It is seldom necessary to include the energy equation in sloshing studies, although a time-dependent **b** would lead to additional terms in Eqs. (6.71) and (6.72). Since only the mass conservation and momentum equations are required for sloshing studies, the body force acceleration **b** can be a function of time.

6. Coefficients listed as zero are neglected. Coefficient $\pi_2 = 1$ provides no model laws.

7. Coefficients $\pi_6 = \pi_7 = D/H$ show that the ratio D/H must be the same in the model and full-size systems, which implies geometric similarity. The normalized body force acceleration of Table 6.2 is $\mathbf{b}^* = \mathbf{G}(t)/g$. Since g is the same in both the model and full size, the components of **G** must be equal. However, the time scales may be different. This will be determined in the next step.

8. The measured variables are pressure, velocity, displacement, and time. The normalized variables of Table 6.2 yield

$$\frac{(P - P_i)_m}{(P - P_i)_f} = \frac{\Delta P_m}{\Delta P_f} = \frac{D_m}{D_f} , \qquad \frac{t_m}{t_f} = \frac{\Delta t_m}{\Delta t_f} = \frac{D_m/\sqrt{H_m}}{D_f/\sqrt{H_f}} = \sqrt{\frac{D_m}{D_f}}$$

$$\frac{V_m}{V_f} = \frac{\Delta V_m}{\Delta V_f} = \sqrt{\frac{H_m}{H_f}} = \sqrt{\frac{D_m}{D_f}} , \qquad \frac{(x, y, z)_m}{(x, y, z)_f} = \frac{D_m}{D_f}$$

It is seen that the ratio of the model pressure disturbance to full-size pressure disturbance is scaled directly as the length scale D_m/D_f. Velocity and time are scaled as $\sqrt{D_m/D_f}$. This means that although the acceleration components of **G** must be the same in both the model and full-size systems, the time scales are different and the shaker table should be programmed accordingly. If one acceleration component in full scale was

$$G_{xf} = b_x \sin 2\pi f_f t_f$$

the corresponding scale model component would be

$$G_{xm} = b_x \sin 2\pi (f_f \sqrt{D_f/D_m}) t_m$$

or the correct model frequency would be

$$f_m = f_f \sqrt{D_f/D_m}$$

Displacements are also scaled according to the length scale D_m/D_f.

The scale model laws obtained in this example are recognized as Froude scaling. The results of two experiments by Aslam et al. [6.11] are shown in Fig. 6.9. Small-size dimensions were 1/15 of full size, and natural periods of the second antisymmetric mode are seen to be 0.67 and 2.5 s, respectively, giving the ratio $t_m/t_f = 3.87$, which verifies the time-scale ratio, $\sqrt{D_f/D_m} = \sqrt{15} = 3.87$.

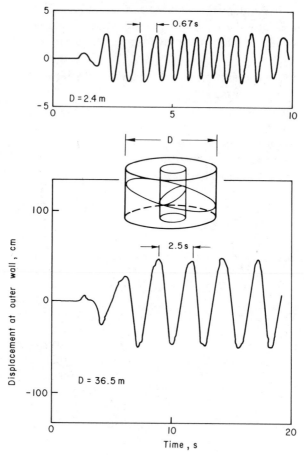

Figure 6.9 Second Antisymmetric Slosh Mode, Annular Tank [11]

If the containing cylinder was flexible, it would be necessary to write and normalize the equations for cylindrical shell dynamics in addition to the fluid equations. Additional model coefficients would be obtained.

6.15 Unsteady Steam Discharge from a Pipe

The possibility of pipe failure sometimes makes it desirable to predict the unsteady discharge properties and reaction forces resulting from the postulated rupture of a pipe attached to a vessel containing high-pressure saturated steam. The full-size pipe has length L, diameter D, and initial steam pressure $P_i = P_0$, as indicated in Fig. 6.10. The rupture is assumed to be instantaneous and circumferential. Consider a full-size system which has the following example parameters:

Figure 6.10   Unsteady Steam Discharge from a Pipe

$P_0 = 6.87\,\text{MPa}$ $\mathcal{H} = 100\,\text{W/m}^2\text{-K}$

$\rho_0 = 35.37\,\text{kg/m}^3$ $f = 0.015$

$C_0 = \sqrt{kg_0 P_0 / \rho_0} = 488\,\text{m/s}$ $D = 1.0\,\text{m}$

$\kappa_0 = 60\,\text{MW/K-m}$ $A = 0.785\,\text{m}^2$

$\nu_0 = 5.58 \times 10^5\,\text{m}^2/\text{s}$ $g = 9.8\,\text{m/s}^2$

$c_{p0} = 6\,\text{kJ/kg-K}$ $v_f = 0.0013\,\text{m}^3/\text{kg}$

$T_0 = 284°\text{C} = 557\,\text{K}$ $v_g = 0.0278\,\text{m}^3/\text{kg}$

$L = 100\,\text{m}$ $v_{fg} = 0.0265\,\text{m}^3/\text{kg}$

$\theta = -\pi/2$ $h_{fg} = 1518\,\text{kJ/kg}$

$k \approx 1.3\,\text{for steam}$ $V_i = 0$

The decompression of saturated steam in a pipe will cause its state to drop into the saturated mixture region. The quantities β and c_p for a saturated mixture are infinite. However, the ratio β/c_p occurs in several model coefficients of Table 6.3. If we employ β and c_p from Table 2.1 with the two-phase state equation

$$v = v_f + \frac{v_{fg}}{h_{fg}}(h - h_f)$$

we obtain

$$\frac{\beta}{c_p} = \frac{1}{v}\left(\frac{\partial v}{\partial h}\right)_p = \frac{1}{v}\frac{v_{fg}}{h_{fg}}$$

in the saturated region. Saturated steam with $v = v_g$ yields $\beta/c_p = 6.3 \times 10^{-4}\,\text{kg/kJ}$. The steps of Table 6.7 are summarized next.

1. The boundary is determined by the pipe wall of diameter D and length L.

2. The phenomenon is one-dimensional flow.

3. The space interval is $L_r = L$. Since a sudden rupture occurs, the problem is dominated by propagation effects. Therefore, the response time is

$$\Delta t = \frac{L}{C_0}$$

4. Saturated steam usually can be treated like a perfect gas in problems like this. Therefore, the pressure disturbance at the ruptured end will correspond to a sudden decrease from P_0 to approximately the critical value of Eq. (2.59),

$$\frac{P_c}{P_0} = \left(\frac{2}{k+1}\right)^{k/(k-1)} = 0.546$$

This is not strictly true, as will be shown in Chapter 8, but it is an acceptable estimate for model laws. The pressure disturbance, therefore, is approximately

$$\Delta P = P_0\left(1 - \frac{P_c}{P_0}\right) \cong 0.454 P_0$$

The approximate density disturbance for, say, isentropic decompression, is

$$\Delta \rho = \rho_0 - \rho_c = \rho_0\left(1 - \left(\frac{P_c}{P_0}\right)^{1/k}\right) = 0.37\rho_0$$

and from Eq. (2.116), as in Example 6.3,

$$\Delta T = T_0\left(\frac{\Delta P}{P_0} - \frac{\Delta \rho}{\rho_0}\right) = (0.454 - 0.37)T_0 = 0.084 T_0$$

An approximate change in velocity is

$$\Delta V \cong C_0$$

It is necessary to estimate a reference wall heat transfer rate q'_r from

$$q'_r = \frac{q''_r P_w L}{L} = q''_r P_w = q''_r \pi D$$

where $q''_r = \mathcal{H}\,\Delta T$. Thus,

$$q'_r = \mathcal{H}\pi D\,\Delta T = (100\ \text{W/m}^2\text{-K})\pi(1.0\ \text{m})(0.084)(507\ \text{K})$$

$$= 13,380\ \text{W/m}$$

5. The appropriate model coefficients of Table 6.3 are, for the parameters given,

$$\pi_1 = \frac{0.454}{k} = 0.35 \qquad\qquad \pi_{11} = 4.8 \times 10^{-4}$$

$$\pi_2 = 1$$

$$\pi_2^0 = 0 \qquad\qquad \pi_{13} = 0.34 \frac{L}{D} \frac{T_0}{C_0} \mathcal{H} \frac{v_{fg}}{h_{fg}} = 6 \times 10^{-5}$$

$$\pi_3 = 0.37 \qquad\qquad \pi_{14} = \frac{fL}{D} = 1.5$$

$$\pi_4 = 9.3 \times 10^{-7} \qquad\qquad \pi_{15} = 0.65 P_0 \frac{v_{fg}}{h_{fg}} = 0.08$$

$$\pi_6 = \pi_1 = 0.35 \qquad\qquad \pi_{16} = \frac{gL}{C^2} \sin\theta = 0.004$$

$$\pi_9 = 0.084 \qquad\qquad \pi_{17} = \pi_{18} = 0 \quad (c_p \to \infty \text{ in saturated mixture})$$

$$\pi_{10} = 0.454 P_0 \frac{v_{fg}}{h_{fg}} = 0.054$$

6. Coefficients π_{11}, π_{13}, π_{16}, π_{17}, and π_{18} usually are small enough to neglect. Coefficients π_1 and π_6 contain only the ratio of specific heats k. Coefficients π_2, π_3, and π_9 resulted in rational numbers and do not influence scaling laws. Coefficients π_{10} and π_{15} yielded the group $P_0 v_{fg}/h_{fg}$, and π_{14} resulted in fL/D.

7. A scale model should have the following nondimensional groups identical to the full-size system: k, $P_0 v_{fg}/h_{fg}$, and fL/D. If the same friction factor f could be expected in a model pipe, then the model should have the same L/D ratio as full size. The use of steam would satisfy the requirement of equal k values in both. If steam was used at full size pressure P_0 in the scale model, then the model coefficient $P_0 v_{fg}/h_{fg}$ would be satisfied. Such a

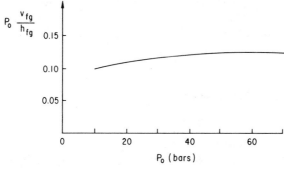

Figure 6.11 The Saturated Steam-Water Property $P_0 v_{fg}/h_{fg}$

scale model could be of smaller geometry than full size, but sometimes full-size pressure is not available in a laboratory.

If it is desirable to run a scale model at reduced pressure, $P_0 v_{fg}/h_{fg}$ must still be the same. This requirement might restrict the test to high pressure. However, Fig. 6.11 shows that this model coefficient is not a strong function of pressure. Therefore, it would be possible to operate a small-scale steam system at reduced pressure without introducing appreciable error.

8. Pressure, velocity, density, space, and time data from a well-instrumented small-scale test would be scaled up to make full-size predictions from the normalized variables of Table 6.2:

$$\frac{(P-P_i)_f}{(P-P_i)_m} = \frac{\Delta P_f}{\Delta P_m} = \frac{P_{if}}{P_{im}}, \qquad \frac{x_f}{x_m} = \frac{L_f}{L_m}$$

$$\frac{V_f}{V_m} = \frac{\Delta V_f}{\Delta V_m} = \frac{C_{0f}}{C_{0m}}, \qquad \frac{t_f}{t_m} = \frac{\Delta t_f}{\Delta t_m} = \frac{L_f}{L_m}\frac{C_m}{C_f}$$

$$\frac{(\rho-\rho_i)_f}{(\rho-\rho_i)_m} = \frac{\Delta \rho_f}{\Delta \rho_m} = \frac{\rho_{if}}{\rho_{im}}$$

Suppose it also was desirable to simplify the describing equations for a theoretical solution. The list of model coefficients in step 5 shows which terms in Eqs. (6.64) to (6.68) can be neglected. The equations obtained are

$$\frac{\partial P}{\partial t} + V\frac{\partial P}{\partial x} + \frac{\rho C^2}{g_0}\frac{\partial V}{\partial x} = \frac{\beta}{g_0 c_p}\frac{f}{4}\frac{P_w}{A}\frac{kPV^3}{2}$$

$$\frac{\partial V}{\partial t} + V\frac{\partial V}{\partial x} + \frac{g_0}{\rho}\frac{\partial P}{\partial x} + \frac{f}{4}\frac{P_w}{A}\frac{V^2}{2} = 0$$

$$\frac{\partial P}{\partial t} + V\frac{\partial P}{\partial x} - \frac{C^2}{g_0}\left(\frac{\partial \rho}{\partial t} + V\frac{\partial \rho}{\partial x}\right) = \frac{\beta}{c_p}\frac{f}{4}\frac{P_w}{A}\frac{\rho V^3}{2g_0}$$

which could be solved numerically by the method of characteristics discussed in Chapter 8.

6.16   Relief Tank Pressurization

Suppose that it is necessary to discharge a high-pressure gas vessel through a relief valve into a vented tank, which is expected to decrease environmental noise. If the relief tank pressure transient, and especially the maximum pressure, is to be determined from a scale model test, the model laws would be required.

Consider a case of gas at pressure P_0, temperature T_0, density ρ_0, and ratio of specific heats k in a pressure vessel of volume $\mathcal{V}_0$. Sudden discharge occurs through a nozzle of area A_n into a relief tank of volume $\mathcal{V}$. The relief tank is vented to the atmosphere at pressure P_∞ and temperature T_∞ through a frictionless vent pipe of area A_v. The system is shown in Fig. 6.12. The steps of Table 6.7 are employed to provide the following summary analysis:

1. The system actually involves two subsystems, namely the pressure vessel and the relief tank. It is possible that the pressure vessel could be almost discharged before the relief tank reached its maximum pressure. Whether the pressure vessel or relief tank determines the maximum pressure will be discussed further in step 3.

2. The phenomenon to be studied involves the pressurization rate of a volume which undergoes simultaneous charging and discharging. If we can neglect shock wave activity in the relief tank from the sudden release, it is expected that both vessels can be approximated by spatially uniform systems.

3. The space interval in the relief tank corresponds to its volume so that

$$\Delta \mathcal{V} = \mathcal{V}$$

The response time Δt can be estimated as the time required for the relief tank pressure to reach its maximum value at its initial pressure rate. Critical discharge flow will occur because the relief tank pressure is less than the critical value corresponding to $P_c/P_0 = 0.528$ from Eq. (2.59). If we write the mass discharge rate as $\dot{m}_{in} = G_c(P_0, h_0)A_n$, the initial pressure rate of Eq. (2.88) becomes

$$\left.\frac{dP}{dt}\right|_i = \frac{kA_n}{\mathcal{V}} f(k) \sqrt{kg_0 \frac{P_0^3}{\rho_0}}$$

$$f(k) = \left(\frac{2}{k+1}\right)^{[(k+1)/(k-1)]/2}$$

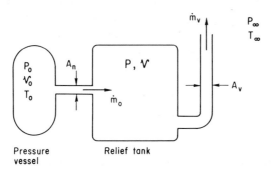

Figure 6.12   Relief Tank Pressurization

The pressure increase to the maximum value $\Delta P = P_{max} - P_\infty$ must be estimated. If we assume critical discharge also through the vent, which is valid if $P_{max}/P_\infty < P_c/P = 0.528$, we can set $dP/dt = 0$ in Eq. (2.88) for an adiabatic constant volume system, with $h_{0,\,in} = h_{0,\,out}$, $\dot{m} = G_c A$, and G_0 from Eq. (2.60), to obtain

$$P_{max} = \frac{A_n}{A_v} P_0$$

so that

$$\Delta P = \frac{A_n}{A_v} P_0 - P_\infty$$

The estimated relief tank response time is then

$$\Delta t \cong \frac{\Delta P}{(dP/dt)_i} = \frac{[(A_n/A_v)P_0 - P_\infty](\mathcal{V}/A_n)}{kf(k)\sqrt{kg_0 P_0^3/\rho_0}}$$

This response time is based on the assumption that discharge from the pressure vessel remains constant at its initial value, as if it were an infinite reservoir. It could be that the vessel discharge happens in a time shorter than Δt. If so, then the vessel time response, estimated from $\Delta t_0 \cong P_0/(dP/dt)_0$, would control the transient, and a better estimate of ΔP in the relief tank would be $(dP/dt)_i \Delta t_0$. This consideration yielded

$$\left.\frac{dP}{dt}\right|_0 = -\frac{kA_b}{\mathcal{V}_0} f(k)\sqrt{kg_0 \frac{P_0^3}{\rho_0}}$$

for the pressure vessel and

$$\Delta t_0 = -\left.\frac{P_0}{dP/dt}\right|_0 = \frac{\mathcal{V}_0 P_0}{kA_n f(k)\sqrt{kg_0 P_0^3/\rho_0}}$$

The ratio of response times is

$$\frac{\Delta t}{\Delta t_0} = \frac{\mathcal{V}}{\mathcal{V}_0}\left(\frac{A_n}{A_0} - \frac{P_\infty}{P_0}\right)$$

The rest of this analysis is based on a case for which the relief tank response time Δt is much shorter than the response time of the vessel, Δt_0. Therefore, Δt is the response time to be used in the model coefficients for a study of the relief tank transient.

4. Other full-size property changes are approximated by

$$\Delta M \cong G_c(P_0, h_0) A_n \, \Delta t = \frac{1}{k} \rho_0 \mathcal{V} \left(\frac{A_n}{A_v} - \frac{P_\infty}{P_0} \right)$$

where G_c was obtained from Eq. (2.60), and

$$\Delta U \cong G_c(P_0, h_0) h_0 A_n \, \Delta t = \frac{\mathcal{V} P_0}{k-1} \left(\frac{A_n}{A_v} - \frac{P_\infty}{P_0} \right)$$

where $h_0 = kP_0/(k-1)\rho_0$.

5. The model coefficients of Table 6.6 are summarized below with

$$\beta_\infty = 1/T_\infty, \qquad \dot{m}_r = G_c(P_0, h_0)A_n, \qquad h_r = [k/(k-1)]P_0/\rho_0$$

$$q_r = 0, \qquad P_i = P_\infty, \qquad \mathcal{V}_i = \mathcal{V}, \qquad U_i = P_\infty \mathcal{V}/(k-1)$$

$$M_i = \rho_\infty \mathcal{V}, \qquad P = \rho RT, \qquad c_p/R = k/(k-1), \qquad C_r^2 = kg_0 RT_\infty.$$

Equations (6.95)–(6.97) and Table 6.6 show that π_{20}, π_{21}, and π_{31} do not have to be calculated because they are coefficients of $d\mathcal{V}^*$, which is zero in a constant-volume system. It follows that

$$\pi_{19} = \frac{\dot{m}_r \, \Delta t}{\Delta M} = \frac{G_c A_n \, \Delta t}{G_c A_n \, \Delta t} = 1 \qquad\qquad \pi_{26} = \frac{P_i \mathcal{V}_i}{\Delta U} = (k-1)\,\frac{P_\infty/P_0}{A_n/A_0 - P_\infty/P_0}$$

$$\pi_{22} = \frac{\dot{m}_r h_{0r} \, \Delta t}{\Delta U} = 1 \qquad\qquad \pi_{27} = \frac{\Delta P \mathcal{V}_i}{\Delta U} = (k-1)$$

$$\pi_{23} = 0$$

$$\pi_{28} = \frac{c_{pr}}{\beta_r}\,\frac{\Delta M}{\Delta U} = 1.0$$

$$\pi_{24} = \frac{U_i}{\Delta U} = \frac{P_\infty/P_0}{A_n/A_0 - P_\infty/P_0}$$

$$\pi_{29} = \frac{g_0 c_{pr} \, \Delta P \mathcal{V}_i}{\Delta U \beta_r C_r^2} = 1$$

$$\pi_{25} = \frac{M_i}{\Delta M} = k\,\frac{(P_\infty/P_0)(T_0/T_\infty)}{A_n/A_0 - P_\infty/P_0}$$

$$\pi_{30} = \frac{g_0 c_{pr} \, \Delta P \, \Delta \mathcal{V}}{\Delta U \beta_r C_r^2} = 1$$

6. The π values of step 5 which reduce to ratios of parameters are π_{24}, π_{25}, π_{26}, and π_{27}. The associated model coefficients which must be the same in full-scale and a scale model relief tank are

$$\pi_{24} = \frac{P_\infty/P_0}{A_n/A_0 - P_\infty/P_0} \qquad\qquad \pi_{25} = k\,\frac{T_0}{T_\infty}\,\pi_{24}$$

$$\pi_{26} = (k-1)\pi_{24} \qquad\qquad \pi_{27} = k-1$$

7. The scale model could be designed with any volume $\mathcal{V}$ since it does not appear in the model coefficients. If it were not possible to lower P_∞ so that a lower vessel pressure P_0 could be used, the area ratio A_n/A_v could be adjusted to keep π_{24} the same as full size.

8. The ratio of scale model to full-size relief tank pressure and time would correspond to

$$\frac{P_s}{P_f} = \frac{\Delta P_s}{\Delta P_f} = \left(\frac{P_{0s}}{P_{0f}}\right) \frac{(A_n/A_v)_s - (P_\infty/P_0)_s}{(A_n/A_v)_f - (P_\infty/P_0)_f}$$

$$\frac{t_s}{t_f} = \frac{\Delta t_s}{\Delta t_f} = \left(\frac{\mathcal{V}_s}{\mathcal{V}_f}\frac{A_{nf}}{A_{ns}}\right) \frac{(A_n/A_v)_s - (P_\infty/P_0)_s}{(A_n/A_v)_f - (P_\infty/P_0)_f} \sqrt{\frac{T_{0f}}{T_{0s}}}$$

When scale model data $P_s(t_s)$ is recorded, the ratios P_s/P_f and t_s/t_f provide scale-up laws for a prediction of the full-size pressure transient.

If vessel and relief tank pressures are the same in full size and a model test, a *segment model* could be designed in which volumes and areas could be reduced by the same factor and yet give identical pressure behavior with time. If this analysis is extended to two-phase liquid-vapor discharge and venting, a small-segment model again could be designed to give full-size $P(t)$ data. Moreover, the equation for P_{max} shows that relief tank pressure from numerous tests should correlate as $f(A_n/A_v)$. Figure 6.13 gives results from early pressure suppression tests which show that a segment model does indeed correlate pressure with the vent-to-vessel nozzle area ratio.

EXAMPLE 6.6: SCALE MODEL TEST FOR TANK PRESSURE
Determine scale model laws for predicting the time-dependent pressure in a vented relief tank while it is charged with air from a pressurized vessel. The full-size system is shown in Fig. 6.12 with vessel pressure $P_0 = 6.87$ MPa (1000 psia), temperature $T_0 = 284°C$ (543°F) $= 557$ K (1003 R), density $\rho_0 = 43$ kg/m³ (2.7 lbm/ft³), vessel volume $\mathcal{V}_0 = 1000$ m³ (35, 287 ft³), nozzle area $A_n = 0.5$ m² (5.4 ft²), relief tank volume $\mathcal{V} = 500$ m³ (17, 644 ft³) vented to the atmosphere at $P_\infty = 0.101$ MPa (14.7 psia) through a vent with negligible friction and flow area $A_v = 5$ m² (54 ft²).
The analysis of Section 6.16 yielded

$$\frac{\Delta t}{\Delta t_0} = 0.043 \qquad \text{(step 3)}$$

which shows that the relief tank time response is substantially shorter than that of the vessel. If this was not the case, it would be necessary to repeat the analysis of Section 6.16 for a response time Δt_0 based on the vessel. Numerical values of the model coefficients from step 5 are

$$\pi_{24} = 0.17, \quad \pi_{25} = 0.03, \quad \pi_{26} = 0.069, \quad \pi_{27} = 0.4$$

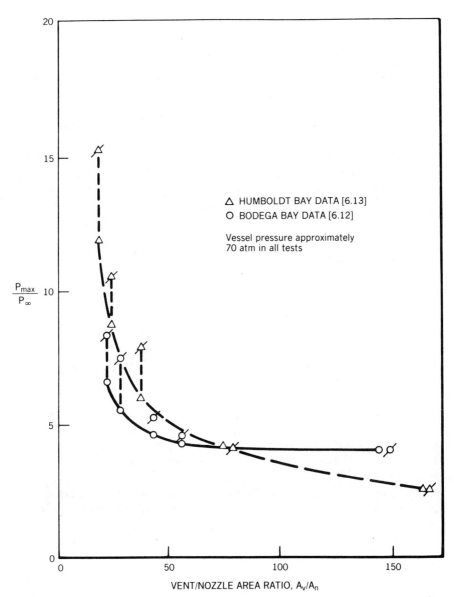

Figure 6.13 Correlation, Maximum Relief Tank Pressure. △ Humboldt Bay Data [13]; ○ Bodega Bay Data [12]

The largest model coefficients are π_{24} and π_{27}. One way to satisfy these requirements is to have the ratios P_∞/P_0, A_n/A_v, and k identical in both the full-size and scale model systems.

6.17   Scale-Modeling Coupled Systems

A scale model experiment requires a system in which the disturbed property can be controlled. However, two or more systems may be coupled so that the disturbance produced by one on the other is not known in advance. Therefore, it is sometimes desirable to scale model several systems together if all scaling laws can be accommodated.

Consider an example system for which a steam relief valve of opening time $t_v = 0.25$ s and a full flow rate $\dot{m} = 90$ kg/s charges an air-filled pipe of length $L = 40$ m and flow area $A = 0.1\,\text{m}^2$, whose discharge end is submerged in a cylindrical water pool of diameter D and undisturbed depth H, both of which are 6 m. Pool pressure and motion during water expulsion and air discharge are to be predicted from a small-scale experiment.

The multidimensional pool action, spatially uniform discharged air bubble, and one-dimensional pipe flow systems are coupled. Pool acoustic and slosh response times are $H/C_L = 0.002$ s and $H/\sqrt{gH} = 0.8$ s. The pipe acoustic time is $L/C_{\text{air}} = 0.1$ s. Comparison with the 0.25 s valve time shows that propagation effects are important in the pipe, and pool acoustics are negligible. Therefore, model coefficients for the pool, pipe, and bubble systems are coupled together. Analyses which follow the steps of table 6.7 for the bubble and pool show that geometrically similar pool action would be achieved if both absolute pressure and temperature of the air were in proportion to the scale factor H_m/H_f, for which the time would scale as $t_m/t_f = \sqrt{H_m/H_f}$. However, absolute temperature cannot be reduced beyond water freezing, which could limit the geometric scale factor.

Analysis of the pipe shows that if absolute temperatures were equal in full and small scales, the pipe length scale would require distortion according to

$$\frac{L_m}{L_f} = \sqrt{\frac{H_m}{H_f}}$$

in order to match pool response time. Moreover, mass and energy flow ratios required by the bubble would be equal to $(H_m/H_f)^{7/2}$, whereas a pipe would provide discharge mass and energy flow ratios scaled as $(H_m/H_f)^3$. Thus, it would be necessary to orifice or otherwise reduce pipe discharge rates in small scale. An orificing procedure by Anderson et al. [14] was successful in modeling gas bubble charging loads on pool walls.

References

6.1 Baker, W. E., P. S. Westine, and F. T. Dodge, *Similarity Methods in Engineering Dynamics*, Hayden, Rochelle Park, N. J. 1973.

6.2 Kline, S. J., *Similitude and Approximation Theory*, McGraw-Hill, New York, 1965.

6.3 Bird, R. B., W. E. Stewart, and E. N. Lightfoot, *Transport Phenomena*, Wiley, New York, 1960.

6.4 Landau, L. D., and E. M. Lifshitz, *Fluid Mechanics*, Addison-Wesley, Massachusetts, 1959.

6.5 Schlichting, H., *Boundary Layer Theory*, Pergamon, New York, 1955.

6.6 Shapiro, A. H., *The Dynamics and Thermodynamics of Compressible Fluid Flow*, Vol. I, Ronald Press, New York, 1953.

6.7 Reynolds, W. C., *Thermodynamics*, McGraw-Hill, New York, 1965.

6.8 Huang, F. F., *Engineering Thermodynamics*, MacMillan, New York, 1976.

6.9 Kreith, F., *Principles of Heat Transfer*, Intext Educational, New York, 1973.

6.10 Moody, F. J., "A Systematic Procedure for Scale Modeling Unsteady Thermofluid Systems," General Electric Report NEDO-25210.

6.11 Aslam, A. M., W. G. Godden, and D. T. Scalise, "Earthquake Sloshing in Annular and Cylindrical Tanks, *ASCE, J. Engineering Mech.*, **105** (1979).

6.12 "Preliminary Hazards Summary Report, Bodega Bay Atomic Park, Unit No. 1," Pacific Gas and Electric Company, 1962.

6.13 Robbins, C. H., "Tests of a Full Scale 1/48 Segment of the Humboldt Bay Pressure Suppression Containment," GEAP-3596, 1960.

6.14 Anderson, W. G., P. W. Huber, and A. A. Sonin, "Small-Scale Modeling of Hydrodynamic Forces in Pressure Suppression Systems: Tests of the Scaling Laws," *Proc., Topical Meeting on Thermal Reactor Safety*, Sun Valley Id., CONF-770708 July 31-August 4, 1977.

6.15 Peebles, T. N., and H. J. Garber, *Chem. Eng. Progress*, **49** (1953) 89–97.

6.16 Lamb, H., *Hydrodynamics*, 6th ed., Dover, New York, 1945.

6.17 ASME 1967 Steam Tables.

Problems

6.1 Missile motion during propulsion in a cylindrical mortar of length L and cross-sectional area A is described by (see Fig. 1.27 and Eq. (1.82))

$$\frac{1}{k} y \frac{d^3y}{dt^3} + \frac{dy}{dt} \frac{d^2y}{dt^2} + \left(\frac{g_0 A}{M} P_\infty + \frac{g_0}{M} F_f + g \right) \frac{dy}{dt} + \frac{g_0}{Mk} y \frac{dF_f}{dt}$$

$$= \frac{g_0}{Mk} (k-1)(q + \dot{m}h_0) H_s \left(1 - \frac{t}{\tau} \right)$$

with initial conditions

$$t = 0, \quad y = 0, \quad \frac{dy}{dt} = 0, \quad \frac{d^2y}{dt^2} = 0$$

where M is the missile mass, y is its upward displacement, k is the gas ratio of specific heats, P_∞ is the ambient pressure, F_f is the wall friction force on the missile (constant), g is the acceleration of gravity, and $(q + \dot{m}h_0)H_s(1 - t/\tau)$ is the energy release rate of the burning propulsion charge in the form of a step with duration τ, which then returns to zero. It is desirable to simplify the differential equation.

(a) Reduce the differential equation to a system of first-order differential equations by introducing

$$V = \frac{dy}{dt}, \quad a = \frac{dV}{dt}, \quad z = \frac{da}{dt} = \frac{d^2V}{dt^2} = \frac{d^3y}{dt^3}$$

(b) Normalize the first-order differential equations and initial conditions with appropriate Δt, Δy, ΔV, and Δa. It is reasonable to pick $\Delta y = L$. The other normalizing parameters can be estimated by assuming that the charge burn time τ is less than the expulsion time, causing gas release with energy $\dot{m}h_0\tau$. This energy release can be employed with the perfect gas equations for an average volume $AL/2$ to give an estimated average propulsion pressure P_{avg}. The average pressure force can be used to estimate Δa, from which ΔV and Δt can be estimated.

(c) Show that for the parameters $k = 1.3$, $F_f = 40$ N (9 lbf), $h_0 = 1115$ kJ/kg (481 B/lbm), $q = 0$ (all energy addition from hot gas formation), $\dot{m} = 50$ kg/s (125 lbm/s), $P_\infty = 101$ kPa (14.7 psia), $\tau = 0.01$ s, $L = 1.0$ m (3.28 ft), $M = 3.0$ kg (6.6 lbm) and $A = 50$ cm^2 (7.75 in^2), the differential equation can be simplified to

$$\frac{1}{k} y \frac{d^3y}{dt^3} + \frac{dy}{dt} \frac{d^2y}{dt^2} = \frac{g_0}{Mk}(k - 1)(\dot{m}h_0)H_s\left(1 - \frac{t}{\tau}\right)$$

6.2 The behavior of the spring-mass-dashpot system (Fig P6.2) is described by

$$\text{DEs:} \quad \frac{dV}{dt} + bV + \frac{Kg_0}{M}x = 0, \quad V = \frac{dx}{dt}$$

$$\text{IC:} \quad t = 0, \quad x = x_0, \quad V = V_0$$

The system is to be studied in a suitable scale model.

(a) Normalize the variables with Δx, Δt, and ΔV where Δx is the maximum amplitude x_0, ΔV can be estimated by converting all the initial spring energy $Kx_0^2/2$ into kinetic energy $M(\Delta V)^2/2g_0$, and Δt is the approximate time for M to travel distance x_0 at ΔV.

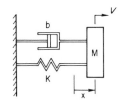

Figure P6.2

(b) Verify that the only model parameter is

$$\pi = b\sqrt{\frac{M}{Kg_0}}$$

and that the full-size variables ( )$_f$ can be obtained from values measured in a scale model test ( )$_s$ through the scale-up equations

$$x_f = \frac{x_{0f}}{x_{0s}}x_s \,, \quad t_f = t_s\sqrt{\frac{M_f}{M_s}\frac{K_s}{K_f}}\,, \quad V_f = V_s\frac{x_{0f}}{x_{0s}}\sqrt{\frac{M_f}{M_s}\frac{K_s}{K_f}}$$

6.3 The design of scale model experiments to study the behavior of single gas bubbles rising in liquid, and single liquid drops moving through gas require a reference length L for use in Table 6.3 and Eqs. (6.69)–(6.72). An obvious reference length is the bubble or drop stable diameter D. Although Example 6.1 gave an estimate for D of a gas bubble in liquid, another analysis may be more appropriate.

It can be shown from potential flow methods in Chapter 9 that the surface pressure on a sphere of radius R moving through stationary liquid of density ρ at velocity U_s (Fig. P6.3) is

$$P - P_\infty = \frac{\rho U_s^2}{2g_0}\left(1 - \frac{9}{4}\sin^2\theta\right)$$

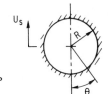

Figure P6.3 ρ

(a) Integrate the pressure over the right hemisphere to show that the outward force is $6R^2U_s^2\rho/2g_0$. (This integration is not difficult, but must be done carefully.)

(b) The right and left hemispheres have equal outward forces which are held together by a surface tension force $2\pi R\sigma$ when a stable radius is attained. Show that the stable radius for a gas bubble in liquid, $\rho = \rho_L$, is

$$R_b = \frac{4g_0\pi\sigma}{6\rho_L U_s^2}$$

(compare with Example 6.1). Then show that the stable radius for a liquid drop in gas is

$$R_d = \frac{4g_0\pi\sigma}{6\rho_g U_s^2}$$

(c) Consider a gas bubble rising by buoyancy in liquid, and a liquid drop falling in gas under the influence of gravity. Let the drag force be expressed as $\rho C_d \pi R^2 U_s^2/2g_0$ (as in Example 6.1), and show that the stable bubble and drop sizes are

$$R_b = R_d \approx 0.88\sqrt{\frac{C_d g_0 \sigma}{g\rho_L}}$$

6.4 The spreading rate of a very viscous liquid on a flat horizontal surface is to be studied both analytically and in a scale model test. The liquid of density ρ and kinematic viscosity ν has the initial shape of a vertical cylinder of radius R_0 and height H_0. Scale model laws for three-dimensional spreading flow can be obtained from Table 6.3 and Eqs. (6.69)–(6.72), but these require an estimate of a velocity change ΔV typical of spreading.

Consider a simple analysis for estimating ΔV in which the liquid cylinder spreads from R_0 and H_0 to a flattening cylinder $R(t)$ and $H(t)$, where R and H always correspond to the initial volume. Let the rate of change of gravitational potential energy be roughly equal to the viscous power outflow as the bottom surface exerts an approximate friction force $F_f \approx \tau\pi R^2$ at the typical velocity $V \approx dR/dt$, where the shear stress is of the form $\tau = \mu(du/dy) \approx \mu V/H$, and y is measured upward from the horizontal surface. Show that the initial velocity for very viscous fluid is

$$V_i = \Delta V = \frac{2gH_0^3}{\nu R_0}, \qquad \nu = \frac{g_0\mu}{\rho}$$

If the liquid is nonviscous, the maximum limiting velocity of spreading corresponds to the dam break front velocity of $2\sqrt{gH_0}$ in Chapter 5. Show that for the viscous analysis here, parameters must satisfy the Reynolds number criterion

$$\frac{\sqrt{gH_0}\,H_0}{\nu} < \frac{R_0}{H_0}$$

6.5 The spreading rate of an ultraviscous liquid on a horizontal flat surface is to be studied in a scale model. The liquid has properties $\rho = 5\,\text{g/cc}\ (312\,\text{lbm/ft}^3) = \text{constant}$, $\nu = 8.0\,\text{m}^2/\text{s}\ (86\,\text{ft}^2/\text{s}) = \text{constant}$, $\beta = 0$, $C \to \infty$, with uniform temperature, and initially is in the form of a cylinder with dimensions $R_0 = H_0 = 1.0\,\text{m}\ (3.28\,\text{ft})$.

(a) Determine appropriate scale model laws. One possible reference velocity change ΔV was considered in Problem 6.4.

Answer:

π_6, π_7, and π_8 contain ν^2/gH_0^3 and R_0/H_0.

(b) If a 1/4 scale model test is to be performed, show that a fluid must be used with a viscosity of $1.0\,\text{m}^2/\text{s}\ (10.8\,\text{ft}^2/\text{s})$.

(c) Also show that for a 1/4 scale model test, 1 s of time in small scale corresponds to 2 in full size.

6.6 If the fluid in Problem 6.4 was glycerine, $\nu = 0.001\,\text{m}^2/\text{s}$, show that the ΔV derived for a high-viscosity fluid exceeds the advancing front velocity resulting from a dam break. Then use the dam break front velocity $2\sqrt{gH_0}$ for ΔV and show that the describing equations for the spreading rate on a flat surface reduce to the inviscid form

$$\nabla \cdot \mathbf{V} = 0, \qquad \frac{\partial \mathbf{V}}{\partial t} + \mathbf{V} \cdot \nabla \mathbf{V} + \frac{g_0}{\rho}\nabla P + g\mathbf{n}_y = 0$$

6.7 Scale modeling an unsteady submerged fluid jet involves a multidimensional flow field for which model laws can be obtained from Tables 6.3 and Eqs. (6.69) and (6.72). An estimated reference length L is required. If a model study is to determine the jet penetration length into stagnant fluid, L should be chosen accordingly.

A crude estimate of L for jet penetration could be made by equating the kinetic energy of a moving jet increment of length δ to the work done by the surrounding fluid as it brings the increment to rest. Let the initial velocity be V_0 at discharge from an opening of diameter D_0, and let the force of the surrounding fluid on the jet increment be $F_f = \tau_w \pi D_0 \delta$, where τ_w is the wall shear stress, estimated for flow in a pipe as $(f/4)\rho V_0^2/2g_0$, and f is an ordinary friction factor (usually between 0.01 and 0.1) which depends on the pipe Reynolds number and pipe relative roughness. Show that a reasonable reference length for studying jet penetration is $L \approx D_0/f$.

6.8 It is desirable to use a scale model for studying both the penetration and diffusion of an unsteady liquid jet submerged in identical liquid. The full-size average jet discharge velocity is V_0 from an orifice of diameter D_0. The density and kinematic viscosity of the liquid are ρ and ν. Also, $C \to \infty$ and $\beta \to 0$.

(a) Use Table 6.3 and Eqs. (6.69) and (6.70) with the reference pressure change equal to the dynamic value $\rho V_0^2/2g_0$, the reference length of Problem 6.7, and reference velocity change V_0 to show that mass conservation and momentum effects would be approximately modeled by preserving the model coefficients

$$\pi_7 = \frac{gD_0}{V_0^2}\frac{1}{f} \quad \text{and} \quad \pi_8 = \frac{f}{V_0 D_0/\nu}$$

Recall that f depends on the Reynolds number. These results show that both the Froude number gD_0/V_0^2 and the Reynolds number $V_0 D_0/\nu$ must be the same in the scale model and full size.

(b) Show that for a 1/9 scale model test, kinematic viscosity should be 1/27 of that in full size.

6.9 The penetration properties of an on-off submerged water jet discharge into surrounding water are to be estimated by a solution to the describing equations. Employ Table 6.3 and Eqs. (6.69)–(6.72) to obtain the simplified describing equations. Only mass conservation and momentum equations are needed, which require model coefficients π_1 through π_8. Water jet properties are

$$V_0 = \begin{cases} 100 \text{ cm/s (3.28 ft/s)}, & 0 \le t \le \tau \\ 0, & \tau < t \end{cases}$$

$\tau = 1.0$ s, $\rho = 1.0$ g/cc (62.4 lbm/ft^3), $D_0 = 10$ cm (0.328 ft), $\nu = 10^{-6}$ m^2/s (1.08 × 10^{-5} ft^2/s).

Answer:

$\nabla \cdot \mathbf{V} = 0$, $\partial \mathbf{V}/\partial t + \mathbf{V} \cdot \nabla \mathbf{V} + (g_0/\rho)\nabla P + \mathbf{b} = 0$.

6.10 A geometrically similar scale model test is to help observe natural convection circulation patterns in a rectangular fluid volume of length L, width W, and height H caused by local heating from an end wall at temperature T_h. The average fluid temperature in the volume is T_∞, and its other properties are α, β, and ν. Consider a reference change in temperature $\Delta T = T_h - T_\infty$ and an associated change in density from Eq. (2.116) at essentially constant pressure.

(a) Show that if buoyant and viscous forces are equated on a vertical fluid column of horizontal width δ and temperature difference ΔT, the rise velocity which permits steady heat conduction from the surface can be expressed by

$$\Delta V = \sqrt{\alpha g H \beta \, \Delta T/2\nu} = \sqrt{\alpha g H \, \Delta\rho/2\nu\rho}, \quad \alpha = \kappa/\rho c_p$$

(b) Show that for liquid flows, the dominant model coefficients of Table 6.3 are

$$\pi_t = Pr(Ra)^{-1/4}(\beta \Delta T)^{-1}, \quad \pi_9 = \frac{\Delta T}{T_\infty}$$

where $Pr = g_0 \mu c_p / \kappa$ and $Ra = Gr\,Pr$ with $Gr = g\beta H^3 \Delta T / \nu^2$.

(c) Show that scale model and full-size times and velocities are related to the scale factor according to

$$\frac{V_s}{V_f} = \frac{t_s}{t_f} = \left(\frac{H_s}{H_f}\right)^{1/5}$$

6.11 A large cylindrical tank of diameter $D = 6\,m$ and height $H = 6\,m$ initially contains stationary fluid of density ρ_0, kinematic viscosity ν, temperature $T_0 = 40°C$, and pressure $P_\infty = 1.0\,atm$. Similar fluid at temperature $T_h = 90°C$ begins to discharge into the tank at velocity $V_h = 10\,m/s$ from a submerged pipe of diameter $D_p = 10\,cm$. It is necessary to find a discharge pipe position and orientation which gives the most uniform temperature distribution in the least amount of time. A scale model study is being considered. Specify such a test when the fluid is air, $\nu = 2 \times 10^{-5}\,m^2/s$, and when the fluid is water, $\nu = 6 \times 10^{-7}\,m^2/s$.

6.12 Equations (1.19), (1.22), and (1.24) yield the differential equation for draining of low-viscosity liquid from a general tank in the form

$$\text{DE:} \quad \frac{dy}{dt} + \frac{C_d A_d}{A(y)} \sqrt{2g(y - y_d)} = 0$$

$$\text{IC:} \quad y(0) = H$$

where y is liquid elevation, y_d is the discharge elevation, C_d is a flow coefficient, A_d is the drain area, and $A(y)$ is the tank cross-sectional area at elevation y. Obtain scaling laws for a geometrically similar test by introducing normalized variables $y^* = y/\Delta y$ and $t^* = t/\Delta t$ which are $O(1)$. (Suggestion: $\Delta y = H - y_d$ and $\Delta t = M_i/\dot{m}_i$, where M_i is the total mass that will drain and $\dot{m}_i$ is the initial gravity draining rate.) Show that for a geometrically similar test, no model coefficients must be retained, and that full-size and scale model times and elevations are related by

$$y_f = y_s \frac{H_f}{H_s}, \quad t_f = t_s \sqrt{\frac{(H - y_d)_f}{(H - y_d)_s}}$$

6.13 A 1/9 small scale, geometrically similar model of a dam rupture and water cascading through a downstream village is to be photographed on movie film at an appropriate speed so that when it is shown at theater speed of 24 frames/s it will appear to be a full-size rupture of a 27 m (89 ft) high dam. If water is used in the scale model, show that the scene should be photographed at 72 frames/s. (Hint: Obtain scale model laws from Table 6.3 and Eqs. (6.69) and (6.70) with a reference velocity $\Delta V = 2\sqrt{gH}$, which corresponds to the advancing front of a dam rupture in Chapter 5.)

6.14 Gas discharges at critical flow into the atmosphere from a reservoir at pressure $P_0 = 10.1$ MPa (1470 psia), density $\rho_0 = 120$ kg/m^3 (7.48 lbm/ft^3), and $k = 1.3$ through a horizontal uniform pipe of diameter $D = 0.1$ m (0.328 ft) and length $L = 100$ m (328 ft). A plug valve at the discharge end is activated to close in a time of $\tau = 0.10$ s. Employ Table 6.3 and Eqs. (6.64)–(6.68) to determine the appropriate equations for predicting unsteady pressure, density, and velocity of fluid in the pipe so that the force on the valve can be determined. (See Example 6.3.)

6.15 The decompression transient of a gas vessel is to be demonstrated in a scale model test. The model volume $\mathcal{V}_s$ will be one fifth of the full-size vessel volume $\mathcal{V}_f$. Discharge will occur from a nozzle of area A_f. Gas initial temperature is T_{0f}. Use the perfect gas idealization and determine the model laws for an accurate test. Note that if $P_\infty/P_0 \leq P_c/P_0$ from Eq. (2.59), the discharge flow rate is critical, which would provide a convenient reference mass discharge rate.

Answer:

The model coefficients show that only the ratio of specific heats k must be the same in both the model and full-size systems. Also, show that times are related by

$$t_f = t_s \frac{A_s}{A_f} \frac{\mathcal{V}_f}{\mathcal{V}_s} \sqrt{\frac{R_{gs}T_s}{R_{gf}T_f}}$$

6.16 Consider the decompression of a vessel containing saturated liquid. Initial properties are mass M_i, energy U_i, pressure P_i, enthalphy h_i and discharge rate $\dot{m}_i$. The volume $\mathcal{V}$ is constant. Show that the describing state equations (6.95)–(6.97) can be approximated as $dU \approx (U/M)\,dM$ for saturated water discharge.

6.17 The equations for nonshocking hydrostatic wave propagation in a horizontal channel, Eqs. (4.1) and (4.2), are

$$\frac{\partial y}{\partial t} + V \frac{\partial y}{\partial x} + y \frac{\partial V}{\partial x} = \frac{\partial (BV)}{\partial x} - \frac{y - B}{D} V \frac{dD}{dx}$$

$$\frac{\partial V}{\partial t} + V \frac{\partial V}{\partial x} + g \frac{\partial y}{\partial x} = -\frac{g_0}{\rho} \frac{\partial P_\infty}{\partial x}$$

where V is the horizontal liquid velocity, y is the liquid elevation above a reference horizontal plane at $y = 0$, $B(x)$ is the floor elevation relative to $y = 0$, and $D(x)$ is the channel width. The liquid initially is at rest with $V_i = 0$ and $y_i = H$. The channel length is L. Consider cases where P_∞ is uniform.

(a) Normalize the describing equations with independent variables $t^* = t/\Delta t$ and $x^* = x/\Delta x$ and the dependent variables $y^* = (y - H)/\Delta y$ and $V^* = (V - V_i)/\Delta V$. Also, let $B^* = (B - B_0)/\Delta B$ and $D^* = (D - D_0)/\Delta D$. Show that the normalized equations and model coefficients can be written as follows:

$$\frac{\partial y^*}{\partial t^*} + \pi_1 V^* \frac{\partial y^*}{\partial x^*} + (\pi_2 + \pi_1 y^*) \frac{\partial V^*}{\partial x^*}$$

$$- (\pi_3 + \pi_4 B^*) \frac{\partial V^*}{\partial x^*} - \pi_5 V^* \frac{dB^*}{dx^*}$$

$$+ \frac{\pi_6 + \pi_8 y^* - \pi_9 + \pi_{10} B^*}{1 + \pi_7 D^*} V^* \frac{dD^*}{dx^*} = 0$$

and

$$\frac{\partial V^*}{\partial t^*} + \pi_1 V^* \frac{\partial V^*}{\partial x^*} + \pi_{11} \frac{\partial y^*}{\partial x^*} = 0$$

with

$$\pi_1 = \frac{\Delta V \, \Delta t}{\Delta x}, \quad \pi_2 = \frac{H \, \Delta t \, \Delta V}{\Delta y \, \Delta x}, \quad \pi_3 = \frac{B_0 \, \Delta t \, \Delta V}{\Delta y \, \Delta x}$$

$$\pi_4 = \frac{\Delta B \, \Delta t \, \Delta V}{\Delta y \, \Delta x} = \pi_5, \quad \pi_6 = \frac{H \, \Delta t \, \Delta V \, \Delta D}{\Delta y \, \Delta x D_0}$$

$$\pi_7 = \frac{\Delta D}{D_0}, \quad \pi_8 = \frac{\Delta t \, \Delta V \, \Delta D}{\Delta x D_0}, \quad \pi_9 = \frac{B_0 \, \Delta t \, \Delta V \, \Delta D}{\Delta y \, \Delta x D_0}$$

$$\pi_{10} = \frac{\Delta B \, \Delta t \, \Delta V \, \Delta D}{\Delta y \, \Delta x D_0}, \quad \pi_{11} = \frac{g \, \Delta y \, \Delta t}{\Delta V \, \Delta x}$$

(b) Consider the case of small disturbance propagation in which $H = 1.0 \, \text{m} \, (3.28 \, \text{ft})$, $L = 10 \, \text{m} \, (32.8 \, \text{ft})$, and the left wall begins moving sinusoidally to a velocity $V(0, t) = V_0 \sin 2\pi f t$. Recall from Chapter 4 that the speed of a hydrostatic wave is $\sqrt{gH}$, and show that if the liquid elevation in the channel is to be studied an appropriate reference time Δt is $1/f$ whenever $f \ll \sqrt{gH}/L$; show that if this condition is not satisfied, an appropriate Δt is $L/\sqrt{gH}$.

(c) Consider the case where $V_0 = 0.1 \text{ m/s}$ (0.328 ft/s) and $f = 2 \text{ Hz}$. Use Eq. (4.47) to estimate the disturbed reference level Δy in terms of $V = V_0$. Show that for $\Delta x = L$, the describing equations of part (a) can be simplified because π_1 is small, which causes the convective terms to be negligible.

(d) Consider additional specializations to part (c) where the wave channel has a flat bottom, $B_0 = \Delta B = 0$, and the width varies according to $D = D_0 + eF(x)$, where e is an amplitude and $F(x)$ is a nondimensional function whose variation is within 0.0 and 1.0. Show that the describing equations can be simplified to

$$\frac{\partial y}{\partial t} + H\frac{\partial V}{\partial x} + H\frac{F'(x)}{F(x)}V = 0, \qquad \frac{\partial V}{\partial t} + g\frac{\partial y}{\partial x} = 0$$

6.18 Consider a hydrostatic wave tank with a flat bottom, $B = 0$, and parallel walls, $D = \text{constant}$. Follow the procedure of Problem 6.17, but use a wall frequency of $f = 0.01 \text{ Hz}$, and show that since V_0/C is small the describing equations can be simplified to

$$\frac{\partial y}{\partial t} + V\frac{\partial y}{\partial x} + H\frac{\partial V}{\partial x} = 0, \qquad \frac{\partial V}{\partial t} + V\frac{\partial V}{\partial x} + g\frac{\partial y}{\partial x} = 0$$

6.19 It is hoped that a particular case of fluid structure interaction (FSI) can be studied in a scale model test. Suppose that a full-scale system (Fig. P6.19) involves a cylindrical tank of diameter D, water height

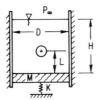

Figure P6.19

H, and rapid collapse of a steam bubble of initial radius R_i at elevation L. Air above the water is at P_∞. The tank walls are essentially rigid, but the bottom is flexible and can be simulated by a mass M with a spring constant K. A model test involves an evacuated glass sphere which is shattered to simulate sudden bubble collapse. Determine if a $1/10$ scale model test is feasible. If it is, provide model laws and scale-up formulas. Full-size values are $D = H = 4 \text{ m}$ (13.1 ft), $L = 2 \text{ m}$ (6.56 ft), $R_i = 20 \text{ cm}$ (0.656 ft), $M = 2500 \text{ kg}$ (5500 lbm), $K = 600 \text{ kN/m}$ (41122 lbf/ft), and $P_\infty = 1.0 \text{ atm}$.

6.20 Superheated steam at $T_s = 773°K$ (931°F) is discharged into saturated
water at 1 atm and $T_\infty = 373$ K (212°F). Stable bubbles of $R_i = 0.5$ cm
(0.197 in) radius form immediately. It is necessary to determine the
time-dependent size of a typical bubble as it cools to pool temperature. Since a bubble rises by buoyancy, a continuous flow of pool
water washes its boundary so that convection is expected to dominate
the heat transfer. An estimated overall heat transfer coefficient is
$\mathcal{H} = 30$ W/m²-K (5.3 B/h-ft²-°F). Useful steam and water properties
are

$$\rho_{gi} = 0.28 \text{ kg/m}^3 \ (0.017 \text{ lbm/ft}^3), \qquad k = c_p/c_v = 1.3$$

$$c_{pg} = 2.0 \text{ J/g-K} \ (0.48 \text{ B/lbm-°F}), \qquad c_{vg} = 1.5 \text{ J/g-K} \ (0.36 \text{ B/lbm-°F}),$$

$$R_g/c_{pg} = (k-1)/k = 0.23, \qquad R_g/c_{vg} = k - 1 = 0.3$$

$$h_g = 2675 \text{ J/g} \ (1150 \text{ B/lbm}), \qquad h_{fg} = 2258 \text{ J/g} \ (971 \text{ B/lbm})$$

(a) Compare bubble collapse times controlled by heat transfer Δt_q
and inertia Δt_I and show that Δt_q is slower and therefore dominates the problem. (Δt_I can be estimated from a Rayleigh bubble
collapse time, $R_i\sqrt{\rho_L/3kg_0P_\infty}$, in Chapter 9).
(b) Select an appropriate reference vaporization or condensation
mass flow rate $\dot{m}$ and show by Table 6.6 and Eqs. (6.95)–(6.97)
that the bubble transient size can be predicted by

Mass: $$-\dot{m} + \frac{dM}{dt} = 0$$

Energy: $$P_\infty \frac{d\mathcal{V}}{dt} + \frac{dU}{dt} - \dot{m}h_g + q = 0$$

State: $$\frac{dU}{dt} = \frac{U}{M}\frac{dM}{dt} + \frac{c_{pg}}{\beta_g}\frac{M}{\mathcal{V}}\frac{d\mathcal{V}}{dt}$$

7 ONE-DIMENSIONAL BULK AND WATERHAMMER FLOWS

This chapter gives formulations and solutions for analyzing unsteady liquid flows in pipes. Bulk and waterhammer solutions are discussed. Interesting and sometimes non-intuitive pipe flow effects are presented, which include boundary conditions associated with pumps, orifices, ideal nozzles, valve opening or closure transients, and flow area or fluid property discontinuities. The almost unlimited amplification of pressure signals in a tapered pipe and the filtering effect of pipe friction on various disturbance frequencies also are treated.

Although paper-and-pencil solutions are emphasized for clarity and understanding, waterhammer in multiple-pipe systems is more readily solved with machine computation. The method of characteristics is applied to the waterhammer equations to obtain formulations which are easily programmed. A procedure is given for synthesizing and solving pipe networks with a variety of boundary conditions.

7.1 Restrictions for Bulk Flow

The equations for one-dimensional pipe flows are summarized in Eqs. (6.64) to (6.68) with model coefficients listed in Table 6.3. Unsteady bulk liquid flows discussed in this chapter are incompressible and thermal conductivity is neglected.

Consider a pipe or other flow passage of length L where the flow initially is stagnant. A velocity disturbance ΔV creates a pressure disturbance ΔP, which can be estimated from the Bernoulli equation. An approximate time response is the transit time of a liquid particle, $L/\Delta V \gg L/C$. If heating occurs, the temperature disturbance above the steady value T_0 is ΔT. It follows that the important parameters for unsteady bulk liquid flow are

$$
\left.\begin{array}{ll}
C_r = \infty , & b = g \\[2mm]
\rho_r = \rho = \text{constant} & c_{pr} = c_p \\[2mm]
\beta_r = 0 , & \Delta P = \dfrac{\rho(\Delta V)^2}{2g_0} \\[4mm]
\kappa_r = 0 , & \Delta\rho = \dfrac{g_0}{C_r^2}\Delta P = 0 \\[4mm]
V_0 = 0 , & \Delta t = \dfrac{L}{\Delta V}
\end{array}\right\} \tag{7.1}
$$

If the flow is not initially stagnant, ΔV becomes the velocity disturbance above the initial V_0.

7.2 Bulk Flow Formulation

The model coefficients were obtained from Table 6.3. It was found that π_1, π_3, π_4, π_{10}, π_{11}, π_{13}, and π_{15} are zero. The remaining coefficients are

$$
\left.\begin{array}{ll}
\pi_2 = 1 & \pi_{14} = \dfrac{fL}{D} \\[4mm]
\pi_2^0 = 0 & \pi_{16} = \dfrac{gL}{(\Delta V)^2}\sin\theta \\[4mm]
\pi_6 = \dfrac{1}{2} & \pi_{17} = \dfrac{q'L}{\Delta V \rho T_0 A c_p} \\[4mm]
\pi_9 = \dfrac{\Delta T}{T_0} & \pi_{18} = \dfrac{L(\Delta V)^2}{g_0 T_0 c_p}
\end{array}\right\} \tag{7.2}
$$

If we employ the hydraulic diameter $D(x) = 4A(x)/P_w(x)$ in Eqs. (6.64) to (6.68), the equations for bulk liquid flows are

$$
\frac{\partial}{\partial x}(AV) = 0 \tag{7.3}
$$

$$
\frac{\partial V}{\partial t} + V\frac{\partial V}{\partial x} + \frac{g_0}{\rho}\frac{\partial P}{\partial x} + \frac{f|V|V}{2D(x)} + g\sin\theta = 0 \tag{7.4}
$$

and

$$\frac{\partial T}{\partial t} + V \frac{\partial T}{\partial x} = \frac{q'_{in} - q'_{out}}{\rho c_p A(x)} + \frac{f}{2g_0 D(x) c_p} V^3 \tag{7.5}$$

When the friction heating term $fV^3/2g_0 D(x)c_p$ is negligible, Eq. (7.5) corresponds to Eq. (3.62) for heated or cooled liquid flows.

Pressure and velocity are independent variables described by Eqs. (7.3) and (7.4) alone if ρ and f are relatively constant. This means that P and V can be obtained for most problems without use of the energy conservation principle of Eq. (7.5). However, once V is obtained, it can be used in Eq. (7.5) to obtain an unsteady temperature solution as discussed in Chapter 3. The bulk flow emphasis in this chapter is on pressure and velocity solutions of Eqs. (7.3) and (7.4).

7.3   Incompressible Accelerating Flows

Consider the flow passage shown in Fig. 7.1. Equation (7.3) is integrated to give $AV = f(t)$. Since $V(0, t) = V_0(t)$ and $A(0) = A_0$ at the pipe entrance, we have

$$V(x, t) = \frac{A(0)}{A(x)} V(0, t) = \frac{A_0}{A(x)} V_0(t) \tag{7.6}$$

The derivatives $\partial V/\partial t$ and $\partial V/\partial x$ are obtained from Eq. (7.6) and employed in (7.4) to give

$$\frac{A_0}{A(x)} \frac{dV_0}{dt} - \frac{A_0^2}{A(x)^3} V_0^2 \frac{dA}{dx} + \frac{g_0}{\rho} \frac{\partial P}{\partial x}$$

$$+ \frac{f}{2D(x)} \frac{A_0^2}{A(x)^2} |V_0|V_0 + g \sin \theta = 0 \tag{7.7}$$

If Eq. (7.7) is integrated with respect to x from 0.0 to x and $P = P(0, t)$ to $P(x, t)$, it follows that

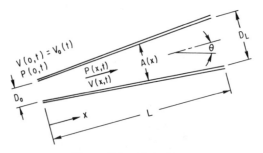

Figure 7.1   Flow Passage

$$P(x, t) - P(0, t) = \frac{\rho}{g_0} \left[\frac{V_0^2}{2} \left(1 - \frac{A_0^2}{A(x)^2} \right) - (g \sin \theta)x \right]$$

$$- \frac{\rho}{g_0} \left[A_0 \frac{dV_0}{dt} \int_0^x \frac{dx}{A(x)} + \frac{fA_0^2}{2} |V_0|V_0 \int_0^x \frac{dx}{A(x)^2 D(x)} \right] \qquad (7.8)$$

Equation (7.8) relates the local time-dependent pressure to entrance pressure and velocity and to the pipe geometry. Usually the entrance pressure and velocity are not independent. However, if pressures $P(0, t)$ and $P(L, t)$ at the entrance and exit boundaries are known, then Eq. (7.8) yields a differential equation for entrance velocity.

Consider the case of a conical pipe with diameter

$$D(x) = \left(1 + K_c \frac{x}{D_0} \right) D_0 \qquad (7.9)$$

where K_c is the diameter gradient

$$K_c = \frac{D_L - D_0}{L} \qquad (7.10)$$

If we employ $A(x) = \pi D(x)^2/4$, Eq. (7.8) becomes

$$P(x, t) = P(0, t) - \frac{\rho g}{g_0} x \sin \theta$$

$$+ \frac{\rho}{g_0} \left\{ \left[1 - \left(\frac{D_0}{D(x)} \right)^4 \right] \left[\frac{V_0^2}{2} - \frac{f}{4K_c} \frac{|V_0|V_0}{2} \right] - \frac{xD_0}{D(x)} \frac{dV_0}{dt} \right\} \qquad (7.11)$$

Suppose the entrance or discharge pressure disturbances are imposed. The flow could reverse direction, making it necessary to retain the term $|V_0|V_0$. However, if the unsteady flow always is in one direction, say from left to right, then $|V_0|V_0$ can be replaced by V_0^2.

Consider cases where the inlet and discharge pressures, $P(0, t)$ and $P(L, t)$, are known, and V_0 is always to the right. Equation (7.11) yields

$$\frac{dV_0}{dt} - \left(1 - \frac{f}{4K_c} \right) \left[1 - \left(\frac{D_0}{D_L} \right)^4 \right] \frac{D_L}{D_0} \frac{1}{L} \frac{V_0^2}{2}$$

$$= \left\{ \frac{g_0}{\rho} [P(0, t) - P(L, t)] - gL \sin \theta \right\} \frac{D_L}{D_0 L} \qquad (7.12)$$

If there is an initial velocity V_{0i}, we have the initial condition

$$V_0(0) = V_{0i} \qquad (7.13)$$

Moreover, if a step change occurs in either the inlet or outlet pressure, we have

$$P(0, t) - P(L, t) = \Delta P = \text{constant} \tag{7.14}$$

The corresponding solution to Eq. (7.12) is summarized in Eqs. (7.15) to (7.20).

GENERAL BULK FLOW IN A CONICAL PIPE, CONSTANT INLET AND OUTLET PRESSURES

$$V_0(t) = V_{ss} \frac{B - e^{-t/\tau}}{B + e^{-t/\tau}} \tag{7.15}$$

$$V_{ss} = \sqrt{\frac{b}{a}}, \qquad \text{steady-state velocity} \tag{7.16}$$

$$B = \frac{1 + V_{0i}\sqrt{a/b}}{1 - V_{0i}\sqrt{a/b}} \tag{7.17}$$

$$a = \left(1 - \frac{f}{4K_c}\right)\left[\left(\frac{D_0}{D_L}\right)^4 - 1\right]\frac{D_L}{2LD_0}, \qquad K_c = \frac{D_L - D_0}{L} \tag{7.18}$$

$$b = \left(\frac{g_0}{\rho}\Delta P - gL\sin\theta\right)\frac{D_L}{D_0 L}, \qquad \Delta P = P(0, t) - P(L, t) = \text{constant} \tag{7.19}$$

$$\tau = \frac{1}{2\sqrt{ab}}, \qquad \text{response time} \tag{7.20}$$

Consider a case where the pipe is attached through an ideal nozzle to a liquid reservoir at constant pressure P_0. The entrance pressure, obtained from the Bernoulli equation, is

$$P(0, t) = P_0 - \frac{V_0^2}{2g_0}\rho \tag{7.21}$$

and the corresponding solution for $V(0, t)$ is summarized in Eqs. (7.22)–(7.28).

BULK FLOW IN A CONICAL PIPE, IDEAL NOZZLE ATTACHMENT TO RESERVOIR AT CONSTANT PRESSURE P_0, AND CONSTANT DISCHARGE PRESSURE P_L

$$V_0(t) = V_{ss1} \frac{B_1 - e^{-t/\tau_1}}{B_1 + e^{-t/\tau_1}} \tag{7.22}$$

$$V_{ss1} = \sqrt{\frac{b_1}{a_1}}, \qquad \text{steady-state velocity} \qquad (7.23)$$

$$B_1 = \frac{1 + V_{0i}\sqrt{a_1/b_1}}{1 - V_{0i}\sqrt{a_1/b_1}} \qquad (7.24)$$

$$b_1 = \left(\frac{g_0}{\rho}\Delta P - gL\sin\theta\right)\frac{D_L}{D_0 L} \qquad (7.25)$$

$$a_1 = \left[\left(1 - \frac{f}{4K_c}\right)\left(\frac{D_0}{D_L}\right)^4 + \frac{f}{4K_c}\right]\frac{D_L}{2LD_0}, \qquad K_c = \frac{D_L - D_0}{L} \qquad (7.26)$$

$$\Delta P = P_0 - P_L = \text{constant} \qquad (7.27)$$

$$\tau_1 = \frac{1}{2\sqrt{a_1 b_1}}, \qquad \text{response time} \qquad (7.28)$$

If the pipe is uniform with $D_0 = D_L$, a limiting process is required for constants a and a_1 in Eqs. (7.18) and (7.26), which yields

$$\left.\begin{array}{l} a = \dfrac{f}{2D_0} \\[3mm] a_1 = \dfrac{f}{2D_0} + \dfrac{1}{2L} \end{array}\right\} \quad \text{as } K_c \to 0 \quad \text{uniform pipe} \qquad (7.29)$$

EXAMPLE 7.1: UNSTEADY DISCHARGE FROM CONICAL PIPE

A frictionless horizontal, conical pipe of length $L = 5$ m (16.4 ft) is attached to a cold-water reservoir at $P_0 = 15$ atm (220 psia). The pipe initially is filled with stagnant water at pressure P_0, and the discharge end is suddenly opened to another reservoir at pressure $P_L = 14$ atm (206 psia). Determine the discharge volume rate $\dot{V}(t)$ as a function of time for the following cases: (1) a *diverging* pipe with $D_0 = 0.1$ m (0.33 ft) and $D_L = 0.2$ m (0.66 ft); (2) a *converging* pipe with $D_0 = 0.2$ m (0.66 ft) and $D_L = 0.1$ m (0.33 ft); and (3) a *uniform* pipe with $D_L = D_0 = 0.1$ m (0.33 ft).

It is assumed that bulk flow will occur with negligible propagation effects. This assumption will be checked after results are obtained.

Equations (7.22) through (7.28) were employed with the following parameters:

Parameter	Case 1 (diverging)		Case 2 (converging)		Case 3 (uniform)	
b_1, m/s² (ft/s²)	40.4	(133)	10.1	(33)	20.2	(66)
a_1, 1/m (1/ft)	0.0125	(0.004)	0.8	(0.24)	0.1	(0.03)
V_{ss}, m/s (ft/s)	56.8	(186)	3.55	(11.6)	14.2	(46.6)
τ_1, s	0.704		0.176		0.35	
B_1,	1.0		1.0		1.0	

Equation (7.22) gives the entrance velocity $V_0(t)$, from which the volume flow rate $\dot{V} = A_0 V_0(t)$ was obtained. Results are shown in Fig. 7.2. It is seen that the diverging

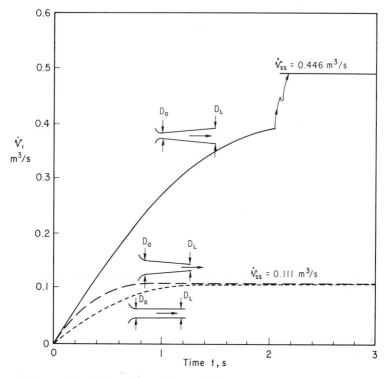

Figure 7.2 Volume Flow Rates in Example Non uniform Pipes

pipe has almost four times the steady-state volume flow rate as either the converging or uniform pipes. Actually, all of the asymptotic steady discharge velocities $V(L, t)$ are equal. Therefore, the pipe with the largest discharge area passes the highest volume flow rate. The time response τ_1 is longest for the diverging pipe.

This bulk flow solution is valid if propagation effects are negligible. The acoustic response time for $C = 1220$ m/s (4000 ft/s) in water is

$$t_p = \frac{L}{C} = \frac{5 \text{ m}}{1220 \text{ m/s}} = 0.0041 \text{ s}$$

The bulk flow response time is much greater than t_p for each case, which justifies the present analysis.

Although the volume rate from the diverging pipe is substantially higher than it is from the converging or uniform pipes of Example 7.1, if entrance pressure falls below the liquid saturation pressure then vapor will form and eventually cause flow choking at the entrance.

7.4 Pipe Reaction Forces

A design concern often is the pipe reaction force created by an unsteady flow. Figure 7.3 shows a segment of straight, nonuniform pipe of length L between two elbows adjoining other segments at angles θ_1 and θ_2. The three axes shown are not necessarily in the same plane. A dotted CV is drawn around the fluid. It is assumed that the elbow fluid volumes are small enough to neglect storage rates. A smaller CV is shown in the pipe segment for the purpose of expressing the stored momentum $\rho A(z)V\,dz$. Pressure forces P_1A_1 and P_2A_2 act on the CV at the elbows. Total forces exerted by the elbows in the z direction are F_1 and F_2, whereas F_{11} and F_{12} are elbow forces exerted in the directions of the bounding segments. The force F_s represents all other forces exerted on the fluid by the pipe segment, which may include friction, orifices, valves, or other fittings. The two heavy dashed lines at the elbows are normal to the segment axis. The momentum principle in the z direction for the entire CV is given by

$$\dot{m}_2 V_2 \cos\theta_2 - \dot{m}_1 V_1 \cos\theta_1 + \frac{d}{dt}\int_0^L \rho A(z)V\,dz$$
$$= g_0[F_1 - F_2 - F_s + (P_1A_1 - F_{11})\cos\theta_1 - (P_2A_2 - F_{22})\cos\theta_2]$$

Next, the momentum principle is written for each elbow CV in a direction normal to the z axis. Since there is no pressure force or momentum flow components normal to z at the heavy dashed boundaries, we obtain

$$g_0(P_1A_1 - F_{11}) = \dot{m}_1 V_1$$

$$g_0(P_2A_2 - F_{22}) = \dot{m}_2 V_2$$

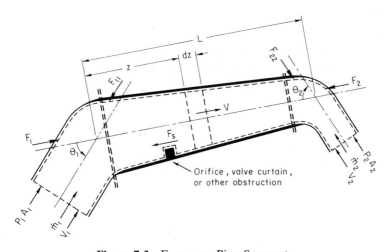

Figure 7.3 Forces on Pipe Segment

The total axial force acting on fluid contained by the pipe segment and elbows at each end is $F_1 - F_2 - F_s$. The corresponding equal and opposite force on the pipe is

$$R_b = -(F_1 - F_2 - F_s) = -\frac{1}{g_0}\frac{d}{dt}\int_0^L \rho A(z)V\,dz \qquad (7.30)$$

where the subscript b refers to a pipe segment *bounded* by two elbows. It is seen from Eq. (7.30) that the reaction force on a bounded segment of pipe depends on the storage rate of momentum and goes to zero whenever the flow is steady.

If the right elbow of Fig. 7.3 is straight, or is absent with discharge from an open pipe, F_2 vanishes and the reaction force becomes

$$R_{\text{open}} = -\left[(P_2 - P_\infty)A_2 + \frac{\dot{m}_2 V_2}{g_0} + \frac{1}{g_0}\frac{d}{dt}\int_0^L \rho A(z)V\,dz\right] \qquad (7.31)$$

If the flow is steady, R_{open} becomes the classical jet thrust.

Equations (7.30) and (7.31) give axial forces on any straight pipe segment, and apply for either bulk or propagative flows. The time-dependent forces obtained for each segment can be employed as forcing functions in a dynamic analysis of a piping system.

EXAMPLE 7.2: VALVE CLOSURE AND PIPE FORCES
Consider a segment of uniform pipe with length L and flow area A, bounded by two elbows at each end. The pipe contains liquid of density ρ flowing at velocity V_i. A valve must be specified for this system which decreases the flow to zero in a time period τ. Valve closure can be programmed to reduce velocity according to

$$V = \begin{cases} V_i\left[1 - \left(\dfrac{t}{\tau}\right)^n\right], & \dfrac{t}{\tau} < 1 \\[2ex] 0, & \dfrac{t}{\tau} > 1 \end{cases}$$

Determine the maximum force resulting for $n = 0.5$, 1.0, and 2.0.

Equation (7.30) gives the reaction force R_b as

TABLE 7.1 Maximum Force on a Pipe Segment during Velocity Reduction $V/V_i = 1 - (t/\tau)^n$

n	$\dfrac{t}{\tau}$	$[R_b(\rho A V_i L/g_0 \tau)^{-1}]_{\max}$
1/2	0	∞
1	1	1
2	1	2.0

$$\frac{R_b}{(\rho A V_i L / g_0 \tau)} = n\left(\frac{t}{\tau}\right)^{n-1}$$

for which maximum values are given in Table 7.1. The maximum force comes from $n = 1/2$, and the smallest from $n = 1$. It was assumed that bulk flow dominates this problem. This assumption is valid only if $L/C \ll \tau$. A valve which reduces flow velocity according to $n = 1/2$ should be avoided because of the large reaction force it imposes on the pipe.

7.5 Boundary Conditions

Whether a pipe flow is bulk or propagative, devices like orifices, valves, and pumps impose local differences between upstream and downstream fluid properties. The properties can be related by an algebraic or differential equation, by tables, or by graphs. Wherever these devices are placed, they introduce boundary conditions which affect unsteady flows in upstream and downstream pipe segments.

Each flow device discussed here is assumed to have a negligible storage rate of fluid, relative to the upstream and downstream pipe segments.

Figure 7.4 shows a general liquid flow device with upstream and downstream flow properties, related by the quasi-steady mass and energy conservation principles

$$\dot{m} = \rho A_u V_u = \rho A_d V_d \tag{7.32}$$

and

$$\frac{V_d^2 - V_u^2}{2g_0} + \frac{P_d}{\rho} - \frac{P_u}{\rho} + \frac{\Delta P_L}{\rho} = \frac{\mathscr{P}_{in} - \mathscr{P}_{out}}{\dot{m}} \tag{7.33}$$

Elevation changes across the device are neglected. When pressure loss occurs, it is given in the form

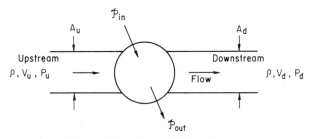

Figure 7.4   General Flow Device

$$\frac{\Delta P_L}{\rho} = K_L \frac{V^2}{2g_0} \tag{7.34}$$

where K_L is a loss coefficient corresponding to velocity V, which usually is V_u or V_d. The mechanical power terms $\mathscr{P}_{in}$ and $\mathscr{P}_{out}$ apply to turbomachinery like pumps or turbines. The head-flow characteristic of a pump or turbine can be written in terms of the pressure change as $\Delta P = f(\dot{m})$, for which the power is expressed by

$$\mathscr{P} = \left(\rho \frac{g}{g_0} \right) H \dot{\mathscr{V}} = \frac{\dot{m} \, \Delta P}{\rho} = \frac{\dot{m} f(\dot{m})}{\rho} \tag{7.35}$$

Boundary conditions for liquid flows through several devices are given by Eqs. (7.36)–(7.40), which are based on Eqs. (7.32) through (7.35).

BOUNDARY CONDITIONS FOR COMMON FLOW DEVICES

Ideal Pipe Attachment to Pressure Vessel

$$P_0 = P + \frac{V^2 \rho}{2g_0} \tag{7.36}$$

Orifice or Valve in a Uniform Pipe

$$P_1 - P_2 = K_L \frac{|V|V}{2g_0} \rho$$

$$V_1 = V_2 \tag{7.37}$$

Orifice or Valve at Discharge End of Pipe

$$P_1 - P_\infty = K_L \frac{V^2}{2g_0} \rho \tag{7.38}$$

Smooth Area Change

$$P_1 - P_2 = \frac{V_2^2 - V_1^2}{2g_0} \rho$$

$$A_1 V_1 = A_2 V_2 \tag{7.39}$$

Centrifugal Pump

$$P_2 - P_1 = \rho \frac{g}{g_0} H_{p0} - \rho \alpha \frac{g}{g_0} A_2^2 V_2^2$$

$$A_1 V_1 = A_2 V_2 \tag{7.40}$$

Boundary conditions where either pressure or velocity are specified are given by the general forms

$$P(0, t) = P_0(t) \quad \text{or} \quad V(0, t) = V_0(t) \tag{7.41}$$

and

$$P(L, t) = P_L(t) \quad \text{or} \quad V(L, t) = V_L(t) \tag{7.42}$$

A valve boundary condition often will include opening or closing transients, which determine the loss coefficient K_L of Eq. (7.37) as a function of time. Many valves can be approximated by an orifice with minimum flow area $a_v(t)$. The flow is assumed to proceed without losses from the valve entrance to $a_v(t)$ and on to the vena-contracta area $C_c a_v(t)$, where it is followed by an abrupt expansion loss to the full pipe area A. The corresponding loss coefficient is approximated by

$$K_L \cong \begin{cases} \left(\dfrac{A}{C_c a_v(t)} - 1 \right)^2 , & \text{velocity in pipe of area } A \\[4mm] \left(1 - \dfrac{C_c a_v(t)}{A} \right)^2 , & \text{velocity at vena-contracta} \end{cases} \tag{7.43}$$

where the contraction coefficient C_c is about 0.65 for a sharp-edged orifice and 1.0 for a gently curved entrance.

Flow characteristics which give flow rate and pressure loss are seldom available for commercial valves, except for some control valves. However, some valve flow coefficients are given by the manufacturer, and are usually defined by a flow coefficient

$$C_f = \frac{\dot{V}}{\sqrt{\Delta P_L}}$$

Figure 7.5 gives a graph of C_f, normalized by $C_{f,\text{full}}$, the value at full opening, in terms of the valve opening time or stroke for a typical globe

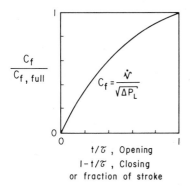

t/τ , Opening
$1-t/\tau$, Closing
or fraction of stroke

Figure 7.5 Flow Coefficient, Typical Globe Valve

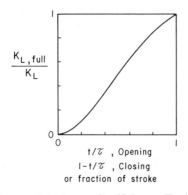

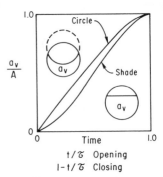

Figure 7.6 Loss Coefficient, Typical Globe Valve

Figure 7.7 Gate Valve Flow Area Transients

valve. If we employ ΔP_L and K_L from Eqs. (7.34) and (7.43), with the volume rate $\dot{V} = AV$, we find that the corresponding loss coefficient is

$$\frac{K_{L,\text{full}}}{K_L} = \left(\frac{C_f}{C_{f,\text{full}}}\right)^2 \tag{7.44}$$

where $K_{L,\text{full}}$ is the fully open valve pressure loss coefficient. Figure 7.6 gives $K_{L,\text{full}}/K_L$ in terms of the fraction of valve opening. Time scales also are shown in Fig. 7.6 for constant stroke speed, where τ is the full-stroke time.

Figure 7.7 gives the flow are ratio a_v/A for gate valves with circular and flat-edge disks as functions of time for constant stroke speed. Loss coefficients are obtained from Eq. (7.43).

A centrifugal pump operating at constant speed has a head-flow characteristic like the one shown in Fig. 7.8, where H_p is the head increase from

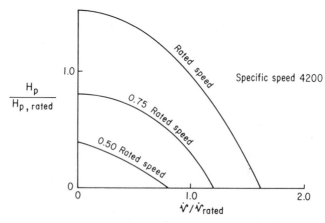

Figure 7.8 Centrifugal Pump Head-Flow-Speed Characteristic

suction to discharge, and $\dot{V}$ is the volume rate. Most pump characteristics can be approximated with the help of Eq. (7.40), by a quadratic equation of the form

$$\frac{1}{\rho}\frac{g_0}{g}\Delta P_p = H_p = H_{p0} - \alpha\dot{V}^2 \tag{7.45}$$

where H_{p0} is the shutoff head at zero flow.

Valves, nozzles, area changes, orifices, and centrifugal pumps provide boundary conditions which include upstream and downstream flow properties. A positive displacement pump provides a periodic velocity boundary condition. These and other devices can occur in various combinations in piping systems.

EXAMPLE 7.3: CENTRIFUGAL PUMP FLOW ACCELERATION

The straight pipe segment of Fig. 7.9 has a centrifugal pump at one end and an orifice of area a_0, which is followed by a quick-opening valve at the other. The pump is running with its suction line in a body of water at atmospheric pressure. The pipe is full, but the flow initially is zero. The pump characteristic follows Eq. (7.45) with $H_{p0} = 100$ m (328 ft) and $\alpha = 10$ s^2/m^5 (0.026 s^2/ft^5), with $\dot{V}$ given in m^3/s (ft^3/s). Sudden opening of the valve allows flow acceleration, and it is desirable to estimate the steady-state fluid velocity V_{ss}, response time τ, and the pipe initial and steady-state reaction forces for the parameters $L = 20$ m (65.6 ft), $A = 0.1$ m^2 (1.08 ft^2), and $C_c a_0 = 0.05$ m^2 (0.538 ft^2).

This problem is solved for the assumption of bulk flow. Friction and other pressure losses are assumed to be negligible compared with that of the orifice. Equation (7.40) is used to express the pressure rise across the pump in terms of velocity V, which yields

$$P_2 - P_\infty = \rho\frac{g}{g_0}H_{p0} - \rho\alpha\frac{g}{g_0}A^2V^2$$

The momentum principle for fluid in the pipe gives

$$P_2 - P_3 = \frac{\rho L}{g_0}\frac{dV}{dt}$$

and the orifice pressure loss of Eqs. (7.34) and (7.43) with $a_v = a_0$ is

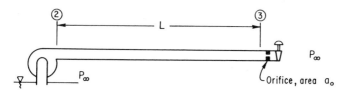

Figure 7.9 Centrifugal Pump Flow Acceleration

$$P_3 - P_\infty = \left(\frac{A}{C_c a_0} - 1\right)^2 \frac{V^2}{2g_0} \rho$$

Thus, the full problem becomes

$$\text{DE:} \qquad \frac{dV}{dt} + aV^2 = b$$

where

$$a = \left[\frac{1}{2}\left(\frac{A}{C_c a_0} - 1\right)^2 + g\alpha A^2\right]\frac{1}{L}, \qquad b = \frac{gH_{p0}}{L}$$

$$\text{IC:} \qquad t = 0, \qquad V = 0$$

This form of differential equation already was discussed in Section 7.3, for which the steady velocity is given by

$$V_{ss} = \sqrt{\frac{gH_{p0}}{(1/2)(A/C_c a_0 - 1)^2 + g\alpha A^2}}$$

and the response time is

$$\tau = \frac{L}{2\sqrt{gH_{p0}[(1/2)(A/C_c a_0 - 1)^2 + g\alpha A^2]}}$$

The open pipe reaction force is obtained from Eq. (7.31) with $P_2 = P_\infty$, $V_2 = VA/a_0 C_c$, and $\dot{m}_2 = \rho AV$, which gives

$$\frac{g_0 \tau^2}{\rho A L^2} R_{\text{open}} = -\left(\frac{A}{C_c a_0}\right)\left(\frac{V_{ss}\tau}{L}\right)^2 \tanh^2 \frac{t}{2\tau} - \frac{V_{ss}\tau/2L}{\cosh^2(t/2\tau)}$$

Steady fluid velocity in the pipe is seen to increase with the pump shutoff head H_{p0}. The response time is proportional to pipe length L and decreases as the orifice becomes smaller. The initial pipe force is obtained from

$$\left.\frac{g_0 \tau^2}{\rho A L^2} R_{\text{open}}\right|_{t/\tau = 0} = -\frac{V_{ss}\tau}{2L} = -0.168$$

whereas the steady-state reaction force, due to the discharging liquid jet, is

$$\left.\frac{g_0 \tau^2}{\rho A L^2} R_{\text{open}}\right|_{t/\tau \to \infty} = -\left(\frac{A}{C_c a_0}\right)\left(\frac{V_{ss}\tau}{L}\right)^2 = -0.227$$

Results for the parameters given are

V_{ss} $= 25.7\,\text{m/s}$ (84 ft/s)
τ $= 0.262\,\text{s}$
$R_{\text{open}} = -98\,\text{kN}$ (22030 lbf) initially at $t = 0$
$R_{\text{open}} = -132\,\text{kN}$ ($-29,674\,\text{lbf}$) at steady state, $t > \tau$

These forces are both exerted on the pipe in a direction toward the pump.

Problems so far have involved slow disturbances for which bulk flow analysis is justified. However, a large class of problems results in propagative flows under the general heading of *waterhammer*.

7.6   Restrictions for Waterhammer Flows

Waterhammer flows are characterized by short disturbance times with $t_d \leq L/C$, constant sound speed C, small Mach number V/C, and small density changes. Heating effects are unimportant, and temperature changes are negligible. Liquid flows and slightly compressed gas flows can be treated by waterhammer theory. If a velocity disturbance ΔV is imposed, the associated impact pressure disturbance of Section 1.11 is $\Delta P \approx \rho C \, \Delta V / g_0$. Thus, the important waterhammer parameters are

$$\left.\begin{aligned}
C_r &= C = \text{constant}\,, & \Delta t &= \frac{L}{C} \\[2mm]
\kappa_r &= 0\,, & \Delta P &= \frac{\rho C \, \Delta V}{g_0} \\[2mm]
\frac{\Delta \rho}{\rho} &\approx 0\,, & \frac{\Delta V}{C} &\approx 0 \\[2mm]
b &= g\,, & q' &= 0 \\[2mm]
\beta_r &= \beta \approx 0\,, & c_{pr} &= c_p
\end{aligned}\right\} \qquad (7.46)$$

7.7   Waterhammer Formulation

The one-dimensional flow model coefficients of Table 6.3, based on the parameters of Eq. (7.46), are

$$\left.\begin{aligned}
\pi_1 &= \pi_2 = \pi_6 = \frac{\Delta V}{C} \approx 0\,, & \pi_{14} &= \frac{fL}{D} \\[2mm]
\pi_2^0 &= \frac{V_0}{C} \approx 0\,, & \pi_{15} &= \frac{\beta C^2}{2 g_0 c_p} \\[2mm]
\pi_3 &= \pi_4 = 0\,, & \pi_{16} &= \frac{gL}{C^2} \sin \theta \approx 0 \\[2mm]
\pi_9 &= \frac{\Delta T}{T_0} = 0\,, & \pi_{17} &= \frac{q'L}{\rho T_0 A C c_p} = 0 \\[2mm]
\pi_{10} &= \frac{\beta C \, \Delta V}{g_0 c_p} \approx 0\,, & \pi_{18} &= \frac{C^2}{g_0 T_0 c_p} \\[2mm]
\pi_{11} &= \pi_{13} = 0\,,
\end{aligned}\right\} \qquad (7.47)$$

It follows that for low Mach number V_0/C, an adiabatic pipe, and negligible elevation effects, the one-dimensional mass conservation and momentum formulations of Eqs. (6.64) and (6.65) become

$$\frac{\partial P}{\partial t} + \frac{\rho C^2}{g_0} \frac{1}{A} \frac{\partial}{\partial x} (AV) = 0 \tag{7.48}$$

$$\frac{\partial V}{\partial t} + \frac{g_0}{\rho} \frac{\partial P}{\partial x} + \frac{f}{2D} |V|V = 0 \tag{7.49}$$

Equations (7.48) and (7.49) are waterhammer equations for rigid pipes with continuous, rather than abrupt, area transitions in the flow direction and with significant wall friction. The effect of pipe nonrigidity can be incorporated in the developments of this chapter by employing a modified fluid-and-pipe sound speed C_{fp}, given by

$$C_{fp} = C\sqrt{\frac{1}{1 + DY_L/\delta Y_p}} \tag{7.50}$$

where Y_L is the liquid elastic modulus, $\rho C^2/g_0$, Y_p is the modulus of the pipe, and δ is the pipe wall thickness.

7.8 General Solutions, Uniform, Frictionless Pipes

If the conditions of uniform area and negligible wall friction are introduced, Eqs. (7.48) and (7.49) reduce to

$$\frac{\partial P}{\partial t} + \frac{\rho C^2}{g_0} \frac{\partial V}{\partial x} = 0 \tag{7.51}$$

and

$$\frac{\partial V}{\partial t} + \frac{g_0}{\rho} \frac{\partial P}{\partial x} = 0 \tag{7.52}$$

Either V or P can be eliminated to obtain

$$\frac{\partial^2 P}{\partial t^2} - C^2 \frac{\partial^2 P}{\partial x^2} = 0 \tag{7.53}$$

or

$$\frac{\partial^2 V}{\partial t^2} - C^2 \frac{\partial^2 V}{\partial x^2} = 0 \tag{7.54}$$

The procedure described in Section 4.3 for hydrostatic waves is applied to Eq. (7.53) to obtain the general solution

$$P(x, t) = F_L\left(t + \frac{x}{C}\right) + F_R\left(t - \frac{x}{C}\right) \tag{7.55}$$

Partial integration of Eqs. (7.51) and (7.52) yields

$$V(x, t) = -\frac{g_0}{\rho C}\left[F_L\left(t + \frac{x}{C}\right) - F_R\left(t - \frac{x}{C}\right)\right] \tag{7.56}$$

The functions F_L and F_R are arbitrary, although they must be twice differentiable.

The propagation diagram of Fig. 7.10 shows basic features of the solution. The slopes of characteristic lines are $dt/dx = \pm C$. The numerical

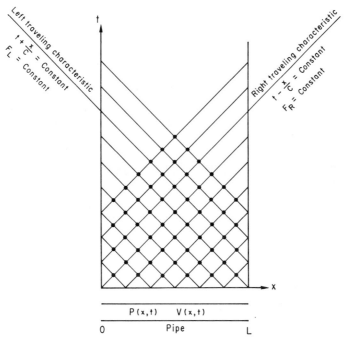

Figure 7.10 Propagation Diagram for Waterhammer

values of functions F_L and F_R are determined at their points of origination and remain constant on their respective characteristics. Initial pressure and velocity are known at all mesh points where left and right traveling characteristics originate at $t = 0$. Boundary conditions at $x = 0$ and $x = L$ also are known. Equations (7.55) and (7.56) are used with initial and boundary conditions to determine the originating values of F_L and F_R. The pressure and velocity at a given location x and time t are always obtained from the addition or subtraction of F_L and F_R in Eqs. (7.55) and (7.56). The solution procedure begins at $t = 0$ and advances along the characteristic lines to obtain P and V at each internal mesh point in the t, x propagation diagram.

The simplest boundary conditions involve specified values of velocity V or pressure P. However, many practical boundary conditions involve algebraic relationships between P and V. Flexible wall boundary conditions sometimes are in the form of a differential equation for P and V.

A simple paper-and-pencil procedure is discussed next for solving most uniform, frictionless pipe waterhammer problems with known initial conditions and either pressure, velocity, or functions of pressure and velocity boundary conditions. The procedure is similar to that of Section 4.6 for hydrostatic waves. Its development is a bit tedious, but it is organized in Table 7.2 for ease in solving problems.

Initial conditions are known in the form

$$P(x, 0) = P_i(x_i) \qquad \text{and} \qquad V(x, 0) = V_i(x_i) \tag{7.57}$$

for which F_L and F_R are obtained from Eqs. (7.55) and (7.56) as

$$F_{Li}(x_i) = \frac{1}{2}\left(P_i(x_i) - \frac{\rho C V_i(x_i)}{g_0}\right)$$

$$F_{Ri}(x_i) = \frac{1}{2}\left(P_i(x_i) + \frac{\rho C V_i(x_i)}{g_0}\right) \tag{7.58}$$

Either pressure or velocity boundary conditions can be specified from the general forms

$$B_0(t) = nP(0, t) + (n - 1)\frac{\rho C V(0, t)}{g_0}, \qquad n = 0 \text{ or } 1 \tag{7.59}$$

and

$$B_L(t) = mP(L, t) + (m - 1)\frac{\rho C V(L, t)}{g_0}, \qquad m = 0 \text{ or } 1 \tag{7.60}$$

If $n = 0$, $B_0(t)$ is a function of specified velocity $V(0, t)$, whereas if $n = 1$, $B_0(t)$ is a function of specified pressure $P(0, t)$. The boundary function

$B(L, t)$ is similarly specified in terms of either pressure $P(L, t)$, if $m = 1$, or velocity $V(L, t)$, if $m = 0$. It follows from Eqs. (7.55), (7.56), (7.59), and (7.60) that the functions

$$F_R(0, t_0) = \frac{B_0(t_0) - F_L(0, t_0)}{2n - 1}, \qquad \text{Left BC}$$

$$(7.61)$$

$$F_L(L, t_L) = B_L(t_L) - (2m - 1)F_R(L, t_L), \qquad \text{Right BC}$$

determine the originating values of $F_R(0, t_0)$ at the left boundary where the time is $t = t_0$ and of $F_L(L, t_L)$ at the right boundary where the time is $t = t_L$. Since F_R and F_L remain constant on characteristic lines spanning the diagram of Fig. 7.11, we can write

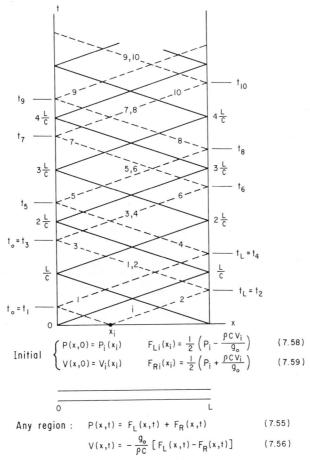

$$\text{Initial} \begin{cases} P(x,0) = P_i(x_i) & F_{Li}(x_i) = \frac{1}{2}\left(P_i - \frac{\rho c V_i}{g_0}\right) & (7.58) \\ V(x,0) = V_i(x_i) & F_{Ri}(x_i) = \frac{1}{2}\left(P_i + \frac{\rho c V_i}{g_0}\right) & (7.59) \end{cases}$$

$$\text{Any region :} \quad P(x,t) = F_L(x,t) + F_R(x,t) \qquad (7.55)$$

$$V(x,t) = -\frac{g_0}{\rho c}\left[F_L(x,t) - F_R(x,t)\right] \qquad (7.56)$$

Figure 7.11 Solution Procedure for Waterhammer in Uniform, Frictionless Pipes

TABLE 7.2 Solution Procedure: Waterhammer in Uniform, Frictionless Pipes (refer to Fig. 7.11)

Initial conditions

$$F_{Li}(x_i) = F_{L1} = \frac{1}{2}\left[P_i(x_i) - \frac{\rho C V_i(x_i)}{g_0}\right], \quad (x_i = x + Ct)$$

$$F_{Ri}(x_i) = F_{R2} = \frac{1}{2}\left[P_i(x_i) + \frac{\rho C V_i(x_i)}{g_0}\right], \quad (x_i = x - Ct)$$

Pressure $P = F_L + F_R$

Velocity $V = -\dfrac{g_0}{\rho C}(F_L - F_R)$

Boundary functions

Left, $x = 0$: $B_0(t) = nP(0,t) + (n-1)\dfrac{\rho C V(0,t)}{g_0}$

Right, $x = L$: $B_L(t) = mP(L,t) + (m-1)\dfrac{\rho C V(L,t)}{g_0}$

Specified boundary pressure: n or $m = 1$
Specified boundary velocity: n or $m = 0$

F_R Components

F_R Components	This Column is Cumulative	This Column is Cumulative
$F_{R1} = F_{R1,2} = F_{R4} = -\dfrac{1}{2n-1}F_{Li}(Ct-x)$	$+\dfrac{1}{2n-1}B_0\left(t-\dfrac{x}{C}\right)$	—
	$+$	
$F_{R5} = F_{R5,6} = F_{R8} = -\dfrac{2m-1}{(2n-1)^2}F_{Li}(Ct-x-2L)$	$\dfrac{2m-1}{(2n-1)^2}B_0\left(t-\dfrac{x}{C}-\dfrac{2L}{C}\right)$	$-\dfrac{1}{2n-1}B_L\left(t-\dfrac{x}{C}-\dfrac{L}{C}\right)$
	$+$	$-$
$F_{R9} = F_{R9,10} = F_{R12} = -\dfrac{(2m-1)^2}{(2n-1)^3}F_{Li}(Ct-x-4L)$	$\dfrac{(2m-1)^2}{(2n-1)^3}B_0\left(t-\dfrac{x}{C}-\dfrac{4L}{C}\right)$	$\dfrac{2m-1}{(2n-1)^2}B_L\left(t-\dfrac{x}{C}-\dfrac{3L}{C}\right)$
	$+$	$-$
$F_{R13} \vdots = F_{R13,14} = F_{R16} = -\dfrac{(2m-1)^3}{(2n-1)^4}F_{Li}(Ct-x-6L)$	$\dfrac{(2m-1)^3}{(2n-1)^4}B_0\left(t-\dfrac{x}{C}-\dfrac{6L}{C}\right)$	$\dfrac{(2m-1)^2}{(2n-1)^3}B_L\left(t-\dfrac{x}{C}-\dfrac{5L}{C}\right)$
		$-$
		$\dfrac{(2m-1)^3}{(2n-1)^4}B_L\left(t-\dfrac{x}{C}-\dfrac{7L}{C}\right)$

F_R Components

$$F_{R3} = F_{R3,4} = F_{R6} = \left(\frac{2m-1}{2n-1}\right)F_{Ri}(-Ct+x+2L)$$

$$F_{R7} = F_{R7,8} = F_{R10} = \left(\frac{2m-1}{2n-1}\right)^2 F_{Ri}(-Ct+x+4L)$$

$$F_{R11} = F_{R11,12} = F_{R14} = \left(\frac{2m-1}{2n-1}\right)^3 F_{Ri}(-Ct+x+6L)$$

$$F_{R15} \vdots = F_{R15,16} = F_{R18} = \left(\frac{2m-1}{2n-1}\right)^4 F_{Ri}(-Ct+x+8L)$$

F_L Components

F_L Components	*This Column is Cumulative*	*This Column is Cumulative*
$F_{L2} = F_{L1,2} = F_{L3}\ = -(2m-1)F_{Ri}(-Ct-x+2L)$	—	$+B_L\left(t+\dfrac{x}{C}-\dfrac{L}{C}\right)$ $+$
$F_{L6} = F_{L5,6} = F_{L7}\ = -\dfrac{(2m-1)^2}{2n-1}F_{Ri}(-Ct-x+4L)$	$-\left(\dfrac{2m-1}{2n-1}\right)B_0\left(t+\dfrac{x}{C}-\dfrac{2L}{C}\right)$ $-$	$\left(\dfrac{2m-1}{2n-1}\right)B_L\left(t+\dfrac{x}{C}-\dfrac{3L}{C}\right)$ $+$
$F_{L10} = F_{L9,10} = F_{L11}\ = -\dfrac{(2m-1)^3}{(2n-1)^2}F_{Ri}(-Ct-x+6L)$	$\left(\dfrac{2m-1}{2n-1}\right)^2 B_0\left(t+\dfrac{x}{C}-\dfrac{4L}{C}\right)$ $-$	$\left(\dfrac{2m-1}{2n-1}\right)^2 B_L\left(t+\dfrac{x}{C}-\dfrac{5L}{C}\right)$ $+$
$F_{L14} \;\vdots\; = F_{L13,14} = F_{L15}\ = -\dfrac{(2m-1)^4}{(2n-1)^3}F_{Ri}(-Ct-x+8L)$	$\left(\dfrac{2m-1}{2n-1}\right)^3 B_0\left(t+\dfrac{x}{C}-\dfrac{6L}{C}\right)$	$\left(\dfrac{2m-1}{2n-1}\right)^3 B_L\left(t+\dfrac{x}{C}-\dfrac{7L}{C}\right)$

F_L Components	*This Column is Cumulative*	*This Column is Cumulative*
$F_{L4} = F_{L3,4} = F_{L5}\ = \left(\dfrac{2m-1}{2n-1}\right)F_{Li}(Ct+x-2L)$	$-\left(\dfrac{2m-1}{2n-1}\right)^2 B_0\left(t+\dfrac{x}{C}-\dfrac{4L}{C}\right)$ $-$	$B_L\left(t+\dfrac{x}{C}-\dfrac{L}{C}\right)$ $+$
$F_{L8} = F_{L7,8} = F_{L9}\ = \left(\dfrac{2m-1}{2n-1}\right)^2 F_{Li}(Ct+x-4L)$	$\left(\dfrac{2m-1}{2n-1}\right)^3 B_0\left(t+\dfrac{x}{C}-\dfrac{6L}{C}\right)$ $-$	$\left(\dfrac{2m-1}{2n-1}\right)B_L\left(t+\dfrac{x}{C}-\dfrac{3L}{C}\right)$ $+$
$F_{L12} = F_{L11,12} = F_{L13}\ = \left(\dfrac{2m-1}{2n-1}\right)^3 F_{Li}(Ct+x-6L)$	$\left(\dfrac{2m-1}{2n-1}\right)^4 B_0\left(t+\dfrac{x}{C}-\dfrac{8L}{C}\right)$ $-$	$\left(\dfrac{2m-1}{2n-1}\right)^2 B_L\left(t+\dfrac{x}{C}-\dfrac{5L}{C}\right)$ $+$
$F_{L16} \;\vdots\; = F_{L15,16} = F_{L17}\ = \left(\dfrac{2m-1}{2n-1}\right)^4 F_{Li}(Ct+x-8L)$	$\left(\dfrac{2m-1}{2n-1}\right)^5 B_0\left(t+\dfrac{x}{C}-\dfrac{10L}{C}\right)$	$\left(\dfrac{2m-1}{2n-1}\right)^3 B_L\left(t+\dfrac{x}{C}-\dfrac{7L}{C}\right)$

$$t_0 = t - \frac{x}{C}, \qquad t_L = t - \frac{L-x}{C} \tag{7.62}$$

to express $F_R(x, t)$ and $F_L(x, t)$ anywhere on the t, x diagram of Fig. 7.11. The solution procedure has been formulated from Eqs. (7.61) and arranged in Table 7.2 so that it is relatively easy to obtain F_L and F_R at any location x and time t when initial conditions are given, and boundary values of P or V are specified as functions of time. The first column in Table 7.2 gives F_R and F_L for the regions identified in Fig. 7.11. The second column is given in terms of the initial condition functions F_{Li} or F_{Ri}. The third and fourth columns involve boundary functions B_0 and B_L. These terms are cumulative and must be added *up to and including* a given F_R or F_L being calculated. If we wanted F_{R9}, we would have

$$F_{R9} = -\frac{(2m-1)^2}{(2n-1)^3} F_{Li}(Ct - x - 4L) + \frac{1}{2n-1} B_0\left(t - \frac{x}{C}\right)$$

$$+ \frac{2m-1}{(2n-1)^2} B_0\left(t - \frac{x}{C} - \frac{2L}{C}\right) + \frac{(2m-1)^2}{(2n-1)^3} B_0\left(t - \frac{x}{C} - \frac{4L}{C}\right)$$

$$- \frac{1}{2n-1} B_L\left(t - \frac{x}{C} - \frac{L}{C}\right) - \frac{2m-1}{(2n-1)^2} B_L\left(t - \frac{x}{C} - \frac{3L}{C}\right)$$

Several examples of this procedure follow.

EXAMPLE 7.4: LINEAR PRESSURE REDUCTION AT PIPE END
Consider a pipe of length L which is initially closed and contains liquid at pressure P_0. Pressure at the right end $x = L$ is reduced linearly, providing the boundary condition

$$P(L, t) = P_0 - Kt$$

The left end at $x = 0$ remains closed with

$$V(0, t) = 0$$

Initial conditions are

$$P(x, t) = P_0, \qquad V(x, 0) = 0$$

Determine the left-end pressure $P(0, t)$.

The procedure of Table 7.2 and Fig. 7.11 are employed in this problem. The left boundary function for specified velocity requires $n = 0$, so that

$$B_0(t) = -\frac{\rho C V(0, t)}{g_0} = 0$$

whereas the right boundary function for specified pressure requires $m = 1$, for which

$$B_L(t) = P(L, t) = P_0 - Kt$$

Table 7.2 yields the following functions:

$$F_{Li}(x_i) = \frac{P_0}{2}, \qquad F_{Ri}(x_i) = \frac{P_0}{2}$$

$$F_{L1}(x, t) = \frac{P_0}{2}$$

$$F_{R2}(x, t) = \frac{P_0}{2}$$

$$F_{R1}(x, t) = F_{R1,2}(x, t) = F_{R4}(x, t) = \frac{P_0}{2}$$

$$F_{L2}(x, t) = F_{L1,2}(x, t) = F_{L3}(x, t)$$
$$= \frac{P_0}{2} - K\left[t + \frac{x - L}{C}\right]$$

$$F_{R3}(x, t) = F_{R3,4}(x, t) = F_{R6}(x, t)$$
$$= \frac{P_0}{2} - K\left[t - \frac{x + L}{C}\right]$$

$$F_{L4}(x, t) = F_{L3,4}(x, t) = F_{L5}(x, t)$$
$$= \frac{P_0}{2} - K\left[t + \frac{x - L}{C}\right]$$

$$F_{R5}(x, t) = F_{R5,6}(x, t) = F_{R8}(x, t)$$
$$= \frac{P_0}{2} - K\left[t - \frac{x + L}{C}\right]$$

Pressure at any x, t is obtained from

$$P(x, t) = F_L(x, t) + F_R(x, t)$$

Specifically, for the left end at $x = 0$ we have

$$P(0, t) = \begin{cases} P_1 = P_0 & 0 < t < \dfrac{L}{C} \\[2mm] P_3 = P_0 - 2K\left(t - \dfrac{L}{C}\right) & \dfrac{L}{C} < t < \dfrac{2L}{C} \\[2mm] P_5 = P_0 - 2K\left(t - \dfrac{L}{C}\right) & \dfrac{2L}{C} < t < \dfrac{3L}{C} \\[2mm] P_7 = P_0 - \dfrac{4KL}{C} & \dfrac{3L}{C} < t < \dfrac{4L}{C} \\[2mm] P_9 = P_0 - \dfrac{4KL}{C} & \dfrac{4L}{C} < t < \dfrac{5L}{C} \\[2mm] P_{11} = P_0 - 2K\left(t - \dfrac{3L}{C}\right) & \dfrac{5L}{C} < t < \dfrac{6L}{C} \\[2mm] \vdots \end{cases}$$

This solution shows that pressure at the left end decreases at twice the rate imposed at the right end, then holds constant for a wave round trip, then continues decreasing.

––––––––––––

Sometimes a boundary condition exists as a function of both pressure and velocity, as given in Eqs. (7.36) through (7.40). When this is the case, substitution of Eqs. (7.55) and (7.56) for pressure and velocity yields left and right boundary functions of the general forms

$$\text{Left BC:} \qquad F_R(0, t_0) = f_R[F_L(0, t_0)] \qquad (7.63a)$$

$$\text{Right BC:} \qquad F_L(L, t_L) = f_L[F_R(L, t_L)] \qquad (7.63b)$$

which are employed to obtain a tabular development similar to Table 7.2, but depending on the functions f_R and f_L instead of B_0 and B_L.

EXAMPLE 7.5: LIQUID DISCHARGE, $V(P)$ BOUNDARY

A frictionless pipe of length L and flow area A is attached at the left end, $x = 0$, to a vessel at pressure P_0 via an ideal nozzle. A partially open valve at the right end , $x = L$, determines the discharge flow rate by control of the throat area a_v. Liquid properties are ρ and C, and atmospheric pressure is P_∞. The valve in this case can be treated like another ideal nozzle, so the initial throat velocity is expressed from the Bernoulli equation in the form $V_{vi} = \sqrt{2g_0(P_0 - P_\infty)/\rho}$. Moreover, initial velocity in the pipe is $V_i = a_{vi}V_{vi}/A$. The valve with area a_{vi} begins to open or close with a linear area-time function to a final value a_{vf} according to

$$\frac{a_v}{A} = \begin{cases} \dfrac{a_{vi}}{A} + \left(\dfrac{a_{vf} - a_{vi}}{A}\right)\dfrac{t}{\tau}, & t < \tau \\[2ex] \dfrac{a_{vf}}{A}, & t \geq \tau \end{cases}$$

where τ is the opening or closing time. Velocity and pressure transients in the pipe are required.

The vessel boundary condition corresponds to that of Eq. (7.36), for which

$$P_0 - P(0, t) - \frac{V(0, t)^2}{2g_0}\rho = 0$$

and the discharge-end boundary condition of Eq. (7.38) is

$$P(L, t) - P_\infty = \phi(t)\frac{V(L, t)^2}{2g_0}\rho$$

$$\phi(t) = \left(\frac{A}{a_v}\right)^2 - 1$$

Substitution of Eqs. (7.55) and (7.56) yields boundary functions for any region in the forms of Eqs. (7.63a) and (7.63b) as

$$F_{Rj}(0, t_0) = f_R[F_{Lj}(0, t_0)] = F_{Lj}(0, t_0) - \frac{\rho C^2}{g_0}$$

$$+ \sqrt{\left(\frac{\rho C^2}{g_0}\right)^2 - 4\left(\frac{\rho C^2}{g_0}\right)\left[F_{Lj}(0, t_0) - \frac{P_0}{2}\right]}$$

$$t_0 = t - \frac{x}{C}, \qquad j = 1, 3, 5, \ldots$$

$$F_{Lk}(L, t_L) = f_L[F_{Rk}(L, t_L)] = F_{Rk}(L, t_L) + \frac{\rho C^2}{\phi(t_L)g_0}$$

$$- \sqrt{\left(\frac{\rho C^2}{\phi(t_L)g_0}\right)^2 + 4\frac{\rho C^2}{\phi(t_L)g_0}\left[F_{Rk}(L, t_L) - \frac{P_\infty}{2}\right]}$$

$$t_L = t - \frac{L - x}{C}, \qquad k = 2, 4, 6, \ldots$$

Since initial velocity in the pipe is V_i, the initial pressure is

$$P_i = P_0 - \frac{\rho V_i^2}{2g_0}$$

Thus, the initial conditions of Eq. (7.58) yield

$$F_{Li} = F_{Li}(x_i) = \frac{1}{2}\left(P_i - \frac{\rho C V_i}{g_0}\right) = \frac{1}{2}\left(P_0 - \frac{\rho V_i^2}{2g_0} - \frac{\rho C V_i}{g_0}\right)$$

$$F_{Ri} = F_{Ri}(x_i) = \frac{1}{2}\left(P_i + \frac{\rho C V_i}{g_0}\right) = \frac{1}{2}\left(P_0 - \frac{\rho V_i^2}{2g_0} + \frac{\rho C V_i}{g_0}\right)$$

both of which are constant over the range of x_i in Fig. 7.11.

Unsteady properties in the pipe are obtained from successive $F_R(0, t_0)$ and $F_L(L, t_L)$ calculations for regions of Fig. 7.11 in the sequence 1, (1, 2), 4, (3, 4), 5, (5, 6), 8, ... and 2, (1, 2), 3, (3, 4), 6, (5, 6), 7, (7, 8), 10, ... Several steps of the computation are outlined below.

$$F_{L1}(0, t_0) = F_{Li}(x_i) = \frac{1}{2}\left(P_0 - \frac{\rho V_i^2}{2g_0} - \frac{\rho C V_i}{g_0}\right)$$

$$F_{R1}(0, t_0) = f_R[F_{L1}(0, t_0)]$$

$$= F_{Li} - \frac{\rho C^2}{g_0} + \sqrt{\left(\frac{\rho C^2}{g_0}\right)^2 - 4\left(\frac{\rho C^2}{g_0}\right)\left(F_{Li} - \frac{P_0}{2}\right)}$$

$$F_{R1}(x, t) = F_{R1}(0, t_0)|_{t_0 = t - x/C} = F_{R1,2}(x, t)$$

$$= F_{R4}(x, t) = f_R[F_{L1}(0, t_0)]|_{t_0 = t - x/C}$$

$$F_{R4}(L, t_L) = F_{R4}(x, t)\Big|_{\substack{x=L \\ t=t_L}}$$

$$F_{L4}(L, t_L) = f_L[F_{R4}(L, t_L)] = F_{R4}(L, t_L)$$

$$+ \frac{\rho C^2}{\phi(t_L)g_0} - \sqrt{\left(\frac{\rho C^2}{g_0}\right)^2 + 4\left(\frac{\rho C^2}{\phi(t_L)g_0}\right)\left[F_{R4}(L, t_L) - \frac{P_0}{2}\right]}$$

$$F_{L4}(x, t) = F_{L3,4}(x, t) = F_{L5}(x, t)$$

$$= f_L[F_{R4}(L, t_L)]\big|_{t_L = t - (L - x)/C}$$

$$F_{L5}(0, t_0) = F_{L5}(x, t)\Big|_{\substack{x=0 \\ t=t_0}}$$

$$F_{R5}(0, t_0) = f_R[F_{L5}(0, t_0)]$$

The sequence continues with

$$F_{R5}(x, t) = F_{R5,6}(x, t) = F_{R8}(x, t) = f_R[F_{L5}(0, t_0)]\big|_{t_0 = t - x/C}$$

$$F_{R8}(L, t_L) = F_{R8}(x, t)\Big|_{\substack{x=L \\ t=t_L}}$$

$$F_{L8}(L, t_L) = f_L[F_{R8}(L, t_L)]$$

$$F_{L8}(x, t) = F_{L7,8}(x, t) = F_{L9}(x, t) = F_{L8}(L, t_L)\big|_{t_L = t - (L - x)/C}$$
$$\vdots$$

The other sequence is obtained as

$$F_{R2}(L, t_L) = F_{Ri}(x_i) = \frac{1}{2}\left(P_0 - \frac{\rho V_i^2}{2g_0} + \frac{\rho C V_i}{g_0}\right)$$

$$F_{L2}(L, t_L) = f_L[F_{R2}(L, t_L)]$$

$$= F_{Ri} + \frac{\rho C^2}{\phi(t_L)g_0}$$

$$- \sqrt{\left(\frac{\rho C^2}{\phi(t_L)g_0}\right)^2 + 4\left(\frac{\rho C^2}{\phi(t_L)g_0}\right)\left(F_{Ri} - \frac{P_0}{2}\right)}$$

$$F_{L2}(x, t) = F_{L2}(L, t_L)\big|_{t_L = t - (L - x)/C} = F_{L1,2}(x, t)$$

$$= F_{L3}(x, t) = f_L[F_{R2}(L, t_L)]\big|_{t_L = t - (L - x)/C}$$

$$F_{L3}(0, t_0) = F_{L3}(x, t)\Big|_{\substack{x=0 \\ t=t_0}}$$
$$\vdots$$

Consider the special case of a pipe with an initially closed valve so that

$$P(x, 0) = P_0 , \qquad V(x, 0) = 0$$

A sudden, complete pipe rupture at $x = L$ is obtained by setting $a_{vi} = 0$ and $a_v = a_{vf} = A$, which makes $\phi(t) = 0$ in the discharge boundary condition, reducing it in the limit to

$$F_{Lk}(L, t_L) = -F_{Rk}(L, t_L) + P_\infty$$

Discharge velocity was calculated from

$$V = -\frac{g_0}{\rho C}[F_{Lk}(L, t_L) - F_{Rk}(L, t_L)]\big|_{k=2,4,6,\ldots}$$

to obtain

$$V_2 = 0.65 \text{ m/s}, \qquad 0 < t \le \frac{L}{C}$$

$$V_4 = 0.65 \text{ m/s}, \qquad \frac{L}{C} < t \le \frac{2L}{C}$$

$$V_6 = 1.95 \text{ m/s}, \qquad \frac{2L}{C} < t \le \frac{3L}{C}$$

$$V_8 = 1.95 \text{ m/s}, \qquad \frac{3L}{C} < t \le \frac{4L}{C}$$
$$\vdots$$

The computation continues, with the discharge velocity increasing in steps whenever a wave returns from the vessel. Results are shown in Fig. 7.12. Equations (7.22)

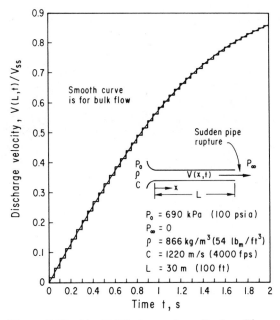

Figure 7.12 Liquid Discharge from Broken Pipe

through (7.28) were used for bulk flow solution to this problem, which also is shown in Fig. 7.12. This comparison shows that the fine structure of the flow transient is displayed by a waterhammer solution, whereas the coarse structure is obtained from bulk flow.

The computation of Example 7.5 is simplified because each region of Fig. 7.11 contains uniform P and V. All changes in P and V occur in steps at the boundaries. This observation is used next to describe a simplified, approximate solution to a class of waterhammer problems.

7.9   Approximate Waterhammer Solutions

Whenever each region of Fig. 7.11 contains uniform F_L and F_R, and consequently uniform P and V, the solution procedure of Table 7.2 is simplified considerably. This simplification arises whenever the boundary conditions can be treated as a series of step changes which occur at the arrival of each reflection of the initial disturbance. If continuous boundary conditions are approximated by a series of step changes, then, although the solution becomes approximate, it often is adequate for estimates.

Consider a uniform, frictionless pipe of length L. Let an initiating disturbance occur at the right end, as shown in Fig. 7.13. If the disturbed property is known as a function of time, as in Fig. 7.13(b), it is replaced by a series of steps occuring at time intervals of $2L/C$. If the boundary condition is algebraic, it is applied in steps whenever a reflection arrives. Regions in Fig. 7.13(a) are numbered according to $1, 2, 3, \ldots$, corresponding to the value of tC/L at the triangular region apex. The initially undisturbed region is designated by i.

Initial conditions are P_i and V_i, which give F_{Ri} from Eq. (7.58), from which $F_{R1} = F_{Ri}$. The boundary condition at $x = L$ yields $F_{L1} = F_{L2}$ in terms of the arriving F_{R1}. The boundary condition at $x = 0$ yields $F_{R2} = F_{R3}$ in terms of the arriving F_{L2}, and so on. The procedure is summarized in Table 7.3, with reference to Fig. 7.13. The procedure just described is another paper-and-pencil solution. It can be extended to include effects of lumped friction if desired [7.2]. When many time intervals are required, or when many cases must be investigated, PC assistance is desirable. However, if one is going to use a PC, the method of characteristics would be a better choice for obtaining accuracy and flexibility. This method is discussed in Section 7.21.

EXAMPLE 7.6: LIQUID PRESSURE AT CLOSING VALVE
A straight, uniform, frictionless pipe of length $L = 61$ m is attached by an ideal nozzle to a vessel at pressure $P_0 = 1000$ kPa. Liquid of density $\rho = 1000$ kg/m³ and

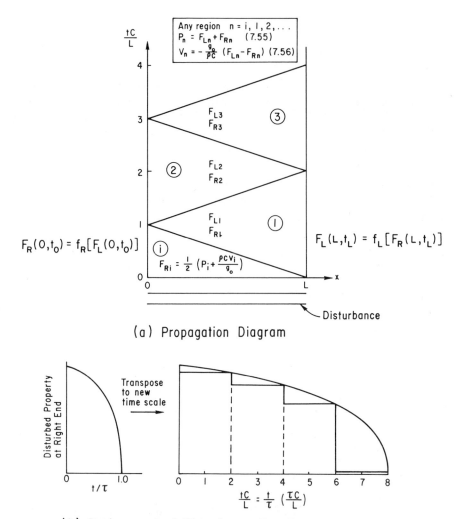

Any region $n = i, l, 2, \ldots$

$$P_n = F_{L_n} + F_{R_n} \quad (7.55)$$

$$V_n = -\frac{g_o}{\rho C} (F_{L_n} - F_{R_n}) \quad (7.56)$$

$$F_R(0, t_0) = f_R[F_L(0, t_0)]$$

$$F_{Ri} = \frac{1}{2} \left(P_i + \frac{\rho C V_i}{g_o} \right)$$

$$F_L(L, t_L) = f_L[F_R(L, t_L)]$$

(a) Propagation Diagram

Disturbance

Transpose to new time scale

Disturbed Property at Right End

$$\frac{tC}{L} = \frac{t}{\tau} \left(\frac{\tau C}{L} \right)$$

(b) Replacement of Disturbance Function with Steps

Figure 7.13 Approximate Waterhammer Solution

sound speed $C = 1220$ m/s flows steadily through the pipe where it is discharged to zero ambient pressure. A valve at the discharge end closes with a valve-to-pipe area ratio

$$\frac{a_v}{A} = 1 - \left(\frac{t}{\tau} \right)^n$$

where τ is the closure time and n is a positive exponent. Determine the pressure $P(L, t)$ just upstream from the valve for a closure time $\tau = 1.0$ s and values of $n = 0.5, 1.0,$ and 2.0. Assume that the valve flow contraction coefficient is $C_c = 1.0$.

The procedure of Fig. 7.13 and Table 7.3 was followed to solve this problem.

TABLE 7.3 Procedure for Approximate Waterhammer Analysis, Uniform, Frictionless Pipe, Disturbance at Right End (see Figure 7.13)

1. Start with a t, x propagation diagram as shown. Use Eqs. (7.55) and (7.56) to express $F_R(0, t)$ for the left boundary condition in the form of Eq. (7.63a).
2. If the right disturbed property boundary condition is given as $P(t)$ or $V(t)$, plot it on a graph with the time scale tC/L, and then approximate it as a series of steps, each of duration $\Delta(tC/L) = 2$. Use Eqs. (7.55) and (7.56) to express $F_L(L, t)$ in the form of Eq. (7.63b).
3. Obtain F_{Ri}, and $F_{R1} = F_{Ri}$ from initial conditions, Eq. (7.58).
4. Obtain $F_{L1} = F_{L2} = f_L[F_{R1}]$, Eq. (7.63b).
5. Obtain $F_{R2} = F_{R3} = f_R[F_{L2}]$, Eq. (7.63a).
6. Repeat for $F_{L3} = F_{L4}$, $F_{R4} = F_{R5}$, etc., for the required time period.
7. Determine pressures or velocities in each required region from $P = F_L + F_R$ and $V = -(g_0/\rho C)(F_L - F_R)$, Eqs. (7.55) and (7.56), respectively.

1. The left boundary condition of Eq. (7.36) is

$$P_0 - P - \frac{V^2 \rho}{2g_0} = 0$$

for which Eqs. (7.55) and (7.56) were employed to obtain

$$F_R = F_L - r + r\sqrt{1 - \frac{4}{r}\left(F_L - \frac{P_0}{2}\right)}$$

$$r = \frac{\rho C^2}{g_0} = 1488 \text{ MPa}, \qquad \frac{P_0}{2} = 0.5 \text{ MPa}$$

2. The right boundary condition for a valve closure, Eq. (7.38), is

$$P(L, t) - P_\infty = K_L \frac{V(L, t)^2}{2g_0} \rho$$

where $K_L = (A/a_v - 1)^2$ from Eq. (7.43). It follows that

$$F_L = F_R + \frac{r}{K_L} - \frac{r}{K_L}\sqrt{1 + \frac{4F_R}{r/K_L} - \frac{P_\infty}{r/K_L}}$$

The disturbance imposed is the valve area transient, shown in Fig. 7.14 for each case to be considered. The approximate steps are shown dotted on the tC/L time scale, as in Fig. 7.13.

3. The initial condition corresponds to $P_i = 0$, for which the Bernoulli equation yields $V_i = 44.76$ m/s. Thus, Eq. (7.58) gives

$$F_{Ri} = \frac{\rho C V_i}{2g_0} = 27.3 \text{ MPa}$$

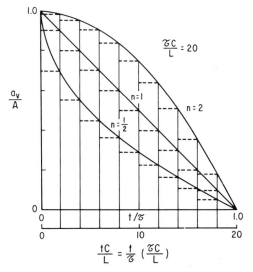

Figure 7.14 Approximation Steps, Valve Area Disturbance

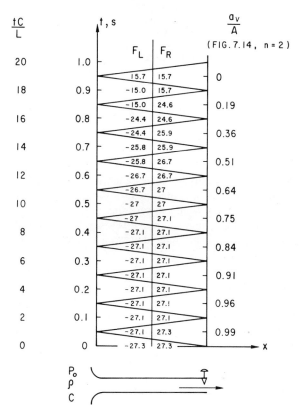

Figure 7.15 Approximate Waterhammer Solution, Valve Closure

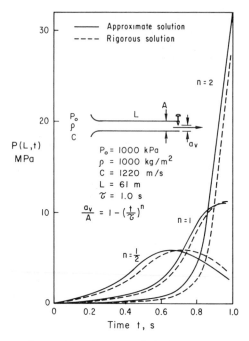

Figure 7.16   Approximate Waterhammer Solution, Pressure at Closing Valve

4, 5, 6. These steps are carried out in Fig. 7.15, which shows the F_L and F_R obtained for $n = 2$.

7. Pressure at the valve is shown in Fig. 7.16, which also gives $P(L, t)$ for cases with $n = 0.5$ and 1.0.

Results show that the valve closure corresponding to $n = 2$ reaches the highest pressure of 31.4 MPa. The reason is that most of the closure comes toward the end of the stroke in a short fraction of the closure time. Still, the pressure magnitude is below the full waterhammer value of $P = \rho C V_i/g_0 = 54.6$ MPa. The dashed lines were calculated by a rigorous procedure, described in Section 7.21.

Various boundary conditions have been considered at either end of a given pipe segment. However, there are instances where a boundary condition separates two pipes, and the upstream and downstream solutions are coupled. Analyses of systems which are coupled through a boundary condition often display features which can be incorporated into a design to alter the effects of waterhammer. Several of these are summarized in the next sections.

7.10 Disturbance Arrival at Discontinuity

A known pressure disturbance $P_0 - P_\infty$ is assumed to travel leftward in a uniform pipe whose properties are denoted by subscript I in Fig. 7.17. It arrives at a discontinuity where the flow area and fluid properties abruptly change to values designated by subscript II. The undisturbed fluid is moving leftward with steady velocity V_I in section I, and for which continuity gives $V_{II} = A_I V_I / A_{II}$ in section II. It is useful to know how such discontinuities affect the transmitted and reflected disturbances. When an area change occurs, two-dimensional effects will be present close to the discontinuity. However, propagation remains essentially one-dimensional about one or two diameters away, for which this analysis applies.

Equations (7.55) and (7.56) with the boundary conditions of (7.39) were

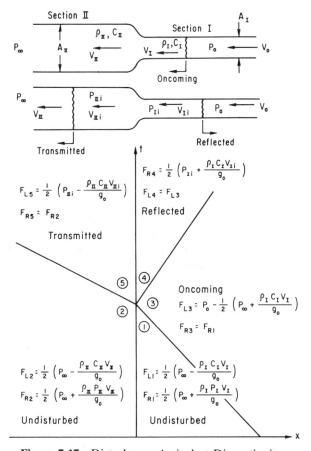

Figure 7.17 Disturbance Arrival at Discontinuity

employed to obtain F_L and F_R in each region of the propagation diagram of Fig. 7.17. It was found that between the transmitted and reflected waves

$$P_{Ii} = P_{IIi} = P_i$$

for low-Mach-number flows. Therefore, if the initial pressure is P_∞ and a pressure disturbance $P_0 - P_\infty$ arrives, the transmitted and reflected pressures are equal, given by

$$\frac{P_i - P_\infty}{P_0 - P_\infty} = 2\left(1 + \frac{A_{II}}{A_I}\frac{\rho_I}{\rho_{II}}\frac{C_I}{C_{II}}\right)^{-1} \tag{7.64}$$

It is seen from Eq. (7.64) that a pressure disturbance reflected at a dead end with $A_{II}/A_I = 0$ is twice the oncoming disturbance. Furthermore, if a disturbance in gas arrives at a liquid surface such that

$$\frac{\rho_{II}C_{II}}{\rho_I C_I} = \frac{\rho_L C_L}{\rho_g C_g} \to \infty$$

then both the reflected and transmitted pressure disturbances are twice the original. A firecracker above the surface of a lake will transmit twice the pressure disturbance into the water. (It would sound twice as loud to the fish.)

Conversely, if a disturbance traveling in liquid arrived at a gas with

$$\frac{\rho_{II}C_{II}}{\rho_I C_I} = \frac{\rho_g C_g}{\rho_L C_L} \approx 0$$

then a small pressure disturbance is transmitted to the gas. The explosion of a submerged firecraker could scarcely be heard by a fisherman in a boat directly above, although he might note the mysterious appearance of stunned fish floating on the surface. Figure 7.18 shows what can happen if acoustic disturbances arrive at a two-fluid muffler. The transmitted signal is substantially diminished whether it passes from gas to liquid to gas, or vice versa. This leads to the practical conclusion that condensation or trapped gas in instrument lines should be avoided.

Equation (7.64) can be used to show that a single-step expansion followed by a contraction, or vice versa, as shown in Fig. 7.19, can alter a transmitted disturbance and render it tolerable downstream. Figure 7.20 shows the character of transmitted disturbances for expansion and contraction area ratios of $1/4$ and $4/1$. It is seen that an expansion or contraction section is useful for dividing a disturbance into a series of smaller disturbances spread over a period of time determined by length L. A series of expansions and contractions is a variation of this concept used in some straight-through internal combustion engine mufflers.

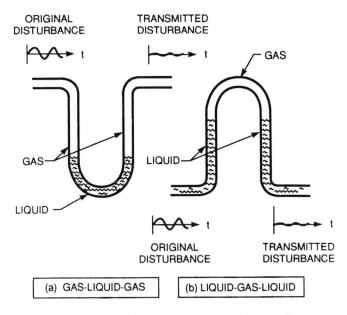

ORIGINAL
DISTURBANCE

TRANSMITTED
DISTURBANCE

GAS

GAS

LIQUID

LIQUID

ORIGINAL
DISTURBANCE

TRANSMITTED
DISTURBANCE

(a) GAS-LIQUID-GAS   (b) LIQUID-GAS-LIQUID

Figure 7.18   Two-Fluid Mufflers. (a) Gas-Liquid-Gas; (b) Liquid-Gas-Liquid

Suppose that a pressure disturbance is expected in a flow passage which muct be reduced to a smaller area. Intuitively, one might expect that several step area reductions, rather than a single full reduction, would result in a less severe transmitted pressure. But waterhammer is not always intuitive. A one-step area reduction to a closed end corresponds to

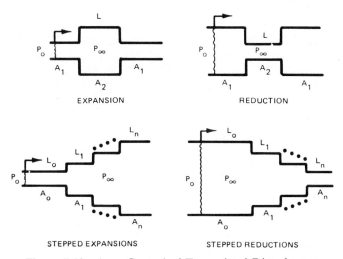

EXPANSION

REDUCTION

STEPPED EXPANSIONS

STEPPED REDUCTIONS

Figure 7.19   Area Control of Transmitted Disturbances

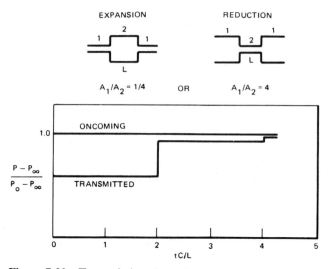

Figure 7.20 Transmission through Expansion or Reduction

$$\frac{P_1 - P_\infty}{P_0 - P_\infty} = 2$$

from Eq. (7.64), which shows that the oncoming pressure doubles. Now suppose that a two-step area reduction is employed with $A_1/A_0 = 1/2$ and $A_2/A_1 = 0$. It follows by repeated use of Eq. (7.64) that the end pressure is

$$\frac{P_2 - P_\infty}{P_0 - P_\infty} = 2^2\left(\frac{1}{1 + 1/2}\right)\left(\frac{1}{1 + 0}\right) = 2.7$$

which shows that successive area reductions can create higher pressure disturbances than a single reduction! Many consecutive reduction steps can be employed to achieve extremely high pressure transmission.

7.11 Orifice Effects

Consider the pipe which contains an orifice in Fig. 7.21. The pressure loss boundary condition is expressed by Eq. (7.37). Equations (7.55) and (7.56) were employed with (7.37) to obtain the transmitted and reflected presure disturbances shown in Fig. 7.21. It is seen that if no orifice exists corresponding to $K_L = 0$, reflected and transmitted disturbances are equal to the oncoming value, as expected. Also, if $K_L \rightarrow \infty$, for a full pipe closure, the transmitted disturbance is zero, and the reflected disturbance is twice the oncoming value. The loss coefficient of Eq. (7.43) for a case with $C_c = 0.6$ was used to show that an orifice-to-pipe diameter ratio of about 0.1 would

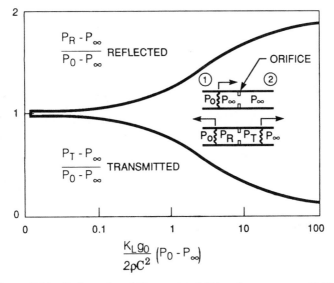

Figure 7.21 Reflected and Transmitted Disturbances at an Orifice

be required to attenuate transmission of a 5-atm disturbance to 2.5 atm. Therefore, an orifice must be small to be effective in attenuating acoustic signals.

Figure 7.21 also shows that for a given orifice, if the disturbance $P_0 - P_\infty$ increases, the fraction of it which is transmitted decreases. This can be observed by noting that low-level sounds can be heard by passengers through a slightly opened window of an automobile, whereas the loud noise of a passing diesel tractor-trailer rig is heard at a comfortable level.

7.12 Orifice in Pipe End Wall

Orifices in the previous section had fluid on both the upstream and downstream sides. It was shown that small orificing is required in order to substantially reduce the transmitted disturbance. However, orificing the end of a pipe, connected to a constant pressure environment as shown in Fig. 7.22, can be used to reduce reflected disturbances. The orifice pressure loss of Eq. (7.38) was employed to obtain the reflected disturbance, which is graphed in Fig. 7.22. It is seen that if no orifice is present, with $K_L = 0$, then zero pressure disturbance is reflected from the open pipe end. Also, if $K_L \to \infty$, corresponding to a closed end, the reflected pressure is again twice the oncoming value. But, perhaps amazingly, Fig. 7.22 shows that if an orifice is specified such that $(K_L g_0 / 2\rho C^2)(P_0 - P_\infty) = 1.0$, a step pressure disturbance and reflection will be identical! An observer in the pipe would not sense a reflected signal, a result which also corresponds to a long pipe

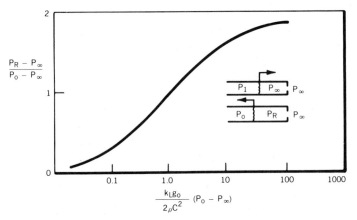

Figure 7.22 Reflected Disturbance at Pipe End Orifice

extension without an orifice. This idea can be useful when a piece of equipment must be tested for response to an abrupt, sustained pressure increase.

7.13 Gas Cushion Boundary

A liquid-filled pipe can be attached to a region of gas, such as a trapped bubble or surge tank, to reduce waterhammer pressure. It is useful to obtain characteristics for a cushion like the one shown in Fig. 7.23.

The liquid-gas boundary displacement Z from its unstressed position is assumed to be small relative to the pipe length. If the initial gas volume is $\mathcal{V}_i$ at pressure P_i, adiabatic, uniform compression by an applied pressure disturbance $P' = P - P_i$ causes the volume to change as

$$P' = P_i\left[\left(\frac{\mathcal{V}_i}{\mathcal{V}}\right)^k - 1\right] \tag{7.65}$$

Equation (7.65) is valid provided that propagation time in the gas is short relative to that in the liquid column. Let the gas volume be related to displacement of the gas-liquid interface by

$$\mathcal{V} = \mathcal{V}_i - sZ \tag{7.66}$$

The constant s could be, for example, the pipe cross-sectional area if a quantity of gas was trapped at the end of a closed pipe. If $\mathcal{V}$ is eliminated from Eqs. (7.65) and (7.66) and the result is differentiated, it follows that with $dZ/dt = V$,

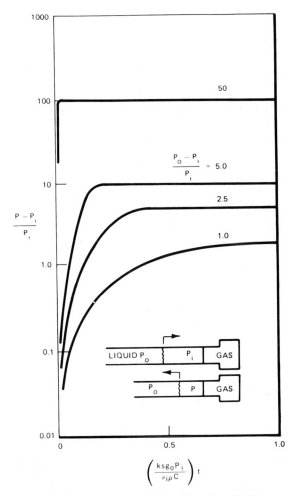

Figure 7.23   Step Reflection from Gas Cushion

$$\frac{dP'}{dt} = \frac{ksP_i}{\mathscr{V}_i}\left(\frac{P'}{P_i}+1\right)^{(k+1)/k} V$$

Consider an oncoming rightward pressure disturbance

$$P_0'(x,t) = P_0(x,t) - P_\infty = F_R\left(t - \frac{x}{C}\right)$$

Equations (7.65) and (7.66) are employed to obtain

$$\frac{d(P'/P_i)}{dt} = -\left(\frac{ksg_0 P_i}{\mathscr{V}_i\rho C}\right)\left(\frac{P'}{P_i}+1\right)^{(k+1)/k}\left(\frac{P'}{P_i}-2\frac{P_0'}{P_i}\right) \qquad (7.67)$$

If the time of disturbance arrival is considered zero, an appropriate initial conditions is $P'(0) = 0$.

Results for an oncoming step disturbance are shown in Fig. 7.23. All reflected pressure disturbances eventually would reach twice the oncoming value if the liquid column is long. However, if the liquid column is short, the time delay of a reflected pressure rise would permit substantial relief. The pressure rise time can be extended by increasing the initial gas volume $\mathcal{V}_i$. This kind of device is used to reduce waterhammer effects in processes with rapid valve opening or closing.

7.14   Spring-Mass Boundary

Consider a pipe which has spring-mass properties associated with axial motion, like a segment between elbow attachments to other pipes. When a waterhammer disturbance enters or leaves a pipe segment, pipe axial motion will affect the propagating disturbance.

Figure 7.24 shows a pressure disturbance $P_0' = P_0 - P_\infty$ approaching a

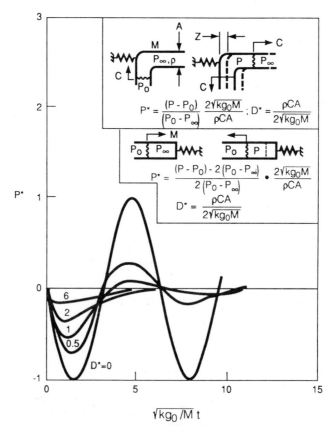

Figure 7.24 Step Pressure Disturbance at Spring-Mass Boundary

horizontal pipe segment from the downward side. Pipe axial motion begins when the disturbance enters the horizontal segment. Leftward pipe motion begins to reduce the pressure disturbance.

If M is the pipe mass and K_p is the equivalent spring constant for axial motion, displacement Z and velocity dZ/dt are related by Newton's law

$$M \frac{d^2 Z}{dt^2} = g_0(P'A - K_p Z) \tag{7.68}$$

Equations (7.55), (7.56), and (7.68) provide the full problem for P' in the form

DE: $\quad \dfrac{d^2 P'}{dt^2} + \left(\dfrac{\rho C A}{2M}\right) \dfrac{dP'}{dt} + \left(\dfrac{K_p g_0}{M}\right) P' = \dfrac{d^2 P_0'}{dt^2} + \left(\dfrac{K_p g_0}{M}\right) P_0'$ (7.69)

ICs: $\quad P'(0) = P_0'(0)$

$$\frac{dP'(0)}{dt} = \frac{dP_0'(0)}{dt} - \left(\frac{\rho C A}{2M}\right) P_0'(0) \tag{7.70}$$

Equation (7.69) describes a linearly damped spring-mass system with a forcing function. The quantity $\rho C A/2M$ acts like a damping coefficient and increases as the pipe mass decreases. A small value of M permits less restrained leftward acceleration and a consequent reduction of P' at first. The subsequent time history of P' depends on the disturbance nature as well as on pipe motion parameters. This problem is an example of fluid structure interaction.

The case of an oncoming step disturbance

$$P_0'(t) = P_0' = P_0 - P_\infty = \text{constant}$$

was considered, for which the solution is graphed in Fig. 7.24. The case

$$D^* = \frac{\rho C A}{2\sqrt{K_p g_0 M}} = 2.0$$

corresponds to critical damping, whereas $D^* > 2$ is overdamped and $D^* < 2$ is underdamped.

A similar solution also was obtained for the closed pipe sketched in Fig. 7.24.

It can be shown from Fig. 7.24 that larger mass M or a stiffer pipe with a higher K_p results in less damping and higher oscillatory pressure amplitudes in the elbow. The least disturbance is transmitted in a pipe whose product $K_p M$ is small.

7.15 Frictional Attenuation

A disturbance at one end of a long pipe sometimes is distorted by friction as it travels. Such distortions can be troublesome in instrument lines. A time scale shift between the disturbance and its measurement can be corrected by the known line acoustic delay time. However, frictional effects can attenuate certain harmonics, thus distorting the measured signal. It is useful to know how much attenuation of various harmonics should be expected.

Consider a fluid line of uniform area and zero initial velocity. It is assumed that the friction parameter π_{14} in Eq. (7.47) is substantial. If V is eliminated from Eqs. (7.48) and (7.49) with $dA/dx = 0$ and the average value of $f|V|/D$ is employed, we obtain

$$\frac{\partial^2 P'}{\partial t^2} - C^2 \frac{\partial^2 P'}{\partial x^2} + R_f \frac{\partial P'}{\partial t} = 0 \qquad (7.71)$$

where the resistance constant R_f is given by

$$R_f = \frac{\bar{f}|V|}{D} \qquad (7.72)$$

Consider a harmonic pressure disturbance of the form

$$P'(0, t) = P_0 \cos \omega t \qquad (7.73)$$

in a long line without reflections. A steady solution at any x is given by

$$\frac{P'(x, t)}{P_0} = \exp(-h_1 x) \cos\left[\omega\left(t - \frac{x}{C}\right) + \frac{x\omega}{C} - h_2 x\right] \qquad (7.74)$$

where

$$\begin{Bmatrix} h_1 \\ h_2 \end{Bmatrix} = \frac{\omega}{C}\left[1 + \left(\frac{R_f}{\omega}\right)^2\right]^{1/4} \begin{Bmatrix} \cos\dfrac{\theta}{2} \\ \sin\dfrac{\theta}{2} \end{Bmatrix} \qquad (7.75)$$

with

$$\theta = \tan^{-1}\left(-\frac{R_f}{\omega}\right) \qquad (7.76)$$

It follows that the amplitude attenuation is expressed by

$$\frac{P_{max}}{P_0} = \exp(-h_1 x) \qquad (7.77)$$

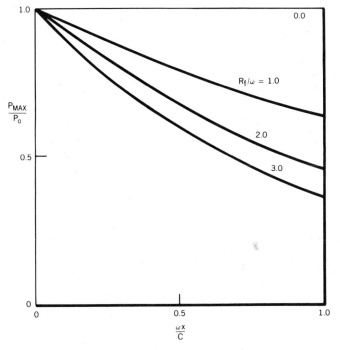

Figure 7.25 Friction Attenuation

The graph of P_{max}/P_0 in Fig. 7.25 shows that amplitudes of harmonic signals diminish with increasing distance from the source. Although it is not obvious in Fig. 7.25, it can be shown that for a given R_f, higher harmonics, which correspond to smaller R_f/ω, attenuate more than lower harmonics. This formulation should be useful in specification of instrument lines to avoid signal distortion by friction.

7.16 Continuous Area Variation

Velocity is eliminated from Eqs. (7.48) and (7.49) for a flow section of known area $A(x)$ and negligible friction to yield

$$\frac{\partial^2 P}{\partial t^2} - C^2 \frac{\partial^2 P}{\partial x^2} - \frac{C^2}{A} \frac{dA}{dx} \frac{\partial P}{\partial x} = 0 \qquad (7.78)$$

with the assumed entrance condition

$$P(0, t) = P_0(t) \qquad (7.79)$$

Since the propagation speed is unaffected by geometry, a general solution might be expected of the form

$$P(x, t) = F(x)F_R\left(t - \frac{x}{C}\right)$$

It can be shown that if the pipe diameter varies linearly with flow area as

$$\frac{A(x)}{A_0} = \left[1 + \frac{1}{2}\left(\frac{1}{A}\frac{dA}{dx}\right)_0 x\right]^2 \tag{7.80}$$

the pressure solution is given by

$$P(x, t) = \left[1 + \frac{1}{2}\left(\frac{1}{A}\frac{dA}{dx}\right)_0 x\right]^{-1} P_0\left(t - \frac{x}{C}\right) \tag{7.81}$$

If the area decreases in the direction of propagation (that is, if $dA/dx < 0$), Eq. (7.81) shows that a pressure disturbance will increase as the denominator approaches zero. If the diameter varies as $D(x)/D_0 = 1 - x/L$, with area $A(x)/A_0 = (1 - x/L)^2$ in a cone of length L, the pressure at $x/L = 0.9$ will be 10 times the entering pressure signal! This observation shows that extremely high pressures can be obtained from relatively small disturbances entering a tapered section. It also helps to explain why the old-fashioned ear horns worked for people hard of hearing.

7.17 Waterhammer Force at Pipe Area Change

Suppose that in Fig. 7.26, a *decompressive* disturbance $P_2 - P_1 > 0$ moves in a pipe of area A_1. It arrives at an abrupt area transition to a pipe of area A_2, where it undergoes simultaneous reflection and transmission as discussed in Section 7.10. It is sometimes necessary to predict the resulting pipe forces. Equation (7.30) was employed with the results of Section 7.10 to obtain the following pipe forces on bounded segments:

$$\frac{F_{1,\text{onc}}}{(P_2 - P_1)A_1} = 1, \qquad \text{oncoming}$$

$$\frac{F_{1,\text{refl}}}{(P_2 - P_1)A_1} = \frac{A_2/A_1 - 1}{A_2/A_1 + 1}, \qquad \text{reflected} \tag{7.82}$$

$$\frac{F_{2,\text{trans}}}{(P_2 - P_1)A_2} = \frac{2}{A_2/A_1 + 1}, \qquad \text{transmitted}$$

The total force of transmission and reflection is

$$\frac{F_{\text{combined}}}{(P_2 - P_1)A_1} = \frac{3 - A_1/A_2}{1 + A_1/A_2} \tag{7.83}$$

These results are shown in Fig. 7.26. It is seen that if A_2/A_1 is large, the combined force can achieve three times the oncoming value. A compressive disturbance corresponds to $P_2 - P_1 < 0$ in Fig. 7.26, for which the force direction is opposite to that shown.

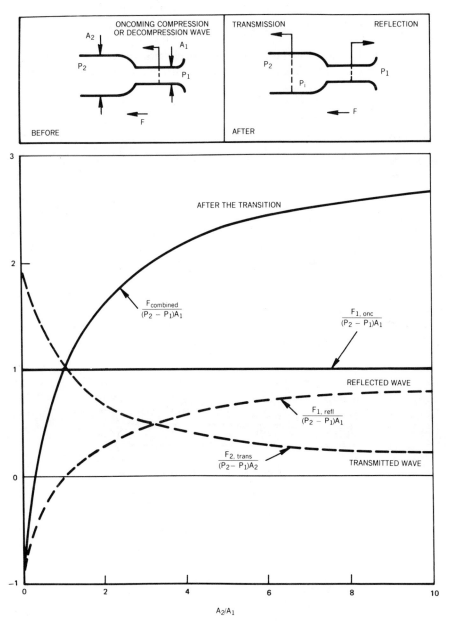

Figure 7.26 Waterhammer Force at Pipe Area Change

7.18   Column Separation

When sudden full or partial closure of a valve occurs on flowing liquid whose temperature is T, a pressure rise $\Delta P = \rho C \, \Delta V/g_0$ occurs on the upstream side. A decompression of $-\Delta P$ would occur downstream of the valve, unless liquid pressure would drop below the saturation value $P_{sat}(T)$, for which case pressure would remain at $P_{sat}(T)$ and vapor flashing would occur.

Consider the case of sudden full valve closure on a liquid flowing at velocity V_i in Fig. 7.27. The downstream column of liquid separates from the valve, decelerates, reverses its motion, and impacts the valve in the opposite direction. It can be shown from mass conservation and momentum that if the vapor readily condenses, the liquid column ideally rebounds back to the valve with velocity equal to its initial value. Thus, the impact pressure rise $\rho C V_i/g_0$ also would occur on the downstream side. Column separation can be avoided by slowing the valve closure to keep downstream pressure higher than the saturation value.

Swing check valves (see Fig. 7.29) are designed to prevent flow reversal if upstream pressure falls below the downstream value. However, flow can sometimes reverse before the valve fully closes. Since a check valve flow area diminshes quickly during the later part of its swing, upstream impact pressure and downstream column separation can result, causing several valve openings and slams in succession.

EXAMPLE 7.7: PREVENTION OF COLUMN SEPARATION
Water of density $\rho = 1000 \, \text{kg/m}^3$ and sound speed $C = 1220 \, \text{m/s}$ discharges from a reservoir at pressure $P_0 = 1101 \, \text{kPa}$. Discharge is through an ideal nozzle connected to a frictionless pipe of length $L = 100 \, \text{m}$ and flow area $A = 0.1 \, \text{m}^2$ into surroundings

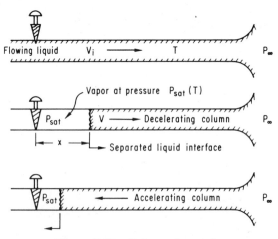

Figure 7.27   Column Separation

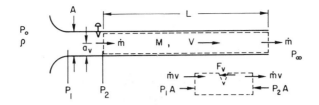

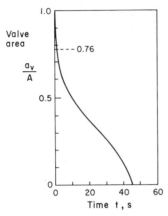

Figure 7.28 Valve Closure to Prevent Column Separation

at pressure $P_\infty = 101$ kPa. The water velocity, based on Bernoulli's equation, is $V_i = 44.76$ m/s. The liquid temperature is such that its saturation pressure is $P_{sat} = 1.0$ kPa, which is less than P_∞. A motor-operated valve is installed just downstream of the nozzle as shown in Fig. 7.28. Specify the most rapid valve closure area ratio a_v/A as a function of time, which will prevent column separation. Also, determine the minimum closure time, the maximum force resulting on the valve, and the waterhammer force for instant valve closure and resulting column separation. The valve contraction coefficient is $C_c = 1.0$.

Column separation can be prevented if $P_2 > P_{sat}$. The dotted CV downstream of the valve is employed to write the momentum principle for the limiting case of $P_2 = P_{sat} = $ constant, which gives

$$\frac{dV}{dt} = \frac{g_0}{\rho L}(P_{sat} - P_\infty)$$

Therefore, velocity in the pipe which just prevents column separation is obtained from integration, which yields

$$V = V_i - \frac{g_0}{\rho L}(P_\infty - P_{sat})t = 44.8 \text{ m/s} - (1.0 \text{ m/s}^2)t(\text{s})$$

The minimum closure time corresponds to $V = 0$, or

$$t_{closure} = \frac{\rho L V_i}{g_0(P_\infty - P_{sat})} = 44.8 \text{ s}$$

The ideal entrance yields $P_0 = P_1 + \rho V^2/2g_0$, and the valve boundary condition of Eq. (7.37) with (7.43) and $P_2 = P_{sat}$ gives

$$\frac{a_v}{A} = \frac{1}{\sqrt{2g_0(P_0 - P_{sat})/\rho V^2 - 1} + 1}$$

which is graphed as a function of time in Fig. 7.28. The initial value $a_v/A = 0.76$ means that the valve could be closed abruptly from $a_v/A = 1$ and finish its closure stroke as shown without column separation.

The valve force obtained from momentum considerations is given by

$$F_v = (P_1 - P_2)A = \left(P_0 - \frac{V^2}{2g_0}\rho - P_{sat}\right)A$$

It follows that the maximum force occurs at full closure, for which

$$F_{v,max} = (P_0 - P_{sat})A = 110 \text{ kN}$$

Furthermore, the instant valve closure waterhammer force which would occur first on the upstream side, then on the downstream side from column separation, is given by

$$F_{v,wh} = \Delta P_{wh}A = \frac{\rho C V_i A}{g_0} = 5461 \text{ kN}$$

7.19 Swing Check Valve

Response of a swing check valve, like the one shown schematically in Fig. 7.29, depends on both its dynamic properties and adjacent flow properties. Complex check valve models can be found in the literature [7.12], most of

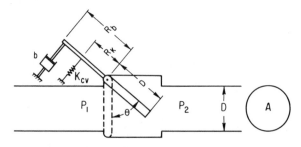

Figure 7.29 Swing Check Valve Model

which contain empirical formulations. The model discussed here is simple but adequate for estimating transient flows.

Fluid flows in the pipe from left to right through an orifice formed by the pipe wall and swinging disk. Flow reversal closes the valve. The rotating disk assembly has a moment of inertia I about the hinge, which includes an effect of the liquid inertia. A spring of elastic constant K_{cv} and a linear dashpot of damping constant b are included to provide a range of control for the valve response. The disk weight W exerts a closure torque for the horizontal orientation shown, and the fluid force on the disk is approximated by

$$F_d \approx A(P_1 - P_2) \approx AK_L \frac{|V - D\dot{\theta}/2|(V - D\dot{\theta}/2)}{2g_0} \tag{7.84}$$

An angular momentum equation for the disk is given by

$$\frac{I}{g_0} \frac{d^2\theta}{dt^2} = F_d \frac{D}{2} \cos\theta - bR_b \frac{d\theta}{dt} + K_{cv}R_k(\theta - \phi) - W \frac{D}{2} \sin\theta \tag{7.85}$$

Angle ϕ is employed to give residual torque from the spring when the disk is seated. Equation (7.85) can be solved by a stepwise procedure, during which F_d is calculated at each step. Equations (7.37) and (7.43) yield $P_1 - P_2$, and the area ratio A/a_v can be expressed in the form

$$\frac{A}{a_v} = \frac{1}{1 - \cos\theta} \tag{7.86}$$

The ideal check valve has rapid response and closes gently, reaching full closure simultaneously at the instant of zero fluid velocity. Realistically, the disk inertia causes its motion to lag behind fluid velocity and sometimes it slams destructively. Selection of coefficients b and K_{cv} sometimes can be determined from analyses of a particular piping system to minimize closure impact loads.

7.20 Ideal Pipe Branch Boundary

Pipe segments are joined to form branches in common piping networks. The ideal multiple branch of N uniform, frictionless pipes coming together in Fig. 7.30 provides a pressure-velocity boundary condition for each of the N pipes in a waterhammer analysis. All flows are shown outward (positive) from the common attachment for generality, which means that one or more of the velocities are inward (negative). Mass conservation for incompressible liquid yields one boundary condition in the form

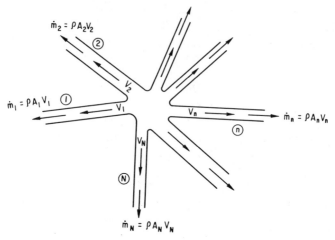

Figure 7.30   Ideal Multiple Branch

$$\sum_{n=1}^{N} A_n V_n = 0 \tag{7.87}$$

The quasi-steady pressure-velocity relationship between any two branches has the general form

$$P_n - P_m = (b_n V_n^2 + b_m V_m^2) \frac{\rho}{2g_0}$$

where b_n and b_m are constants which include branch loss coefficients. Dynamic pressure terms $\rho V^2/2g_0$ usually can be neglected at pipe junctions in waterhammer analysis, so another boundary condition is

$$P_1 \approx P_2 \approx \cdots \approx P_n \approx \cdots \approx P_N = P \tag{7.88}$$

These boundary conditions later are combined with characteristic equations to obtain formulations which are appropriate for waterhammer solutions.

7.21   Waterhammer Solutions by the Method of Characteristics

The method of characteristics for waterhammer solutions is appropriate when it is desirable to include the effects of friction, pipe area changes with various boundary conditions, or several pipes in a piping network. The formulation is suitable for synthesizing simple or complex piping systems and for writing short PC programs to obtain solutions.

The procedure of Section 5.8 is followed, whereby Eqs. (7.48) and (7.49) are first multiplied by unknown constants λ_1 and λ_2, respectively. The derivative forms

$$\frac{dP}{dt} = \frac{\partial P}{\partial t} + \frac{dx}{dt}\frac{\partial P}{\partial x} \quad \text{and} \quad \frac{dV}{dt} = \frac{\partial V}{\partial t} + \frac{dx}{dt}\frac{\partial V}{\partial x}$$

are introduced to eliminate $\partial P/\partial t$ and $\partial V/\partial t$, and then the resulting coefficients of $\partial P/\partial x$ and $\partial V/\partial x$ are made zero by proper choice of the path dx/dt. Thus, the characteristic forms of Eqs. (7.48) and (7.49) are

$$\frac{dP}{dt} + \frac{\rho C}{g_0}\frac{dV}{dt} + B_R = 0 \quad \text{on} \quad \frac{dx}{dt} = C \tag{7.89}$$

$$\frac{dP}{dt} - \frac{\rho C}{g_0}\frac{dV}{dt} + B_L = 0 \quad \text{on} \quad \frac{dx}{dt} = -C \tag{7.90}$$

where

$$B_R = \frac{\rho C^2}{g_0}\frac{1}{A}\frac{dA}{dx}V + \frac{\rho C}{g_0}\frac{f}{D}\frac{|V|V}{2} \tag{7.91}$$

$$B_L = \frac{\rho C^2}{g_0}\frac{1}{A}\frac{dA}{dx}V - \frac{\rho C}{g_0}\frac{f}{D}\frac{|V|V}{2} \tag{7.92}$$

If elevation terms are thought to be important, the term $f|V|V/2D$ in Eqs. (7.91) and (7.92) should be replaced by $f|V|V/2D + g\sin\theta$. Equations (7.89) and (7.90) are integrated on path lines $dx/dt = \pm C$ in the t, x plane for a pipe section shown in Fig. 7.31. The computational field is divided into equal time and space increments, where Δx and Δt are related by

$$\Delta x = C\,\Delta t \tag{7.93}$$

Characteristic lines pass through general mesh points i, j shown with black dots. Equations (7.89) and (7.90) can be integrated by a simple forward difference method, which gives

$$P_{i,j} - P_{i-1,j-1} + \frac{\rho C}{g_0}(V_{i,j} - V_{i-1,j-1}) + B_{Ri-1,j-1}\,\Delta t = 0$$

$$\text{on} \quad \frac{dx}{dt} = C \tag{7.94}$$

and

$$P_{i,j} - P_{i+1,j-1} - \frac{\rho C}{g_0}(V_{i,j} - V_{i+1,j-1}) + B_{Li+1,j-1}\,\Delta t = 0$$

$$\text{on} \quad \frac{dx}{dt} = -C \tag{7.95}$$

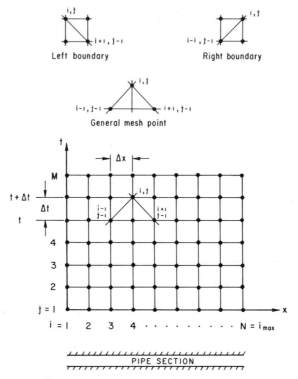

Figure 7.31   Computational Mesh, Pipe

All flow and geometric properties are known at each mesh point i at time t. It follows that simultaneous solution of Eqs. (7.94) and (7.95) give $P_{i,j}$ and $V_{i,j}$ in the forms

$$P_{i,j} = \frac{1}{2}\left(P_{i-1,j-1} + P_{i+1,j-1}\right) + \frac{1}{2}\frac{\rho C}{g_0}\left(V_{i-1,j-1} - V_{i+1,j-1}\right)$$

$$- \frac{1}{2}\left(B_{Ri-1,j-1} + B_{Li+1,j-1}\right)\Delta t \qquad (7.96)$$

and

$$V_{i,j} = \frac{1}{2}\left(V_{i-1,j-1} + V_{i+1,j-1}\right)$$

$$+ \frac{1}{2}\frac{g_0}{\rho C}\left[P_{i-1,j-1} - P_{i+1,j-1} + \left(B_{Li+1,j-1} - B_{Ri-1,j-1}\right)\Delta t\right] \qquad (7.97)$$

for a general mesh point.

Boundary conditions normally are specified as time-dependent pressure, velocity, or pressure-velocity functions like those in Section 7.5.

The process of solution for P and V at all mesh points in a single pipe segment involves a sweep of all mesh points at each time t, indexed by j in Fig. 7.31. Values of P and V are known at the initial condition, corresponding to $i = 1, 2, \ldots, N$, $j = 1$. The sweep begins at time corresponding to $j = 2$. The left boundary condition at $i = 1$, with Eq. (7.95), yields P and V at $i = 1$. The sweep continues from $i = 2$, $j = 2$ and runs to $i = N - 1$, where

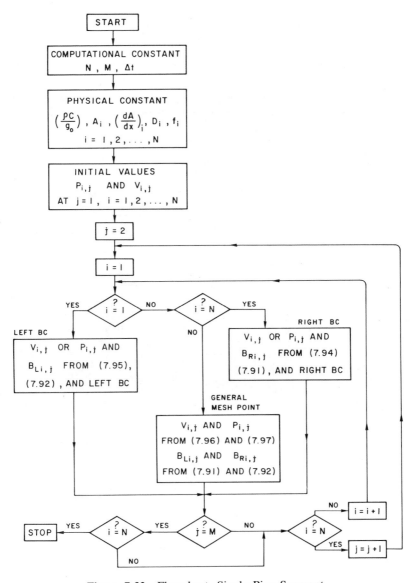

Figure 7.32 Flowchart, Single Pipe Segment

$P_{i,j}$ and $V_{i,j}$ are obtained from Eqs. (7.96) and (7.97) for general mesh points. Then the right boundary condition at $i = N$, $j = 2$, with Eq. (7.94), yields P and V at $i = N$. Values of $B_{Li,j}$ and $B_{Ri,j}$ are obtained at each i, j during a sweep. Sweeps are repeated for $j = 3, \ldots, M$, and the problem is solved. Figure 7.32 gives a simple flowchart for the computation described.

EXAMPLE 7.8: VALVE PRESSURE FROM CHARACTERISTICS

The problem described in Example 7.6, involving the pressure transient at a valve during closure on a 44.76-m/s water flow in a 61-m pipe, is solved in this example by the method of characteristics for the valve area-time function

$$\frac{a_v}{A} = 1 - \left(\frac{t}{\tau}\right)^n \quad \text{with } n = 0.5, 1.0, \text{ and } 2.0$$

Pressure and velocity at a general mesh point are calculated from Eqs. (7.96) and (7.97) with $B_R = B_L = 0$ for a uniform, frictionless pipe, and $\rho C/g_0 = 1.22$ MPa/(m/s). The left boundary condition at $i = 1$ is the vessel pipe attachment through an ideal nozzle, from which Eqs. (7.36) and the left traveling characteristic of (7.95) yield

$$V_{i,j} = -C + \sqrt{1 + \frac{2V_{i+1,j-1}}{C} - \frac{2g_0}{\rho C^2}(P_{i+1,j-1} - P_0)}$$

with $2g_0/\rho C^2 = 0.00134$ MPa^{-1} and $C = 1220$ m/s. Furthermore, when V_{ij} is calculated, Eq. (7.95) is employed to obtain the corresponding $P_{i,j}$ as

$$P_{i,j} = P_{i+1,j-1} + \frac{\rho C}{g_0}(V_{i,j} - V_{i+1,j-1})$$

The right boundary condition at $i = N$ corresponds to a valve, Eq. (7.38), the loss coefficient K_L of (7.43), and the right traveling characteristic of (7.94), which yield

$$V_{i,j} = -\frac{C}{K_L} + \frac{C}{K_L}\sqrt{1 + \frac{2K_L}{C}V_{i-1,j-1} + \frac{2g_0 K_L}{\rho C^2}P_{i-1,j-1}}$$

and

$$P_{i,j} = P_{i-1,j-1} - \frac{\rho C}{g_0}(V_{i,j} - V_{i-1,j-1})$$

with

$$K_L = \left(\frac{A}{a_v C_c} - 1\right)^2, \qquad C_c = 1.0$$

Results are shown as dashed lines in Fig. 7.16, which confirms the earlier results from the simplified procedure of Section 7.9.

7.22 Characteristic Boundary Equations

A *characteristic boundary equation* (CBE) is obtained from simultaneous solution of right or left traveling characteristic equations with quasi-steady boundary conditions like those of Eqs. (7.36)–(7.40) and Sections 7.18–7.20. An example for an ideal nozzle attachment to a vessel and a valve with a specified closure were discussed in Example 7.8. The rest of this section gives a summary of characteristic boundary equations for a collection of useful boundary conditions which should be employed at the boundary mesh points in a piping system calculation. The left end of a pipe segment corresponds to the minimum index i on that segment. The right end corresponds to the maximum index i on that segment. All velocities are considered positive to the right.

1. Ideal Nozzle at Left End

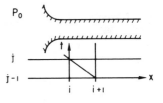

$$V_{i,j} = -C$$

$$+ C\sqrt{1 + \frac{2V_{i+1,j-1}}{C} - \frac{2g_0}{\rho C^2}(P_{i+1,j-1} - P_0)}$$

$$P_{i,j} = P_{i+1,j-1} + \frac{\rho C}{g_0}(V_{i,j} - V_{i+1,j-1})$$

2. Ideal Nozzle at Right End

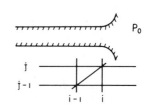

$$V_{i,j} = C$$

$$- C\sqrt{1 - \frac{2V_{i-1,j-1}}{C} - \frac{2g_0}{\rho C^2}(P_{i-1,j-1} - P_0)}$$

$$P_{i,j} = P_{i-1,j-1} - \frac{\rho C}{g_0}(V_{i,j} - V_{i-1,j-1})$$

3. Orifice, Right or Left Flow

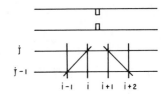

Test for direction of flow:

$$P_{i-1,j-1} - P_{i+2,j-1} + \frac{\rho C}{g_0}(V_{i-1,j-1} + V_{i+2,j-1}) \quad \begin{cases} \geq 0, & \text{rightward} \\ < 0, & \text{leftward} \end{cases}$$

In the following equation, use upper sign within braces for rightward flow, lower sign for leftward flow:

$$V_{i,j} = V_{i+1,j} = \begin{Bmatrix} - \\ + \end{Bmatrix} \frac{2C}{K_L}$$

$$\begin{Bmatrix} + \\ - \end{Bmatrix} \frac{2C}{K_L} \sqrt{1 \begin{Bmatrix} + \\ - \end{Bmatrix} \frac{K_L}{2C}(V_{i-1,j-1} + V_{i+2,j-1}) \begin{Bmatrix} + \\ - \end{Bmatrix} \frac{g_0 K_L}{2\rho C^2}(P_{i-1,j-1} - P_{i+2,j-1})}$$

$$P_{i,j} = P_{i-1,j-1} - \frac{\rho C}{g_0}(V_{i,j} - V_{i-1,j-1})$$

$$P_{i+1,j} = P_{i+2,j-1} + \frac{\rho C}{g_0}(V_{i+1,j} - V_{i+2,j-1})$$

4. Orifice at Right End of Pipe

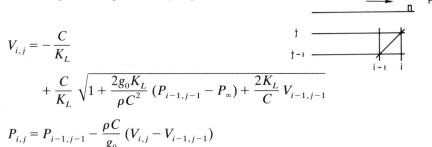

$$V_{i,j} = -\frac{C}{K_L}$$

$$+ \frac{C}{K_L}\sqrt{1 + \frac{2g_0 K_L}{\rho C^2}(P_{i-1,j-1} - P_\infty) + \frac{2K_L}{C}V_{i-1,j-1}}$$

$$P_{i,j} = P_{i-1,j-1} - \frac{\rho C}{g_0}(V_{i,j} - V_{i-1,j-1})$$

5. Orifice at Left End of Pipe

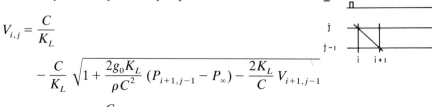

$$V_{i,j} = \frac{C}{K_L}$$

$$- \frac{C}{K_L}\sqrt{1 + \frac{2g_0 K_L}{\rho C^2}(P_{i+1,j-1} - P_\infty) - \frac{2K_L}{C}V_{i+1,j-1}}$$

$$P_{i,j} = P_{i+1,j-1} + \frac{\rho C}{g_0}(V_{i,j} - V_{i+1,j-1})$$

6. *Branch of N Segments* (*Section* 7.20). Each velocity component is positive if outward. Subscript n designates the boundary mesh point i value for P_n and V_n in segment n. Subscript n_0 is the next mesh point index $i+1$ outward from the branch in segment n. All values of P_n are equal to P. Flow area of branch n is A_n.

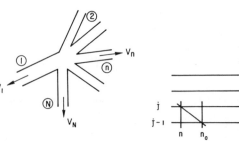

$$P = P_{n,j} = \frac{\sum_{n=1}^{N} A_n \left(P_{n_0,j-1} - \frac{\rho C}{g_0} V_{n_0,j-1} \right)}{\sum_{n=1}^{N} A_n}$$

$$V_{n,j} = V_{n_0,j-1} + \frac{g_0}{\rho C} (P - P_{n_0,j-1})$$

7. Centrifugal Pump

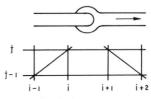

Test for direction of flow:

$$P_0 + (P_{i-1,j-1} - P_{i+2,j-1}) + \frac{\rho C}{g_0} (V_{i-1,j-1} + V_{i+2,j-1}) \quad \begin{cases} \geq 0, & \text{rightward} \\ < 0, & \text{leftward} \end{cases}$$

In the following equation, use upper sign within braces if rightward flow, lower sign if leftward:

$$V_{i+1,j} = \begin{Bmatrix} - \\ + \end{Bmatrix} a_1 \begin{Bmatrix} + \\ - \end{Bmatrix} a_1$$

$$\cdot \left[1 \begin{Bmatrix} + \\ - \end{Bmatrix} \left(a_3 + a_2(V_{i-1,j-1} + V_{i+2,j-1}) - a_4(P_{i+2,j-1} - P_{i-1,j-1}) \right) \right]^{1/2}$$

$$V_{i,j} = \frac{A_{i+1}}{A_i} V_{i+1,j}$$

$$P_{i,j} = P_{i-1,j-1} - \frac{\rho C}{g_0} (V_{i,j} - V_{i-1,j-1})$$

$$P_{i+1,j} = P_{i+2,j-1} + \frac{\rho C}{g_0} (V_{i+1,j} - V_{i+2,j-1})$$

$$b_1 = \rho \frac{g}{g_0} A_i^2 \alpha, \quad P_0 = \rho \frac{g}{g_0} H_{p_0}, \quad \alpha, H_{p_0} \text{ from Eq. (7.45)}$$

$$b_2 = \frac{\rho C}{2g_0 b_1}, \quad b_3 = 1 + \frac{A_{i+1}}{A_i}$$

$$a_1 = b_2 b_3, \quad a_2 = \frac{2}{b_2 b_3^2}, \quad a_3 = \frac{P_0}{b_1 b_2^2 b_3^2}, \quad a_4 = \frac{1}{b_1 b_2^2 b_3^2}$$

8. Swing Check Valve

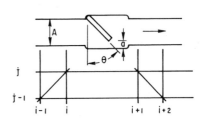

Test for direction of flow:

$$P_{i-1,j-1} - P_{i+2,j-1} + \frac{\rho C}{g_0}(V_{i-1,j-1} + V_{i+2,j-1}) \quad \begin{cases} \geq 0, & \text{rightward} \\ < 0, & \text{leftward} \end{cases}$$

In the following equation, use upper sign within braces if rightward flow, lower sign if leftward:

$$V_{i,j} = V_{i+1,j} = \begin{Bmatrix} - \\ + \end{Bmatrix} \frac{2C}{K_L} \begin{Bmatrix} + \\ - \end{Bmatrix} \frac{2C}{K_L} \left[1 \begin{Bmatrix} + \\ - \end{Bmatrix} \frac{K_L}{2C}(V_{i-1,j-1} + V_{i+2,j-1})\right.$$

$$\left. \begin{Bmatrix} + \\ - \end{Bmatrix} \frac{g_0 K_L}{2\rho C^2}(P_{i-1,j-1} - P_{i+2,j-1})\right]^{1/2}$$

$$P_{i,j} = P_{i-1,j-1} - \frac{\rho C}{g_0}(V_{i,j} - V_{i-1,j-1})$$

$$P_{i+1,j} = P_{i+2,j-1} + \frac{\rho C}{g_0}(V_{i+1,j} - V_{i+2,j-1})$$

$$K_L = K_L(t) \cong \left(\frac{A}{a_v} - 1\right)^2 = \left(\frac{\cos\theta}{1 - \cos\theta}\right)^2$$

Time integration of θ (See Fig. 7.29 and Eqs. (7.84)–(7.86) with $\Delta\omega/\Delta t \approx d^2\theta/dt^2$):

$$\Delta\omega = \frac{g_0}{I}\left[\frac{AD}{2}(P_{i,j} - P_{i+1,j})\cos\theta + K_{cv}R_k(\theta + \phi) - bR_b\omega - \frac{WD}{2}\sin\theta\right]\Delta t$$

$$\Delta\theta = \omega\,\Delta t, \qquad 0 \leq \theta \leq \frac{\pi}{2}$$

9. *Column Separation at Full Valve Closure.* If $P_{i,j} < P_{sat}(T)$ is obtained from computation, then reset $P_{i,j} = P_{sat}(T)$ until column returns and let $V_{i,j}$ be the interface velocity. When the interface again arrives at closed valve, the boundary condition again becomes $V_{i,j} = 0$. The column displacement is

$$x_c = \sum \Delta x_c, \qquad \Delta x_c = V_{i,j}\,\Delta t$$

Column returns when $x_c = 0$ again.

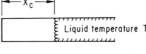

10. *Specified Velocity or Pressure*

$P_{i\max,j}$ or $P_{i,j} = f(t)$ prescribed

or $V_{i\max,j}$ or $V_{i,j} = g(t)$ prescribed

The characteristic boundary equations summarized in this section are used as building blocks in the synthesis procedure of piping systems described next.

7.23   Synthesis of Piping Systems for Waterhammer Solution

Waterhammer programs are available for the solution of complex piping systems with options for multiple pipes of various diameter and length, branches, pumps, valves, and other devices which create disturbances or otherwise affect the system response (see, for example, references [7.1], [7.2], [7.14], [7.16], and [7.27]). The method of synthesizing a piping network discussed in this section involves computational steps to be joined with a straightforward procedure for processing. Separate pipe segments are divided into increments of equal size, separated by sequentially numbered mesh points. Boundary mesh points at the ends of each segment are identified and calculated from appropriate boundary conditions. A computational sweep yields P and V at all mesh points every time step until the solution is complete.

Consider the example of multiple boundaries and pipe sections in Fig. 7.33, which contains reservoirs, a check valve, a gate valve, a pump, and a

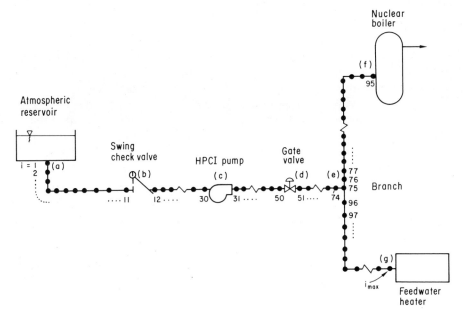

Figure 7.33 Example System, Multiple Boundaries

branch. Each boundary is identified with a letter in the figure. The shortest pipe segment should be divided into at least 10 increments. All segments were divided into increments of equal length. Example mesh point numbers are shown at the boundaries of each pipe segment. The computational mesh of Fig. 7.31 is employed for each pipe segment, with segment a, b containing mesh points $i = 1, 2, \ldots, 11$, segment b, c containing $i = 12, 13, \ldots$, $30, \ldots$. The values of index i at boundaries are flagged so that appropriate boundary conditions will be employed for computing P and V at these mesh points. For example, the indices $i = 11, 12$ are flagged because they designate boundaries of the check valve.

Once the computational mesh points $i = 1, \ldots i_{max}$ are assigned to the entire piping system to be analyzed, initial values of P and V are obtained at each i, probably from a steady state computation. One or more disturbances are specified at boundaries, and a computational sweep proceeds from $i = 1$ to i_{max} to obtain P and V at the end of the first time step Δt. Equations (7.96) and (7.97) are used for P and V at general mesh points, and the characteristic boundary equations of Section 7.22, or other appropriate formulations, are used to obtain P and V at i values which are flagged at boundaries. This procedure makes it possible to synthesize simple or complex piping systems.

Synthesis and computation of the example piping system shown in Fig. 7.33 is summarized by the following steps:

1. Sketch the entire piping system to be analyzed.

2. Denote each boundary with a different letter. Boundaries occur whenever a pipe segment terminates at a reservoir, pump, valve, branch, orifice, area change, or other device.

3. Prescribe an incremental pipe length ΔL which fits an integer number of times, N, in each of the pipe segments, with $N \geq 10$ for the shortest pipe segment.

4. Assign index i to increment boundaries, shown as black dots in Fig. 7.33, which actually are mesh points, in each pipe segment. Index i begins with $i = 1$, and increases sequentially as $2, 3, \ldots, i_{max}$. When a pipe branch occurs, continue assigning i along one branch to its termination, then return to the branch and continue assigning i to another branch until its termination is reached, and so on.

5. Flag all i index values at boundaries. Example values of i at boundaries in Fig. 7.33 are $i = 1$ (reservoir), $11, 12$ (check valve), $30, 31$ (pump), $50, 51$ (gate valve), $74, 76, 96$ (branch), 95 (Boiler), i_{max} (feedwater heater).

6. Assign appropriate characteristic boundary equations from Section 7.22 to each boundary index i which is flagged in step 5.

7. Perform a computational sweep from $i = 1$ to i_{max} to calculate $P_{i,j}$ and $V_{i,j}$ at the next time, using Eqs. (7.96) and (7.97) for general mesh points i, and appropriate characteristic boundary equations from Section 7.22 for $P_{i,j}$ and $V_{i,j}$ at i values of steps 5 and 6.

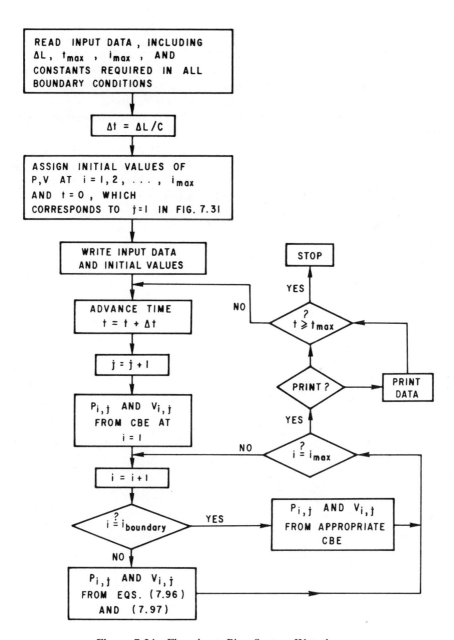

Figure 7.34 Flowchart, Pipe System Waterhammer

8. Store or list the $P_{i,j}$ and $V_{i,j}$ values obtained in step 7 and repeat step 7 for the next time $t + \Delta t$. Continue repeating steps 7 and 8 until the analysis is complete.

A general flowchart for the computation is shown in Fig. 7.34.

The specific example problem of Fig. 7.33 involves a centrifugal pump trip and speed coastdown, water flow reversal driven from the high pressure nuclear boiler, check valve closure, and column separation.

The high pressure coolant injection (HPCI) system is designed to supply water to the nuclear core for emergency cooling if boiler water is lost during an accident. The HPCI system is tested periodically by starting the pump to ensure its availability if needed. When the HPCI pump speed increases, it raises external water from atmospheric pressure to boiler pressure, at which time the gate valve opens, and foreward flow into the nuclear boiler begins. However, if an HPCI pump overspeed trip occurs and the pump begins to coast down, the discharge pressure will decrease. Consequently, the higher boiler pressure could cause the foreward flow to decrease until backflow is achieved, which could cause the swing check valve to slam closed. Column

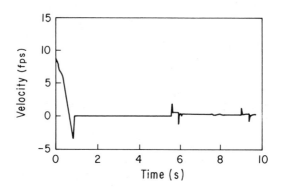

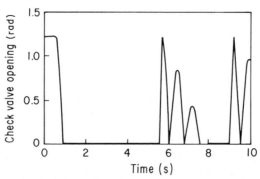

Figure 7.35 Water Velocity and Check Valve Opening

separation could occur at the suddenly closed check valve because of the backward-flowing water inertia between the valve and reservoir.

Pump coastdown speed data were recorded and used as the input disturbance to the system of Fig. 7.33. Figure 7.35 shows the calculated water velocity in the discharge pipe and the check valve opening. Note that the water velocity decreases to zero and reverses in less than one second, and then stops as the check valve closure is complete. Figure 7.36 shows measured and calculated pressures in the suction and discharge pipes. The several pressure spikes between 0.0 and 3.0 s are caused by reflections from that part of the system between (c) and (f), Fig. 7.33. The reverse flow stopped by the check valve causes column separation in the section between (a) and (b). The vapor cavity grows to about 4.0 ft, and then the water column rebounds, impacting the closed check valve at about 6.0 s, and again at 9.0 s, knocking it open each time as shown in Fig. 7.35, and further evidenced by the smaller pressure spikes in Fig. 7.36.

This example is offered to demonstrate that the MOC solution to waterhammer problems, and a catalog of appropriate boundary conditions,

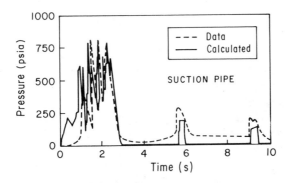

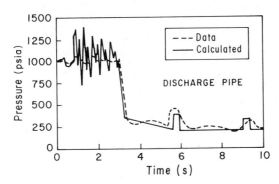

Figure 7.36 Pump Suction and Discharge Pipe Pressures

can be readily employed to synthesize complex piping system transients. The procedure described in this chapter is suited to PC computation.

References

7.1   Fabic, S., "Computer Program WHAM for Calculations of Pressure, Velocity, and Force Transients in Liquid Filled Piping Networks," Report No. 67-49-R, Kaiser Engineering, 1967.

7.2   Streeter, V. L., and E. B. Wylie, *Hydraulic Transients*, McGraw-Hill, New York, 1967.

7.3   Parmakian, J., *Waterhammer Analysis*, Dover, New York, 1963.

7.4   Chen, W. L., Y. W. Shin, and D. H. Thompson, "Analysis of One-dimensional Pressure Pulse Propagation in a Closed Fluid System," *Nucl. Eng. and Design*, **31** (1974).

7.5   Shin, Y. W., and W. L. Chen, "Numerical Fluid-Hammer Analysis by the Method of Characteristics in Complex Piping Networks," *Nucl. Eng. and Design*, **33** (1975).

7.6   Youngdahl., C. K., C. A. Kot, and R. A. Valentin, "Pressure Transient Analysis in Piping Systems Including the Effects of Plastic Deformation and Cavitation," ASME/CSMG Pressure Vessel and Piping Conference, Paper No. 78-PVP-56, 1978.

7.7   Jones, S. E., and D. J. Wood, "The Effect of Axial Boundary Motion on Pressure Surge Generation," *ASME Trans.JBE*, (1972).

7.8   Zielke, W., "Frequency-Dependent Friction in Transient Pipe Flow," *ASME Trans.JBE*, **90** (1968).

7.9   Tarantive, F. J., and W. T. Rouleau, "Waterhammer Attenuation with a Tapered Line," *ASME Trans.JBE*, (1969).

7.10 Serkiz, A. W., "Waterhammer in U. S. Nuclear Power Plants," ASME Paper No. 87-PVP-16.

7.11 Uffer, R. A., "Prevention of Power Plant Waterhammer," ASME Paper No. 87-PVP-19.

7.12 Thorley, A. R. D., "Check Valve Behavior Under Transient Flow Conditions, A State of the Art Review," *ASME J Fluids Engineering*, **111**(1), (1989).

7.13 Kim, H. T., "Waterhammer Formulation for Pipe Network with Check Valves and Its Applications," *Proc. 3rd International Topical Meeting on Nuclear Power Plant Thermal Hydraulics and Operations*, Seoul, Korea, November, 1988.

7.14 *PISCES, Transient Hydraulic Analysis of Systems and Components*, Pisces International, Physics International Scientific Codes and Engineering Services, 2700 Merced Street, San Leandro, California.

7.15 Wiggert, D. C., et al. "Coupled Transient Flow and Structural Motion in Liquid-Filled Piping Systems," *Proc. 5th International Conference on Pressure Surges*, Hanover, F. R. G., September, 1986.

7.16 Chaudhry, M. H., *Applied Hydraulic Transients*, Van Nostrand, New York, 1979.

7.17   Joung, Il-bok, and Y. S. Shin, "A New Model on Transient Wave Propagation in Fluid Flow Tubes," *ASME J. Pressure Vessel Technology*, **109** (1987).

7.18   Stepanoff, A. J., *Centrifugal and Axial Flow Pumps*, 2nd Ed., Wiley, New York, 1957.

7.19   Wiggert, D. C., R. S. Otwell, and F. J. Hatfield, "The Effect of Elbow Restraint on Pressure Transients," *ASME J. Fluids Engineering*, **107**(3), (1985).

7.20   *Fluid Transients in Fluid-Structure Interaction*, ASME Special Publication FED-Vol. 30, 1985, Papers by Schwirian, R. E., "Multidimensional Water-hammer Analysis Using a Node-Flow Link Approach;" Wiggert, D. C., F. J. Hatfield, and S. Stuckenbruck, "Analysis of Liquid and Structural Transients in Piping by the Method of Characteristics;" Simpson, A. R., and E. B. Wylie, "Problems Encountered in Modeling Vapor Column Separation."

7.21   Rothe, P. H., D. H. Evans, and C. J. Crowley, "Blowdown with Check Valve Slam," ASME Special Publication, *Fluid Transients and Structural Interactions in Piping Systems*, June 1981.

7.22   Wylie, E. B., and V. L. Streeter, *Fluid Transients*, McGraw-Hill, New York, 1978.

7.23   Wiggert, D. C., C. S. Martin, M. Naghash, and P. V. Rao, "Modeling of Transient Two-Component Flow Using a Four-Point Implicit Method," ASME Special Publication FED-Vol. 4, June, 1983.

7.24   Wiggert, D. C., and M. J. Sundquist, "Fixed Grid Characteristics for Pipeline Transients," *ASCE J. Hydraulic Division*, **103** (1977).

7.25   Brown, F. T., "The Transient Response of Fluid Lines," *ASME J. Basic Engineering* **84** (1962).

7.26   D'Souza, A. F., and R. Oldenburger, "Dynamic Response of Fluid Lines," *ASME J. Basic Engineering*, **86** (1964).

7.27   Brown, R. J., "Water-Column Separation at Two Pumping Plants," *ASME J. Basic Engineering*, **90** (1968).

7.28   Wiggert, D. C., and M. J. Sundquist, "The Effect of Gaseous Cavitation on Fluid Transients," *ASME J. Fluids Engineering*, **101** (1979).

7.29   Wylie, E. B., and V. L. Streeter, "Column Separation in Horizontal Pipelines," *9th IAHR Symp. on Fluid Mechanics*, Fort Collins, Colorado, June 1978.

Problems

7.1   Show that for unsteady bulk flow, the case of a uniform pipe with $D_0 = D_L = D$ causes Eq. (7.12) to become

$$\frac{dV_0}{dt} + \frac{f}{D}\frac{V_0^2}{2} = \frac{g_0}{\rho L}\left[P(0, t) - P(L, t)\right] - g\sin\theta$$

7.2   Use the results of Problem 7.1 for a uniform, vertical pipe with downward flow ($\theta = -\pi/2$) to show that:

(a) Pressure will be constant at every point in the pipe during steady flow if the velocity is

$$V_0 = \sqrt{\frac{2Dg}{f}}$$

(b) Pressure will be constant at every point in the pipe if the liquid is introduced at the top with velocity

$$V_0(t) = \sqrt{\frac{2Dg}{f}} \, \tanh \sqrt{\frac{gf}{2D}} \, t$$

7.3 The discharge of a centrifugal pump is connected to a uniform, horizontal, frictionless pipe of length L which discharges into the atmosphere at pressure P_∞. The pump suction draws water from a lake at atmospheric pressure and has a head-flow characteristic described by Eq. (7.40). Use Eq. (7.12) to show that if the pump is running and a valve suddenly opens to permit discharge, water velocity in the pipe can be predicted from a solution to the problem

DE: $\quad \dfrac{dV_0}{dt} + (\alpha g A^2) V_0^2 = g H_{p0}$

IC: $\quad t = 0, \quad V_0 = 0$

7.4 A horizontal, frictionless pipe wall is corrugated so that its cross-sectional area varies as

$$A(x) = A_0 + B \sin 2\pi \, \frac{x}{\lambda}, \qquad B = \text{constant}$$

where $B < A_0$ and λ is the corrugation wavelength.

(a) Use Eq. (7.8) to show that if the pipe length is any multiple

$$L = \left(\frac{4n+1}{4} \right) \lambda, \qquad n = 0, 1, 2, \ldots$$

the differential equation describing liquid velocity between $x = 0$ and L, with end pressures $P(0, t)$ and $P(L, t)$, is

$$P(0, t) - P(L, t) = \frac{\rho L}{g_0} \left[\frac{4}{(4n+1)\pi} \frac{A_0}{\sqrt{A_0^2 - B^2}} \tan^{-1} \sqrt{\frac{A_0 - B}{A_0 + B}} \right] \frac{dV_0}{dt}$$

(b) Show that if the pipe is uniform with $B = 0$, the differential equation reduces to

$$P(0, t) - P(L, t) = \frac{\rho L}{g_0} \frac{dV_0}{dt}$$

7.5 A suddenly applied pressure of 1010 kPa (147 psia) is exerted at one end of a 50-m (164-ft) long, 0.2-m (0.66-ft) diameter, horizontal, water-filled pipe. The friction factor is $f = 0.01$.

(a) Show that the maximum velocity is 24 m/s (78.7 ft/s).

(b) Approximately how long will it be until the maximum velocity of part (a) is approached?

Answer:

0.6 s

7.6 A vessel of liquid (Fig. P7.6) is pressurized to P_0. A uniform, frictionless pipe of diameter D, horizontal length L_h, and vertical length L_v is connected to the vessel as shown, A fast-acting valve

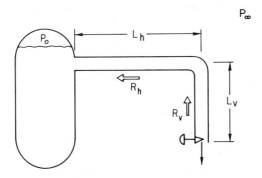

Figure P7.6

opens at the end, and unsteady discharge begins. Neglect gravity effects and show that the reaction forces R_h and R_v on the horizontal and vertical pipe sections are given by

$$\frac{R_h}{(P_0 - P_\infty)A} = -4\,\frac{L_h}{L}\,\frac{e^{-t/\tau_1}}{(1 + e^{-t/\tau_1})^2}$$

$$\frac{R_v}{(P_0 - P_\infty)A} = -4\,\frac{L_v}{L}\,\frac{e^{-t/\tau_1}}{(1 - e^{-t/\tau_1})^2} - 2\,\frac{(1 - e^{-t/\tau_1})^2}{(1 + e^{-t/\tau_1})^2}$$

$$\tau_1 = \frac{L}{\sqrt{2g_0(P_0 - P_\infty)/\rho}}\,, \qquad L = L_h + L_v$$

7.7 Liquid of density ρ and sound speed C is flowing at steady velocity V_∞ and pressure P_∞ in a frictionless, uniform pipe of area A which is attached to a vessel at pressure P_0. Discharge is into the atmosphere at pressure P_∞. An instant-acting valve at the discharge end partially closes to create a minimum flow area A_v. Assume that the valve is simulated by an ideal nozzle, and show that the waterhammer pressure is given by

$$\frac{P - P_\infty}{\rho C^2/g_0} = \frac{V_\infty}{C} + \frac{1}{B}\left(1 - \sqrt{1 + 2B\frac{V_\infty}{C}}\right), \qquad B = \left(\frac{A}{A_v}\right)^2 - 1$$

7.8   Show from the result in Problem 7.7 that in the limit when $A_v \to A$, the pressure P becomes P_∞. Also show that when $A_v \to 0$, $P - P_\infty$ becomes the waterhammer pressure $\rho V_\infty C/g_0$.

7.9   Consider a liquid of density ρ and sound speed C flowing in a pipe of area A at velocity V_∞ and pressure P_∞. An instant acting valve suddenly reduces the minimum area from A to A_v. If the valve is simulated by an orifice with a pressure loss given by Eq. (7.37), show that the net force of liquid on the valve is given by

$$\frac{F}{(\rho C^2/g_0)A} = 2\frac{V_\infty}{C} + \frac{4}{K_L}\left(1 - \sqrt{1 + K_L\frac{V_\infty}{C}}\right)$$

Assume that column separation does not occur and that the upstream and downstream pipes are very long.

7.10 A step pressure wave of 10 atm (147 psia) sweeps through a long pipe of 10 cm (3.94 in) diameter filled with water. What orifice diameter D_0 in the pipe end wall will permit the increased pipe pressure to remain at 10 atm following reflection? Take $\rho = 1000 \text{ kg/m}^3$ (62.4 lbm/ft^3), $C = 1380 \text{ m/s}$ (4526 ft/s), and use an orifice coefficient $C_c = 0.6$.

Answer:

$D_0 = 1.6 \text{ cm}$ (0.63 in)

7.11 It is necessary to protect a pressure-sensing instrument from occasional abrupt pressure steps. If the originating signal is a pressure step $P_0 - P_\infty = 1.0 \text{ atm}$ (14.7 psi) and the instrument line contains oil with $\rho = 800 \text{ kg/m}^3$ (50 lbm/ft^3) and $C = 1200 \text{ m/s}$, (3936 ft/s) estimate the orifice size required to reduce the transmitted pressure signal to 0.25 atm (3.68 psi). Use $C_c = 0.6$ for the orifice.

Answer:

$D_{orf}/D_{pipe} = 0.05$

7.12 An air column somehow became trapped in the high point of an oil-filled instrument tube, which is intended to convey pressure signals. If an oncoming step pressure P_0 is propagating through the oil, what fraction will be transmitted through oil after it passes through the air column? Take $\rho_{air} = 1.22 \text{ kg/m}^3$ (0.076 lbm/ft^3), $C_{air} = 335 \text{ m/s}$ (1099 ft/s), $\rho_{oil} = 800 \text{ kg/m}^3$ (50 lbm/ft^3), $C_{oil} = 1200 \text{ m/s}$ (3936 ft/s).

Answer:

$$\Delta P_2 / \Delta P_0 = 0.0017$$

7.13 Consider an instrument tube similar to the one in Problem 7.12, but normally filled with air. An oil column somehow is trapped in the low point. Show that with the same oil and air properties, the fraction of an oncoming step pressure which is transmitted is 0.0017.

7.14 The liquid in a pipe of length L is initially undisturbed with pressure and velocity $P_i = V_i = 0$. A piston at the left end begins to oscillate according to

$$x_p = x_0 \sin \omega t$$

where x_0 is its amplitude and ω is the oscillation frequency. The right end is closed. Use Table 7.2 to show that the resulting pressure on the piston for several wave transit times (see Fig. 7.11 for regions) is given by

$$P_1^* = \cos \omega t , \qquad 0 < t < \frac{L}{C}$$

$$P_3^* = \cos \omega t , \qquad \frac{L}{C} < t < \frac{2L}{C}$$

$$P_5^* = \cos \omega t + \cos \omega \left(t - \frac{2L}{C} \right) + \cos \omega \left(t - \frac{4L}{C} \right), \qquad \frac{2L}{C} < t < \frac{3L}{C}$$

$$P_7^* = \cos \omega t + 2 \cos \omega \left(t - \frac{2L}{C} \right), \qquad \frac{3L}{C} < t < \frac{4L}{C}$$

$$P_9^* = \cos \omega t + \cos \omega \left(t - \frac{2L}{C} \right) - 2 \cos \omega \left(t - \frac{4L}{C} \right)$$

$$+ \cos \omega \left(t - \frac{6L}{C} \right), \qquad \frac{4L}{C} < t < \frac{5L}{C}$$

where $P^* = P(0, t)/(\rho C x_0 \omega / g_0)$. Show that if $\omega = \pi C / n L$, pressure on the piston will increase by $2\rho C x_0 \omega / g_0$ every time interval of $2L/C$ after $t = 2L/C$.

7.15 Liquid is flowing by gravity in a pipeline of length $L = 10 \text{ km}$ $(32,800 \text{ ft})$ upstream of a shutoff valve. It is necessary to close the valve at the end of the day without creating substantial pressure. If the sound speed and density of the liquid are $C = 1000$ m/s (3280 ft/s) and $\rho = 800$ kg/m³ (50 lbm/ft³), show that the maximum valve closure time which will result in full waterhammer pressure is 20 s.

7.16 The valve closure in Problem 7.15 is programmed to close in 40 s, causing a linear reduction in velocity from $V_i = 10$ m/s (32.8 ft/s). Neglect friction and use the procedure of Table 7.2 to show that the waterhammer pressure at the valve corresponds to the sketch in Fig. P7.16. Oil pressure at the upstream reservoir is 40 kPa gage (5.8 psia) and both ambient and initial pipe pressures are zero gage.

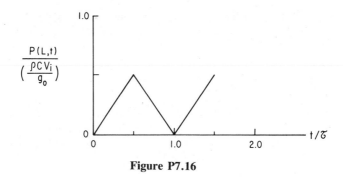

Figure P7.16

7.17 Follow the approximate waterhammer solution procedure of Example 7.6 to verify the pressure obtained for a closing valve shown in Fig. 7.16 for $n = 1$ or 1/2.

7.18 Water at 10 MPa pressure, 1000 kg/m³ density, and 1220 m/s sound speed initially fills a vessel and attached pipe of length $L = 61$ m and area $A = 0.1$ m². The water initially is stationary because a valve at the discharge end is closed. The valve opens with an area-time function

$$\frac{A_v}{A} = \left(\frac{t}{\tau}\right)^n$$

where $\tau = 1.0$ s is the full opening time. Its contraction coefficient is $C_c = 1.0$, and discharge is to $P_\infty = 1.0$ atm. The vessel-pipe attachment is through an ideal nozzle. An approximate waterhammer analysis can be performed since $\tau \gg 2L/C = 0.164$ s. Follow the procedure of Example 7.6 to show that for $n = 1$, pressure corresponding to the t, x diagram of Fig. 7.13 is given by

Region	Pressure (MPa)	Region	Pressure (MPa)
1	1.8	i	5.0
3	2.9	2	9.9
5	2.5	4	9.7
7	2.0	6	9.4
9	1.5	8	8.9
11	0.9	10	8.2
13	0.5	12	7.5
15	0.2	14	6.6
17	0.1	16	5.8
19	0.0	18	5.1

7.19 Verify from Eqs. (6.64)–(6.68) and the restrictions of Section 7.6 that the describing equations for waterhammer in uniform, frictionless pipes with elevation changes are

$$\frac{\partial P}{\partial t} + \frac{\rho C^2}{g_0} \frac{\partial V}{\partial x} = 0$$

$$\frac{\partial V}{\partial t} + \frac{g_0}{\rho} \frac{\partial P}{\partial x} + g \sin \theta = 0$$

Also show that if the elevation effect is absorbed in the pressure term by introducing

$$P^+ = P + \left(\rho \frac{g}{g_0} \sin \theta \right) x$$

the describing equations reduce to the forms given by Eqs. (7.51) and (7.52) with P replaced by P^+. Finally, show that general solutions for P and V are given by

$$P(x, t) = F_L\left(t + \frac{x}{C} \right) + F_R\left(t - \frac{x}{C} \right) - \left(\frac{g}{g_0} \sin \theta \right) x$$

$$V(x, t) = -\frac{g_0}{\rho C} \left[F_L\left(t + \frac{x}{C} \right) - F_R\left(t - \frac{x}{C} \right) \right]$$

7.20 A general solution to the linear wave equations of (7.53) or (7.54) can be obtained from a simple axis rotation. Let the x, t axes be nondimensionalized with arbitrary length L and time τ and rotated counterclockwise through an angle α. Introduce this axis rotation into Eq. (7.53) with the choice $\alpha = \pi/4$, integrate the result twice, and show that the solution of Eq. (7.55) can be obtained.

7.21 Use Eqs. (7.87), (7.88), the left traveling characteristic of (7.95) for uniform, frictionless pipes ($B_L = 0$), and the sketch for a branch of N segments in Section 7.22 to derive the corresponding $P_{n,j}$ and $V_{n,j}$. Remember that the fluid velocity is positive outward from the branch in any given pipe segment.

7.22 Show that if the characteristic boundary equation (CBE) for pressure in the branch of N segments of Section 7.22 is applied to the two-pipe area transition branch, Section 7.10, Fig. 7.17, with $P_\infty = 0$, $P_{n0,j-1} = P_0$, and $V_{n0,j-1} = -V_0$, the transmitted and reflected pressure $P_{n,j}$ is given by

$$P_{n,j} = \frac{2P_0}{1 + A_{II}/A_I}$$

Hint: Note that $\rho C V_0/g_0 = P_0$ and that V_0 is negative.

7.23 Consider the multistep area reduction shown in Fig. 7.19, containing liquid with P_∞ initial pressure. If an oncoming disturbance P_0 arrives, show that at the nth step, the transmitted pressure is

$$\frac{P_n - P_\infty}{P_0 - P_\infty} = 2^n \prod_{j=1}^{n} \left(1 + \frac{A_j}{A_{j-1}}\right)^{-1}$$

8 ONE-DIMENSIONAL LARGE-AMPLITUDE PRESSURE WAVES

This chapter includes many formulations and graphs to help estimate unsteady, compressible flow properties and forces resulting from large-amplitude disturbances propagating through pipes. Flow disturbances include pipe ruptures and valve opening or closing at various rates. Most examples involve perfect gas, saturated liquid/vapor mixtures, or mixtures of incompressible liquid and perfect gas.

Simple pressure wave propagation is formulated for cases where disturbances travel in one direction only. A method is described for estimating the distortion of a large-amplitude disturbance as it propagates. Compound wave propagation is formulated by the method of characteristics (MOC), which differs from the constant density treatment in Chapter 7 because fluid density here depends on pressure. The unusual properties of moving normal shocks are developed and applied to valve closure transients, and cases where fluid is charged into a pipe. An approximate, relatively accurate treatment of shocks which form during a MOC computation is described.

8.1 Uniform Pipes, Simple Wave Propagation

Consider an ideal, uniform, frictionless, adiabatic, horizontal pipe like that shown in Fig. 8.1. If the fluid is compressible and nonconducting, its properties are determined by the reduced forms of Eqs. (6.13), (6.11), and (6.15):

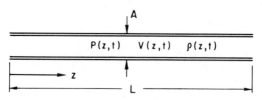

Figure 8.1   Uniform Pipe and Properties

Mass Conservation

$$\frac{\partial P}{\partial t} + V \frac{\partial P}{\partial z} + \frac{\rho C^2}{g_0} \frac{\partial V}{\partial z} = 0 \tag{8.1}$$

Momentum Creation

$$\frac{\partial V}{\partial t} + V \frac{\partial V}{\partial z} + \frac{g_0}{\rho} \frac{\partial P}{\partial z} = 0 \tag{8.2}$$

Energy Conservation

$$T\left(\frac{\partial s}{\partial t} + V \frac{\partial s}{\partial z}\right) = 0 \tag{8.3}$$

Equation (8.3) shows that for the case being considered, the entropy of a fluid particle, traveling at velocity V, remains unchanged. Moreover, if all fluid particles entering the pipe have the same entropy, the unsteady flow is everywhere insentropic. It follows that for a simple compressible substance, the fluid density $\rho(P, s)$ can be expressed as a function of pressure only. That is,

$$\rho = \rho(P) \qquad \text{isentropic flow} \tag{8.4}$$

Furthermore, since the speed of sound can be expresed as the thermodynamic property $C = \sqrt{g_0(\partial P/\partial \rho)_s}$, we can write

$$C = C(P) \qquad \text{isentropic flow} \tag{8.5}$$

A procedure similar to that used for large-amplitude hydrostatic waves in Section 5.1 is employed to obtain a solution of Eqs. (8.1) and (8.2). It is assumed that pressure can be expressed as a function of velocity only, or $P = P(V)$, which yields the solution

$$V = V_i \pm g_0 \int_{P_i}^{P} \frac{dP}{\rho(P)C(P)} \tag{8.6}$$

with

$$z = (V \pm C)t + f(V) \tag{8.7}$$

The function $f(V)$ is arbitrary, and must be determined from boundary conditions of a given problem. Equation (8.6) is a unique relationship between local velocity and pressure for simple waves. A given value of V determines the corresponding pressure P, density $\rho(P)$, and the sound speed $C(P)$. It follows that Eq. (8.7) is a relationship giving either V or P as a function of z and t. Furthermore, constant values of P and V yield straight characteristic lines when Eq. (8.7) is plotted on a t, z diagram, as shown in Fig. 8.2. The $+$ sign is for right traveling, and the $-$ sign is for left traveling simple waves.

State equations for any simple compressible substance can be used in Eqs. (8.6) and (8.7). The case of a perfect gas at constant entropy yields

$$\rho = \rho_i \left(\frac{P}{P_i} \right)^{1/k} \tag{8.8}$$

and

$$C = \sqrt{\frac{kg_0 P}{\rho}} = \sqrt{\frac{kg_0 P_i}{\rho_i}} \sqrt{\left(\frac{P}{P_i} \right)^{(k-1)/k}} = C_i \left(\frac{P}{P_i} \right)^{(k-1)/2k} \tag{8.9}$$

It follows from Eq. (8.6) that

$$V = V_i \pm \frac{2C_i}{k-1} \left[\left(\frac{P}{P_i} \right)^{(k-1)/2k} - 1 \right] \tag{8.10}$$

and from (8.9) that

$$C = C_i \pm \frac{k-1}{2}(V - V_i) \tag{8.11}$$

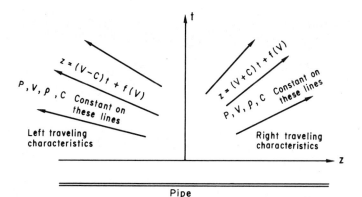

Left traveling characteristics

Right traveling characteristics

Pipe

Figure 8.2 t, z Diagram for Simple Waves

where the subscript i denotes initial values. Fluid velocity V is considered positive to the right on a t, z diagram.

Hydrostatic wave disturbances of Chapter 5 sometimes involved a moving vertical wall in a wave tank. One-dimensional compressible flows in this chapter are confined to a pipe or other flow passage, with common disturbances caused by piston motion, valve opening or closure, or fluid charging.

8.2 Accelerating Piston Withdrawal, Simple Waves

Figure 8.3 shows a long cylinder of stationary gas at initial properties $V_i = 0$, $P = P_i$, $\rho = \rho_i$, and $C = C_i$. A leakless piston begins leftward motion at acceleration a_p, with velocity $V = a_p t$ and displacement $z_p = a_p t^2/2$. It follows that the velocity of a fluid particle adjacent to the piston is

$$V = -V_p = -a_p t \qquad (8.12)$$

whereas its displacement is

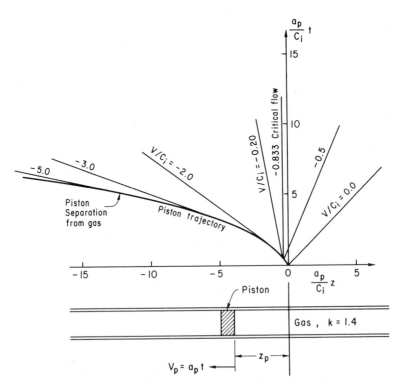

Figure 8.3 Accelerating Piston Withdrawal from Gas

$$z = -z_p = -\frac{a_p t^2}{2} \tag{8.13}$$

Disturbance propagation in the gas is rightward, for which the + sign is employed in Eqs. (8.7), (8.10), and (8.11). Elimination of z and t from Eqs. (8.7), (8.12), and (8.13) yields

$$f(V) = \frac{k}{2}\frac{V^2}{a_p} + C_i \frac{V}{a_p}$$

for which Eqs. (8.7) and (8.11) give

$$z = \left(\frac{k+1}{2}V + C_i\right)t + \frac{k}{2a_p}V^2 + \frac{C_i V}{a_p} \tag{8.14}$$

Several characteristic lines are shown in Fig. 8.3 for $k = 1.4$. The critical flow condition $(\partial z/\partial t)_v = 0$, at which the leftward gas velocity is readily obtained from Eq. (8.14) as

$$V = -C = -\frac{2C_i}{k+1} \quad \text{at} \quad t = \frac{2C_i}{(k+1)a_p}, \quad z = -\frac{1}{2}\left(\frac{2}{k+1}\right)^2 \frac{C_i^2}{a_p} \tag{8.15}$$

When gas pressure on the piston becomes zero, the piston separates from the gas and the maximum gas front velocity is obtained from Eq. (8.10) as

$$V = -\frac{2C_i}{k-1} \quad \text{at} \quad t = \frac{2C_i}{(k-1)a_p}, \quad z = -\frac{1}{2}\left(\frac{2}{k-1}\right)^2 \frac{C_i^2}{a_p} \tag{8.16}$$

The tail, or decompression front, follows the rightward path $z = C_i t$, or

$$z = C_i t, \quad \text{tail location} \tag{8.17}$$

The slopes of characteristic lines in the t, z plane are obtained from Eq. (8.14) as $(\partial z/\partial t)_v = (k+1)V/2 + C_i$. Since fluid velocity V increases in a negative direction, the characteristic lines, which do not cross each other, produce an *expansion fan*.

EXAMPLE 8.1: MISSILE EXPULSION
A cylindrical missile of mass M is to be propelled from a long cylinder of compressed gas as shown in Fig. 8.4. The gas is initially compressed to pressure P_i with sound speed C_i. A vacuum exists on the right. Therefore, the only force is $P_p A$ acting on the left side, where P_p is gas pressure on the missile. Thus, the missile motion is governed by

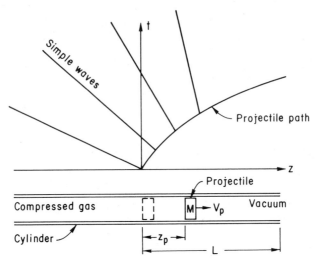

Figure 8.4 Projectile Propulsion

$$P_p A = \frac{M}{g_0} \frac{dV_p}{dt}$$

It is assumed that no leakage occurs between the missile and the frictionless cylinder wall. The compressed gas region is long so that only simple waves propagate leftward from the missile. Determine the missile speed when it leaves the cylinder at $z = L$. Use the parameters

$$
\begin{array}{ll}
P_i = 2\,\text{MPa} \ (288\,\text{psia}) & M = 5\,\text{kg} \ (11.0\,\text{lbm}) \\
A = 0.1\,\text{m}^2 \ (1.07\,\text{ft}^2) & k = 1.4 \\
C_i = 356\,\text{m/s} \ (1168\,\text{fps}) & L = 8\,\text{m} \ (26.2\,\text{ft})
\end{array}
$$

Equation (8.10) with $V_i = 0$ and the $-$ sign for leftward disturbance propagation relates velocity and pressure at the missile, which is combined with the missile equation of motion to obtain P_p in the form

$$\frac{P_p}{P_i} = P^* = \left[1 + \frac{2k}{k+1} \frac{P_i A g_0}{C_i M} t\right]^{-2k/(k+1)}$$

The corresponding missile velocity is

$$V_p = \frac{dz_p}{dt} = -\frac{2C_i}{k-1}\left[(1 + t^*)^{-(k-1)/(k+1)} - 1\right]$$

and the missile displacement is given by

$$z_p = \frac{C_i^2 M}{kP_i A g_0}\left(\frac{k+1}{k-1}\right)\left\{t^* - \left(\frac{k+1}{2}\right)\left[(1 + t^*)^{2/(k+1)} - 1\right]\right\}$$

where

$$t^* = \left(\frac{2k}{k+1}\right)\frac{P_iAg_0}{C_iM}t$$

It follows from the numbers given that the expulsion time is $t = 0.03$ s and the missile expulsion velocity is $V_p = 420$ m/s (1378 ft/s).

8.3 Piston Acceleration into Gas, Simple Waves

Suppose that the piston of Fig. 8.3 moves with constant acceleration b_p into the gas with the velocity and displacement of fluid adjacent to the piston given by

$$V = V_p = b_p t , \qquad z = z_p = (1/2)b_p t^2 \qquad (8.18)$$

An analysis similar to that of Section 8.2 yields the characteristic lines

$$z = \left(\frac{k+1}{2}V + C_i\right)t - \frac{k}{2b_p}V^2 - \frac{C_i}{b_p}V \qquad (8.19)$$

The slope of a characteristic line in the z, t plane is

$$\frac{dz}{dt} = \frac{k+1}{2}V + C_i$$

The first characteristic leaves the piston with a slope or propagation speed of $dz/dt = C_i$. Since the piston is accelerating, subsequent characteristics leave with increased propagation speed, eventually overtaking the earlier characteristics. Figure 8.5 shows a velocity profile at several times for piston acceleration into gas. If a profile becomes vertical anywhere so that

$$\left(\frac{\partial z}{\partial V}\right)_t = 0 \qquad (8.20)$$

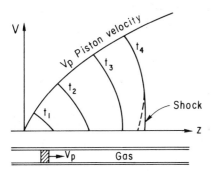

Figure 8.5 Piston Acceleration and Shock Formation

it means that the fluid velocity being propagated on adjacent characteristic lines are equal. This condition only implies parallel characteristic lines. However, if the velocity profile also has an inflection point with

$$\left(\frac{\partial^2 z}{\partial V^2}\right)_t = 0 \qquad (8.21)$$

like that shown in Fig. 8.5, it means that a characteristic line has overtaken and would cross an earlier one. This condition corresponds to the formation of a moving shock wave. If the condition of Eq. (8.20) occurs at the frontmost characteristic of a propagating disturbance, Eq. (8.21) is not required, although a frontal shock forms.

The case of constant piston acceleration into stationary gas gives

$$\left(\frac{\partial z}{\partial V}\right)_t = \left(\frac{k+1}{2}\right)t - \frac{k}{b_p}V - \frac{C_i}{b_p} = 0$$

and

$$\left(\frac{\partial^2 z}{\partial V^2}\right)_t = -\frac{k}{b_p} \neq 0$$

The second condition makes no sense, although the first condition at the wave front where $V = 0$ yields the time of shock formation

$$t_{shock} = \left(\frac{2}{k+1}\right)\frac{C_i}{b_p} \qquad (8.22)$$

Equation (8.19) with $V = 0$ gives the location where the shock forms as

$$z_{shock} = \left(\frac{2}{k+1}\right)\frac{C_i^2}{b_p} \qquad (8.23)$$

Wherever a shock appears, simple waves overtake it and reflect back into the disturbed fluid, which creates compound waves. Therefore, the solution presented in this section applies only for simple wave propagation in one direction until the formation of a shock. Equation (8.19) was used to obtain the normalized velocity profiles of Fig. 8.6 for a gas with $k = 1.4$.

EXAMPLE 8.2: PRESSURE ON ACCELERATING PISTON
A piston advances at constant acceleration $b_p = 10,000 \, \text{m/s}^2$ (32,800 ft/s²) into stagnant gas, $k = 1.4$, at initial properties $P_i = 1.0 \, \text{atm}$ (14.7 psia) and $C_i = 356 \, \text{m/s}$ (1168 ft/s). Determine pressure on the piston when a shock first appears and the location of shock formation.

It follows from Eqs. (8.22) and (8.23) that the shock forms at time

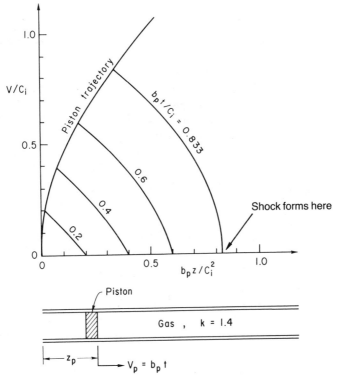

Figure 8.6 Piston Acceleration and Velocity Profiles

$$t_{\text{shock}} = 0.0297 \text{ s}$$

at location

$$z_{\text{shock}} = 10.56 \text{ m} \ (34.6 \text{ ft})$$

If we put piston velocity $V_p = b_p t$ into Eq. (8.10), we can solve for pressure on the piston in the form

$$\frac{P}{P_i} = \left[1 + \left(\frac{k-1}{2} \right) \frac{b_p t}{C_i} \right]^{2k/(k-1)}$$

It is found that at the time of shock formation, pressure on the piston is

$$P = 297 \text{ kPa} \ (43.2 \text{ psia})$$

8.4 Sudden Pipe Rupture and Discharge

The rupture of a high-pressure gas pipe normally raises questions about how fast the gas discharges, the rate of receiver pressurization, the source vessel depressurization rate, and both pipe reaction and gas jet impingement forces. Answers to all of these questions depend on a determination of the discharge velocity, pressure, and density.

Consider the frictionless, uniform, closed pipe of Fig. 8.7, which initially is filled with stagnant gas at properties P_i, ρ_i, and C_i. An instantaneous, full pipe rupture is postulated at the left end. The boundary condition here does not correspond to a prescribed piston velocity as before. However, if a piston were withdrawn at infinite acceleration, Eq. (8.15) shows that critical flow would occur at location

$$z = -\frac{1}{2}\left(\frac{2}{k+1}\right)^2 \frac{C_i^2}{a_p}\bigg|_{a_p \to \infty} = 0 \qquad (8.24)$$

that is, at the discharge plane. Critical flow corresponds to $V = -C$ from Eq. (8.15).

Simple waves will propagate in the pipe until they reach the other end, and reflected disturbances begin to propagate back toward the rupture. Characterics in the ruptured pipe correspond to $a_p \to \infty$ in Eq. (8.14) giving

$$z = \left(\frac{k+1}{2}V + C_i\right)t \qquad \text{pipe rupture} \qquad (8.25)$$

Note that all the simple characteristics originate at $z = t = 0$ and form an *expansion fan*, as they propagate with slopes ranging from $dz/dt = C_i$ for $V = 0$ to 0.0 for $V = -C$.

The condition $V = -C$ is employed to obtain pressure and density in the discharge plane from Eqs. (8.8) and (8.10) as

$$\frac{P}{P_i} = \left(\frac{2}{k+1}\right)^{2k/(k-1)} \qquad \text{at discharge} \qquad (8.26)$$

and

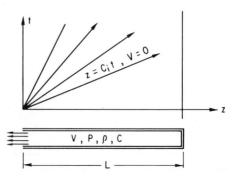

Figure 8.7 Pipe Rupture Discharge, Simple Waves

$$\frac{\rho}{\rho_i} = \left(\frac{2}{k+1}\right)^{2/(k-1)} \qquad \text{at discharge} \qquad (8.27)$$

It follows from Eqs. (8.27) and (8.15) that the critical discharge mass flux is

$$G = \rho V = -\rho_i C_i \left(\frac{2}{k+1}\right)^{(k+1)/(k-1)} \qquad (8.28)$$

The minus sign means that discharge is to the left.

EXAMPLE 8.3: UNSTEADY CRICITAL GAS FLOW
A frictionless pipe is closed at one end by a fast-acting valve, and the other end is attached with a perfect nozzle to a gas vessel with $k = 1.4$. The valve suddenly opens to the full pipe area. Determine the ratio of the initial gas discharge mass flux to that of steady state for unchanging vessel properties.

The steady-state critical mass flux for a perfect gas is obtained from Eq. (2.60). If we put $C_i = \sqrt{kg_0 P_i/\rho_i}$ into Eq. (8.28), disregarding the $-$ sign, the ratio of initial to steady-state discharge mass fluxes for $k = 1.4$ is

$$\frac{G_{init}}{G_{ss}} = \left(\frac{2}{k+1}\right)^{(k+1)/2(k-1)} = 0.58$$

This result shows that the initial discharge rate from a suddenly opened pipe is only about 60 percent of the steady-state critical flow rate.

8.5 Disturbance Signal Distortion

Signal distortion can occur during propagation through a long instrument pipe which contains compressible fluid. It is advisable to estimate the amount of distortion in order to avoid serious data interpretation errors.

Consider the long tube of Fig. 8.8, filled with stagnant compressible fluid at pressure P_i. A velocity disturbance imposed at the left end $z = 0$ has the general form $V(0, t) = V_0(t)$, with the constraint that it starts from zero (that is, $V_0(0) = 0$), and shocks do not occur. It is desirable to compare the disturbance at fixed locations z_1 and z_2 in order to estimate the amount of distortion.

Equation (8.7) is written at z_1 in the form

$$t_1 = \frac{z_1 - f(V_1)}{V_1 + C(V_1)}$$

Since we are considering cases in which shocks do not form, an instantaneous value of velocity V_1 at z_1 propagates and will reach z_2 at time

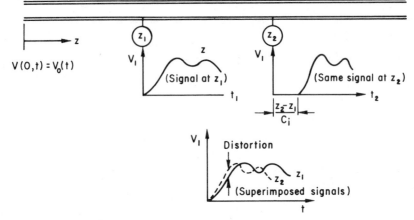

Figure 8.8   Distortion of Velocity Disturbance

$$t_2 = \frac{z_2 - f(V_1)}{V_1 + C(V_1)}$$

Since the velocity disturbance starts from $V_0(0) = 0$ and does not create shocks, the disturbance front propagates at speed C_i. It follows that the profiles at z_1 and z_2 in Fig. 8.8 can be compared if they are superimposed with common starting times. That means that time t_2 must be shifted by the amount $(z_2 - z_1)/C_i$, which gives

$$\Delta t = \left[t_2 - \frac{z_2 - z_1}{C_i} \right] - t_1 = (z_2 - z_1) \left[\frac{1}{V_1 + C(V_1)} - \frac{1}{C_i} \right] \qquad (8.29)$$

The time interval Δt gives the velocity distortion during propagation from z_1 to z_2. Note that if V_1 is small, $C(V_1) \approx C_i$, and virtually no distortion would occur. Suppose that a time trace of $V_1(t_1)$ at z_1 is given. Use of Eq. (8.29) makes it possible to superimpose the trace $V_1(t_1)$ and the distorted velocity trace as seen by an observer when it arrives at z_2.

If it is desirable to estimate the distortion without superimposing it on a signal trace, the fractional distortion of V_1 can be written as

$$\frac{\Delta V_1}{V_1} \approx \frac{dV_1}{dt_1} \frac{\Delta t}{V_1} = \frac{dV_1}{dt_1} \frac{z_2 - z_1}{V_1} \left[\frac{1}{V_1 + C(V_1)} - \frac{1}{C_i} \right] \qquad (8.30)$$

Equation (8.30) is easy to use of $V_1(t_1)$ is expressed analytically. The maximum distortion $\Delta V_{1,\text{max}}$ can be obtained from the condition $d(\Delta V_1/V_1)/dt_1 = 0$.

EXAMPLE 8.4: DISTORTION OF SINUSOIDAL SIGNAL
A velocity disturbance at location z_1 in a long tube of initially stagnant gas is described by

$$V_1(t_1) = V(z_1, t_1) = V_0 \sin 2\pi f t_1$$

Obtain an expression for the velocity distortion and determine its maximum value during propagation to another location z_2. Use the data

$$
\begin{array}{ll}
k = 1.4 & f = 10 \text{ Hz} \\
C_i = 448 \text{ m/s} & z_1 = 1 \text{ m} \\
V_0 = 6.28 \text{ m/s} & z_2 = 11 \text{ m}
\end{array}
$$

Equations (8.11) and (8.30) yield

$$\frac{\Delta V_1}{V_1} \approx -\frac{2\pi f}{V_0}\left(\frac{V_0}{C_i}\right)^2 (z_2 - z_1)\left(\frac{k+1}{2}\right) \cdot \left[\frac{\cos 2\pi f t_1}{((k+1)/2)(V_0/C_i)\sin 2\pi f t + 1}\right]$$

Note that the velocity distortion is slight for small values of f, V_0/C_i, or $z_2 - z_1$. Differentiation of $\Delta V_1/V_1$ with respect to time shows that for small V_0/C_i maximum distortion occurs when $\sin 2\pi f t \approx 0$, or

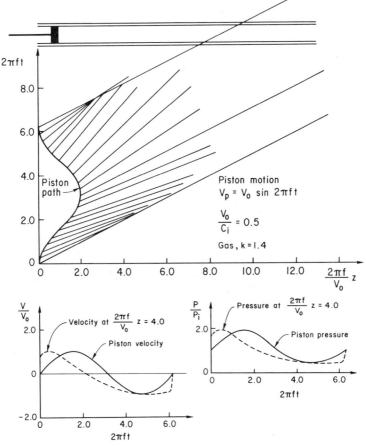

Figure 8.9   Distortion of Sinusoidal Disturbance

$$\frac{\Delta V_{1,max}}{V_1} \approx \frac{2\pi f}{V_0} \left(\frac{V_0}{C_i}\right)^2 (z_2 - z_1)\left(\frac{k+1}{2}\right)$$

The given parameters yield a maximum fractional distortion of

$$\frac{\Delta V_{1,max}}{V_1} \approx 0.0023$$

If distortion of a pressure signal is to be determined, Eq. (8.6) can be employed to obtain $\Delta P/P$ from a velocity distortion analysis.

Another method for estimating signal distortion in pipes filled with compressible fluid requires the construction of a t, z diagram like that of Fig. 8.2, in order to obtain the velocity-time traces at locations z_1 and z_2. Figure 8.9 gives such a construction for a piston at the left end with velocity $V_p = V_0 \sin 2\pi f t$ and $V_0/C_i = 0.5$. The graphs of V/V_0 and P/P_i as functions of $2\pi f t$ show how these traces are distorted from their generated form at the piston and give their arrival at a fixed location z.

8.6   Solution by the Method of Characteristics (MOC)

Simple wave solutions do not apply when disturbances overlap from opposite directions and compound waves are formed. However, the *method of characteristics* (MOC) can be used to obtain solutions if either simple or compound waves are involved.

Equations (6.64)–(6.68) for one-dimensional flow of a nonconducting fluid ($\kappa = 0$) in a rigid pipe ($\partial A/\partial t = 0$) are rewritten here as

Mass Conservation

$$\frac{\partial P}{\partial t} + V \frac{\partial P}{\partial z} + \frac{\rho C^2}{g_0} \frac{\partial V}{\partial z} = F_1 \tag{8.31}$$

Momentum Creation

$$\frac{\partial V}{\partial t} + V \frac{\partial V}{\partial z} + \frac{g_0}{\rho} \frac{\partial P}{\partial z} = F_2 \tag{8.32}$$

Energy Conservation

$$\frac{\partial \rho}{\partial t} + V \frac{\partial \rho}{\partial z} - \frac{g_0}{C^2}\left(\frac{\partial P}{\partial t} + V \frac{\partial P}{\partial z}\right) = F_3 \tag{8.33}$$

where

$$F_1 = -\frac{\rho C^2}{g_0} \frac{1}{A} \frac{dA}{dz} V + \frac{\beta C^2}{g_0 c_p} \mathscr{F} \tag{8.34}$$

$$F_2 = -\frac{f P_w}{4A} \frac{|V|V}{2} - g \sin \theta \tag{8.35}$$

$$F_3 = -\frac{\beta}{c_P} \mathscr{F} \tag{8.36}$$

and

$$\mathscr{F} = \frac{q'_{in} - q'_{out}}{A} + \frac{f P_w}{4A} \frac{\rho V^3}{2g} \tag{8.37}$$

A procedure similar to that described in Section 5.8 is followed, except that this case involves three equations instead of two. Equations (8.31) to (8.33) are multiplied by the unknown functions λ_1, λ_2, and λ_3 and then are added. Pressure, velocity, and density can be represented functionally as $P(z, t)$, $V(z, t)$, and $\rho(z, t)$ and differentiated to obtain

$$\frac{\partial \phi}{\partial t} = \frac{d\phi}{dt} - \frac{dz}{dt} \frac{\partial \phi}{\partial z}, \qquad \phi = P, V, \rho$$

from which $\partial P/\partial t$, $\partial V/\partial t$, and $\partial \rho/\partial t$ can be eliminated. The coefficients of derivatives $\partial P/\partial z$, $\partial V/\partial z$, and $\partial \rho/\partial z$ are set equal to zero, which leaves the equation

$$\left(\lambda_1 - \frac{g_0}{C^2} \lambda_3 \right) \frac{dP}{dt} + \lambda_2 \frac{dV}{dt} + \lambda_3 \frac{d\rho}{dt} = F_1 \lambda_1 + F_2 \lambda_2 + F_3 \lambda_3 \tag{8.38}$$

where λ_1, λ_2, and λ_3 must satisfy the three simultaneous, homogeneous equations

$$\begin{bmatrix} V - \dfrac{dz}{dt} & \dfrac{g_0}{\rho} & -\dfrac{g_0}{C^2} \left(V - \dfrac{dz}{dt} \right) \\[2ex] \dfrac{\rho C^2}{g_0} & V - \dfrac{dz}{dt} & 0 \\[2ex] 0 & 0 & V - \dfrac{dz}{dt} \end{bmatrix} \begin{bmatrix} \lambda_1 \\ \lambda_2 \\ \lambda_3 \end{bmatrix} = \begin{bmatrix} 0 \\ 0 \\ 0 \end{bmatrix} \tag{8.39}$$

Three solutions are obtained for the path dz/dt, corresponding to λ_1, λ_2, and λ_3 as

$$\frac{dz}{dt} = V + C, \qquad \frac{\lambda_2}{\lambda_1} = \frac{\rho C}{g_0}, \qquad \lambda_3 = 0$$

$$\frac{dz}{dt} = V - C, \qquad \frac{\lambda_2}{\lambda_1} = -\frac{\rho C}{g_0}, \qquad \lambda_3 = 0$$

$$\frac{dz}{dt} = V, \qquad \lambda_1 = \lambda_2 = 0, \qquad \lambda_3 \neq 0$$

It follows that Eqs. (8.31)–(8.33) become ordinary differential equations to be integrated on specific paths in the t, z plane according to

Right Traveling Characteristic

$$dP + \frac{\rho C}{g_0} dV = F \, dt \quad \text{on} \quad \frac{dz}{dt} = V + C \quad (8.40)$$

Left Traveling Characteristic

$$dP - \frac{\rho C}{g_0} dV = G \, dt \quad \text{on} \quad \frac{dz}{dt} = V - C \quad (8.41)$$

Fluid Particle Path

$$d\rho - \frac{g_0}{C^2} dP = H \, dt \quad \text{on} \quad \frac{dz}{dt} = V \quad (8.42)$$

where

$$\left. \begin{aligned} F &= F_1 + \frac{\rho C}{g_0} F_2 \\[2mm] G &= F_1 - \frac{\rho C}{g_0} F_2 \\[2mm] H &= F_3 \end{aligned} \right\} \quad (8.43)$$

The sound speed $C = C(P, \rho)$ is obtained from an appropriate state equation in terms of P and ρ. Simultaneous solution of Eqs. (8.40)–(8.43) gives P, V, and ρ at all z and t.

This formulation is based on continuous fluid properties. It does not describe behavior across a shock, although it applies to fluid on either side.

A computational procedure for obtaining P, V, and ρ is described with the help of Fig. 8.10. The t, z plane is divided into a mesh of equal time increments Δt, and equal space increments Δz. Properties are known at each mesh point at arbitrary time t. Right and left traveling characteristics arriving at a general mesh point $z, t + \Delta t$ designated ④, come from points ⓡ, ⓛ, and ⓟ of the previous time t, and simultaneous solution of Eqs. (8.40)–(8.43) gives properties at ④.

Values of fluid properties at points ⓡ, ⓛ, and ⓟ of Fig. 8.10 are interpolated for subcritical flows from mesh points ①, ②, and ③ as

SUBCRITICAL FLOW

$$\phi_R = \phi_2 + (\phi_1 - \phi_2) \frac{\Delta z_R}{\Delta z} \quad (8.44)$$

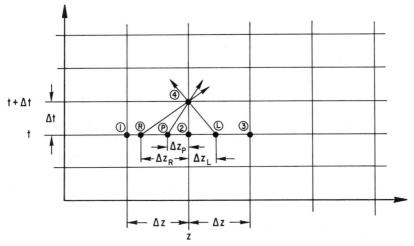

Figure 8.10   Method of Characteristics Grid, Subcritical Flow Procedure

$$\phi_L = \phi_2 + (\phi_3 - \phi_2)\,\frac{\Delta z_L}{\Delta z} \tag{8.45}$$

$$\phi_p = \phi_2 + (\phi_1 - \phi_2)\,\frac{\Delta z_p}{\Delta z} \tag{8.46}$$

$$\phi = P, V, \rho, C$$

where, based on Eq. (5.92) for large-amplitude hydrostatic waves,

$$\frac{\Delta z_R}{\Delta z} = \frac{V_2 + C_2}{\Delta z/\Delta t + V_2 + C_2 - (V_1 + C_1)} \tag{8.47}$$

$$\frac{\Delta z_L}{\Delta z} = \frac{-V_2 + C_2}{-V_2 + C_2 + (V_3 - C_3) + \Delta z/\Delta t} \tag{8.48}$$

$$\frac{\Delta z_p}{\Delta z} = \frac{V_2}{\Delta z/\Delta t + V_2 - V_1} \tag{8.49}$$

The finite difference forms of Eqs. (8.40)–(8.42) for computation of P, V, and ρ at point ④ become

$$P_4 - P_R + \left(\frac{\rho C}{g_0}\right)_R (V_4 - V_R) = F_R\,\Delta t \tag{8.50}$$

$$P_4 - P_L - \left(\frac{\rho C}{g_0}\right)_L (V_4 - V_L) = G_L\,\Delta t \tag{8.51}$$

$$\rho_4 - \rho_p - \left(\frac{g_0}{C^2}\right)_p (P_4 - P_p) = H_p\,\Delta t \tag{8.52}$$

Equations (8.50) and (8.51) can be solved simultaneously for P_4 and V_4, to get

$$V_4 = \frac{P_R - P_L + (\rho C/g_0)_R V_R + (\rho C/g_0)_L V_L + (F_R - G_L)\Delta t}{(\rho C/g_0)_R + (\rho C/g_0)_L} \quad (8.53)$$

$$P_4 = \left\{\left(\frac{\rho C}{g_0}\right)_R P_L + \left(\frac{\rho C}{g_0}\right)_L P_R + \left(\frac{\rho C}{g_0}\right)_R \left(\frac{\rho C}{g_0}\right)_L (V_R - V_L)\right.$$
$$\left. + \left[\left(\frac{\rho C}{g_0}\right)_L F_R + \left(\frac{\rho C}{g_0}\right)_R G_L\right]\Delta t\right\} \cdot \left[\left(\frac{\rho C}{g_0}\right)_R + \left(\frac{\rho C}{g_0}\right)_L\right]^{-1}$$
$$(8.54)$$

The solution for ρ_4 is obtained by putting P_4 into Eq. (8.52).

If a flow is supercritical with $V > C$, characteristic lines correspond to Fig. 8.11. The characteristic from ⓡ and the path line from ⓟ have the same interpolation equation as given by Eqs. (8.44), (8.46), (8.47), and (8.49). However, the characteristic from ⓛ has interpolation properties

$$\phi_L = \phi_2 + (\phi_1 - \phi_2)\frac{\Delta z_L}{\Delta z}, \qquad \phi = P, V, \rho, C \quad (8.55)$$

with

$$\frac{\Delta z_L}{\Delta z} = \frac{V_2 - C_2}{V_2 - C_2 + (V_1 - C_1) + \Delta z/\Delta t} \quad (8.56)$$

The computations described will be stable if the points ⓡ, ⓛ, and ⓟ in Figs. 8.10 or 8.11 remain within ① and ③. This condition is the Courant stability criterion, whereby $\Delta z_R/\Delta z$, $\Delta z_L/\Delta z$, and $\Delta z_P/\Delta z$ are all less than 1.0. It follows from Eqs. (8.47)–(8.49) and (8.56) that $\Delta t/\Delta z$ should satisfy the criterion

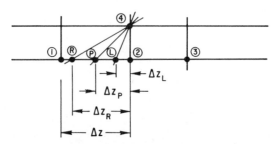

Figure 8.11   Method of Characteristics Grid, Supercritical Flow Procedure

$$\textit{Subcritical} \qquad\qquad\qquad \textit{Supercritical}$$

$$\frac{\Delta t}{\Delta z} < \begin{cases} \dfrac{1}{V_1 + C_1} \\[2mm] \dfrac{1}{C_3 - V_3} \\[2mm] \dfrac{1}{V_1} \end{cases} \quad \text{or} \quad \frac{1}{C_1 - V_1} \qquad (8.57)$$

That is, $\Delta t/\Delta z$ should be less than the smallest slope of all characteristic lines.

8.7 Appropriate Boundary Conditions

Boundary conditions given for unsteady liquid flows and waterhammer applications in Chapter 7 can be revised for unsteady compressible flows. Four of the most common boundary elements for compressible pipe flows are sketched in Figs. 8.12–8.15. These are, respectively, an ideal pipe attachment to a reservoir, a pipe flow area transition, an ideal valve or nozzle, and a specified mass charging rate. Each boundary condition employs the principle of energy conservation to show that the quasi-steady stagnation enthalpy is constant between points upstream and downstream of the boundary element. That is,

$$h_0 = h(P, \rho) + \frac{V^2}{2g_0} = \text{constant}$$

at any instant across the boundary element. The ideal attachment of Fig. 8.12 has isentropic flow from the reservoir into the pipe. The ideal area transition of Fig. 8.13 is restricted to nonshocking isentropic flow, with mass flow rate continuity expressed by $\dot{m}_1 = \dot{m}_2$. The valve or nozzle of Fig. 8.14

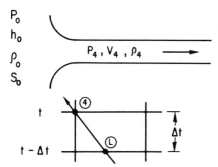

Figure 8.12 Isentropic Pipe Attachment to Reservoir. $h_0 = h\ (P_4, \rho_4) + V_4^2/2g_0$ (Energy), $\rho_4 = \rho_4\ (P_4, s_0)$ (Isentropic Flow), $P_4 = P_L + (\rho C/g_0)_L\ (V_4 - V_L) + G_L \Delta t$ (Left Traveling Characteristic), P_4, V_4, ρ_4 Are Unknowns

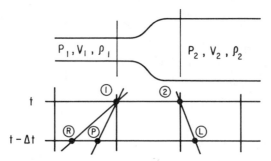

Figure 8.13 Pipe Area Transition. $\rho_1 A_1 V_1 = \rho_2 A_2 V_2$ (Mass Conservation), $h(P_1, \rho_1) + V_1^2/2g_0 = h(P_2, \rho_2) + V_2^2/2g_0$ (Energy), $s(P_1, \rho_1) = s(P_2, \rho_2)$ (Isentropic Flow), $P_1 = P_R - (\rho C/g_0)_R (V_1 - V_R) + F_R \Delta t$ (Right Traveling Characteristic), $\rho_1 = \rho_P + (g_0/C^2)_P (P_1 - P_P) + H_P \Delta t$ (Path line), $P_2 = P_L + (\rho C/g_0)_L (V_2 - V_L) + G_L \Delta t$ (Left Traveling Characteristic), $P_1, \rho_1, V_1, P_2, \rho_2, V_2$ Are Unknowns

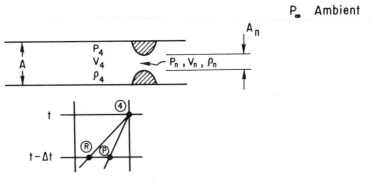

Figure 8.14 Subcritical or Critical Flow through a Valve or Nozzle. $\dot{m} = \rho_4 V_4 A_4 = \rho_n V_n A_n$ (Mass Conservation), $h_0 = h(P_4, \rho_4) + V_4^2/2g_0 = h(P_n, \rho_n) + V_n^2/2g_0$ (Energy), $s(P_4, \rho_4) = s(P_n, \rho_n)$ (Isentropic Flow), $P_4 = P_R - (\rho C/g_0)_R (V_4 - V_R) + F_R \Delta t$ (Right Traveling Characteristic), $\rho_4 = \rho_P + (g_0/C^2)_P (P_4 - P_P) + H_P \Delta t$ (Path Line), If

$$V_n \begin{cases} <C_n(P_n, \rho_n), P_n = P_\infty \ (Subcritical) \\ >C_n(P_n, \rho_n), V_n = C_n(P_n, \rho_n) \ (Critical) \end{cases}, P_4, V_4, \rho_4, P_n, V_n, \rho_n \text{ Are Unknowns}$$

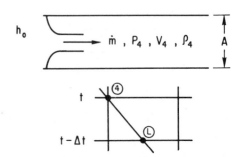

Figure 8.15 Specified Charging Rate. $\dot{m} = \rho_4 A V_4$ (Mass Conservation), $h_0 = h(P_4, \rho_4) + V_4^2/2g_0$ (Energy), $\dot{m}$ Specified, h_0 Known, $P_4 = P_L + (\rho C/g_0)_L (V_4 - V_L) + G_L \Delta t$ (Left Traveling Characteristic), P_4, V_4, ρ_4 Are Unknowns

is treated as an ideal, isentropic nozzle whose throat area A_n can be a given function of time. The mass charging rate $\dot{m}$ of Fig. 8.15 is specified. Simultaneous equations which incorporate these boundary conditions with the MOC equations also are given in Figs. 8.12 through 8.15.

If the fluid can be approximated as a perfect gas, algebraic state equations are easily employed in the characteristic boundary conditions of Figs. 8.12 through 8.15. If the fluid is a saturated liquid-vapor mixture or other simple compressible substance, the state may have to be obtained from tabulated or computer retrievable properties. Although some of the simultaneous equations for a given boundary condition can be combined, iterative solutions for P, V, and ρ at a boundary are straightforward. Solutions are described for several important cases in the next section.

8.8 Unsteady Blowdown from Gas-Filled Pipes

Consider the case of stagnant gas in a uniform pipe attached by an ideal nozzle to a reservoir with stagnation properties P_0 and ρ_0, as shown in Fig. 8.16. The pipe either ruptures or a valve opens fully at distance L from the reservoir, permitting discharge to ambient conditions. A solution for local pressure and velocity is wanted.

The left boundary condition is described by the pipe attachment to a reservoir of Fig. 8.12. If we write

$$h = \frac{k}{k-1} \frac{P}{\rho}$$

for static enthalpy and $P/\rho^k = P_0/\rho_0^k$ for isentropic flow, the energy equation yields

$$\frac{k}{k-1} \frac{P_0}{\rho_0} \left[1 - \left(\frac{P_4}{P_0} \right)^{(k-1)/k} \right] = \frac{V_4^2}{2g_0}$$

which was solved simultaneously at each time step for P_4 and V_4 with the left traveling characteristic of Eq. (8.51).

The right boundary condition corresponds to a fully open valve in the pipe of Fig. 8.14 with $A_4 = A_n$, for which the sonic condition

$$V_4 = C_4 = \sqrt{\frac{k g_0 P_4}{\rho_4}}$$

was solved simultaneously with Eqs. (8.50) and (8.52) for V_4, P_4, and ρ_4. Pressure, velocity, and density were obtained at all mesh points between the reservoir and discharge boundaries every time step from Eqs. (8.52)–(8.54).

A calculated time- and space-dependent pressure surface for gas discharge from a frictionless, uniform, unheated pipe is shown in Fig. 8.16.

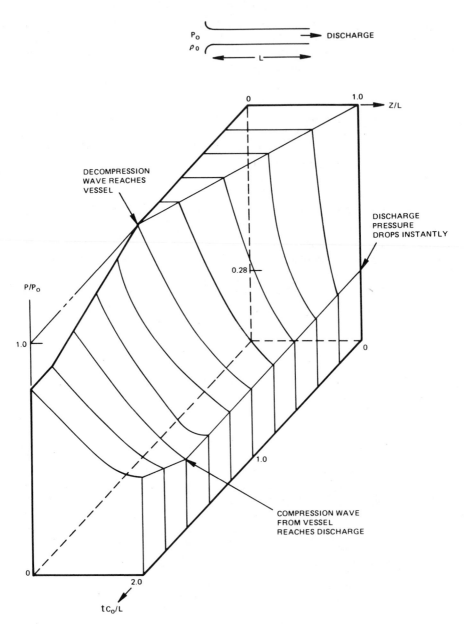

Figure 8.16 Pressure-Space-Time Surface, Pipe Rupture

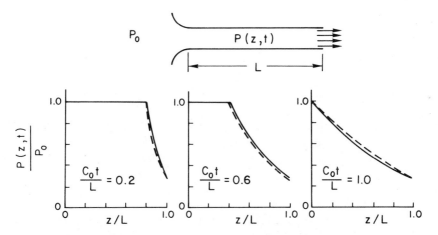

Figure 8.17   Comparison, Simple Wave Theory and Method of Characteristics. ——— Simple Wave Theory; – – – Method of Characteristics

Pressure, time, and space variables were normalized so that the results apply to pipes of any length and any vessel stagnation conditions for a gas with $k \approx 1.4$.

EXAMPLE 8.5: SIMPLE WAVE THEORY AND THE MOC

Consider the pipe rupture and gas discharge of Fig. 8.16. Show that simple wave theory can be employed to predict the time-dependent pressure profile during the time $t = L/C_0$ while the initial decompression travels from the ruptured end to the reservoir.

Equation (8.7) is applied to the right end $z = L$, where the boundary condition for critical discharge is $V = C$. Since disturbance propagation is to the left, the negative sign is used, which gives $f(V) = L$ for all time. It follows from Eq. (8.7), the sound speed C of Eq. (8.11), and the initial condition $V_i = 0$, that

$$\frac{V}{C_i} = \frac{2}{k+1}\left(1 - \frac{L-z}{C_i t}\right)$$

The simple wave pressure profile obtained from Eq. (8.10) becomes

$$\frac{P}{P_i} = \left(\frac{2}{k+1} + \frac{k-1}{k+1}\frac{L-z}{C_i t}\right)^{2k/(k-1)}$$

The comparison given in Fig. 8.17 shows that the simple wave and MOC solutions give the same result. When the disturbance arrives at the reservoir, a return wave is propagated to the right, and simple wave theory is not valid for simultaneous right and left traveling disturbances.

Other applications of the MOC are given later in this chapter. However, it is appropriate to discuss moving normal shocks first, since they can form during the propagation of a continuous disturbance.

8.9   Moving Normal Shocks, General Fluid

Shocks are discontinuities in flow properties which propagate relative to the fluid. The first appearance of a developing shock was discussed in Section 8.3. Equations (8.1)–(8.3) or (8.31)–(8.33) are based on continuous flow properties. Although they apply on either side of a shock, they do not describe property changes across the shock.

Moving shock relationships can be obtained by analysis of a discontinuity moving rightward at speed S relative to the pipe in Fig. 8.18. Properties in the undisturbed fluid are designated by subscript x. The disturbance has occurred at the left, and shocked properties are designated by subscript y. The disturbance has occurred at the left, and shocked properties are designated by subscript y. Mass, momentum, and energy principles across the shock are written for the dotted CV in which storage terms are negligible. Therefore, we have

Mass

$$\dot{m} = \rho_y A(S - V_y) = \rho_x A(S - V_x)$$

Momentum

$$\dot{m}(V_y - V_x) = g_0(P_y - P_x)A$$

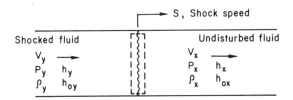

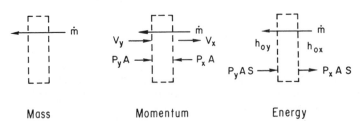

Figure 8.18   Moving Normal Shock Formulation

Energy

$$\dot{m}(h_{0y} - h_{0x}) + (P_x - P_y)AS = 0$$

If we substitute $h_0 = h + V^2/2g_0$, rearrangement yields

$$\rho_x(S - V_x) = \rho_y(S - V_y) \tag{8.58}$$

$$\rho_x(S - V_x)^2 - \rho_y(S - V_y)^2 = g_0(P_y - P_x) \tag{8.59}$$

$$h_x - h_y + \frac{1}{2}\frac{P_y - P_x}{\rho_y}\left(\frac{\rho_y}{\rho_x} + 1\right) = 0 \tag{8.60}$$

Equations (8.58)–(8.60) relate properties across a moving normal shock for a simple compressible fluid. State equations for the fluid are required for a complete analysis.

EXAMPLE 8.6: SPEED OF SMALL SHOCKS

It is postulated that a shock of vanishingly small strength (that is, $P_y = P_x + dP$) travels at sound speed in the fluid. Verify this from eqs (8.58)–(8.60) for moving normal shocks.

Let the undisturbed properties ρ_x, V_x, h_x, and P_x be written as ρ, V, h, and P, respectively. Also, consider an infinitesimal disturbance $P_y = P + dP$, which causes density, velocity, and enthalpy disturbances $\rho_y = \rho + d\rho$, $V_y = V + dV$, and $h_y = h + dh$. Substitution into Eq. (8.60) yields

$$dh - \frac{1}{\rho}dP = 0$$

from which the Gibbs equation of Section 2.7 shows that $T\,ds = 0$; that is, the entropy is constant across an infinitesimal propagating disturbance. Moreover, if second-order differentials are neglected, Eqs. (8.58) and (8.59) yield

$$dV = (S - V)\,d\rho\,, \qquad s = \text{constant}$$

and

$$\rho(S - V)\,dV = g_0\,dP\,, \qquad s = \text{constant}$$

If we note that the term $S - V$ is disturbance velocity relative to the moving fluid and eliminate $\rho\,dV$, we obtain

$$S - V = \sqrt{g_0\frac{dP}{d\rho}}\bigg|_{s=\text{constant}} = \sqrt{g_0\left(\frac{\partial P}{\partial \rho}\right)_s} = C$$

Comparison with Eq. (2.45) shows that a shock of vanishingly small strength travels at the sound speed relative to the fluid.

———————

Moderate and strong shocks cause nonisentropic property changes. The entropy increase $\Delta s = s_y - s_x$ across a shock is related to the energy dissipation and can be determined from an integration of Eq. (2.17) in the form

$$\Delta s = s_y - s_x = \int_x^y \frac{1}{T}\left(dh - \frac{1}{\rho}\,dP\right) \tag{8.61}$$

Since entropy is an exact differential, the integral of Eq. (8.61) does not depend on the process path through the shock.

8.10 Moving Shocks in a Perfect Gas

If we employ the perfect gas state equations of Table 2.2 in Eqs. (8.58)–(8.60), moving normal shock property ratios can be expressed as follows:

$$\frac{\rho_y}{\rho_x} = \left(\frac{k+1}{k-1}\frac{P_y}{P_x}+1\right)\left(\frac{k+1}{k-1}+\frac{P_y}{P_x}\right)^{-1} \tag{8.62}$$

$$\frac{T_y}{T_x} = \frac{P_y}{P_x}\left(\frac{\rho_x}{\rho_y}\right) = \sqrt{\frac{C_y}{C_x}} \tag{8.63}$$

$$\frac{S-V_x}{C_x} = \sqrt{\frac{k+1}{2k}\frac{P_y}{P_x}+\frac{k-1}{2k}} \tag{8.64}$$

$$\frac{V_y-V_x}{C_x} = \left(\frac{P_y}{P_x}-1\right)\sqrt{\frac{2}{k(k-1)}}\left(\frac{k+1}{k-1}\frac{P_y}{P_x}+1\right)^{-1} \tag{8.65}$$

The entropy change of Eq. (8.61) is

$$\frac{s_y-s_x}{R} = \frac{\Delta s}{R} = \frac{k}{k-1}\ln\frac{T_y}{T_x}-\ln\frac{P_y}{P_x} \tag{8.66}$$

These properties are graphed in Fig. 8.19 for a gas of $k = 1.4$. The dotted curve labeled $0.1(R_s/P_x A)$ is the shock force exerted on the pipe segment in which it is traveling, discussed in Section 8.16.

EXAMPLE 8.7: PROPERTIES OF A BLAST WAVE
An explosion occurs at sea level, creating a moving shock which has the pressure ratio $P_y/P_x = 5$ when it arrives at a given location. If the undisturbed air is stationary with $T_x = 20°C$ (68°F) = 293 K, $R = 287$ J/kg-K (53.3 ft-lbf/lbm-°F), and $C_x = 343$ m/s (1125 ft/s), determine the temperature and velocity of the shocked gas and find the shock speed. Then determine the air temperature which would have resulted if it was compressed isentropically to P_y.
    Figure 8.19 shows that for $V_x = 0$ and the shock pressure ratio given, $T_y/T_x = 1.75$, $V_y/C_x = 1.35$, and $S/C_x = 2.1$, from which we obtain

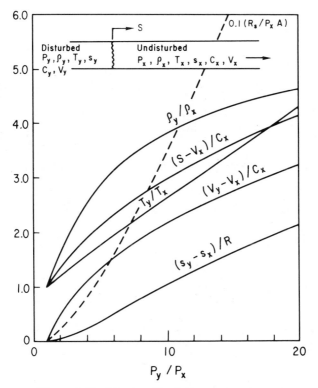

Figure 8.19 Moving Shock Properties, $k = 1.4$

$T_y = 513\,\text{K} = 240°\text{C}\ (463°\text{F})$ shock temperature

$V_y = 463\,\text{m/s}\ (1519\,\text{ft/s})$ shocked gas velocity

$S = 720\,\text{m/s}\ (2362\,\text{ft/s})$ shock speed

The environment created by this relatively small shock is astounding! The temperature is high enough to melt or otherwise damage many common substances, including human skin. The shocked gas velocity far exceeds that of a tornado and could destroy many structures.

If the undisturbed air was compressed isentropically to shock pressure, the temperature ratio would be

$$\frac{T_y}{T_x} = \left(\frac{P_y}{P_x}\right)^{(k-1)/k} = 1.584$$

from which

$$T_y = 464\,\text{K} = 191°\text{C}\ (376°\text{F})$$

The gas temperature from isentropic compression is seen to be substantially less than the 238°C (460°F) shock temperature.

(a) Present formulation   (b) Valve closure application

Figure 8.20   Valve Closure Application of Fig. 8.19. (*a*) Present Formulation; (*b*) Wave Closure Application

Moving shocks are sometimes caused by sudden valve closure on fluid flow in a pipe. It is possible to use the moving normal shock equations already developed or Fig. 8.19 to determine resulting shock properties for this case.

EXAMPLE 8.8: GAS SHOCK FROM SUDDEN VALVE CLOSURE
Air is discharging from a pipe at sonic speed $V = C = 500$ m/s (1640 ft/s) and pressure $P = 6.0$ MPa (882 psia). Sudden valve closure stops the flow. Determine the shock pressure exerted on the valve and the shock velocity.

Figure 8.20(*a*) shows a sketch with the shock and gas velocities employed in the formulations of this section. Figure 8.20(*b*) shows how to interpret these formulations for valve closure on a moving gas. Here we have $P_x = P = 6.0$ MPa (882 psia), $C_x = C = 500$ m/s (1640 ft/s), $V_x = -V = -100$ m/s (328 ft/s) and $V_y = 0$ at the closed valve. Therefore,

$$\frac{V_y - V_x}{C_x} = -\frac{-500}{500} = 1.0$$

from which Fig. 8.19 gives a shock pressure of

$$\frac{P_y}{P_x} = 3.6, \qquad P_y = 21.6 \text{ MPa (3175 psia)}$$

Moreover, Fig. 8.19 also gives

$$\frac{S - V_x}{C_x} = \frac{S - (-500)}{500} = 1.8$$

from which the shock speed relative to the pipe is

$$S = (1.8)(500) + (-500) = 400 \text{ m/s (1312 ft/s)}$$

_____

8.11 Characteristics and Approximate Shocks

The MOC is based on continuous flow properties, and has no specific feature for the development and tracking of shocks. Incorporation of moving shock boundary conditions into the MOC is inefficient and cumbersome. Methods which make use of artificial dissipation have been attempted without satisfactory results. However, the MOC has an inherent numerical dissipation mechanism which displays shock formation and properties with reasonable accuracy for many engineering applications. Shocks appear in the form of steep pressure gradients over about three to five mesh points of thickness, regardless of the total number of spatial mesh lines. Relatively uniform pressures and velocities are found to occur upstream and downstream from the approximate forward moving shock, shown by the sample computation in Fig. 8.21 at two different times. It is informative to compare these results from the MOC with calculations from moving normal shock theory.

The pressure ratio P_y/P_x was obtained from the MOC solution for a typical unsteady flow where an approximate shock developed and moved into stationary gas. This pressure ratio was used to compute other properties from Eqs. (8.62) through (8.65) for moving shocks. Shock speed from the MOC solution was obtained from the computed pressure profile locations at the times shown in Fig. 8.21. It is seen from the results in Table 8.1 that for a moderate shock pressure ratio of 4.08, approximate shock properties from the MOC closely agree with results from the moving shock equations. Additional, similar computations support the conclusion that errors introduced by approximate shocks in the MOC are generally small and proportional to the shock pressure ratio.

Figure 8.22 demonstrates a potential problem with the MOC. When the flow is supersonic to the right, point L is to the left of z_i, t and the point z_i, $t + \Delta t$ is independent of downstream conditions. Therefore, the MOC

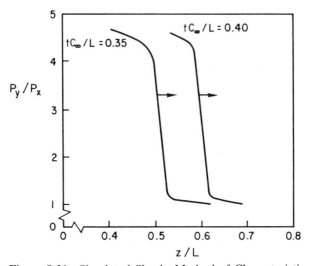

Figure 8.21 Simulated Shock, Method of Characteristics

TABLE 8.1   Moderate Shock Simulation with the MOC

Property	Predicted by MOC	Calculated by Shock Equations	Percentage Error, MOC
P_y/P_x	4.08	—	—
V_y/C_x	1.17	1.15	2
ρ_y/ρ_x	2.69	2.53	6
C_y/C_x	1.23	1.27	3
S/C_x	1.80	1.92	6

cannot recognize shock waves moving leftward against flows exceeding sonic speed to the right. Neither can shocks move to the right if a leftward flow exceeds sonic speed. However, a simple procedure overcomes this difficulty. Shock speed in gas is obtained from Eq. (8.64), and its movement between vertical mesh lines at z_i and z_{i-1} is set to occur in the delayed time interval

$$\Delta t_s = \frac{\Delta z}{S}$$

This method of advancing a shock against sonic or supersonic flow excludes the need for imposing moving shock boundary conditions at every time step, but does not automatically conserve mass. Mass conservation across the shock region in Fig. 8.23 is ensured by writing

$$\dot{m}_{i+1} - \dot{m}_i + \frac{d}{dt}(M_i + M_{i+1}) = 0$$

Since flow rate and stored mass can be written as $\dot{m} = \rho AV$ and $M = \rho A\,\Delta z$, and since $dz_i/dt = -S$ and $dz_{i+1}/dt = S$, we obtain

$$V_i = \frac{\rho_{i+1}V_{i+1}A_{i+1} - S(\rho_i A_i - \rho_{i+1}A_{i+1})}{\rho_i A_i} \qquad (8.67)$$

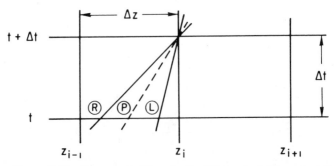

Figure 8.22   Supersonic Flow in the Method of Characteristics

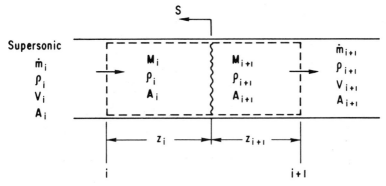

Figure 8.23 Special Mass Conservation Model

Equation (8.67) guarantees mass conservation across a shock which moves against oncoming sonic or supersonic flow. Although the MOC applies upstream and downstream, the moving shock can be tracked by the procedure just described.

8.12 Comparison with Pressure Data

Figure 8.24 gives a comparison of pressures predicted from a MOC calculation and measured during a safety valve steam discharge into a long pipe with the end submerged in water. Pertinent data is given in Table 8.2. The water expulsion was simulated in the MOC with a boundary condition based on the analysis of Section 1.10. It was necessary to apply the water column velocity at its original boundary mesh point rather than to follow a moving gas-liquid interface boundary. This requirement yields a higher predicted maximum pressure because the gas can only move at water column velocity at the fixed boundary, whereas it actually can move faster as it fills the void

TABLE 8.2 Data, Safety Valve Steam Discharge into Submerged Pipe

Valve flow rate (steady state)	69 kg/s (152 lbm/s)
Valve opening time	0.02 s
Steam pressure in reservoir	6.9 MPa (1014 psia)
Steam density in reservoir	35 kg/m^3 (2.18 lbm/ft^3)
Pipe flow area	295 cm^2 (46 in^2)
Pipe length	22.6 m (74 ft)
Initial air pressure	101 kPa (14.7 psia)
Initial air density	1.14 kg/m^3 (0.071 lbm/ft^3)
Initial air temperature	38°C (100°F)
Initial air sound speed	354 m/s (1161 ft/s)
Submerged length	4.88 m (16 ft)

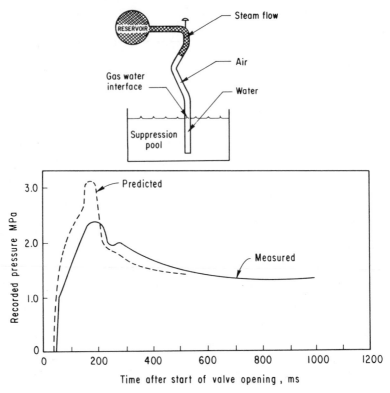

Figure 8.24   Comparison of Model with Measured Pressure

created by water motion. The overall calculation follows experimental trends reasonably well.

Figure 8.25 gives a comparison of the MOC solution with experimental measurements of Edwards [20] for blowdown of a steam-water mixture from a ruptured pipe. State equations from Table 2.4 and the critical flow properties of Fig. 2.20 for bubbly saturated steam-water mixtures were employed for the pipe of a 4 m (13.12 ft) length and 7 cm (2.76 in) diameter initially containing subcooled water at 7070 kPa (1030 psia) whose saturation pressure was 3330 kPa (486 psia). A glass rupture disk initiated the blowdown. The pressure measurement is for a position 15 cm (5.9 in) upstream from discharge. The MOC solution is seen to give a close prediction.

Time-dependent fluid properties can impose substantial reaction forces on the associated piping. The next few sections give procedures and specific cases which can be used to estimate unsteady flow forces on piping systems caused by a sudden pipe rupture or valve operation.

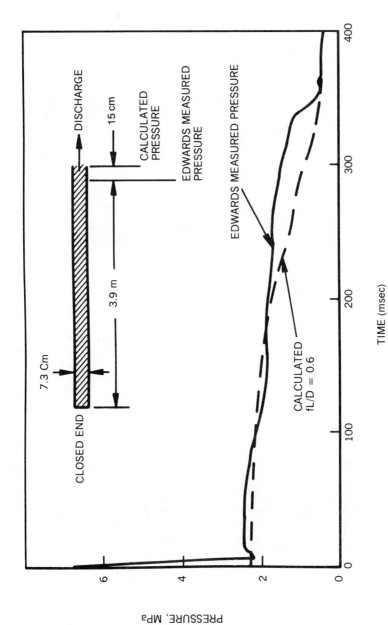

Figure 8.25 Comparison with Edwards Experiment [8.20]

463

8.13   Forces from Instantaneous Pipe Rupture

Consider a uniform pipe of length L and area A which initially contains fluid at pressure P_0 and enthalpy h_0. If the pipe contains liquid, P_0 and h_0 determine the pressure P_{sat} at which vapor would begin to form during isentropic decompression.

A sudden, complete circumferential rupture of the pipe is assumed to occur so that wave propagation initially is into the pipe in one direction. Therefore, simple wave theory of Section 8.1 was applied to pipes containing steam-water mixtures in order to obtain the initial discharge properties. Density and sound speed were determined by isentropic pressure changes, Eqs. (8.4) and (8.5), based on state equations of Tables 2.3 and 2.4 for ideal liquids and saturated liquid-vapor mixtures.

Initial discharge velocity $V_{dis} = C_{dis}$, pressure P_{dis}, and density ρ_{dis} were obtained by employing subcooled and saturated state properties for water in Eq. (8.6). Results are given in Fig. 8.26 where P_0 is the initial pressure, P_{sat}

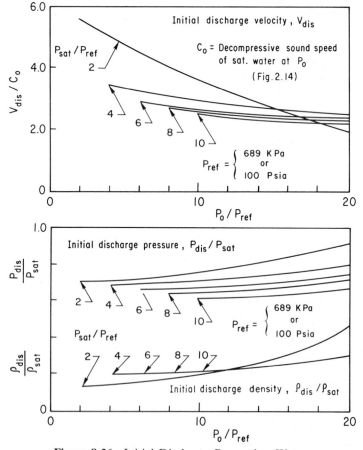

Figure 8.26   Initial Discharge Properties, Water

is the saturation pressure, ρ_{sat} is the saturated water density corresponding to its temperature, and C_0 is the decompressive sound speed of Fig. 2.14 for saturated water at the initial pressure P_0.

EXAMPLE 8.9: INITIAL WATER DISCHARGE RATE

A pipe contains subcooled water at 6.9 MPa pressure and temperature $T = 204°C$ (400°F). A sudden, full-pipe rupture occurs. Determine the initial mass discharge flux and pressure and compare with the steady state discharge flux and pressure from a pipe which is attached to a vessel containing water at the same state.

Liquid decompression above P_{sat} does not appreciably change the temperature T. Flashing will occur when pressure drops below P_{sat} at $T = 204°C$ (400°F). Steam-water properties give $P_{sat} = 1724$ kPa (253 psia) with a corresponding water density $\rho_{sat} = 860$ kg/m^3 (53.6 lbm/ft^3). The corresponding stagnation enthalpy is $h_0 = 875$ kJ/kg (377 B/lbm).

It follows from Fig. 2.14 that $C_0 = 13$ m/s (42.6 ft/s). Then, for $P_0/P_{ref} = 10$ and $P_{sat}/P_{ref} = 2.5$, Fig. 8.26 gives $V_{dis}/C_0 = 3.5$, $\rho_{dis}/\rho_{sat} = 0.2$, and $P_{dis}/P_{sat} = 0.72$. It follows that

$$G_{initial} = \frac{\rho_{dis}}{\rho_{sat}} \frac{V_{dis}}{C_0} \rho_{sat} C_0 = 7826 \text{ kg/m}^2\text{-s } (1600 \text{ lbm/s-ft}^2)$$

and

$$P_{dis} = \frac{P_{dis}}{P_{sat}} P_{sat} = 1241 \text{ kPa } (180 \text{ psia})$$

If we employ Figs. 2.20(a) and (b) for the steady mass flux and pressure at P_0 and h_0, we obtain

$$G_c = 97,640 \text{ kg/m}^2\text{-s } (19,967 \text{ lbm/s-ft}^2)$$

$$P_c = 1374 \text{ kPa } (200 \text{ psia})$$

It is seen that the initial discharge rate is substantially lower than the steady critical flow value.

––––––––––

The open pipe reaction force of Eq. (7.31) during unsteady discharge is rewritten as

$$R_{open} = R_{bd} + R_u \tag{8.68}$$

where

$$R_{bd} = -\left[(P_2 - P_\infty)A_2 + \frac{\dot{m}_2 V_2}{g_0}\right] \tag{8.69}$$

and

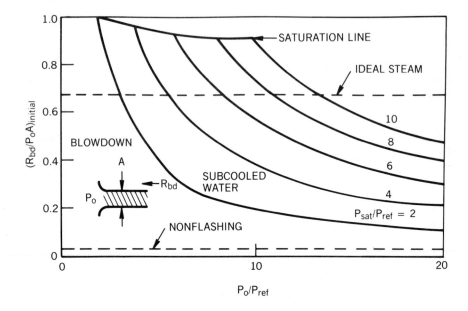

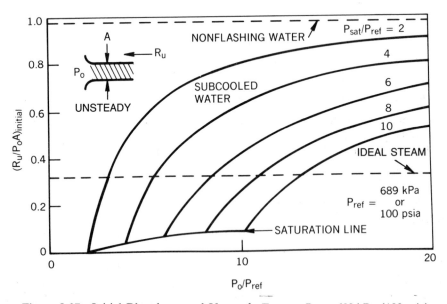

Figure 8.27   Initial Blowdown and Unsteady Forces: $P_{\text{ref}} = 689$ kPa (100 psia)

$$R_u = -\frac{1}{g_0}\frac{d}{dt}\int_0^L V\rho A(z)\,dz \qquad (8.70)$$

Here, R_{bd} is called the *blowdown reaction force*, which is the thrust caused by fluid discharge from an open pipe. Also, R_u is called the *unsteady reaction force* caused by the rate of fluid momentum change in the pipe. The subscript 2 refers to properties at the open end of a pipe discharging fluid. The length of a given pipe segment is L. It is convenient to determine R_{bd} and R_u separately in order to estimate forces on individual bounded pipe segments, and also on the discharging open pipe. Figure 8.27 gives the initial blowdown and unsteady forces for steam-water pipe ruptures.

EXAMPLE 8.10: BLOWDOWN JET IMPINGEMENT FORCE

A very long frictionless pipe of flow area $A = 0.1\,m^2$ ($1.08\,ft^2$) is filled with water at pressure $P_0 = 6.89\,MPa$ (1013 psia) and temperature $T = 230°C$ (446°F), for which the saturation pressure is $P_{sat} = 2.76\,MPa$ (406 psia). A sudden rupture causes fluid discharge and impingement on a wall which is normal to the jet. Determine the jet impingement force on the wall prior to reflected pressure waves in the pipe.

If the momentum principle is applied to a CS drawn around the fluid jet, it is easily shown that the jet impingement force is equal to the blowdown force R_{bd}. Since the ideal pipe is without friction, the initial blowdown force of Fig. 8.27 is exerted until the decompression wave has propagated to the far end of the pipe and returned to the rupture. Therefore, the initial blowdown (and jet impingement) force at $P_0 = 6.89\,MPa$ (1013 psia) and $P_{sat} = 2.76\,MPa$ (406 psia) is obtained as $R_{bd}/P_0A = 0.37$, or

$$F_{jet} = R_{bd} = (0.37)P_0A = 250\,kN\ (56,000\,lbf)$$

An increase in water temperature causes an increase in saturation pressure. It is seen from Fig. 8.27 that if the saturation pressure was increased, both blowdown and impingement forces also would increase.

The MOC was applied to a uniform pipe of length L and area A, attached to a reservoir of ideal steam at pressure P_0, similar to the case shown in Fig. 8.16. Both unsteady and blowdown forces on the entire pipe segment are shown for a range of the pipe friction parameter fL/D in Fig. 8.28. The blowdown force, shown by solid lines, immediately starts at $R_{bd}/P_0A = 0.68$, and then either rises or falls to a steady value, which decreases with increasing fL/D. The dotted unsteady force lines begin at $R_u/P_0A = 0.32$ and decrease to zero when steady state is reached. Use of the steady state friction factor in Fig. B.1 of Appendix B reduces calculated forces below those which would be obtained from an unsteady friction factor.

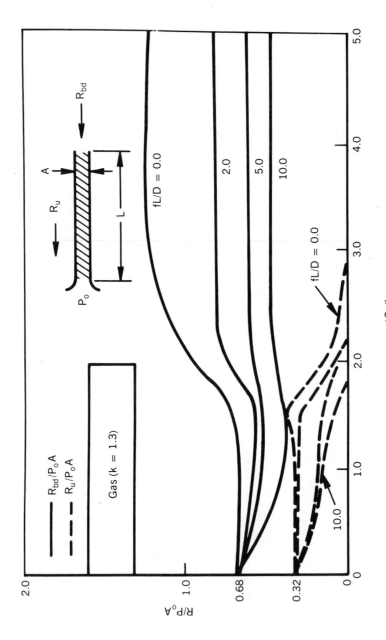

Figure 8.28 Blowdown and Total Unsteady Forces. —— R_{bd}/P_oA; – – – R_u/P_oA

If a pipe has several straight bounded segments between elbows, it is useful to obtain the unsteady force on each segment rather than on the entire pipe. If a bounded pipe segment lies between z_1 and z_2, we can write the total unsteady force of Eq. (8.70) as

$$R_u = -\frac{1}{g_0}\frac{d}{dt}\int_{z_1}^{z_2} \rho AV\, dz = -\frac{1}{g_0}\frac{d}{dt}\left[\int_0^{z_2} \rho AV\, dz - \int_0^{z_1} \rho AV\, dz\right] \tag{8.71}$$

Note that $\int_0^{z_n} \rho AV\, dz$, $n = 1, 2$, is the total momentum of fluid in the pipe between $z = 0$ and $z = z_n$. Figure 8.29 gives the unsteady force in a pressurized gas pipe between the reservoir and position z for sudden discharge from a rupture at $z = L$. An unsteady force does not occur in the pipe segment from $z = 0.0$ to z_n until the disturbance propagates a distance $L - z_n$. The force on a bounded pipe segment extending between z_1 and z_2 is obtained by subtracting the total unsteady force on segment $z = 0$ to z_1 from the total unsteady force on segment $z = 0$ to z_2.

EXAMPLE 8.11: PIPE FORCES FROM RELIEF VALVE BLOW
A closed pipe of length $L = 100$ m, flow area $A = 0.1$ m^2, and negligible friction is attached to a saturated steam reservoir at pressure $P_0 = 13.8$ MPa. A relief valve opens instantly at the discharge end to a flow area equal to that of the pipe. Determine (a) the initial blowdown force, (b) the steady blowdown force and time required to reach it, and (c) the unsteady force history on a straight pipe segment between elbows at distances $z_1 = 40$ m and $z_2 = 60$ m from the reservoir.

(a) Either Fig. 8.27 or 8.28 gives the initial steam blowdown force as $R_{bd}/P_0 A = 0.68$, for which

$$R_{bd} = 0.94 \text{ MN}$$

(b) Figure 8.28 also gives the steady blowdown force as $(R_{bd}/P_0 A)_{\text{steady}} = 1.23$, or

$$R_{b,\text{steady}} = 1.7 \text{ MN}$$

which is reached at an approximate time $tC_0/L = 3.0$. Figure 2.14 gives a sound speed of about 400 m/s in the saturated steam at 13.8 MPa (136 atm), so the time to reach steady blowdown is

$$t = (3.0)L/C_0 = 0.75 \text{ s}$$

(c) The unsteady force history is obtained with the help of Fig. 8.29 at $z_1/L = 0.4$ and $z_2/L = 0.6$. Table 8.3 gives the force history for $P_0 = 13.8$ MPa, $A = 0.1$ m^2, $C_0 = 400$ m/s, and $L = 100$ m. The unsteady force begins to increase at 0.1 s as the decompression front moves into the pipe segment at $z = 60$ m. It rises to 310 kN at 0.15 s when the decompression reaches $z = 40$ m and falls as the front moves out of the segment. Reflection from the reservoir causes a recompression to enter the

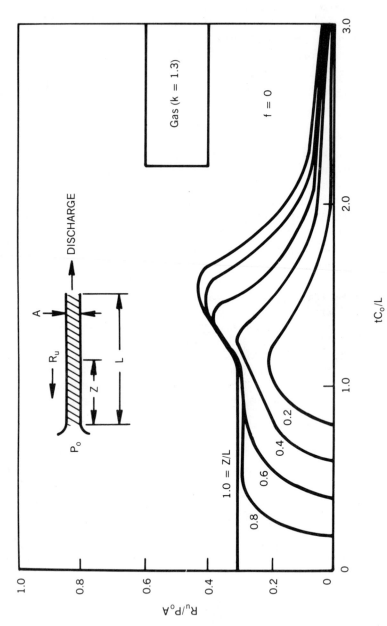

Figure 8.29 Typical Unsteady Force

TABLE 8.3 Unsteady Force History on a Pipe Segment

$\dfrac{tC_0}{L}$	$z_2/L = 0.6$ $\left.\dfrac{R_u}{P_0A}\right\|_2$	$z_1/L = 0.4$ $\left.\dfrac{R_u}{P_0A}\right\|_1$	$\dfrac{R_u}{P_0A} = \left.\dfrac{R_u}{P_0A}\right\|_2 - \left.\dfrac{R_u}{P_0A}\right\|_1$	t (s)	R_u (kN)
0.4	0.00	0.00	0.00	0.10	0
0.5	0.17	0.00	0.17	0.13	235
0.6	0.23	0.00	0.23	0.15	310
0.7	0.27	0.15	0.12	0.18	166
0.8	0.29	0.20	0.09	0.20	124
1.0	0.30	0.25	0.05	0.25	69
1.2	0.34	0.30	0.04	0.30	55
1.4	0.38	0.25	0.13	0.35	179
1.6	0.23	0.17	0.06	0.40	83
1.8	0.15	0.09	0.06	0.45	83
2.0	0.08	0.06	0.02	0.50	28
3.0	0.03	0.02	0.01	0.75	14

segment at 0.3 s, the force increases to 179 kN at 0.35 s, and then decreases until it almost vanishes at 0.75 s when steady blowdown is approached.

When steady blowdown from a pipe rupture is reached, the unsteady force disappears, leaving only the blowdown force acting on the discharging pipe segment. The steady state blowdown force of Eq. (8.69) was graphed in Fig. 8.30 for a steam-water mixture at a range of pressures and enthalpies for a pipe without friction. Ideal steam corresponds to $R_{bd}/P_0A = 1.25$, whereas nonflashing water corresponds to $R_{bd}/P_0A = 2.0$.

8.14 Pipe Forces from Safety Valve Charging

The operation of a safety or pressure relief valve is common in manufacturing, chemical, and energy industries for protection or shutdown purposes. Usually such valves are designed to open quickly and provide a specified discharge rate in order to relieve reservoir pressure. The discharge is sometimes vented through pipes to the atmosphere, but in the case of contaminated fluid it may be vented to quench tanks or other safety containers. The venting pipe system must be designed to withstand forces resulting from the valve charging rate.

Consider the case of a vent pipe initially filled with atmospheric air. It contains a valve which discharges fluid from a pressurized tank, as shown in Fig. 8.31. Energetic fluid expansion into the vent pipe can create a shock in

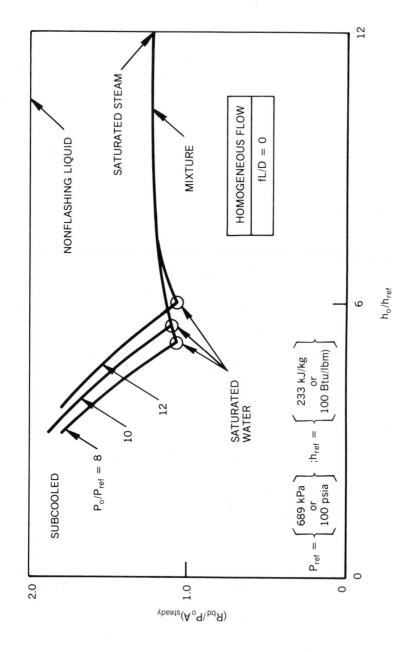

Figure 8.30 Steady Blowdown Force. Homogeneous Flow $fL/D = 0$; $P_{ref} = 689$ kPa (100 psia), $h_{ref} = 233$ kJ/kg (100 B/lbm)

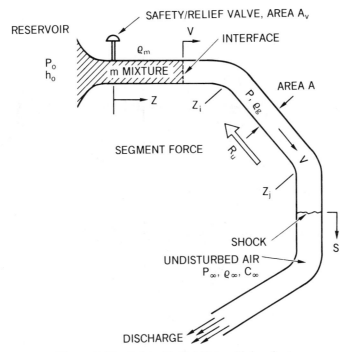

Figure 8.31   Safety/Relief Valve Piping System

the initial air and corresponding unsteady pipe forces. Suppose that the vent pipe total length is L with flow area A, and the initial air properties are P_∞, ρ_∞, and C_∞. If the valve opening time τ is long relative to the pressure wave propagation time L/C_∞, then the valve opening characteristic is important. Otherwise, instant valve opening is a reasonable idealization. The valve opening criterion is, therefore,

Condition for instant valve opening approximation

$$\frac{\tau C_\infty}{L} \ll 1.0 \tag{8.72}$$

8.15   Gas Pressure Relief, Slow Valve Opening

Consider gas pressure relief where the valve opening is important, idealized by the linear ramp mass flow rate

$$\dot{m}_{in} = \begin{cases} \dot{m}_\infty \dfrac{t}{\tau}, & \dfrac{t}{\tau} < 1 \\[2ex] \dot{m}_\infty, & \dfrac{t}{\tau} > 1 \end{cases} \tag{8.73}$$

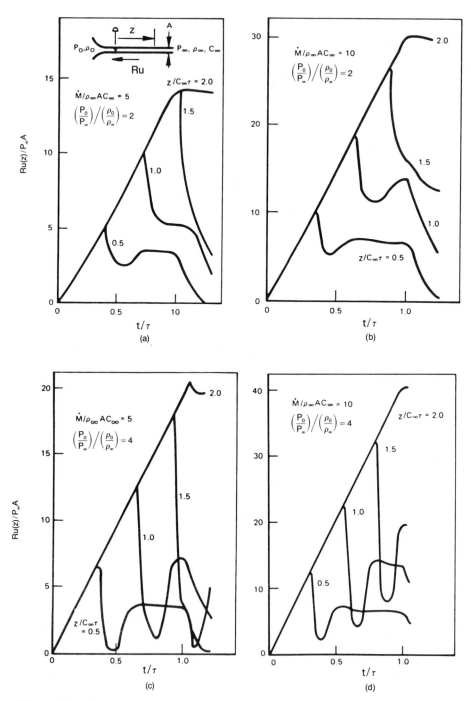

Figure 8.32 Pipe Reaction Force, Valve Flow Acceleration. Gas ($K = 1.3$) (a), (b) $(P_0/P_\infty) \div (\rho_0/\rho_\infty) = 2$; (c), (d) $(P_0/P_\infty) \div (\rho_0/\rho_\infty) = 4$

The MOC and Eq. (8.71) were used to predict segment forces for gas discharge ($k = 1.3$) into pipes initially filled with air at standard atmospheric conditions. Results are summarized in Fig. 8.32, which gives normalized values of the unsteady force $R_u(z)/P_\infty A$ on pipe lengths between the reservoir and distance z. Time is normalized with the valve full opening time τ. The curves cover a range of the normalized steady valve discharge rate $\dot{m}/\rho_\infty A C_\infty$ and gas properties in the tank $(P_0/P_\infty)/(\rho_0/\rho_\infty)$. Each family of curves represents the total unsteady force on a straight pipe between the tank and normalized distance $z/C_\infty\tau$. These curves can be used to estimate segment pipe forces for a range of the parameters involved.

EXAMPLE 8.12: RELIEF VALVE FORCE ON PIPE SEGMENT

A 200-m-long pipe of $0.07\,\text{m}^2$ flow area contains atmospheric air with properties $P_\infty = 101\,\text{kPa}$, $\rho_\infty = 1.22\,\text{kg/m}^3$, and $C_\infty = 341\,\text{m/s}$. A relief valve charges the pipe so that a 150-kg/s steady flow rate is reached linearly in $\tau = 0.1\,\text{s}$, after which it is constant. Source steam is at 6897 kPa pressure and $36\,\text{kg/m}^3$ density. Estimate the force-time history on a pipe segment between two elbows at $z_i = 35\,\text{m}$ and $z_j = 70\,\text{m}$ from the valve. Consider only that period of time prior to reflections from the discharge end.

The valve opening criterion of Eq. (8.72) yields

$$\frac{\tau C_\infty}{L} = 0.18$$

which is judged to be large enough for the valve opening characteristic to be important. Steam charging parameters are given by

$$\frac{\dot{m}}{\rho_\infty A C_\infty} = 5.15$$

$$\frac{P_0/P_\infty}{\rho_0/\rho_\infty} = 2.3$$

The segment force is to be obtained between

TABLE 8.4 Relief Valve Force on Pipe Segment

t/τ	$R_{u,j}/P_\infty A$	$R_{u,i}/P_\infty A$	t (s)	$R_{u,ij} = R_{u,j} - R_{u,i}$ (kN)
0.0	0.0	0.0	0.00	0.0
0.2	2.5	2.5	0.02	0.0
0.4	5.0	5.0	0.04	0.0
0.6	8.1	8.1	0.06	0.0
0.8	11.2	6.0	0.08	36.9
1.0	14.1	5.4	0.10	61.7
1.2	14.4	4.0	0.12	73.7

$$\frac{z_i}{C_\infty \tau} = 1.026$$

and

$$\frac{z_j}{C_\infty \tau} = 2.05$$

Approximate values of $R_u/P_\infty A$ were obtained from Fig. 8.32(a) for the two locations and are tabulated in Table 8.4 as a function of time.

8.16 Gas Pressure Relief, Rapid Valve Opening

Next, we consider cases where a gas relief valve opening time τ is short enough to assume instant opening, or step charging. That is, $\tau C_\infty/L \ll 1.0$. If the valve discharge rate rises immediately to the value $\dot{m}$, a shock wave will move through initial gas in the vent pipe, followed by a density interface as shown in Fig. 8.31. Valve boundary conditions of Fig. 8.14 for gas flow correspond to mass and energy conservation in the forms

$$\dot{m}_{\text{in}} = \rho_{\text{in}} A V_{\text{in}} \tag{8.74}$$

and

$$h_0 = \frac{k_{\text{in}}}{k_{\text{in}} - 1} \frac{P_0}{\rho_0} = \frac{k_{\text{in}}}{k_{\text{in}} - 1} \frac{P_{\text{in}}}{\rho_{\text{in}}} + \frac{V_{\text{in}}^2}{2g_0} \tag{8.75}$$

The interface condition is one of continuous pressure and velocity. Equations (8.74) and (8.75) were employed in the moving shock equations of Section 8.10, and unsteady forces were obtained from Eq. (8.71). The shock pressure P_s/P_∞ and force $R_s/P_\infty A$ are shown in Figs. 8.33 and 8.34. The shock force also is shown as a function of shock pressure by the dotted line in Fig. 8.19. Shock speed $S/C_\infty = S/C_x$ and velocity of gas behind the shock $V/C_\infty = V_y/C_x$ can be obtained from Fig. 8.19 at values of $P_y/P_x = P_s/P_\infty$. The interface moves at velocity $V_I = V_y$, and the corresponding force $R_I/P_\infty A$ is graphed in Fig. 8.35.

EXAMPLE 8.13: RELIEF VALVE SHOCK FORCE IN PIPE
The pipe system, steam source, and initial conditions are the same as those employed in Example 8.12, with the exception that valve time to steady state is $\tau = 0.001$ s. Estimate the shock pressure and force history on that segment between 35 m and 70 m from the valve.
   Equation (8.72) gives

$$\frac{\tau C_\infty}{L} = \frac{(0.001)(341)}{35} = 0.0097 \ll 1$$

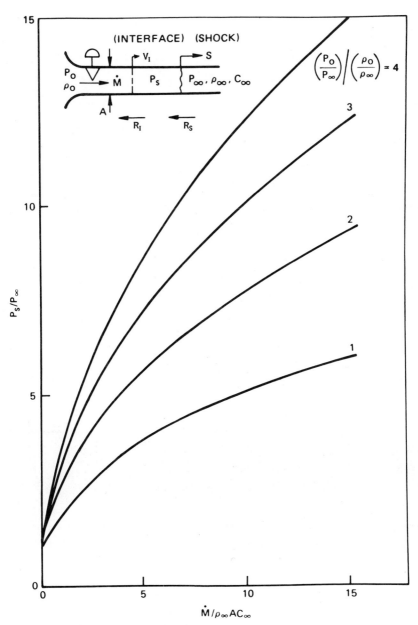

Figure 8.33 Valve Step Charging, Shock Pressure

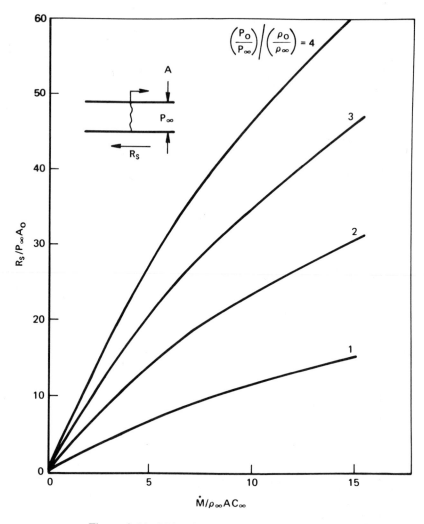

Figure 8.34   Valve Step Charging, Shock Force

which justifies the idealization of instantaneous valve opening. Valve flow parameters correspond to those of example 8.12, or

$$\frac{\dot{m}}{\rho_\infty A C_\infty} = 5.15 \quad \text{and} \quad \frac{P_0/P_\infty}{\rho_0/\rho_\infty} = 2.3$$

Forces, pressures, and speeds obtained from Figs. 8.33–8.35 and 8.19 are given below:

$$
\begin{array}{ll}
R_s/P_\infty A = 17 & R_s = 120 \text{ kN} \\
P_s/P_\infty = 6.3 & P_s = 638 \text{ kPa} \\
R_1/P_\infty A = 0.4 & R_1 = 2.84 \text{ kN} \\
S/C_\infty = 2.28 & S = 777 \text{ m/s} \\
V_1/C_\infty = 1.6 & V_1 = 546 \text{ m/s}
\end{array}
$$

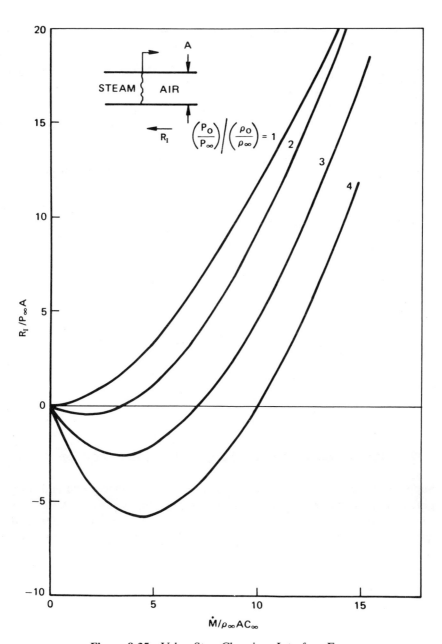

Figure 8.35 Valve Step Charging, Interface Force

TABLE 8.5 Pipe Segment Force-Time History

Time Interval (s)	Shock Force (kN)	Interface Force (kN)	Total Force (kN)
$0 < t < 0.045$	0.0	0.0	0.0
$0.045 < t < 0.064$	120.5	0.0	120.5
$0.064 < t < 0.09$	120.5	2.8	123.3
$0.09 < t < 0.128$	0.0	2.8	2.8
$0.128 < t$	0.0	0.0	0.0

Arrival times of the shock and interface at the segment are obtained from

$$t_s = \frac{z_i}{S} = \frac{35}{777} = 0.045 \text{ s} \qquad \text{shock}$$

$$t_I = \frac{z_i}{V_I} = \frac{35}{546} = 0.064 \text{ s} \qquad \text{interface}$$

and durations of these forces are determined from

$$t_{s,\text{duration}} = \frac{z_j - z_i}{S} = \frac{70 - 35}{777} = 0.045 \text{ s} \qquad \text{shock}$$

$$t_{I,\text{duration}} = \frac{z_j - z_i}{V_I} = \frac{70 - 35}{546} = 0.064 \text{ s} \qquad \text{interface}$$

Table 8.5 gives the force-time history.

8.17 Sudden Liquid-Vapor Discharge into a Gas-Filled Pipe

Suppose that a saturated liquid-vapor mixture is suddenly charged into a pipe containing gas with properties $P_x = P_\infty$, $\rho_x = \rho_\infty$, $C_x = C_\infty$, and $h_x = h_\infty$. The advancing mixture interface is at the same pressure and velocity of the adjacent shocked gas, namely P_y and V_y. The mixture charging rate

$$\dot{m} = A V_y \rho_{my} \tag{8.76}$$

and stagnation enthalpy

$$h_0 = h_{my} + \frac{V_y^2}{2g_0} \tag{8.77}$$

are known, where ρ_{my} and h_{my} refer to the mixture density and static enthalpy in the pipe of area A. If we employ Table 2.4 to write

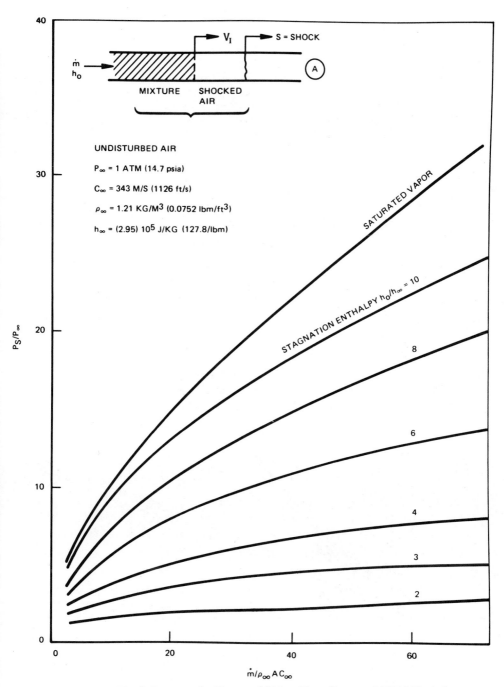

Figure 8.36 Shock Pressure in Terms of Valve Flow Rate and Fluid Enthalpy

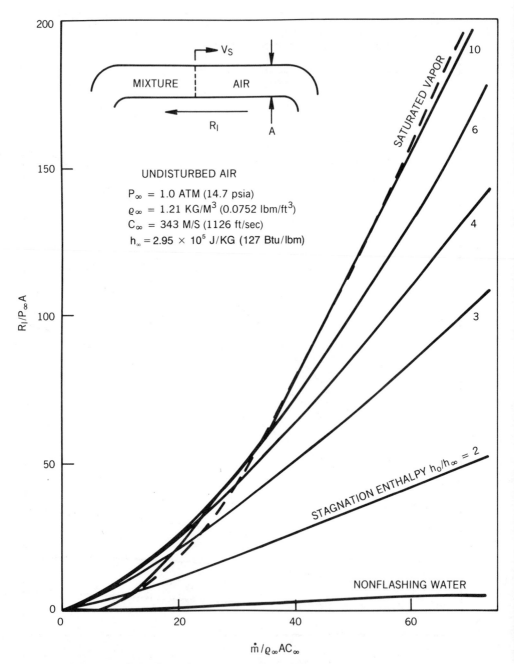

Figure 8.37 Interface Force in Terms of Valve Flow Rate and Fluid Enthalpy. Undisturbed Air: Parameters Same as for Fig. 8.36

$$h_{my} = h_f + \frac{h_{fg}}{v_{fg}} \left(\frac{1}{\rho_{my}} - v_f \right)$$

and eliminate ρ_{my}, we obtain

$$h_0 = h_f + \frac{h_{fg}}{v_{fg}} \left(\frac{AV_y}{\dot{m}} - v_f \right) + \frac{V_y^2}{2g_0} = h_0 \left(P_y, V_y; \frac{\dot{m}}{A} \right) \qquad (8.78)$$

Since h_f, h_{fg}, v_f, and v_{fg} are functions of P_y only, Eq. (8.78) relates P_y and V_y for given values of h_0 and $\dot{m}/A$. Equations (8.76) and (8.78) were employed with (8.62), (8.64), and (8.65) to obtain normal shock properties P_y, V_y, ρ_y, and S, and the mixture density ρ_{my} in terms of h_0 and $\dot{m}/A$. Then Eq. (8.70) was used to obtain the pipe unsteady reaction force caused by the advancing interface. Figure 8.36 gives the shock pressure P_s/P_∞ for ranges of the enthalpy parameter h_0/h_∞ and the charging parameter $\dot{m}/\rho_\infty A C_\infty$. The shock force $R_s/P_x A = R_s/P_\infty A$, shock speed $S/C_x = S/C_\infty$, and mixture front velocity $V_y/C_x = V_y/C_\infty$ can be obtained by entering Fig. 8.19 with $P_y/P_x = P_s/P_\infty$. The mixture-gas interface force $R_I/P_\infty A$ is given in Fig. 8.37. It is seen that both shock and interface forces increase with enthalpy of the charging flow.

Example 8.14: Maximum Allowable Charging Rate

A pressure relief valve in a pipe of area $A = 0.1 \text{ m}^2$ (1.08 ft^2) is designed to prevent the shock and interface forces from exceeding 100 kN ($22{,}400 \text{ lbf}$). Determine the maximum charging rate $\dot{m}$ through a quick-opening relief valve for which neither the shock nor interface force will exceed this limit. The charging fluid comes from a reservoir containing saturated water at 5 MPa (735 psia) pressure. The pipe initially contains air with standard properties $P_\infty = 101 \text{ kPa}$ (14.7 psia), $\rho_\infty = 1.21 \text{ kg/m}^3$ (0.075 lbm/ft^3), $C_\infty = 343 \text{ m/s}$ (1125 ft/s), and $h_\infty = 295 \text{ kJ/kg}$ (684 B/lbm).

Saturated water at 5 MPa (735 psia) has the stagnation enthalpy $h_{fo} = h_f(P_0) = 1150 \text{ kJ/kg}$, which gives

$$\frac{h_{fo}}{h_\infty} = 3.90$$

The allowable shock or mixture interface force parameters are

$$\frac{R_I}{P_\infty A}, \frac{R_s}{P_\infty A} = \frac{100}{(101)(0.1)} = 9.9$$

The allowable shock force gives the shock pressure $P_s/P_\infty = 4.4$ in Fig. 8.19, for which Fig. 8.36 at $h_{fo}/h_\infty = 3.9$ gives $\dot{m}/\rho_\infty A C_\infty = 17.0$. Moreover, the allowable interface force and same enthalpy ratio give $\dot{m}/\rho_\infty A C_\infty = 10$ from Fig. 8.37. It is seen that the interface force gives the smallest allowable charging rate parameter, so the charging rate should not exceed

$$\dot{m} = 10 \rho_\infty A C_\infty = (10)(1.21)(0.1)(343) = 415 \text{ kg/s} \ (913 \text{ lbm/s})$$

Valve flow rate $\dot{m}$ is determined by fluid pressure and enthalpy in the vessel and by the valve minimum flow area A_v. If critical flow (Fig. 2.20) in A_v determines the mass charging rate, then the shock pressure and shock and interface forces can be obtained from Figs. 8.38–8.40. The shock pressure and force reach a relative maximum in the subcooled region, and the interface force reaches an absolute maximum. This behavior results from the fact that as stagnation enthalpy decreases below the saturated liquid

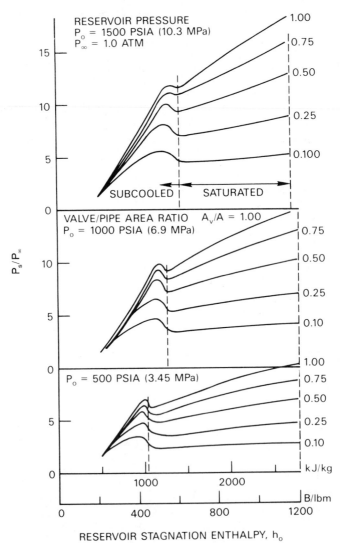

Fig. 8.38 Shock Pressure

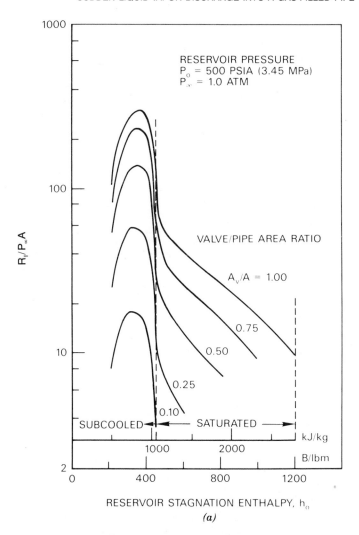

RESERVOIR PRESSURE
P_o = 500 PSIA (3.45 MPa)
P_∞ = 1.0 ATM

VALVE/PIPE AREA RATIO

A_v/A = 1.00

0.75

0.50

0.25

0.10

SUBCOOLED ◄───── SATURATED ─────►

$R_f/P_\infty A$

RESERVOIR STAGNATION ENTHALPY, h_0

(a)

Figure 8.39 The Interface Force

value, mass flux increases, according to Fig. 2.20(*a*), but enthalpy decreases so that the energy discharge rate passes through a maximum at some value of subcooling.

EXAMPLE 8.15: MIXTURE DISCHARGE FORCES

A relief valve is designed to discharge $\dot{m}_g$ = 100 kg/s of saturated steam from a boiler at a pressure of P_0 = 6.9 MPa into a relief pipe of area A = 0.1 m², initially containing atmospheric air at P_∞ = 101 kPa and sound speed C_∞ = 340 m/s. An off-design condition is postulated for which subcooled water at 6.9 MPa and h_0 = 1000 kJ/kg is discharged through the same valve. The shock and interface forces are required for the off-design condition. A straight pipe segment 5 m long between z_i = 10 m and

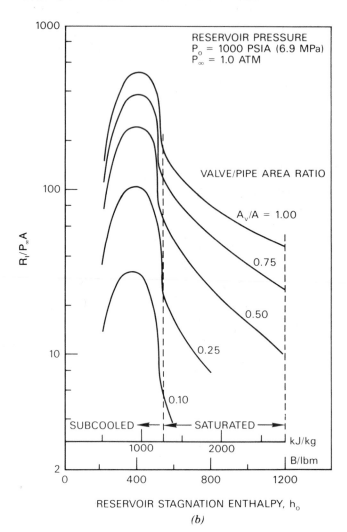

RESERVOIR STAGNATION ENTHALPY, h_o

(b)

Figure 8.39   *(Continued)*

$z_j = 15$ m, measured along the pipe from the boiler, is of particular interest. The times of arrival, duration, and exit from this segment should be obtained for both the shock and mixture interface.

First it is necessary to estimate the valve flow area from the known steam discharge rate. Figure 2.20(*a*) gives the critical mass flux $G_c = 9780$ kg/m²-s for saturated steam at 6.9 MPa. Therefore, the valve flow area is

$$A_v = \frac{\dot{m}_g}{G_c} = \frac{100}{9780} = 0.01 \text{ m}^2$$

It follows that

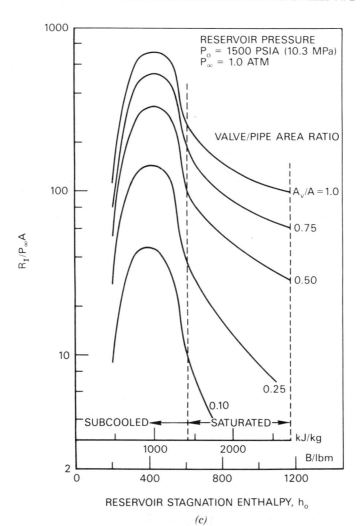

(c)

Figure 8.39 *(Continued)*

$$\frac{A_v}{A} = \frac{0.01}{0.1} = 0.1$$

Although the piping system is already designed for steam discharge, it will be interesting to compare reaction forces and time durations for both the steam and subcooled water cases. Figures 8.38–8.40 with $P_0 = 6.9$ MPa, saturated steam, and $A_v/A = 0.1$ give

$$\frac{R_s}{P_\infty A} = 9.0, \quad \frac{R_I}{p_\infty A} = 0, \quad \frac{P_s}{P_\infty} = 4.2$$

Since $P_\infty A = (101)(0.1) = 10.1$ kN, it follows that

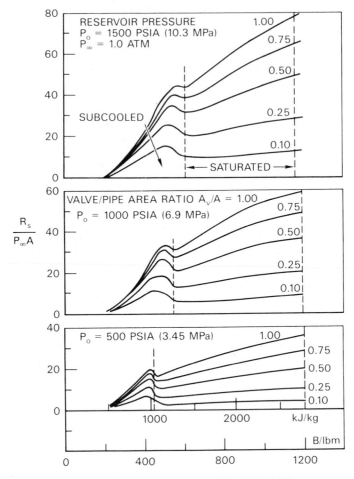

Figure 8.40   The Shock Force

$$
\left.
\begin{aligned}
R_s &= 90.9 \text{ kN} \\
R_I &= \text{negligible} \\
P_s &= 420 \text{ kPa}
\end{aligned}
\right\} \quad \text{steam}
$$

The shock and interface speeds are obtained from Fig. 8.19 at $P_s/P_\infty = 4.2$ as

$$
\frac{S}{C_\infty} = 1.95, \qquad \frac{V_y}{C_\infty} = 1.2
$$

or

$$
\begin{aligned}
S &= 663 \text{ m/s} \\
V_y &= 408 \text{ m/s}
\end{aligned} \quad \text{steam}
$$

The arrival, duration, and exit times of the shock in the pipe segment for steam discharge are

$$
\left.
\begin{aligned}
t_{s,\text{arrival}} &= \frac{z_i}{S} = 0.015 \text{ s} \\[2mm]
t_{s,\text{duration}} &= \frac{z_j - z_i}{S} = 0.0075 \text{ s} \\[2mm]
t_{s,\text{exit}} &= t_{s,\text{arrival}} + t_{s,\text{duration}} = 0.0225 \text{ s}
\end{aligned}
\right\} \quad \text{steam}
$$

Times associated with the steam interface are not considered because the interface force is negligible.

Subcooled water discharge is considered next. Again, Figs. 8.38–8.40 at $P_0 = 6.9$ MPa, $h_0 = 1000$ kJ/kg, and $A_v/A = 0.1$ give

$$
\frac{R_s}{P_\infty A} = 11 , \quad \frac{R_I}{P_\infty A} = 30 , \quad \frac{P_s}{P_\infty} = 4.5
$$

from which

$$
\left.
\begin{aligned}
R_s &= 111 \text{ kN} \\
R_I &= 303 \text{ kN} \\
P_s &= 450 \text{ kPa}
\end{aligned}
\right\} \quad \text{subcooled}
$$

Also, from Fig. 8.19,

$$
\frac{S}{C_\infty} = 2.0 , \quad \frac{V_y}{C_\infty} = 1.25
$$

or

$$
S = 680 \text{ m/s} , \quad V_y = 425 \text{ m/s} \quad \text{subcooled}
$$

The arrival, duration, and exit times of the shock and interface in the pipe segment are

$$
\left.
\begin{aligned}
t_{s,\text{arrival}} &= \frac{10}{680} = 0.0147 \text{ s} \\[2mm]
t_{s,\text{duration}} &= \frac{15 - 10}{680} = 0.00735 \text{ s} \\[2mm]
t_{s,\text{exit}} &= t_{s,\text{arrival}} + t_{s,\text{duration}} = 0.022 \text{ s} \\[2mm]
t_{I,\text{arrival}} &= \frac{10}{425} = 0.024 \text{ s} \\[2mm]
t_{I,\text{duration}} &= \frac{15 - 10}{425} = 0.012 \text{ s} \\[2mm]
t_{I,\text{exit}} &= t_{I,\text{arrival}} + t_{I,\text{duration}} = 0.036 \text{ s}
\end{aligned}
\right\} \quad \text{subcooled}
$$

It is seen that the shock passes through the pipe segment and exits from it at 0.022 s before the interface arrives at 0.024 s. Therefore, the shock and interface forces are applied at different times for different durations without overlapping. If overlap did occur, it would be necessary to add the shock and interface forces for the period of overlap.

———————

The prediction of force magnitudes and time histories on a piping system usually must be considered from the standpoint of piping dynamic response. Although a force magnitude may not overstress a pipe if applied steadily, time dependence of the force sometimes can excite natural frequencies of the pipe system and thereby overstress it. The force predictions in this chapter usually can be employed with pipe dynamic models for a complete unsteady stress analysis.

8.18 Moving Shocks in a Gas-Liquid Mixture

Shock formation in a bubbly mixture of noncondensible gas and liquid can occur in several practical instances, which include the rapid closure of a valve or the sudden arrival of a mixture at a flow area reduction. Even a mixture of saturated liquid and vapor will respond like noncondensible gas in liquid because a shock drives vapor into the superheated state and liquid into the subcooled state.

The homogeneous, bubbly mixture state equations of Tables 2.4 and 2.5 are written as

$$\rho_m = \frac{1}{x_m v_g + (1 - x_m)v_L} \tag{8.79}$$

$$h_m = x_m h_g + (1 - x_m)h_L \tag{8.80}$$

$$h_g = \frac{k}{k-1}\frac{P}{\rho_g} = \frac{k}{k-1}Pv_g \tag{8.81}$$

$$h_L = e_L + Pv_L \tag{8.82}$$

$$C_g = \sqrt{kg_0 Pv_g} \tag{8.83}$$

where mixture quality x_m is constant. These equations were employed in Eqs. (8.58)–(8.60) to obtain the mixture shock equations

$$\frac{\rho_{ym}}{\rho_{xm}} = \left(\frac{k+1}{k-1}\frac{P_y}{P_x} + 1\right)\left(\frac{k+1}{k-1} + \frac{P_y}{P_x}\right)^{-1} \tag{8.84}$$

$$\frac{V_{ym} - V_{xm}}{\sqrt{x_m} C_{gx}} = \left(\frac{P_y}{P_x} - 1\right)\sqrt{\frac{2}{k(k-1)}\left(\frac{k+1}{k-1}\frac{P_y}{P_x} + 1\right)^{-1}} \qquad (8.85)$$

and

$$\frac{1}{1 + b_{mx}}\left(\frac{S_m - V_{xm}}{\sqrt{x_m} C_{gx}}\right) = \sqrt{\frac{k+1}{2k}\frac{P_y}{P_x} + \frac{k-1}{2k}} \qquad (8.86)$$

where

$$b_{mx} = \frac{1 - x_m}{x_m}\frac{\rho_{gx}}{\rho_L} \qquad (8.87)$$

Comparison of Eqs. (8.84)–(8.86) with (8.62), (8.64), and (8.65) shows that gas-liquid mixture shock properties can be obtained with the help of perfect gas shock formulations already derived in Section 8.10 and plotted in Fig. 8.19 for $k = 1.4$.

EXAMPLE 8.16: FLOW STOPPAGE MIXTURE SHOCK
A bubbly mixture of gas and water flowing in a pipe at velocity $V_x = 100\,\text{m/s}$ (328 ft/s) is suddenly stopped by a fast-closing valve. If the oncoming mixture properties are $P_x = 101\,\text{kPa}$ (14.7 psia), $C_{gx} = 334\,\text{m/s}$ (1096 ft/s), $x_m = 0.05$, $\rho_{gx} = 1.26\,\text{kg/m}^3$ (0.078 lbm/ft^3), and $\rho_L = 1000\,\text{kg/m}^3$ (62.32 lbm/ft^3) determine the resulting shock pressure P_y and speed S. Also, compare results with a case of pure gas flow with all oncoming properties the same except $x_m = 1.0$.
   The procedure of Example 8.8 and Fig. 8.20 are employed to obtain

$$\frac{V_{ym} - V_{xm}}{\sqrt{x_m} C_{gx}} = \frac{0 - (-100)}{\sqrt{0.05}\,(334)} = 1.399$$

Comparison of Eqs. (8.85) and (8.65) shows that this calculation corresponds to $(V_y - V_x)/C_x$ for perfect gas flow. Therefore, Fig. 8.19 yields $P_y/P_x = 5$, from which the shock pressure is

$$P_y = 5P_x = 5(101) = 505\,\text{kPa (74 psia)} \qquad \text{mixture}$$

Furthermore, comparison of Eqs. (8.86) and (8.64), and Fig. 8.19 give

$$\frac{S - V_x}{C_x} = 2.1 = \frac{1}{1 + b_{mx}}\left(\frac{S_m - V_{xm}}{\sqrt{x_m} C_{gx}}\right)$$

The mixture property b_{mx} is

$$b_{mx} = \frac{1 - x_m}{x_m}\frac{\rho_{gx}}{\rho_L} = \frac{1 - 0.05}{0.05}\frac{1.26}{1000} = 0.024$$

from which the shock speed is

$$S_m = 60.6\,\text{m/s (199 ft/s)} \qquad \text{mixture}$$

If the oncoming fluid was pure gas with $x_m = 1.0$, we would obtain

$$\frac{V_y - V_x}{C_x} = \frac{0 - (-100)}{334} = 0.299$$

from which Fig. 8.19 gives $P_y/P_x = 1.5$, or a shock pressure of

$$P_y = 152 \text{ kPa } (22 \text{ psia}) \qquad \text{gas}$$

and $(S - V_x)/C_x = 1.2$, or a shock speed of

$$S = (-100) + (1.2)(334) = 300 \text{ m/s } (984 \text{ ft/s}) \qquad \text{gas}$$

It is seen that the 505 kPa mixture shock pressure is substantially higher than the 152 kPa gas shock pressure. Furthermore, the 300-m/s (984 ft/s) gas shock speed is five times higher than the 60.6-m/s (199 ft/s) mixture shock speed.

The gas-liquid mixture shock model of this section is compared with experimental data of Rehder and Pucci [17] in Fig. 8.41. Shock pressure was measured when a bubbly air-water mixture, flowing in a 20-m-long pipe, was abruptly stopped by a 0.01-s valve closure. The comparison shows favorable agreement.

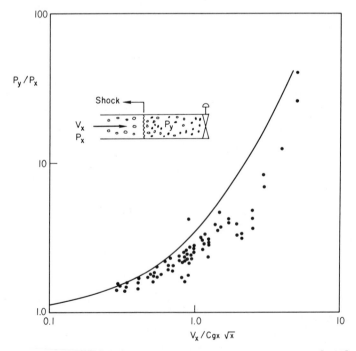

Figure 8.41   Shock Pressue in bubbly Gas/Liquid Mixture [8.17]

References

8.1   Moody, F. J., and P. T. Tran, "The Effect of Vessel Steam/Water Properties on Safety Relief Valve Forces," ASME Paper No. 87-PVP-21.

8.2   *Structural Design of Nuclear Plant Facilities*, Vols. I, II, III, American Society of Civil Engineers, 1973.

8.3   Haupt, R. W., and R. A. Meyer, *Safety Relief Valves*, ASME Special Publication PVP-33, 1979.

8.4   Fabic, S., "Blodwn-2: Westinghouse APD Computer Program for Calculation of Fluid Pressure, Flow, and Resulting Transients during a Loss of Flow Accident," American Nuclear Society Transactions, Vol. 12, No. 1, 1969.

8.5   Hanson, G. H., "Subcooled Blowdown Forces on Reactor System Components," Report IN-1354, Idaho Nuclear Corp., 1970.

8.6   Semprucci, L. B., and B. P. Holbrook, "The Application of RELAP 4/REPIPE to Determine Force-Time Histories on Relief Valve Discharge Piping," *Safety Relief Valves*, R. W. Haupt, and R. A. Meyer, eds., ASME Special Publication PVP-33, 1979.

8.7   "Design Basis for Protection of Light Water Nuclear Power Plants against Effects of Postulated Pipe Rupture," ANSI/ANS-58.2-1980.

8.8   Wheeler, A. J., and F. J. Moody, "A Method for Computing Transient Pressures and Forces in Safety Relief Valve Discharge Lines," *Safety Relief Valves*, R. W. Haupt, and R. A. Meyer, eds., ASME Special Publication PVP-33, 1979.

8.9   Landau, L. D., and E. M. Lifshitz, *Fluid Mechanics*, Addison-Wesley, Massachusetts, 1959.

8.10   Hsiao, W. T., P. Valandani, and F. J. Moody, "A Method to Determine Forces Developed during a Time-Dependent Opening of a Relief Valve Discharging a Two-Phase Mixture," ASME Paper No. 81-WA/NE-15, 1981.

8.11   Von Neumann, J., and R. D. Richtmyer, "A Method for the Numerical Calculation of Hydrodynamic Shocks," *J. Appl. Phys.*, **21** (1950).

8.12   Moody, F. J., "Fluid Reaction and Impingement Loads," Specialty Conference on Structural Design of Nuclear Plant Facilities," Vol. 1, ASCE, 1973.

8.13   Rudinger, G., *Non-Steady Duct Flow*, Dover, New York, 1969.

8.14   Moody, F. J., A. J. Wheeler and M. Ward, "The Role of Various Parameters on Safety Relief Valve Pipe Forces," *Safety Relief Valves*, R. W. Haupt, R. A. Meyer, eds., ASME Special Publication PVP-33, 1979.

8.15   Moody, F. J., "Unsteady Piping Forces Caused by Hot Water Discharge from Suddenly Opened Safety/Relief Valves," *Nuclear Engineering and Design*, **72** (1982).

8.16   McCready, J., et al. "Steam Vent Clearing Phenomena and Structural Response of the BWR Torus (Mark I Containment)," GE Report NEDO-10859, 1973.

8.17   Rehder, W. A., and P. F. Pucci, "The Effects of Air in Damping Water Borne Pressure Pulses," MS Thesis, U. S. Naval Post Graduate School, Monterey, Calif., 1964.

8.18   Wheeler, A. J., and E. A. Siegel, "Measurements of Piping Forces in a Safety Valve Discharge Line, Testing and Analysis of Safety/Relief Valve Performance," ASME-AIChE-ANS Special Publication, A. Sing, and M. Bernstein, eds., 1983.

8.19   Bernstein, M. D., J. A. Werhane, and J. C. Smith, "Recent ASME Code Changes Permit Wider Choice of Safety Valve Location," ASME Paper No. 81-WA/NE-14, 1981.

8.20   Edwards, A. R., and T. P. O'Brien, "Studies of Phenomena Connected with the Depressurization of Water Reactors," *J. Brit. Nuclear Energy Soc.* **9** (1970).

8.21   Moody, F. J., "Pipe Forces Caused by a Sudden Density Change through a Discharging Safety/Relief Valve," *Fluid-Structure Dynamics*, ASME Special Publication PVP-98-7, 1985.

Problems

8.1   Steam discharge suddenly begins through a safety valve into a pipe filled with stagnant air. If the steam front acts like a piston moving into the air with an acceleration of $50,000 \text{ m/s}^2$ $(164,000 \text{ ft/s}^2)$ at what distance will a shock form? Take sound speed in the undisturbed air as 400 m/s (1312 ft/s).

Answer:

2.76 m (9.05 ft)

8.2   Show that if the entropy change across a shock in Eq. (8.66) is set equal to zero, the resulting expression between pressure and temperature is the same as that obtained from Eq. (2.75).

8.3   A valve at the end of a long gas-filled pipe opens instantly, letting gas discharge to the atmosphere. Gas properties initially in the pipe are P_i, ρ_i, and C_i. The valve is simulated by a perfect nozzle of throat area a, and the pipe area is A. Consider a case where critical gas flow occurs in the valve throat (nozzle), and develop a method for determining throat properties P_a, ρ_a, $V_a = C_a$, and the mass flux G_{ca} in terms of initial gas properties and the area ratio a/A. Also show how to determine gas properties P_A, ρ_A, and V_A in the pipe at the valve entrance. Several suggestions follow:

(a) Write the steady energy conservation equation between pipe and valve throat properties $(\ )_A$ and $(\ )_a$, incorporate with perfect gas state equations and an isentropic process for relating pressure and density to show that the valve throat-to-entrance pressure ratio can be written as

$$\frac{P_a}{P_A} = \left[\frac{2}{k+1} + \left(\frac{k-1}{k+1}\right)\left(\frac{V_A}{C_A}\right)^2\right]^{k/(k-1)}$$

Note that if $A \gg a$, pipe velocity $V_A \to 0$, and P_a/P_A becomes the pressure ratio for critical flow through a nozzle given by Eq. (2.59).

(b) Employ the valve-choked condition $V = C$ to show that

$$\frac{V_a}{C_A} = \frac{C_a}{C_A} = \left(\frac{P_a}{P_A}\right)^{1/2}$$

(c) Use steady mass conservation between the valve inlet and throat to show that V_A/C_A can be obtained for the area ratio a/A from

$$\frac{a}{A} = \frac{V_A}{C_A}\left[\frac{2}{k+1} + \left(\frac{k-1}{k+1}\right)\left(\frac{V_A}{C_A}\right)^2\right]^{-(k+2)/2(k-1)}$$

(d) Use Eq. (8.10) to show that P_A can be determined from

$$\frac{V_A}{C_A} = f\left(\frac{a}{A}\right)\Big|_{\text{step (c)}} = \frac{2}{k-1}\left[\left(\frac{P_i}{P_A}\right)^{(k-1)/2k} - 1\right]$$

(e) Finally, show that the critical discharge mass flux from the valve throat is given by

$$\frac{G_{ca}}{\rho_i C_i} = \frac{\left[\frac{2}{k+1} + \left(\frac{k-1}{k+1}\right)\left(\frac{V_A}{C_A}\right)^2\right]^{(k+2)/2(k-1)}}{\left[1 + \left(\frac{k-1}{2}\right)\frac{V_A}{C_A}\right]^{(k+1)/(k-1)}}$$

and that for $V_A = C_A$ ($a/A = 1$), $G_{ca}/\rho_i C_i$ reduces to Eq. (8.28).

8.4 A long tube of flow area A is filled with stationary gas, $V_i = 0$, at pressure $P_i = P_\infty$. A piston at the left end, $x = 0$, undergoes a single compressive cycle with a displacement given by

$$x_p = \begin{cases} x_0 \sin 2\pi ft, & 0 \le t \le \frac{1}{f} \\ 0, & \frac{1}{f} < t \end{cases}$$

(a) Show with the help of Eq. (8.10) that the energy imparted to the gas for small x_0 is $2P_\infty A x_0^2 \pi^2 fk/C_i$.

(b) Part (a) is based on large-amplitude pressure wave relationships, even though x_0 was assumed to be small. Use the waterhammer expression for pressure on the piston, $P_p - P_\infty = \rho C V_p / g_0$ for outgoing waves to show that the corresponding energy imparted to the gas is $\rho C A x_0^2 \pi^2 f / g_0$.

8.5 Show that for either liquid flows or mildly disturbed gas flows with $V \ll C$, the signal distortion of Eq. (8.30) approaches zero.

8.6 Show that for the case of a uniform, frictionless, unheated pipe, liquid flow with constant ρ and C reduces the characteristic solution of Eqs. (8.53) and (8.54) to

$$V_4 = \frac{g_0}{2\rho C} (P_R - P_L) + \frac{1}{2} (V_R + V_L)$$

$$P_4 = \frac{1}{2} (P_L + P_R) + \frac{\rho C}{2 g_0} (V_R - V_L)$$

[Note: These expressions can be obtained from Eqs. (7.96) and (7.97), which apply to waterhammer.]

8.7 Show the the help of Eqs. (8.47) and (8.48) that for liquid flows with $V \ll C$, the characteristic lines in Fig. 8.10 intersect the horizontal grid lines at

$$\Delta z_R = \Delta z_L = C \Delta t$$

Note that in the characteristic waterhammer solution of Chapter 7, Fig. 7.31, Δz_R and Δz_L are equal to the vertical grid spacing.

8.8 Equations (8.62) through (8.66) express property changes across a moving normal shock for which the disturbed pressure P_y is greater than the undisturbed pressure P_x. Someone wants to use these equations to analyze the propagation of a rarefaction wave by employing $P_y < P_x$. He argues that since he can obtain real numerical results from all of the equations, the analysis of a rarefaction wave is justified. Demonstrate that for $P_y < P_x$, Eq. (8.66) gives an entropy *decrease* across a shock which violates the second law and cannot be physically valid.

8.9 A normal shock wave with $P_y / P_x = 10$ moves through stationary atmospheric air, $C_x = 343$ m/s (1125 ft/s) and $T_x = 300$ K (540°R). It arrives at a rigid flat wall, from which it is reflected. Use either Fig. 8.19 or Eqs. (8.62) through (8.65) to show that the pressure exerted on the wall is 52 atm (764 psi). Also show that the oncoming sound speed ratio C_y / C_x is $(T_y / T_x)^{1/2}$.

8.10 A 0.1-m^2 $(1.08$ ft$^2)$ pipe contains subcooled water at 6.9 MPa $(1000$ psia) and 204 C $(400°F)$. Use the results from Example 8.9 to show that the initial blowdown thrust force is 70 kN $(15,736$ lbf). Compare with the steady-state blowdown thrust of Fig. 8.30.

8.11 A long pipe of 10 in^2 area contains water at 1000 psia. Its temperature corresponds to a saturation pressure of 400 psia. A sudden rupture occurs and the discharging fluid is directed against a rigid wall perpendicular to the flow direction. Use Fig. 8.27 to show that the initial impingement force (equal to the initial blowdown force) is 4000 lbf. The saturated liquid enthalpy is 424 B/lbm for $P_{sat} = 400$ psia. Determine the corresponding steady blowdown force.

Answer:

$15,500$ lbf from Fig. 8.30

8.12 The pipe of Problem 8.11 contains a straight 50-ft segment between two $90°$ elbows. What maximum unsteady force do you expect on this segment?

Answer:

6000 lbf, from Fig. 8.27

8.13 A pipe is attached to a vessel of pressurized gas. The friction parameter is $fL/D = 10$. A sudden, complete pipe rupture occurs so that discharge creates forces on the piping system, and the discharging fluid jet impinges on a flat wall perpendicular to the flow direction.

   (a) Show by a simple momentum analysis that the blowdown and impingement forces are equal for a case where the jet forward motion is stopped.

   (b) Show that the impingement force is higher initially than when steady flow is reached.

   (c) Show that the final impingement force is greater than the initial value for $fL/D = 0$ and 2.

8.14 Discharge suddenly occurs from a ruptured pipe which is attached to a gas vessel, $k = 1.3$, pressurized to 10 MPa $(1450$ psia). The total pipe length is 100 m $(328$ ft), and the flow area is 0.1 m^2 $(1.08$ ft$^2)$. Determine the maximum unsteady force on a pipe segment between elbows at 60 and 80 m $(197$ and 262 ft) from the vessel. Assume negligible friction so that Fig. 8.29 can be employed.

Answer:

$R_{u,max} = 260$ kN $(58,448$ lbf)

8.15 A 100-m-long relief pipe of flow area $0.1 \, m^2$ $(1.08 \, ft^2)$ initially contains atmospheric air with $P_\infty = 0.101 \, MPa$ (14.7 psia), $\rho_\infty = 1.22 \, kg/m^3$ $(0.076 \, lbm/ft^3)$, and $C_\infty = 341 \, m/s$ (1118 ft/s). A relief valve at the left end, $z = 0$, occasionally opens, causing the relief discharge of high-pressure steam with properties $P_0 = 6.06 \, MPa$ (879 psia), $\rho_0 = 36.6 \, kg/m^3$ $(2.28 \, lbm/ft^3)$, and a ramp-flat discharge with a valve opening time $\tau = 0.1 \, s$ to the steady discharge rate of $\dot{m} = 208 \, kg/s$ (458 lbm/s). Employ Fig. 8.32 to show that the maximum reaction force on a straight pipe segment between elbows at distances 17 m (55.8 ft) and 34 m (111.6 ft) from the valve is about 63 kN (14, 162 lbf). Show that an analysis based on instant valve opening in Fig. 8.34 yields 140 kN (31, 472 lbf). How can the first analysis be justified?

8.16 The sudden discharge of steam from a safety-relief valve into a long $0.2\text{-}m^2$ $(2.15 \, ft^2)$ pipe filled with atmospheric air, $P_\infty = 101 \, kPa$ (14.7 psia), has the properties

$$\frac{\dot{m}}{\rho_\infty A C_\infty} = 10 \, , \qquad \left(\frac{P_0}{P_\infty}\right)\left(\frac{\rho_0}{\rho_\infty}\right)^{-1} = 3$$

The shock propagating in air, and the steam-air interface both are in the same straight pipe segment for a brief interval of time. Use Figs. 8.34 and 8.35 to show that the total force exerted on this segment is 798 kN (179, 390 lbf).

8.17 A quick-opening valve discharges hot water into a long, $0.1 \, m^2$ $(1.08 \, ft^2)$ pipe initially containing atmospheric air at 101 kPa (14.7 psia). The valve-to-pipe flow area ratio is $A_v/A = 0.5$. If the valve acts like an ideal nozzle and the reservoir is at 6.9 MPa (1000 psia) at what enthalpy will the interface force be largest as it moves through the pipe, and what is its magnitude.

Answer:

$h_0 = 595 \, kJ/kg$ (256 B/lbm) and $R_I = 222 \, kN$ (49, 906 lbf), from Fig. 8.39(b).

8.18 A $0.1\text{-}m^2$ $(1.08 \, ft^2)$ relief valve piping system, initially filled with atmospheric air, is designed for a 202-kN (45, 410-lbf) unsteady force. The system has a valve-to-pipe area ratio of $A_v/A = 0.1$. Use Figs 8.39 and 8.40 to show that the lowest water enthalpy that can be discharged through this piping system from a reservoir at 10.3 MPa (1494 psia) which will not cause the design force to be exceeded by either the shock or interface force is 1400 kJ/kg (602 B/lbm).

8.19 Show that for a vanishingly small shock (sound wave) the propagation speed in a stationary bubbly mixture can be obtained in the form

$$S = \left(1 + \frac{1-x}{x} \frac{\rho_g}{\rho_L}\right) \sqrt{x_m}\, C_g$$

8.20 Show that the shock force in a pipe filled with a stationary bubbly gas-liquid mixture can be expressed as

$$\frac{F}{A} = \frac{1 + b_{mx}}{g_0} \frac{\rho_{xm} x_m C_{gx}^2}{k} \frac{[(k+1)/(k-1)]\dfrac{P_y}{P_x} + 1}{(k+1)/(k-1) + P_y/P_x} \left(\frac{P_y}{P_x} - 1\right)$$

9 MULTIDIMENSIONAL INCOMPRESSIBLE BULK AND WATERHAMMER FLOWS

This chapter contains analyses, graphs, and examples which should be useful in solving a variety of multidimensional unsteady flow problems. Incompressible bulk flow formulations are included for the analysis of submerged fluid discharge, bubble expansion and collapse, and forces exerted by unsteady flows on walls and other structural members. Analyses also are included for liquid sloshing in tanks, the motion of submerged objects, and the stability properties of interfaces which separate two fluids of different densities. Waterhammer analyses are given for propagative disturbances in extended fluid regions. The pressure and forces resulting from sudden pipe discharge, vapor bubble collapse, and other rapid transients can be estimated from the graphs and methods included.

9.1 Incompressible Bulk Flow Formulation

Equations (6.69) to (6.72) describe the behavior of multidimensional flows. Incompressible criteria are $\Delta \rho = 0$, $\beta = 0$, and infinite sound speed, $C \rightarrow \infty$. Therefore, in a fluid region of dimension L, any disturbance time t_d is much greater than the acoustic propagation time L/C, which approaches zero. Hence, conditions for bulk flow are satisfied. Moreover, Eq. (2.116) shows that when ρ is constant, $\Delta T = 0$ during pressure changes.

A pressure disturbance ΔP in bulk flow imposes a velocity disturbance, estimated from Eq. (6.75) as

$$\Delta P = \rho \, \frac{(\Delta V)^2}{2g_0}$$

Moreover, if ΔP acts over an area A to accelerate a fluid region of density ρ and length L, Newton's law in the form

$$A \, \Delta P = \frac{\rho A L}{g_0} \frac{\Delta V}{\Delta t}$$

gives the response time $\Delta t = 2L/\Delta V$. It follows that the model coefficients in Table 6.3 for the mass conservation and momentum formulations, Eqs. (6.69) and (6.70), are

$$\pi_1 = \pi_3 = \pi_4 = \pi_5 = \pi_9 = \pi_{10} = \pi_{11} = 0\,, \qquad \pi_2 = \pi_6 = 2\,,$$

$$\pi_2^0 = 2\,\frac{V_0}{\Delta V}\,, \quad \pi_7 = \frac{4bL}{(\Delta V)^2}\,, \quad \pi_8 = \frac{4\nu}{L\,\Delta V}\,, \quad \pi_{12} = \frac{2\nu\,\Delta V}{g_0 L c_p T}$$

Thus, Eq (6.69) yields the mass conservation principle in the form

$$\nabla \cdot \mathbf{V} = 0 \qquad\qquad (9.1)$$

Furthermore, with Γ obtained from Eqs. (6.30) and (6.31), the momentum formulation of Eq. (6.70) becomes

$$\frac{\partial \mathbf{V}}{\partial t} + \mathbf{V} \cdot \nabla \mathbf{V} + \frac{g_0}{\rho} \nabla P + \mathbf{b} = \nu \nabla^2 \mathbf{V} \qquad\qquad (9.2)$$

The energy principle, Eq. (6.71), gives no additional equations.

9.2 Ideal Flow Approximation

Ideal flow is both *inviscid* ($\nu = 0$) and *irrotational*. The irrotational character of a flow is quantified by considering the rectangular fluid element in Fig. 9.1 at times t and $t + dt$. The difference in vertical velocity components

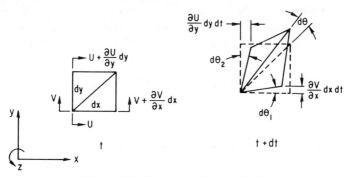

Figure 9.1 Rotation about z Axis

$(\partial v/\partial x)\,dx$ causes rotation of dx about z, and the difference in horizontal velocity components $(\partial u/\partial y)\,dy$ causes rotation of dy about z. Small angles give

$$d\theta_1 \approx \tan d\theta_1 = \frac{\partial v}{\partial x}\,dt\,, \qquad d\theta_2 \approx \tan d\theta_2 = \frac{\partial u}{\partial y}\,dt$$

The average counterclockwise angular displacement of the diagonal is $d\theta \approx (d\theta_1 - d\theta_2)/2$, for which the *rotation* (angular velocity) about z is

$$\omega_z = \frac{d\theta}{dt} = \frac{1}{2}\left(\frac{\partial v}{\partial x} - \frac{\partial u}{\partial y}\right) \tag{9.3}$$

Similarly,

$$\omega_x = \frac{1}{2}\left(\frac{\partial w}{\partial y} - \frac{\partial v}{\partial z}\right), \qquad \omega_y = \frac{1}{2}\left(\frac{\partial u}{\partial z} - \frac{\partial w}{\partial x}\right) \tag{9.4}$$

The rotation vector is, therefore,

$$\boldsymbol{\omega} = \omega_x \mathbf{n}_x + \omega_y \mathbf{n}_y + \omega_z \mathbf{n}_z = \frac{1}{2}\nabla \times \mathbf{V} \tag{9.5}$$

If rotation exists in a fluid, a submerged toothpick will rotate.

A fluid property closely related to rotation is *vorticity*, defined as

$$\boldsymbol{\zeta} = \nabla \times \mathbf{V} \tag{9.6}$$

The transport of vorticity is obtained by applying the vector operation $\nabla \times$ to each term of Eq. (9.2), which gives

$$\frac{d\boldsymbol{\zeta}}{dt} = \frac{\partial \boldsymbol{\zeta}}{\partial t} + \mathbf{V}\cdot\nabla\boldsymbol{\zeta} = \nu\nabla^2\boldsymbol{\zeta} \tag{9.7}$$

The derivative $d\boldsymbol{\zeta}/dt$ is the rate of change of vorticity seen by an observer moving with a fluid particle. The right side of Eq. (9.7) characterizes diffusion of vorticity. Thus, vorticity is transported by convection and diffusion. If the fluid is inviscid ($\nu = 0$), vorticity is convected but not diffused. An *irrotational vortex* is a singularity, outside of which fluid rotation is zero. The two-dimensional velocity field for x, y coordinates with an irrotational vortex at the origin is given by

$$\mathbf{V} = \frac{K}{r}\,\mathbf{n}_\theta \tag{9.8}$$

If we write $\mathbf{n}_\theta = -\sin\theta\,\mathbf{n}_x + \cos\theta\,\mathbf{n}_y$ with $r = \sqrt{x^2 + y^2}$, then

$$\boldsymbol{\zeta} = \nabla \times \mathbf{V} = K\,\frac{(y^2 - x^2) - (y^2 - x^2)}{(x^2 + y^2)^2}\,\mathbf{n}_z$$

which is seen to be zero, as is the rotation, for all points x, y except at the vortex itself where $x = y = 0$.

Equation (9.7) gives $d\zeta/dt = 0$ for an inviscid region where $\nu = 0$. It follows that if $\zeta = 0$ initially, the fluid rotation is forever zero, and the flow is irrotational. Only inviscid, irrotational flows are considered in this chapter.

9.3 The Velocity Potential

Irrotational flow is characterized by $\nabla \times \mathbf{V} = 0$, which implies that $\mathbf{V}$ can be expressed as the gradient of a scalar function ϕ, or

$$\mathbf{V} = \nabla\phi \tag{9.9}$$

because of the vector identity $\nabla \times (\nabla\phi) = 0$. The function ϕ is called the velocity potential, and velocity components in x, y, z are

$$u = \frac{\partial\phi}{\partial x}, \quad v = \frac{\partial\phi}{\partial y}, \quad w = \frac{\partial\phi}{\partial z} \tag{9.10}$$

The velocity component V_s in an arbitrary direction s is obtained from Fig. 9.2 as the projection of $\mathbf{V}$ on a unit vector $d\mathbf{r}/|d\mathbf{r}| = d\mathbf{r}/ds$, where $d\mathbf{r}$ is tangent to s. Then $V_s = \mathbf{V} \cdot d\mathbf{r}/ds$. If we put

$$\mathbf{V} = \frac{\partial\phi}{\partial x}\,\mathbf{n}_x + \frac{\partial\phi}{\partial y}\,\mathbf{n}_y + \frac{\partial\phi}{\partial z}\,\mathbf{n}_z$$

$$\mathbf{r} = x\mathbf{n}_x + y\mathbf{n}_y + z\mathbf{n}_z$$

it follows that

$$V_s = \frac{\partial\phi}{\partial x}\frac{dx}{ds} + \frac{\partial\phi}{\partial y}\frac{dy}{ds} + \frac{\partial\phi}{\partial z}\frac{dz}{ds} = \frac{d\phi}{ds} \tag{9.11}$$

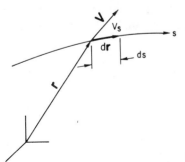

Figure 9.2 Velocity Components in S direction

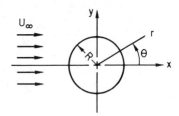

Figure 9.3 Uniform Flow Past Stationary Cylinder

That is, the derivative of ϕ with respect to length in any direction s yields the velocity component in that direction.

EXAMPLE 9.1: VELOCITY ON A SUBMERGED CYLINDER
The two-dimensional velocity potential for ideal uniform flow past the submerged stationary cylinder of radius R in Fig. 9.3 is [9.7]

$$\phi = U_\infty \left(r + \frac{R^2}{r} \right) \cos \theta$$

Determine the velocity on the cylinder surface.
   If ϕ is confined to $r = R$,

$$\phi = 2U_\infty R \cos \theta$$

The derivative of ϕ on the circle is obtained from $d\phi/ds$, where ds is length on the circle, $R\,d\theta$, giving

$$V_s = \frac{d\phi}{ds} = -2U_\infty \sin \theta$$

Note that θ is measured counterclockwise from the positive x axis, and length s increases in a direction opposite to the flow, which explains why V_s is negative.

———————

9.4   Ideal Fluid Formulations

When a flow is irrotational, the velocity expressed by Eq. (9.9) sometimes can be used to obtain ϕ for a flow pattern. The example of uniform flow in the x direction corresponds to

$$\mathbf{V} = U_\infty \mathbf{n}_x = \nabla \phi = \frac{\partial \phi}{\partial x} \mathbf{n}_x + \frac{\partial \phi}{\partial y} \mathbf{n}_y + \frac{\partial \phi}{\partial z} \mathbf{n}_z$$

or $\partial \phi / \partial x = U_\infty$, for which

$$\phi = U_\infty x , \quad \text{uniform left-to-right flow} \tag{9.12}$$

Also, a spherically symmetric point source produces a volume flow rate $\dot{V}$ which is uniform in all directions and creates a velocity field $V_r = \dot{V}/4\pi r^2$, where r is the distance from the source. Since $V_r = \partial\phi/\partial r$,

$$\phi = -\frac{\dot{V}}{4\pi}\frac{1}{r} \qquad \text{spherically symmetric point source} \qquad (9.13)$$

The velocity potentials of Eqs. (9.12) and (9.13) are based on incompressible bulk flow. However, all fluids are somewhat compressible. A local disturbance can be characterized by bulk flow in the immediate neighborhood, but it also can be characterized by propagative flow in a larger region. This can be seen by considering a point source with a disturbance time t_d. Equation (1.2) shows that a bulk flow analysis applies if

$$t_d \gg \frac{r}{C} \qquad \text{or} \qquad r \ll Ct_d \qquad (9.14)$$

That is, the response of a fluid region whose extent is much less than Ct_d is dominated by bulk flow. A much larger region could be dominated by propagative response. Bulk flows are considered in this chapter through section 9.18, and propagative or waterhammer flows are discussed in sections 9.19–9.26.

If $\mathbf{V} = \nabla\phi$ is employed in the bulk flow mass conservation principle of Eq. (9.1), we obtain Laplace's equation

$$\nabla^2\phi = 0 \qquad (9.15)$$

which can be written as follows for common coordinate systems of Fig. 9.4:

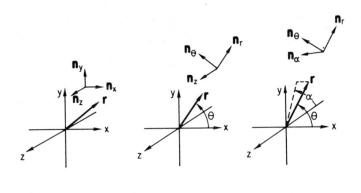

Rectangular   Cylindrical polar   Spherical

Figure 9.4   Common Coordinate Systems

LAPLACE'S EQUATION IN COMMON COORDINATE SYSTEMS (FIG. **9.4**)

Rectangular

$$\nabla^2\phi = \frac{\partial^2\phi}{\partial x^2} + \frac{\partial^2\phi}{\partial y^2} + \frac{\partial^2\phi}{\partial z^2} = 0 \tag{9.16}$$

Cylindrical Polar

$$\nabla^2\phi = \frac{1}{r}\frac{\partial}{\partial r}\left(r\frac{\partial\phi}{\partial r}\right) + \frac{1}{r^2}\frac{\partial^2\phi}{\partial\theta^2} + \frac{\partial^2\phi}{\partial z^2} = 0 \tag{9.17}$$

Spherical

$$\nabla^2\phi = \frac{1}{r^2}\frac{\partial}{\partial r}\left(r^2\frac{\partial\phi}{\partial r}\right) + \frac{1}{r^2\sin\theta}\frac{\partial}{\partial\theta}\left(\sin\theta\frac{\partial\phi}{\partial\theta}\right)$$

$$+ \frac{1}{r^2\sin^2\theta}\frac{\partial^2\phi}{\partial\alpha^2} = 0 \tag{9.18}$$

A solution for ϕ in a fluid region requires ϕ or its normal derivative $\partial\phi/\partial n$ to be specified over the region boundary. Since Laplace's equation is linear, any linear combination of solutions $\phi_0, \phi_1, \phi_2, \ldots, \phi_n$ provides additional solutions.

Consider Fig. 9.5(a) where a sphere of radius R moves along the x axis at velocity $U_s(t)$ in a liquid whose velocity is $U_\infty(t)$. This flow pattern is axially symmetric about x, so α dependence disappears from Eq. (9.18). No flow crosses the spherical wall. Every point on the surface moves in the x direction with velocity $U_s(t)$, whose component normal to the spherical surface is $U_s\cos\theta$, which also is equal to the adjacent normal fluid velocity component. Therefore, boundary conditions are

$$\frac{\partial\phi}{\partial r} = U_s(t)\cos\theta \qquad \text{on} \qquad r = R$$

$$\phi = U_\infty(t)x = U_\infty(t)r\cos\theta \text{ at } r\to\infty$$

$$\phi \text{ is symmetric about the } x \text{ axis}$$

$$\phi \text{ is periodic in } \theta \text{ every } 2\pi \text{ radians}$$

A solution of Eq. (9.18) by separation of variables yields

Sphere Moving at $U_s(t)$ along x in Uniform Flow $U_\infty(t)$

$$\phi(r,\theta,t) = \left[U_\infty(t)\left(r + \frac{R^3}{2r^2}\right) - U_s(t)\frac{R^3}{2r^2}\right]\cos\theta \tag{9.19}$$

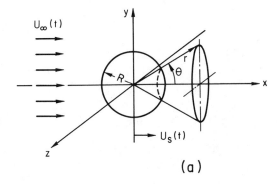

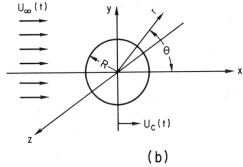

Figure 9.5 Moving Sphere and Cylinder. (*a*) Sphere; (*b*) Cylinder

A similar solution for the cylinder in Fig. 9.5(*b*) moving at velocity $U_c(t)$ is given by

Cylinder Moving at $U_c(t)$ along x in Uniform Flow $U_\infty(t)$

$$\phi(r, \theta, t) = \left[U_\infty(t)\left(r + \frac{R^2}{r}\right) - U_c(t)\,\frac{R^2}{r} \right]\cos\theta \qquad (9.20)$$

The inviscid fluid momentum principle of Eq. (9.2) is simplified with $\nu = 0$. The identity $\mathbf{V}\cdot\nabla\mathbf{V} = \nabla(\mathbf{V}\cdot\mathbf{V}/2) = \nabla(V^2/2)$ and a body force acceleration potential

$$\Omega = b_x x + b_y y + b_z z \qquad (9.21)$$

for which

$$\mathbf{b} = \nabla\Omega \qquad (9.22)$$

are employed in Eq. (9.2) with $\mathbf{V} = \nabla\Phi$ to obtain

$$\nabla\left(\frac{\partial \phi}{\partial t} + \frac{V^2}{2} + \frac{g_0}{\rho} P + \Omega\right) = 0 \qquad (9.23)$$

where

$$V^2 = \nabla\phi \cdot \nabla\phi = \mathbf{V} \cdot \mathbf{V} \qquad (9.24)$$

The integral of Eq. (9.23) yields the unsteady Bernoulli equation

$$\frac{\partial \phi}{\partial t} + \frac{V^2}{2} + \frac{g_0}{\rho} P + \Omega = f(t) \qquad (9.25)$$

where $f(t)$ is a general time-dependent function to be determined from known information in a given problem. If gravity is the only body force acceleration component,

$$\Omega = gy \qquad \text{gravity only} \qquad (9.26)$$

Equations (9.15) and (9.25) give the ideal incompressible, inviscid, irrotational flow formulations for mass conservation and momentum. Often ϕ can be determined from a solution of Eq. (9.15) without momentum, and the result then can be employed in Eq. (9.25) to obtain local pressure.

EXAMPLE 9.2: PRESSURE ON A VIBRATING CYLINDER
A cylinder of radius R is submerged in stationary liquid of density ρ. The cylinder is oscillating in the x direction with a periodic displacement

$$x_c = x_0 \sin \omega t$$

where x_0 is the amplitude. Determine the resulting unsteady pressure on the cylinder at angles $\theta = 0$ and $\pi/2$. Neglect gravity.
  The cylinder velocity is

$$U_c(t) = \frac{dx_c}{dt} = x_0 \omega \cos \omega t$$

Since $U_\infty(t) = 0$, the velocity potential of Eq. (9.20) becomes

$$\phi(r, \theta, t) = -(x_0 \omega \cos \omega t) \frac{R^2}{r} \cos \theta$$

Pressure anywhere in the liquid is obtained from Eq. (9.25), where the velocity term V^2 corresponds to $\mathbf{V} \cdot \mathbf{V}$, in which

$$\mathbf{V} = \frac{\partial \phi}{\partial r} \mathbf{n}_r + \frac{1}{r} \frac{\partial \phi}{\partial \theta} \mathbf{n}_\theta$$

Far away at $r \to \infty$, we have $V = 0$, $P = P_\infty$, and $\phi \to 0$ so that $f(t) = g_0 P_\infty/\rho$. It follows that

$$P^*(r, \theta, t) = \frac{P - P_\infty}{R^2 \omega^2 \rho / 2 g_0} = -\frac{2x_0}{r} \sin \omega t \cos \theta$$

$$- \frac{x_0^2 R^2}{r^4} \cos^2 \omega t \qquad (9.27)$$

Notice that the pressure field vanishes for large r, whereas on the cylinder wall at $\theta = 0$ and $\pi/2$ the required pressures are

$$P^*(R, 0, t) = -2 \frac{x_0}{R} \sin \omega t - \left(\frac{x_0}{R}\right)^2 \cos^2 \omega t$$

and

$$P^*(R, \pi/2, t) = -\left(\frac{x_0}{R}\right)^2 \cos^2 \omega t$$

9.5   The Rayleigh Bubble

Problems involving phenomena like submerged explosions, gas discharge, and vapor bubble collapse can be approximated by a Rayleigh bubble analysis. An ideal spherical bubble of radius $R(t)$ and pressure $P_b(t)$ is submerged in a liquid of density ρ and distant pressure P_∞. The surrounding liquid response is obtained by simulating the bubble wall motion with the point source of Eq. (9.13). That is, the bubble wall velocity dR/dt corresponds to $\partial \phi / \partial r$ at $r = R$, which yields

$$\frac{dR}{dt} = \left. \frac{\partial \phi}{\partial r} \right|_{r=R} = \frac{\dot{V}}{4\pi} \frac{1}{R^2}$$

so that

$$\phi = -R^2 \frac{dR}{dt} \frac{1}{r} \qquad (9.28)$$

Since $\partial \phi / \partial t$ and V both vanish at $r \to \infty$, substitution into Eq. (9.25) with gravity neglected yields $f(t) = g_0 P_\infty / \rho$, and

$$P(r, t) - P_\infty = \frac{\rho}{g_0} \left[\frac{R^2}{r} \frac{d^2R}{dt^2} + \frac{2R}{r} \left(\frac{dR}{dt}\right)^2 - \frac{1}{2} \frac{R^4}{r^4} \left(\frac{dR}{dt}\right)^2 \right] \qquad (9.29)$$

Since $P(R, t) = P_b(t)$, Eq. (9.29) yields the Rayleigh equation for bubble radius as

$$R \frac{d^2R}{dt^2} + \frac{3}{2} \left(\frac{dR}{dt}\right)^2 = \frac{g_0}{\rho} [P_b(t) - P_\infty] \qquad (9.30)$$

Bubble pressure is required before $R(t)$ can be obtained. Useful solutions for various $P_b(t)$ are summarized in the next sections.

9.6 Constant Pressure Bubble Response

If $P_b(t) - P_\infty$ is a positive constant in Eq. (9.30), a bubble of initial radius R_i grows according to

$$R = R_i + \sqrt{\frac{2}{3}\frac{g_0}{\rho}(P_b - P_\infty)}\, t \tag{9.31}$$

However, if $P_b - P_\infty$ is a negative constant, the solution is quite different, and the bubble collapses. This can occur when a vapor bubble suddenly is surrounded by cool liquid, causing rapid vapor condensation and a decreased P_b. If P_{sat} is the saturated vapor pressure at liquid temperature, the limiting case is obtained with $P_b - P_\infty = P_{sat} - P_\infty < 0$ in Eq. (9.30). If initial conditions $R(0) = R_i$ and $dR(0)/dt = 0$ are employed, the first integral of Eq. (9.30) gives

$$\frac{dR}{dt} = \sqrt{\frac{2}{3}\frac{g_0}{\rho}(P_\infty - P_{sat})\left(\frac{R_i^3}{R^3} - 1\right)} \tag{9.32}$$

A second integration yields

$$\frac{t}{R_i}\sqrt{\frac{g_0(P_\infty - P_{sat})}{\rho}} = 0.915[1 - I_{(R/R_i)}(\tfrac{5}{6}, \tfrac{1}{2})] \tag{9.33}$$

where $I_{(R/R_i)}(\tfrac{5}{6}, \tfrac{1}{2})$ is an incomplete Beta function. Full collapse occurs when $R = 0$, or at the normalized time

$$\frac{t}{R_i}\sqrt{\frac{g_0(P_\infty - P_{sat})}{\rho}} = 0.915 \tag{9.34}$$

Equations (9.32) and (9.33) give the curves in Fig. 9.6.

The pressure near a collapsing bubble is obtained by incorporating Eqs. (9.30) and (9.32) into (9.29), for which

$$\frac{P(r, t) - P_\infty}{P_\infty - P_{sat}} = \frac{1}{3}\left(\frac{R_i}{r}\right)\left(\frac{R}{R_i}\right)\left[\left(\frac{R_i}{R}\right)^3 - \left(\frac{R_i}{r}\right)^3 + \left(\frac{R}{R_i}\right)^3\left(\frac{R_i}{r}\right)^3 - 4\right] \tag{9.35}$$

Consider a collapsing bubble where pressure is recorded at the location of its initial radius, $r = R_i$. Equation (9.35) reduces to

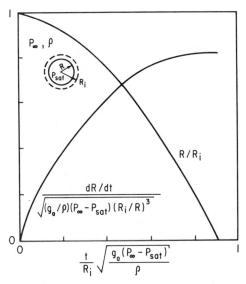

Figure 9.6   Rayleigh Bubble Collapse

$$\frac{P(R_i, t) - P_\infty}{P_\infty - P_{sat}} = \frac{1}{3}\left(\frac{R}{R_i}\right)\left[\left(\frac{R_i}{R}\right)^3 + \left(\frac{R}{R_i}\right)^3 - 5\right] \qquad (9.36)$$

Figure 9.7 obtained from Eqs. (9.33) and (9.36) shows an amazingly high pressure spike outside the bubble before full bubble collapse occurs! This largely unexpected result can be explained by a careful look at the equation. It is seen that a collapsing bubble wall velocity increases with time as R decreases, according to Eq. (9.32). However, the liquid velocity at $r > R$, obtained from Eq. (9.28) as $V(r, t) = \partial\phi/\partial r = (R/r)^2(dR/dt)$, can be shown to decrease in time as R decreases. Therefore, the abrupt pressure rise is caused from liquid deceleration! The rate of pressure rise in nearby liquid is dominated by bulk flow at first. However, it becomes so steep toward the end that it becomes acoustic dominated, which explains the sharp cracking or popping noises near cavitating machinery. Continuous bubble collapse pressures occur near the tip of submerged ship propellers and pump impellers, eventually pitting and disintegrating the metal surfaces.

EXAMPLE 9.3: BUBBLE COLLAPSE TIME
A water vapor bubble of initial radius $R_i = 2$ cm (0.79 in) is suddenly expelled into cold water, $\rho = 1$ g/cc (62.4 lbm/ft³) and $C = 1200$ m/s (3936 ft/s) at 1 atm. (14.7 psia). If instantaneous vapor condensation reduces bubble pressure to zero, determine the collapse time and the largest fluid region for which the bulk flow solution applies.

Equation (9.34) yields the collapse time

$$t_{collapse} = 1.8 \text{ ms}$$

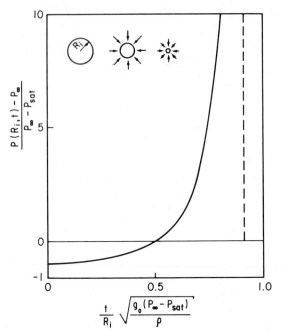

Figure 9.7   Pressure Near Collapsing Bubble

The applicable fluid region for bulk flow, Eq. (9.14), corresponds to the disturbance time

$$t_d \doteq t_{\text{collapse}} \gg r/C \quad \text{or} \quad r \ll C t_{\text{collapse}} = 2.16 \text{ m} \ (7.1 \text{ ft})$$

Therefore, the bubble collapse would impose bulk flow in a region much less than 2 m (6.6 ft), say about 20 cm (7.9 in) radius from the bubble center. Acoustic effects would become increasingly important for regions greater than 20 cm (7.9 in).

The Rayleigh bubble collapse analysis is based on bulk flow in an infinite liquid region. This is justified because the majority of liquid affecting its wall motion lies within the bulk flow region.

9.7   Small-Amplitude Bubble Oscillation

A compressible gas bubble resembles a spring, and the surrounding liquid a mass. Slight compression and release of a gas bubble in pressure equilibrium with surrounding liquid would start an oscillation. Let bubble pressure be isentropically related to its radius by

$$P_b = P_\infty \left(\frac{R_\infty}{R} \right)^{3k} \tag{9.37}$$

where R_∞ is the undisturbed size. Pressure is slightly increased so that initially $R(0) = R_\infty - \epsilon$. An oscillation occurs upon release, based on the small-amplitude solution of Eq. (9.30) (see section C.2 of Appendix C), which yields

$$R(t) = R_\infty - \epsilon \cos 2\pi f t \tag{9.38}$$

with frequency

$$f = \frac{1}{2\pi R_\infty} \sqrt{\frac{3kg_0 P_\infty}{\rho}} \tag{9.39}$$

It is seen that the frequency varies inversely with bubble radius.

9.8 Large-Amplitude Bubble Oscillation

The adiabatic gas pressure of Eq. (9.37) was employed for a bubble undergoing large-amplitude oscillation. Maximum bubble pressure $P_{b,\max}$ occurs at minimum radius $R_{\min}$. Equation (9.30) was solved with a fourth-order Runge-Kutta integration (section C.1 of Appendix C) for one cycle of oscillation, and results are given in Fig. 9.8. The period of oscillation $T = 1/f$ is employed with Eq. (9.39) to express the small-amplitude normalized period as

$$\frac{T}{R_{\min}} \sqrt{\frac{g_0 P_\infty}{\rho}} = \frac{2\pi}{\sqrt{3k}} = 3.06 , \qquad k = 1.4$$

for the limiting case of small $P_{b,\max}$ in Fig. 9.8.

The large-amplitude bubble wall velocity is shown in Fig. 9.9. It has been observed [9.12] that when a high-pressure bubble is discharged deeply into water, it undergoes oscillation with a relatively stable boundary. The reason for this stability is related to the bubble wall acceleration, which is in a direction toward the gas (a stable condition) during most of its oscillation cycle, as noted by the negative slope of the wall velocity in Fig. 9.9.

The pressure field from an oscillating bubble is described by Eq. (9.29). However, at oscillation extremes when $dR/dt = 0$, Eqs. (9.29) and (9.30) yield

$$P(r, t) - P_\infty = \frac{R}{r} (P_b - P_\infty) \tag{9.40}$$

where maximum pressure $P_{b,\max}$ corresponds to minimum radius $R_{\min}$, and

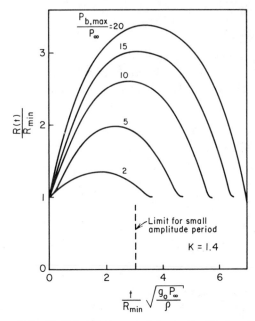

Figure 9.8 Bubble Radius, Adiabatic Oscillating Bubble

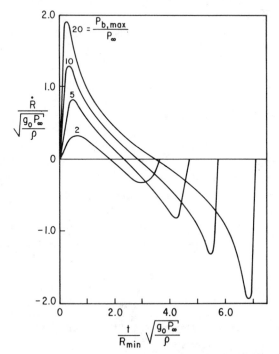

Figure 9.9 Bubble Wall Velocity, Adiabatic Oscillating Bubble

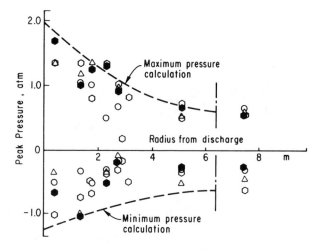

Figure 9.10   Pressure Field from Oscillating Air Bubble

minimum pressure $P_{b,\min}$ corresponds to maximum radius $R_{\max}$. The pressure field of Eq. (9.40) closely matches data taken for the oscillation of a 1.3-m^3 relief valve air bubble in Fig. 9.10.

Since $V = 0$ in Eq. (9.25) at the extremes of bubble oscillation, we may write

$$\frac{\partial}{\partial t}\,\nabla^2\phi + \frac{g_0}{\rho}\,\nabla^2 P = 0$$

However, $\nabla^2\phi = 0$, and the equation for P at instants of zero velocity is

$$\nabla^2 P = 0\,, \qquad V = 0 \tag{9.41}$$

Whenever the pressure satisfies Laplace's equation in a fluid region, P cannot exceed the maximum boundary pressure.

9.9   Surface Tension Effects

When bubbles or droplets are small, surface tension can play an important role in the dynamic response. Pressure difference caused by surface tension across a curved two-fluid interface, is formulated here for the geometry of Fig. 9.11. The differential area section has radii of curvature R_{c1} and R_{c2}. Surface tension forces $\delta F_{\sigma 1} = \sigma\,\delta L_1$ and $\delta F_{\sigma 2} = \sigma\,\delta L_2$ act normal to the sides δL_1 and δL_2 and tangent to the interface itself. A force balance is written for small angles as

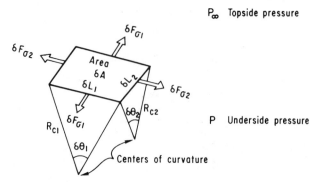

Figure 9.11 Surface Tension Components

$$(P - P_\infty)\, \delta A = 2\left(\delta F_{\sigma 1} \sin \frac{\delta \theta_2}{2} + \delta F_{\sigma 2} \sin \frac{\delta \theta_1}{2} \right)$$

Since $\delta A = \delta L_1\, \delta L_2 = (R_{c1}\, \delta \theta_1)(R_{c2}\, \delta \theta_2)$, it follows that

$$P - P_\infty = \sigma\left(\frac{1}{R_{c1}} + \frac{1}{R_{c2}} \right) \tag{9.42}$$

The radii of curvature are equal in a spherical bubble or droplet so that

$$P - P_\infty = \frac{2\sigma}{R}, \qquad \text{spherical interface} \tag{9.43}$$

Equation (9.43) is significant when liquid boils. Consider liquid at pressure $P_L = P_{\text{sat}}$ and saturation temperature T_{sat}. Small cavities in a solid heating surface can trap noncondensible gas, providing bubble nucleation sites. The gas partial pressure is P_g, and vapor in these cavities is at pressure $P_v = P_{\text{sat}}$. The cavity pressure is the sum $P_v + P_g$, and the pressure difference across an ideal spherical gas-liquid interface at the top of the cavity is

$$P_v + P_g - P_L = \frac{2\sigma}{R}$$

Heating of adjacent liquid to $T > T_{\text{sat}}$ increases pressure P_v according to the Clausius-Clapeyron equation

$$dT = T \frac{v_{fg}}{h_{fg}}\, dP$$

obtained from Eq. (2.116) with the saturation properties $\beta \to \infty$ and $\beta / c_p = v_{fg}/v h_{fg}$. Integration yields

$$T_v - T_{\text{sat}} \approx T_{\text{sat}} \frac{v_{fg}}{h_{fg}} \left(\frac{2\sigma}{R} - P_g \right) \tag{9.44}$$

Equation (9.44) gives the superheat temperature $T_v - T_{sat}$ required for a spherical vapor bubble of radius R to remain in equilibrium. Higher T_v causes bubble growth, and lower T_v causes a bubble to shrink. It is seen that the surface tension is important in boiling whenever cavities satisfy the criterion $R \le 2\sigma/P_g$.

9.10   Flat Wall Simulation near a Bubble

The pressure and velocity fields of a bubble near a flat wall can be simulated by the superposition of two equal bubbles at distances $\pm D$ on the x axis, as shown in Fig. 9.12. Equation (9.28) and the idealization that neither bubble is affected by the other yields the potential functions

$$\phi_1 = R^2 \frac{dR}{dt} \frac{1}{r_1} \quad \text{and} \quad \phi_2 = -R^2 \frac{dR}{dt} \frac{1}{r_2}$$

where

$$r_1 = \sqrt{(x+D)^2 + y^2 + z^2} \quad \text{and} \quad r_2 = \sqrt{(x-D)^2 + y^2 + z^2}$$

The combined velocity potential is $\phi = \phi_1 + \phi_2$. Differentiation of ϕ shows that

$$u = \frac{\partial \phi}{\partial x} = 0 \quad \text{on } x = 0$$

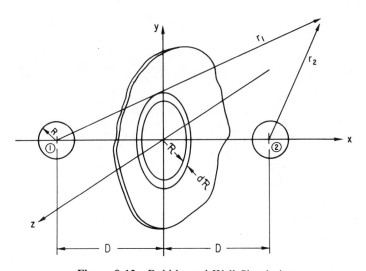

Figure 9.12   Bubble and Wall Simulation

which implies that for this superposition of a bubble and its image, the y, z plane could be replaced by a rigid flat wall. If the distant pressure is P_∞ and gravity is again neglected, Eqs. (9.25) and (9.30) can be employed to express the wall pressure at $r = r_1 = r_2$ on $x = 0$ as

$$P(\mathscr{R}, t) - P_\infty = \frac{2R}{\sqrt{D^2 + \mathscr{R}^2}} \left\{ (P_b - P_\infty) + \frac{\rho}{2g_0} \left(\frac{dR}{dt} \right)^2 \left[1 - \frac{2R^3\mathscr{R}^2}{(D^2 + \mathscr{R}^2)^{5/2}} \right] \right\}$$

$$\mathscr{R} = \sqrt{y^2 + z^2} \tag{9.45}$$

Wall pressure is constant on circles of $\mathscr{R} = $ constant and diminishes either with increasing bubble distance D or larger values of $\mathscr{R}$. Recall that when an oscillating bubble radius passes through zero velocity, the pressure field satisfies Laplace's equation and the corresponding wall pressure cannot exceed bubble pressure. However, Eq. (9.45) shows that $P(\mathscr{R}, t)$ could exceed P_b. This apparent anomaly is a direct result of neglecting the coupling effects between the bubble and its image. When the distance D is large, coupling is not important. Therefore, if we impose the requirement that the maximum wall pressure $P(0, t) \le P_b$ when $dR/dt = 0$, we obtain the minimum permissible wall distance

$$\frac{D}{R} \ge 2 \tag{9.46}$$

That is, Eq. (9.45) gives a reasonable approximation of wall pressure from a nearby bubble if its center is at least twice its maximum radius from the wall.

When $dR/dt \ne 0$, the coefficient of $(dR/dt)^2$ in Eq. (9.45) is about 1.0 for $D/R \ge 2$. This increases the maximum wall pressure to

$$P_{\max}(0, t) - P_\infty = \frac{2R}{D} \left[(P_b - P_\infty) + \frac{\rho}{2g_0} \left(\frac{dR}{dt} \right)^2 \right]$$

$$\le (P_b - P_\infty) + \frac{\rho}{2g_0} \left(\frac{dR}{dt} \right)^2 \tag{9.47}$$

which means that during bubble expansion or contraction, wall pressure can exceed bubble pressure, even if the limiting value of D/R in Eq. (9.46) is satisfied.

EXAMPLE 9.4: BUBBLE PRESSURE EXERTED ON WALL
A submerged bubble oscillation in water, $\rho = 1000$ kg/m^3 (62.4 lbm/ft^3), is simulated in a test by a spherical bladder which is inflated and deflated by a programmed hydraulic fluid so that its radius is

$$R = R_\infty + \Delta R \sin 2\pi f t$$

where $R_\infty = 0.5\,\text{m}$ (1.64 ft), $\Delta R = 0.2\,\text{m}$ (0.66 ft), and the frequency is $f = 10\,\text{Hz}$. This bladder center is a distance $D = 1.0\,\text{m}$ (3.28 ft) from a rigid flat wall. If the distant fluid pressure is $P_\infty = 101\,\text{kPa}$ (14.7 psia), determine the maximum wall pressure at $y = z = 0$ when $dR/dt = 0$ and compare with the absolute maximum pressure during a cycle.

The wall velocity and acceleration are

$$\frac{dR}{dt} = \Delta R 2\pi f \cos 2\pi ft$$

$$\frac{d^2R}{dt^2} = -\Delta R(2\pi f)^2 \sin 2\pi ft$$

Whenever $dR/dt = 0$, we have

$$2\pi ft = \frac{2n+1}{2}\pi, \qquad n = 0, 1, 2, \ldots$$

for which

n	R	d^2R/dt^2
0	$R_\infty + \Delta R$	$-\Delta R(2\pi f)^2$
1	$R_\infty - \Delta R$	$+\Delta R(2\pi f)^2$
—	(repeats for $n = 2, 3, \ldots$)	

Equation (9.30) at $dR/dt = 0$ gives

$$P_b - P_\infty = \frac{\rho}{g_0}\begin{cases} -(R_\infty + \Delta R)\,\Delta R(2\pi f)^2 & \text{Minimum} \\ (R_\infty - \Delta R)\,\Delta R(2\pi f)^2 & \text{Maximum} \end{cases}$$

and on the wall at $\mathscr{R} = 0$,

$$P(0,t) - P_\infty = \frac{2R}{D}(P_{b,\text{max}} - P_\infty)$$

$$= \frac{2(R_\infty - \Delta R)^2}{D}\frac{\rho}{g_0}\Delta R(2\pi f)^2 = 142\,\text{kPa} \;(20.7\,\text{psia})$$

at

$$\frac{dR}{dt} = 0 \quad \text{and} \quad t = \frac{2n+1}{4f}, \qquad n = 0, 1, 3, 5$$

The pressure at any time during a cycle is obtained from Eqs. (9.30) and (9.45) at $\mathscr{R} = 0$ with the corresponding R, dR/dt, and d^2R/dt^2, which give

$$\psi = \frac{P(0,t) - P_\infty}{(2\rho/Dg_0)(2\pi f)^2 R_\infty^3}$$

$$= -\frac{\Delta R}{R_\infty}\left(1 + \frac{\Delta R}{R_\infty}\sin 2\pi ft\right)^2 \sin 2\pi ft + 2\left(\frac{\Delta R}{R_\infty}\right)^2\left(1 + \frac{\Delta R}{R_\infty}\sin 2\pi ft\right)(1 - \sin^2 2\pi ft)$$

Extremes of ψ are found from the condition $dP(0, t)/dt = 0$, which occurs at times corresponding to either

$$\cos 2\pi ft = 0, \qquad 2\pi ft = \frac{2n+1}{2}\pi$$

or

$$\sin 2\pi ft = -\frac{4}{9}\frac{R_\infty}{\Delta R}\left(1 - \sqrt{1 - \frac{9}{16}\left[1 - 2\left(\frac{\Delta R}{R_\infty}\right)^2\right]}\right)$$

The first solution was obtained earlier for $dR/dt = 0$. The second result yields

$$\sin 2\pi ft = -0.238$$

for which

$$P(0, t) - P_\infty = 355 \text{ kPa} \quad (51.7 \text{ psi})$$

Comparison with the 142 kPa (20.7 psi) obtained for $dR/dt = 0$ shows that the velocity effect on wall pressure can be substantial.

———————

The wall pressure force can be obtained by integrating the differential force $dF = [P(\mathcal{R}, t) - P_\infty]\, dA$ on a ring of radius $\mathcal{R}$ and width $d\mathcal{R}$ in the y, z plane shown in Fig. 9.12, where $dA = 2\pi\mathcal{R}\, d\mathcal{R}$. It follows that for the wall pressure of Eq. (9.45)

$$\frac{F}{4\pi R^2[(P_b - P_\infty) + (\rho/2g_0)(dR/dt)^2]}$$

$$= \frac{1}{R}(\sqrt{D^2 + \mathcal{R}^2} - D) - \frac{1}{2}\left(\frac{R}{D}\right)^2 \frac{\mathcal{R}^4}{(D^2 + \mathcal{R}^2)^2}\left(1 + \frac{P_b - P_\infty}{(\rho/2g_0)(dR/dt)^2}\right)^{-1}$$

$$(9.48)$$

The first term on the right of Eq. (9.48) comes from $\partial\phi/\partial t$ in Eq. (9.25) and increases as $\mathcal{R}$ increases. Therefore, the bubble force depends on wall size. The potential function for a point source, $\phi = -R^2(dR/dt)/r$, can be written in terms of the volume flow rate $\dot{V}$ as $\phi = -(\dot{V}/4\pi)/r$. If $\dot{V}$ is constant, the wall pressure force becomes

$$F = -\frac{\dot{V}^2}{16\pi D^2}\frac{\rho}{g_0}\frac{\mathcal{R}^4}{(D^2 + \mathcal{R}^2)^2}, \qquad \dot{V} = \text{constant} \qquad (9.49)$$

which has the infinite wall limit

$$F_\infty = \lim_{\mathcal{R}\to\infty} F = -\frac{\dot{V}^2}{16\pi D^2}\frac{\rho}{g_0} \qquad \dot{V} = \text{constant} \qquad (9.50)$$

Equations (9.49) and (9.50) show that a point source with a constant volume rate exerts a force on the wall which pulls it towards the source! This is true also for a point sink with $\dot{V}$ replaced by $-\dot{V}$!

9.11 Other Geometries by Superposition

The pressure field from a bubble or point source in two- and three-sided corners can be approximated by the superposition shown in Fig. 9.13. The velocity potentials in either case are given by

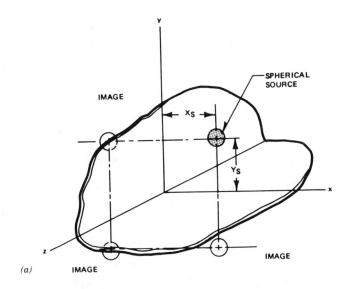

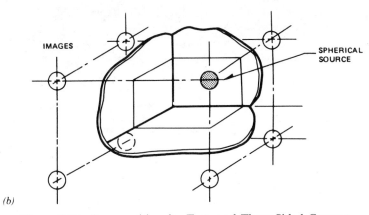

Figure 9.13 Superposition for Two- and Three-Sided Corners

$$\phi = -\frac{\dot{\mathcal{V}}}{4\pi} \sum_{n=1}^{N} \frac{1}{r_n}, \qquad N = \begin{cases} 4, & \text{two-sided corner} \\ 8, & \text{three-sided corner} \end{cases} \qquad (9.51)$$

Geometries of Fig. 9.13 require superposition of a finite number of images because wall symmetry is readily obtained. However, some geometries require many more images because symmetry is not possible. Consider the source between two parallel walls in Fig. 9.14. Source ① must have image ② for symmetry about the right wall, and ③ and ④ for symmetry about the left wall. But ③ and ④ must have ⑤ and ⑥ for symmetry about the right wall, and so on, extending to the superposition of an infinite number of source images. The superposition can be written as

$$\phi \approx -\frac{\dot{\mathcal{V}}}{4\pi} \sum_{n=0}^{\infty} \left(\frac{1}{T_{1n}} + \frac{1}{T_{2n}} + \frac{1}{T_{3n}} + \frac{1}{T_{4n}} \right)$$

$$T_{(1,2)n} = \sqrt{[x(+,-)b - (2n+1)a]^2 + \mathcal{R}^2} \qquad (9.52)$$

$$T_{(3,4)n} = \sqrt{[x(-,+)b + (2n+1)a]^2 + \mathcal{R}^2}$$

$$\mathcal{R}^2 = y^2 + z^2$$

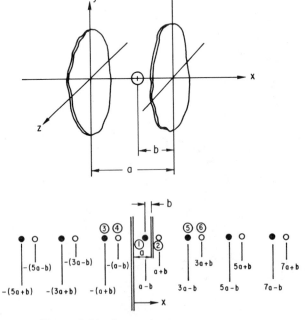

Figure 9.14 Source between Parallel Walls

If n becomes large, each term in the series becomes $1/2an$, which is nonconvergent. This implies that $\partial\phi/\partial t$ and, consequently, the unsteady pressure also are nonconvergent. However, the velocity field

$$\mathbf{V} = \frac{\partial\phi}{\partial x}\,\mathbf{n}_x + \frac{\partial\phi}{\partial\mathcal{R}}\,n_{\mathcal{R}}$$

yields terms involving $\{[x \pm b \mp (2n+1)a]^2 + \mathcal{R}^2\}^{-3/2}$, which are indeed convergent, approaching $1/(2na)^3$ for large n. That is, this superposition of velocity fields is convergent in steady or unsteady flows. Steady pressure fields can be obtained from Eq. (9.25) and the velocity field. Unsteady pressure fields require an approximate limited superposition of the velocity potential time derivative.

Applications involve forces exerted on finite surfaces, which generally can be approximated by a finite number of images. Suppose that it is necessary to estimate the force exerted by a bubble transient on the submerged flat plate in Fig. 9.15. The superposition pattern shown is required in two planes parallel to y, z at $x = \pm x_0$ to satisfy the condition of zero velocity perpendicular to the plate. The black circles are sources and the light circles are sinks, arranged to give uniform pressure at the flat plate edges when the fluid velocity is zero. The velocity potential on the wall $x = 0$, y, z, obtained from superposition, is

$$\phi = -\frac{\dot{\mathcal{V}}}{4\pi}\left[\sum_n\sum_m \frac{2}{\sqrt{(y-y_m)^2+(z-z_n)^2}} - \sum_n\sum_m \frac{2}{\sqrt{(y-y_m)^2+(z-z_n)^2}}\right]$$

(9.53)

$$n, m = 0, \pm 1, \pm 2, \ldots$$

where all combinations of n and m are implied. Source locations correspond to

$$\begin{aligned} z_n &= 2(2n+1)L - z_0, & y_m &= 2(2m+1)H - y_0 \\ z_n &= 4nL + z_0, & y_m &= 4mH + y_0 \end{aligned}$$

(9.54)

Sink locations correspond to

$$\begin{aligned} z_n &= 2(2n+1)L - z_0, & y_m &= 4mH + y_0 \\ z_n &= 4nL + z_0, & y_m &= 2(2m+1)L - y_0 \end{aligned}$$

(9.55)

Large n and m in Eqs. (9.53)–(9.55) yield

$$\phi\big|_{n,m\ \text{large}} \to 0$$

Source-sink clusters can be neglected after $n = m = 5$ with less than 10 percent error in ϕ.

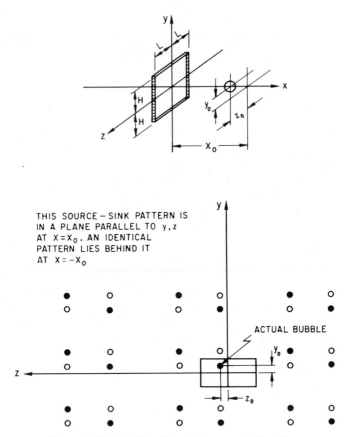

Figure 9.15   Bubble Near Submerged Flat Plate

Other superpositions can be used to construct solutions for sources or bubbles in various geometries, which include straight walls and flat free surfaces.

9.12   Submerged Sparger Discharge

Liquid or gas discharge from the end of a submerged pipe can exert a substantial pressure on nearby walls. A sparger is sometimes employed to reduce local pressure, and usually consists of a straight pipe with closed ends and many small holes in the wall, as shown in Fig. 9.16. This device distributes the discharge with a flow field which approaches that of a finite line source. If the total discharge volume rate is $\dot{\mathcal{V}}$, the velocity potential for a uniform line source on the z axis can be obtained by superimposing many differential point sources. That is, for a line segment $d\eta$ we have

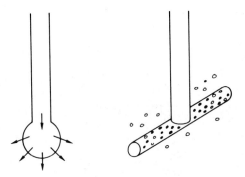

Pipe discharge   Discharge through sparger

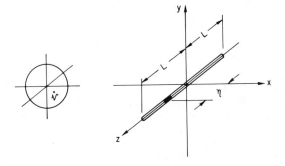

Line source

Figure 9.16 Fluid Discharge from Pipe and Sparger

$$d\phi = -\frac{\dot{V}}{8\pi L}\frac{1}{\zeta}\,d\eta\,, \qquad \zeta = \sqrt{x^2 + y^2 + (z - \eta)^2}$$

The integral over sparger length $2L$ yields ϕ for a uniform line source on z of length $2L$ and total volume rate $\dot{V}$ in the form

Line Source on z

$$\phi_L = -\frac{\dot{V}}{8\pi L}\ln\left[\frac{z + L + \sqrt{(z + L)^2 + r^2}}{z - L + \sqrt{(z - L)^2 + r^2}}\right], \qquad r^2 = x^2 + y^2 \qquad (9.56)$$

If two such line sources are superimposed at $x = \pm x_0$ as

$$\phi = \phi_L(x - x_0, y, z, t) + \phi_L(x + x_0, y, z, t)$$

an infinite flat wall in the y, z plane is simulated. Pressure at the origin is

$$P_L(0,0,0,t) - P_\infty = -\frac{\rho}{g_0}\left(\frac{\partial\phi}{\partial t} + \frac{V^2}{2}\right)$$

$$= \frac{\rho}{g_0}\frac{1}{4\pi L}\frac{d\dot{V}}{dt}\ln\left[\frac{\sqrt{L^2 + x_0^2} + L}{\sqrt{L^2 + x_0^2} - L}\right] \qquad (9.57)$$

Discharge from the end of a pipe is simulated by the superposition of two point sources, Eq. (9.13), of the same volume rate $\dot{V}$ and spacing on $x = \pm x_0$ to obtain,

$$P_s(0,0,0,t) - P_\infty = \frac{2\rho}{g_0}\frac{1}{4\pi x_0}\frac{d\dot{V}}{dt} \qquad (9.58)$$

The ratio of line-to-point source wall pressures at the origin is

$$\frac{P_{L,\max} - P_\infty}{P_{s,\max} - P_\infty} = \frac{x_0}{2L}\ln\frac{\sqrt{1 + (x_0/L)^2} + 1}{\sqrt{1 + (x_0/L)^2} - 1} \qquad (9.59)$$

for equal volume rates.

EXAMPLE 9.5: SPARGER WALL PRESSURE REDUCTION
A discharging gas bubble exerts a 10-atm (147-psi) pressure increase on the closest point of a flat wall 2 m (6.56 ft) away from the bubble center. If the same volume rate is discharged from an equally distant sparger parallel to the wall, what length L will reduce the pressure from 10 atm (147 psi) to 5 atm (74 psi)?

Equation (9.59) is employed with $(P_{L,\max} - P_\infty)/(P_{s,\max} - P_\infty) = 5/10 = 1/2$, giving $x_0/L = 0.23$, so $L = 8.7$ m (28.5 ft). A sparger of 17.4 m (57 ft) total length should reduce the maximum wall pressure from 10 (147 psi) to 5 (74 psi) atm. Such a long sparger should be further analyzed for non-uniform discharge. Other sparger designs consisting of multiple pipes in the form of a cross or fan also are used in practice. Integration shows that although the maximum pressure on a wall is reduced by a sparger, the total force on the wall sometimes is increased!

9.13 Small-Amplitude Liquid Sloshing

The motion of a liquid surface in a tank is important in transportation, earthquake responses, space vehicle low-gravity sloshing motion, and other applications. Sometimes only the sloshing natural frequencies of liquid are required. If container motion or other disturbances can be specified to avoid the natural frequencies, sloshing can be reduced. Several cases are considered for rigid containers where gravity is the restoring force and surface tension effects are negligible.

Sloshing in a rectangular container is indicated in Fig. 9.17. The fluid motion is governed by Laplace's equation $\nabla^2\phi = 0$ with zero normal fluid

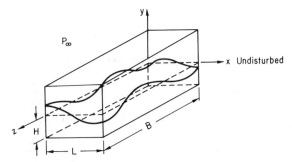

Figure 9.17 Liquid Sloshing in a Rectangular Tank. Disturbed Surface Elevation Above H is Given by $\eta = \eta(x, y, z, t)$.

velocity at the wall boundaries, and a uniform pressure on the free surface. Equation (9.25) relates free surface pressure, velocity, elevation, and ϕ as

$$\frac{\partial \phi}{\partial t} + \frac{V^2}{2} + \frac{g_0}{\rho} P_\infty + gy = f(t) \qquad \text{on } y = \eta$$

Although P_∞ is known, this equation contains both ϕ and η. A kinematic boundary condition introduces no new physics, but yields another equation involving ϕ and η and is obtained by writing the surface elevation as

$$\eta = \eta(x, z, t)$$

Differentiation yields

$$\frac{d\eta}{dt} = \frac{\partial \eta}{\partial x}\frac{dx}{dt} + \frac{\partial \eta}{\partial z}\frac{dz}{dt} + \frac{\partial \eta}{\partial t} \qquad \text{on } y = \eta$$

The motion of a surface particle can be tracked if we write $dx/dt = u = \partial\phi/\partial x$, $dz/dt = w = \partial\phi/\partial z$, and note that $d\eta/dt$ becomes $v = \partial\phi/\partial y$. Initially specified surface displacement η_0 and velocity $(d\eta/dt)_0$ complete the full problem. A product solution for x, y, z geometry

$$\phi(x, y, z, t) = X(x)Y(y)Z(z)T(t)$$

results in the eigensolutions

$$\phi_{n,m} = T_{n,m}(t)[e^{\beta_{n,m}y} + e^{-\beta_{n,m}(2H+y)}] \cos \chi_n x \cos \sqrt{\beta_{n,m}^2 - \chi_n^2}\, z$$

$$\chi_n = \frac{n\pi}{L}, \quad \beta_{n,m} = \sqrt{\left(\frac{n\pi}{L}\right)^2 + \left(\frac{m\pi}{B}\right)^2} \qquad (9.60a)$$

which satisfies Laplace's equation and the wall boundary conditions. The free surface condition contains nonlinear terms, which usually require numerical procedures for solution. However, a small-amplitude solution for ϕ, obtained by superposition of eigensolutions, gives useful information. First, $\phi(x, y, z, t)$ and its derivatives are expanded in Taylor series about $y = 0$ to project their values onto the free surface. Since velocity components involve derivatives of ϕ with respect to x, y, and z, we can absorb $f(t)$ and $g_0 P_\infty/\rho$ into ϕ and then impose the small-amplitude initial condition

$$\eta(x, z, 0) = \eta_0(x, z) = \epsilon f_s(x, z) \tag{9.60b}$$

to obtain the linearized surface conditions transferred back to $y = 0$ in the forms

$$\frac{\partial \phi}{\partial t}(x, y, z, t) + g\eta(x, z, t) = 0 \quad \text{on} \quad y = 0 \tag{9.60c}$$

and

$$\frac{\partial \eta(x, z, t)}{\partial t} = \frac{\partial \phi(x, y, z, t)}{\partial y} \quad \text{on} \quad y = 0 \tag{9.60d}$$

Elimination of η gives the boundary condition,

$$\frac{\partial^2 \phi}{\partial t^2} + g \frac{\partial \phi}{\partial y} = 0, \quad \text{on} \quad y = 0 \tag{9.60e}$$

If an eigensolution of ϕ is substituted from Eq. (9.60a), we obtain

$$\frac{d^2 T_{n,m}}{dt^2} + \omega_{n,m}^2 T_{n,m} = 0 \tag{9.60f}$$

for which the natural circular frequency is

$$\omega_{n,m} = \sqrt{g\beta_{n,m} \tanh(\beta_{n,m}H)},$$
$$\beta_{n,m} = \sqrt{\left(\frac{n\pi}{L}\right)^2 + \left(\frac{m\pi}{B}\right)^2} \qquad n, m = 1, 2, \ldots \tag{9.61a}$$

If sloshing occurs only in one direction, terms in the other direction can be neglected. That is,

$$\beta_{n,m} = \begin{cases} \beta_n = \dfrac{n\pi}{L}, & \text{sloshing in } L \text{ direction only} \tag{9.61b} \\[2mm] \beta_m = \dfrac{m\pi}{B}, & \text{sloshing in } B \text{ direction only} \tag{9.61c} \end{cases}$$

EXAMPLE 9.6: SLOSHING FREQUENCY, RECTANGULAR TANKS
Two rectangular tanks have dimensions $L = 1.0$ m and $B = 0.5$ m, but one has a depth $H = 0.2$ m and the other is very deep. Compare natural frequencies for the

fundamental slosh mode $n = m = 1$ in the L direction only and then in the B direction only.

Equations (9.61a)–(9.61c) give

L direction only, $n = 1$			B direction only, $m = 1$		
$\beta_n = \beta_1 = (\pi/1.0)m^{-1}$			$\beta_m = \beta_1 = (\pi/0.5)m^{-1}$		
H (m)	ω_1 (rad/s)	f_1 (Hz)	H (m)	ω_1 (rad/s)	f_1 (Hz)
0.2	4.14	0.66	0.2	7.22	1.15
∞	5.50	0.88	∞	7.84	1.25

The depth effect on slosh frequency is not strong unless $\beta_{n,m}H$ is less than about 1.0.

If the initial surface elevation $\eta_0(x, z)$ has zero velocity, a superposition of Eq. (9.60a) and a solution of Eq. (9.60f) yield the velocity potential

$$\phi(x, y, z, t) = -\epsilon \sum_m \sum_n I_{n,m} \left[\frac{e^{\beta_{n,m}(H+y)} + e^{-\beta_{n,m}(H+y)}}{e^{\beta_{n,m}H} + e^{-\beta_{n,m}H}} \right]$$

$$\times \cos \frac{n\pi x}{L} \cos \frac{m\pi z}{B} \sin \omega_{n,m}t \qquad (9.62)$$

$$I_{n,m} = \frac{4g}{\omega_{n,m}BL} \int_0^B \int_0^L f_s(x, z) \cos \frac{n\pi x}{L} \cos \frac{m\pi z}{B} \, dx \, dz$$

and the surface is obtained from Eqs. (9.60c) and (9.62) with $y = 0$ as

$$\eta(x, z, t) = \epsilon \frac{1}{g} \sum_m \sum_n \omega_{n,m} I_{n,m} \cos \frac{n\pi x}{L} \cos \frac{m\pi z}{B} \cos \omega_{n,m}t \qquad (9.63)$$

EXAMPLE 9.7: SLOSHING FORCE ON A TANK WALL
A rectangular tank (see Fig. 9.17) has dimensions $B = 1.0$ m, $L = 5.0$ m, and undisturbed depth $H = 2.0$ m. A disturbance causes sloshing of low amplitude $\epsilon = 0.1$ m in the L direction at the fundamental mode. Determine the force exerted on the tank by the water motion.

Equation (9.62) is written for the fundamental mode in x, neglecting z dependence, with $f_s(x) = \cos(\pi x/L)$. It follows that

$$\phi = -2g\epsilon \frac{1}{\omega_1} \left[\frac{e^{\beta_1(H+y)} + e^{-\beta_1(H+y)}}{e^{\beta_1 H} + e^{-\beta_1 H}} \right] \cos \frac{\pi x}{L} \sin \omega_1 t$$

with

$$\omega_1 = \sqrt{\frac{g\pi}{L} \tanh \frac{\pi H}{L}}$$

Components of the water velocity

$$\mathbf{V} = \frac{\partial \phi}{\partial x} \mathbf{n}_x + \frac{\partial \phi}{\partial y} \mathbf{n}_y$$

are obtained from ϕ and employed in the momentum equation

$$\mathbf{F} = \frac{1}{g_0} \frac{d}{dt} \int_{cv} \rho \mathbf{V} \, d\mathcal{V} + \mathbf{W}$$

with $d\mathcal{V} = B \, dx \, dy$ and the weight of water, $W = -\rho(g/g_0) BLH$ to give

$$\mathbf{F} = -\frac{\rho g}{g_0} \left[\epsilon \frac{4LB}{\pi} \tanh \beta_1 H \cos \omega_1 t \, \mathbf{n}_x + BLH\mathbf{n}_y \right]$$

$$= -5291 \cos (2.28)t \, \mathbf{n}_x - 98,000\mathbf{n}_y$$

(**F** is newtons, t is seconds)

Note that for a small-amplitude slosh, the vertical force remains equal to the fluid weight.

If sloshing occurs in the annular tank of inner and outer radii R_i and R, Fig. 9.18, an analysis similar to that for rectangular tanks gives the natural slosh frequencies

$$\omega_{n,m} = \sqrt{g\gamma_{n,m} \tanh \gamma_{n,m} H} \tag{9.64}$$

The wall boundary condition at $r = R$ was applied to obtain the mth eigenvalue $\gamma_{n,m}$ for a cylindrical tank with $R_i = 0$ from the roots of

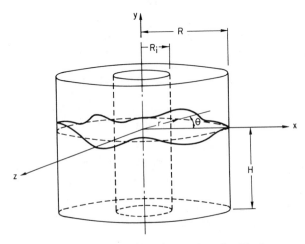

Figure 9.18   Sloshing in an Annular Tank

TABLE 9.1 Roots of Eq. (9.65) for $n = 1$, $R_i = 0$ (Cylindrical Tank)

m	$\gamma_{1,m} R$
1	1.841
2	5.331
3	8.536
4	11.706
5	14.864

$$(\gamma_{n,m} R) J_{n-1}(\gamma_{n,m} R) - n J_n(\gamma_{n,m} R) = 0, \qquad n = 1, 2, \ldots \qquad (9.65)$$

where the index n gives the integer number of periods of θ from 0 to 2π. Several roots for $n = 1$ are given in Table 9.1. The outer and inner wall boundary conditions at $r = R$ and $r = R_i$ were employed to obtain eigenvalues for an annular tank from the roots of

$$\frac{dJ_n(\gamma_{n,m} r)}{dr}\bigg|_{r=R} \frac{dY_n(\gamma_{n,m} r)}{dr}\bigg|_{r=R_i} - \frac{dJ_n(\gamma_{n,m} r)}{dr}\bigg|_{r=R_i} \frac{dY_n(\gamma_{n,m} r)}{dr}\bigg|_{r=R} = 0$$

$$(9.66)$$

EXAMPLE 9.8: SLOSH FREQUENCY IN AN ANNULAR TANK

Determine the sloshing frequency corresponding to a $\cos n\theta$ variation, $n = 1$, and the fundamental radial mode with $m = 1$ in an annular body of water between cylinders of inner and outer radii $R_i = 1.4$ m (4.6 ft) and $R = 4.5$ m (14.8 ft) of depth $H = 2.0$ m (6.56 ft). Compare with the frequency for a cylindrical tank without the inner wall.

The eigenvalue for the annular tank is obtained from the first root of Eq. (9.66), which is $\gamma_{n,m} = \gamma_{1,1} = 0.35$/m (0.107 /ft). The slosh frequency of Eq. (9.64) yields

Annular

$$\omega_{1,1} = 1.44 \text{ rad/s}, \qquad f_{1,1} = 0.22 \text{ Hz}$$

The eigenvalue for the cylindrical tank corresponds to the first root of Table 9.1, or $\gamma_{1,1} = 0.41$/m, for which

Cylindrical

$$\omega_{1,1} = 1.64 \text{ rad/s}, \qquad f_{1,1} = 0.26 \text{ Hz}$$

The slosh frequencies are seen to be slower in an annular tank than in a cylindrical tank of the same outer radius.

9.14 Large-Amplitude Liquid Motion

Numerical procedures have been formulated for tracking large-amplitude free surface motion. One simple procedure uses a stationary mesh like that in Fig. 9.19. Laplace's equation is solved by relaxation for ϕ at all mesh points every time step. Veolcity components are obtained from $\mathbf{V} = \nabla\phi$ at each point where a free surface crosses a mesh line, and the free surface position $\mathbf{r}$ is advanced every time step Δt according to $\Delta\mathbf{r} = \mathbf{V}\,\Delta t$.

The relaxation procedure is formulated for the computational star in Fig. 9.19. Values of ϕ_A, ϕ_B, ϕ_C, and ϕ_D are written from Taylor series expansions about ϕ_0 as

$$\phi_{\binom{A}{C}} = \phi_0(\pm)\left(\frac{\partial\phi}{\partial x}\right)_0\left(\frac{A}{C}\right) + \frac{1}{2!}\left(\frac{\partial^2\phi}{\partial x^2}\right)_0\left(\frac{A^2}{C^2}\right) + O(H^3)$$

and

$$\phi_{\binom{B}{D}} = \phi_0(\pm)\left(\frac{\partial\phi}{\partial y}\right)_0\left(\frac{B}{D}\right) + \frac{1}{2!}\left(\frac{\partial^2\phi}{\partial y^2}\right)_0\left(\frac{B^2}{D^2}\right) + O(H^3)$$

where H is the maximum branch length. Simultaneous solutions for the second derivatives and substitution into Laplace's equation yield the center residual

$$2\text{RES}_0 = (BD + AC)(A + C)(B + D)\phi_0 - AC(A + C)(D\phi_B + B\phi_D)$$
$$- BD(B + D)(C\phi_A + A\phi_C) + O(H)$$

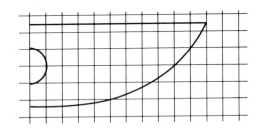

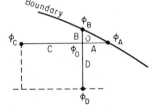

Figure 9.19 Stationary Mesh and Computational Star

which is reduced to an acceptable minimum by successive computational sweeps of all mesh points lying inside the fluid region. Overrelaxation involves computation of $RES_{0,\text{current}}$ at a mesh point and immediate adjustment of ϕ_0 according to

$$\phi_{0,\text{new}} = \phi_{0,\text{current}} - \frac{F_R}{4} RES_{0,\text{current}}$$

The overrelaxation factor F_R is in the range $0.0 < F_R < 0.5$ and can be optimized [9.14].

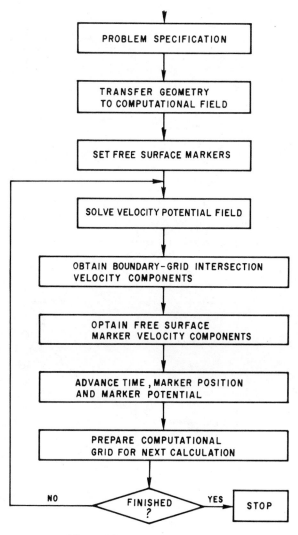

Figure 9.20 Overall Flowchart

Interpolation of ϕ and its derivatives from mesh point values provides the velocity potential, velocity components, and the total time derivative

$$\frac{d\phi}{dt} = \frac{\partial\phi}{\partial t} + \mathbf{V}\cdot\nabla\phi$$

at markers positioned on the free surfaces. The markers are moved distances $u\,\Delta t$, $v\,\Delta t$, and $w\,\Delta t$ during a time step Δt, and a marker value of ϕ changes by an amount $(d\phi/dt)\,\Delta t$. The marker positions display the altered free surface, and values of ϕ are reverse-interpolated to the mesh lines.

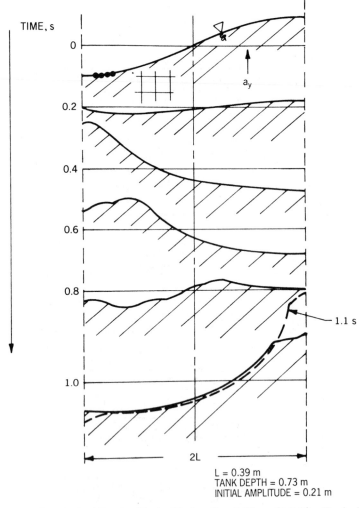

L = 0.39 m
TANK DEPTH = 0.73 m
INITIAL AMPLITUDE = 0.21 m

Figure 9.21 Large-Amplitude Tank Slosh. $L = 0.39\,\mathrm{m}$ (1.3 ft), Tank Depth = 0.73 m (2.4 ft), Initial Amplitude = 0.21 m (0.7 ft)

Then the field is ready for another overrelaxation solution at $t + \Delta t$. A computational flow chart is given in Fig. 9.20. Several two-dimensional example solutions are shown in Figs. 9.21 through 9.24.

Figure 9.21 shows distortion of an initially sinusoidal free surface under the action of gravity. The wave becomes steeper and higher near the walls and flatter through the middle. The wall amplitude after one slosh cycle is almost three times the initial value.

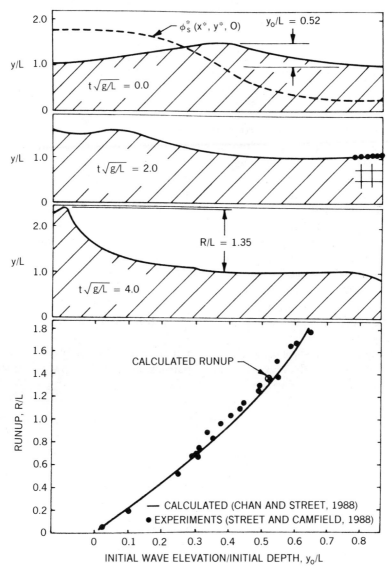

Figure 9.22 Solitary Wave Run-Up on a Wall. Calculated [9.2]. Experiments [9.16]

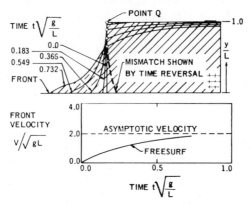

Figure 9.23 Dam Break Problem

Figure 9.22 shows motion of a solitary wave [9.15] and run-up elevation on a vertical wall, which compares well with data. A solitary wave is one of the so-called permanent types, which may have large amplitude and can propagate great distances without significant deformation.

Figure 9.23 shows the early stages of a dam break problem. Point Q adjacent to the dam moves straight downward. A fishing boat behind a suddenly failed dam would gently follow the surface to a lower elevation without moving horizontally relative to the shore! However, a diver searching for treasure at the base of the dam would be swept downstream. The frontal velocity is seen to approach $2\sqrt{gL}$, value obtained from hydrostatic wave theory. Since inviscid flow has no dissipation, problems described in this section are reversible. If the time step is changed from Δt to $-\Delta t$, a problem will run backwards and eventually should reach the initial configuration. Any mismatch gives a measure of truncation and roundoff errors in the computation.

Figure 9.24 shows the growth of a high-pressure bubble below a free liquid surface. It is seen that expansion is mostly upward, ultimately rupturing the top surface.

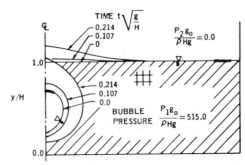

Figure 9.24 Growth of High-Pressure Bubble

9.15 Other Large-Amplitude Methods

Finite difference schemes to solve the mass conservation and momentum laws of Eqs. (9.1) and (9.2) have been formulated which use computational *cells* rather than stars. The marker-and-cell (MAC) method [9.17] employs the pressure difference of two adjacent cells to advance average values of velocity components on each cell boundary from the momentum law. The cell interior pressures are adjusted iteratively, forcing velocity values to satisfy mass conservation. Computational markers are assigned local velocity components. Displacement of the markers during a time step yields an appearance of the velocity field. The example computation shown in Fig. 9.25 was obtained from one of the SOLA [9.10] programs for the growth of a gas bubble being charged into a pool of water.

9.16 Interface Stability

A stability study of the interface between fluids with unequal densities can be employed to predict if interface disturbances grow, if droplets or bubbles can form, or if each of the adjacent fluids can remain separated without

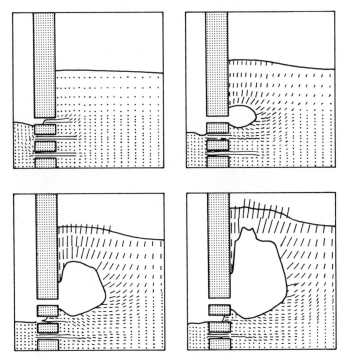

Figure 9.25 Gas Charging into a Water Pool [9.10]

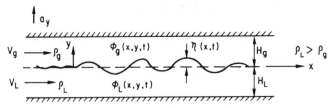

Figure 9.26   Fluid Streams Interface Instability (Helmholtz)

entraining the other. Linearized theory is employed in this section to determine parameter ranges which are consistent with stable and unstable interfaces.

The interface instabilities considered here are in one of the two classifications shown in Figs. 9.26 and 9.27: *Helmholtz instability*, which pertains to adjacent fluid streams with unequal densities, traveling at different velocities parallel to the interface; and *Taylor instability*, which pertains to fluids with unequal densities, undergoing acceleration in a direction normal to the interface.

Fluid densities are designated as ρ_L and ρ_g with $\rho_L > \rho_g$, and the respective velocity potentials are $\phi_L(x, y, t)$ and $\phi_g(x, y, t)$. The fluid streams in Fig. 9.26 have depths H_L and H_g and undisturbed velocities V_L and V_g. The disturbed interface displacement is $\eta(x, t)$. The body force acceleration a_y in Figs. 9.26 and 9.27 may correspond to that of gravity, $a_y = g$, if the bounding walls are resting on the earth with the y coordinate upward. The heavier fluid is below the lighter one in Fig. 9.26, which, for a normal gravity field, can be interpreted as a horizontal liquid stream flowing beneath a gas stream. However, the heavier fluid is above the lighter one in Fig. 9.27, for which one interpretation involves a glass of water turned upside down.

The two-dimensional problems are specified as follows:

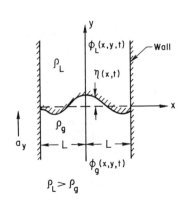

Figure 9.27   Two-Fluid Accelerating Interface Instability (Taylor)

DEs:

$$\frac{\partial^2 \phi_i}{\partial x^2} + \frac{\partial^2 \phi_i}{\partial y^2} = 0, \qquad i = g, L \qquad (9.67a)$$

Kinematic Free Surface BCs

$$\frac{\partial \eta}{\partial t} = \frac{\partial \phi_i}{\partial y} - \frac{\partial \phi_i}{\partial x} \frac{\partial \eta}{\partial x} \qquad \text{on} \quad y = \eta, \quad i = g, L \qquad (9.67b)$$

Dynamic Free Surface BCs

$$\frac{\partial \phi_i}{\partial t} + \frac{V_i^2}{2} + \frac{g_0}{\rho_i} P_i + a_y y = f_i(t) \qquad \text{on} \quad y = \eta, \quad i = g, L \qquad (9.67c)$$

Interface Pressure Difference, Eq. (9.42)

$$P_g - P_L = \frac{\sigma}{R_c} = \frac{\sigma}{[1 + (\partial \eta / \partial x)^2]^{3/2}} \frac{\partial^2 \eta}{\partial x^2} \qquad (9.67d)$$

HELMHOLTZ, FIG. 9.26   **TAYLOR, FIG. 9.27**

Wall BCs

$$\frac{\partial \phi_i}{\partial y} = 0, \quad i, y = \begin{cases} g, H_g \\ L, -H_L \end{cases} \qquad \left. \frac{\partial \phi_i}{\partial x} \right|_{x = \pm L} = 0, \quad i = g, L \qquad (9.67e)$$

Distant BCs

Not required $\nabla \phi_i \big|_{y \to \pm \infty} = 0, \quad i = g, L \qquad (9.67f)$

ICs

$$\nabla \phi_i = V_i, \quad i = g, L \qquad\qquad \nabla \phi_i = 0, \quad i = g, L \qquad (9.67g)$$

$$\eta(x, 0) = \epsilon f_s(x) \qquad (9.67h)$$

Here, the parameter ϵ is a small disturbance amplitude.

A regular perturbation solution (Section C.2 of Appendix C) is formulated by writing ϕ_g, ϕ_L, and η as power series in the small parameter as

$$\phi_g(x, y, t) = \phi_{g0}(x, y, t) + \epsilon \phi_{g1}(x, y, t) + \cdots \qquad (9.67i)$$

$$\phi_L(x, y, t) = \phi_{L0}(x, y, t) + \epsilon \phi_{L1}(x, y, t) + \cdots \qquad (9.67j)$$

$$\eta(x, t) = \epsilon \eta_1(x, t) \qquad (9.67k)$$

The fluid streams interface instability problem of Fig. 9.26 gives $\phi_{g0} = V_g x$, $\phi_{L0} = V_L x$, $f_g(t) = V_g^2/2$, and $f_L(t) = V_L^2/2$ for the ϵ^0 analysis. The ϵ^1 analysis yields ϕ_{g1} and ϕ_{L1} from product solutions of the form $T(t)X(x)Y(y)$, which satisfy Eqs. (9.67a), (9.67e), and (9.67f), but which also contain undetermined time-dependent coefficients and separation constants.

A wave train initial disturbance is assumed of the form

$$\eta_1(x, t) = \cos \omega(t - x/C) = \cos(\omega t - 2\pi x/\lambda) \tag{9.68a}$$

where λ is the wave length, related to the wave speed C and the circular frequency by

$$C = \frac{\lambda \omega}{2\pi} \tag{9.68b}$$

The kinematic conditions of Eq. (9.67b) are employed with (9.68a) to determine the time-dependent coefficients and separation constants of ϕ_{g1} and ϕ_{L1}. Finally, the dynamic boundary conditions of order ϵ^1 from Eq. (9.67c) are employed with (9.67d) and the expressions for ϕ_{g1}, ϕ_{L1}, and η_1 to obtain

$$\frac{C}{\lambda} = \frac{\omega}{2\pi} = F_1 \pm F_2 \tag{9.69a}$$

where

$$F_1 = \frac{V_L \rho_L \alpha_L + V_g \rho_g \beta_g}{\lambda(\rho_L \alpha_L + \rho_g \beta_g)} \tag{9.69b}$$

$$F_2 = \frac{\sqrt{F_3 - F_4}}{(\rho_L \alpha_L + \rho_g \beta_g)} \tag{9.69c}$$

$$F_3 = \left[\frac{2\pi g_0 \sigma}{\lambda^3} + \frac{a_y}{2\pi\lambda}(\rho_L - \rho_g)\right](\rho_L \alpha_L + \rho_g \beta_g) \tag{9.69d}$$

$$F_4 = \frac{\rho_L \rho_g \alpha_L \beta_g}{\lambda^2}(V_g - V_L)^2 \tag{9.69e}$$

$$\alpha_L = \coth \frac{2\pi H_L}{\lambda}, \qquad \beta_g = \coth \frac{2\pi H_g}{\lambda} \tag{9.69f}$$

If Eqs. (9.69a)–(9.69f) show that ω is real, the interface waves of Eq. (9.68a) remain periodic in time and space. However, if ω is complex, Eq. (9.68a) yields unstable interface waves with exponential growth. The only way ω can be complex is if $F_4 > F_3$ in Eq. (9.69c), which can occur for wave lengths which satisfy the following conditions:

UNSTABLE INTERFACE WAVE LENGTHS, FLUID STREAMS INTERFACE INSTABILITY (HELMHOLTZ, FIG. 9.26)

$$\left[\frac{2\pi g_0 \sigma}{\lambda^3} + \frac{a_y}{2\pi\lambda}(\rho_L - \rho_g)\right](\rho_L\alpha_L + \rho_g\beta_g) - \frac{\rho_L\rho_g}{\lambda^2}\alpha_L\beta_g(V_g - V_L)^2 < 0 \tag{9.70a}$$

or the equivalent form

$$(1 - \sqrt{1 - r})\lambda_R \le \lambda \le (1 + \sqrt{1 - r})\lambda_R \tag{9.70b}$$

$$\lambda_R = \frac{2\pi\rho_L\rho_g\alpha_L\beta_g(V_g - V_L)^2}{2a_y(\rho_L\alpha_L + \rho_g\beta_g)(\rho_L - \rho_g)} \tag{9.70c}$$

$$r = \frac{(2\pi)^2 g_0\sigma}{a_y(\rho_L - \rho_g)}\frac{1}{\lambda_R^2} \tag{9.70d}$$

If the interface is unstable, F_2 in Eq. (9.69a) is imaginary, and Eq. (9.68a) yields an amplitude which grows according to $\exp(2\pi F_2 t)$. Thus, the fastest growing unstable wave length results in a maximum value of F_2, corresponding to the condition $dF_2/d\lambda = 0$. Several cases of particular interest are summarized below:

CASES OF PARTICULAR INTEREST, FLUID STREAMS INTERFACE INSTABILITY (HELMHOLTZ, FIG. 9.26)

Case 1: Fastest Growing Wave Length in Range of Eq. (9.70b)

$$\lambda_{\text{fastest growing}} = 2\lambda_R\left[1 \pm \sqrt{1 - (2\pi)^2 \frac{3g_0\sigma}{4a_y}\frac{1}{\lambda_R^2}}\right] \tag{9.71a}$$

Case 2: Equal Stream Velocities; Waves Are Never Unstable

$$C = \sqrt{\frac{(2\pi g_0\sigma/\lambda) + (a_y\lambda/2\pi)(\rho_L - \rho_g)}{\rho_L\alpha_L + \rho_g\beta_g}} \tag{9.71b}$$

$$C = \sqrt{\frac{a_y\lambda}{2\pi}\left(\frac{\rho_L - \rho_g}{\rho_L + \rho_g}\right)}, \qquad \sigma = 0, \qquad H_L, H_g \to \infty \tag{9.71c}$$

$$C = \sqrt{\frac{a_y\lambda}{2\pi}}, \qquad \rho_g \ll \rho_L, \qquad \sigma = 0, \qquad H_L, H_g \to \infty \tag{9.71d}$$

Case 3: Deep Fluids, Negligible Surface Tension

$$\lambda < \frac{(V_g - V_L)^2}{a_y}\frac{2\pi\rho_L\rho_g}{\rho_L^2 - \rho_g^2}, \qquad \text{unstable} \tag{9.71e}$$

Case 4: *Streams at Rest, Liquid over Gas* $(a_y = -g)$

$$\lambda \geq 2\pi\sqrt{\frac{g_0\sigma}{g(\rho_L - \rho_g)}}, \qquad \text{unstable} \qquad (9.71\text{f})$$

Case 5: *Vertical Fluid Streams, or* $a_y = 0$

$$\lambda \geq \frac{2\pi g_0\sigma(\rho_L\alpha_L + \rho_g\beta_g)}{\rho_L\rho_g\alpha_L\beta_g(V_g - V_L)^2}, \qquad \text{unstable} \qquad (9.71\text{g})$$

EXAMPLE 9.9: UNSTABLE WAVES FROM WIND OVER WATER
Wind blows over a deep lake at $V_g = 15$ m/s (49.2 fps) velocity. What length waves
will be unstable, and is there a fastest growing wave length? Take $\rho_L = 1000$ kg/m^3
(62.4 lbm/ft^3), $\rho_g = 1$ kg/m^3 (0.06232 lbm/ft^3), and $\sigma = 0.07$ N/m (0.0048 lbf/ft).
Since $\rho_g \ll \rho_L$, $\alpha_L = \beta_g = 1.0$, and $a_y = g$, Eq. (9.70a) reduces to

$$\frac{2\pi g_0\sigma}{\lambda^3} + \frac{g}{2\pi\lambda}\rho_L - \frac{\rho_g V_g^2}{\lambda^2} < 0$$

which yields two values of λ, for an unstable range

$$1.98\text{ mm (0.078 in)} < \lambda < 142\text{ mm (5.6 in)}$$

Wavelengths outside this range are expected to be stable. The fastest growing wave
length is obtained from Eq. (9.71a) as

$$\lambda_{\text{fastest growing}} = \frac{2\pi\rho_g V_g^2}{g\rho_L}\left[1 \pm \sqrt{1 - \frac{6\pi gg_0\rho_L\sigma}{\rho_g^2 V_g^4}}\right] = \begin{cases} 268\text{ mm} \\ 19.7\text{ mm} \end{cases}$$

The two values obtained are compared with the permissible unstable range of Eq.
(9.70b). It follows that a wavelength of about 19.7 mm (0.78 in) is expected to
dominate the surface. Since this wavelength is unstable, large-amplitude growth
would lead to wave-breaking whitecaps and possible entrainment in the air stream.

———————

Equations (9.67a)–(9.67k) were employed to analyze the two-fluid accelerating interface instability problem of Fig. 9.27. A regular perturbation
solution yields $\phi_{g0} = \phi_{L0} = 0$. Separation of variables was applied to the ϵ^1
(linear) problem to obtain ϕ_{g1} and ϕ_{L1}, which satisfied Eqs. (9.67a), (9.67e),
and (9.67f) for either symmetric or antisymmetric interfaces, except for
undetermined time-dependent coefficients. Equations (9.67b)–(9.67d) were
combined from the ϵ^1 problem to eliminate η_1 and obtain a differential
equation for the time-dependent coefficients of ϕ_{g1} and ϕ_{L1} of the form

$$\ddot{T}_n + \omega_n^2 T_n = 0 \qquad (9.72)$$

where

$$\omega_n^2 = \frac{g_0 \sigma}{2\pi(\rho_L + \rho_g)L^3} F(n)\left[\frac{a_y(\rho_L - \rho_g)L^2}{g_0 \sigma} - F(n)\right] \qquad (9.73a)$$

$$F(n) = \begin{cases} n\pi, & n = 1, 2, \ldots, \quad \text{symmetric} \\ \dfrac{2n+1}{2}\pi, & n = 0, 1, 2, \ldots, \quad \text{antisymmetric} \end{cases} \qquad (9.73b)$$

A real value of the circular frequency ω_n implies a stable oscillation and an imaginary value yields exponentially growing interface disturbances. The fastest growing unstable wave is obtained from the condition $d\omega_n/dn = 0$. Results of the Taylor instability analysis are summarized below:

INTERFACE INSTABILITY, ACCELERATION NORMAL TO INTERFACE (TAYLOR, FIG. 9.27)

Interface Profile

$$\eta_1(x, t) = \begin{cases} \cos n\pi x/L, & n = 1, 2, \ldots, \quad \text{symmetric} \\ \sin(2n+1)\pi x/2L, & n = 0, 1, 2, \ldots, \quad \text{antisymmetric} \end{cases}$$

$$(9.74a)$$

Wavelength

$$\lambda_n = \begin{cases} 2L/n, & n = 1, 2, \ldots, \quad \text{symmetric} \\ 4L/(2n+1), & n = 0, 1, 2, \ldots, \quad \text{antisymmetic} \end{cases}$$

$$(9.74b)$$

Conditions for Instability

$$L\sqrt{\frac{a_y(\rho_L - \rho_g)}{g_0 \sigma}} > \begin{cases} n\pi, & n = 1, 2, \ldots, \quad \text{symmetric} \\ (2n+1)\pi/2, & n = 0, 1, 2, \ldots, \quad \text{antisymmetric} \end{cases}$$

$$(9.74c)$$

or, in terms of wavelength

$$\lambda\sqrt{\frac{a_y(\rho_L - \rho_g)}{g_0 \sigma}} > 2\pi \qquad (9.74d)$$

Fastest Growing Wavelength

$$\lambda_{\text{fastest growing}} = 2\pi\sqrt{\frac{3g_0 \sigma}{a_y(\rho_L - \rho_g)}} \qquad (9.74e)$$

If the wall half-width L does not satisfy the inequality (9.74c), the interface is stable. Also, a stable interface can occur in low-gravity environment, even for large values of L.

EXAMPLE 9.10: WATER IN AN INVERTED TEST TUBE

Estimate the maximum diameter of a 10 cm long, open test tube containing water, which will continue to hold water in it when inverted. The air-water surface tension is 0.07 N/m.

If the water existed as a solid, frictionless plug, ambient pressure would support it in the inverted test tube as its weight would create a slight vacuum at the closed end. The water would also remain in the test tube if the air-liquid interface was stable. However, if the interface was unstable, a long air column would rise upward, displacing water downward, thus emptying the tube. It is necessary, then, to *estimate* the tube maximum diameter for which the interface remains stable.

Equation (9.74c) yields, for

$$\rho_L = 1000 \text{ kg/m}^3 \gg \rho_g \quad \text{and} \quad a_y = g = 9.8 \text{ m/s}^2$$

$$L > \sqrt{\frac{g_0 \sigma}{a_y(\rho_L - \rho_g)}} \begin{cases} n\pi \\ (2n+1)\pi/2 \end{cases}$$

or

$$L > \begin{cases} 8.4 \text{ mm}, & n = 1, \quad \text{symmetric} \\ 4.2 \text{ mm}, & n = 0, \quad \text{antisymmetric} \end{cases}$$

It is expected that if the tube diameter is much more than 4 mm, a slight disturbance would cause the water to run out (see Prob. 9.19).

———

9.17 Acceleration of Submerged Structures

Consider the fluid force on the moving cylinder in a uniform flow, already discussed in Section 9.4. The fluid pressure exerts a force $dF = P(R, \theta)RL\, d\theta$ in Fig. 9.28, which has an x component $dF_x = -dF \cos \theta$. Pressure $P(R, \theta)$ is obtained from Eqs. (9.20) and (9.25). Integration around the cylinder yields

$$F_x = \pi R^2 L \frac{\rho}{g_0} (2\dot{U}_\infty - \dot{U}_c) \tag{9.75}$$

Equation (9.75) is based on ideal flow. Actual flow would exhibit boundary layer separation, wake formation, and a drag force

$$F_D = C_D A(U_\infty - U_c)^2 \frac{\rho}{2g_0}$$

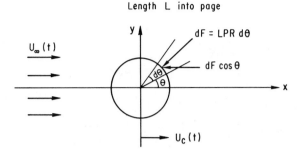

Figure 9.28   Fluid Force on Cylinder

would develop. The force given by Eq. (9.75) often dominates problems involving the acceleration of submerged objects.

EXAMPLE 9.11: A PROPOSED TOW-TANK TEST
It is anticipated that a submerged, stationary cylindrical structure of radius $R = 0.5$ m will be exposed to an unsteady water flow of acceleration $\dot{U}_\infty = 10$ m/s². The supporting braces should accommodate the resulting force. An engineering firm has proposed a test to determine this force. They plan to tow the cylinder in a large tank of water at the expected water acceleration and to measure the force required. They claim that the experiment is justified because the well-understood form drag on an object is the same either for flow past a stationary object or for motion of the object at the same velocity through stagnant fluid. Will the proposed test give useful data for this application?

Equation (9.75) shows that the fluid force per unit length exerted on a stationary cylinder is $2\pi R^2 \rho \dot{U}_\infty / g_0 = 15.7$ kN/m, whereas the force on a cylinder moving at $\dot{U}_c = -\dot{U}_\infty$ in stagnant liquid is $\pi R^2 \rho \dot{U}_\infty / g_0 = 7.85$ kN/m, or only half as much! The reason is that accelerating fluid has a pressure gradient in the direction of its acceleration which creates an additional force similar to that of buoyancy. The proposed experiment will give erroneous results.

Although a factor of 2 applies for a cylinder, the force of accelerating flow past objects with other shapes is always greater than the force of towing at the same acceleration in stagnant fluid. Forces in either case are the same only if the object has zero volume, such as a flat plate.

The form drag force with $C_D \approx 1.0$ is about 2 kN/m when the velocity reaches 2 m/s, which is beginning to rival the acceleration force of Eq. (9.75).

If an external rightward force F_{ex} is applied to the cylinder, we have the equation of motion

$$F_{ex} + F_x = \frac{M_c}{g_0} \dot{U}_c$$

where M_c is the cylinder mass. If F_x is employed from Eq. (9.75),

$$F_{ex} = \frac{M_c + \rho\pi R^2 L}{g_0} \dot{U}_c - \frac{2\rho\pi R^2 L}{g_0} \dot{U}_\infty \qquad (9.76)$$

It is seen that if $U_\infty = $ constant, the fluid force appears to increase the cylinder mass by an additional amount $\rho\pi R^2 L$, which is called the *virtual mass*.

9.18   Virtual Mass

Suppose that an applied force F_{ex} is accelerating a cylinder through stationary liquid. If the cylinder mass is zero, the acceleration force of Eq. (9.75) is

$$F_{ex} = \frac{\rho\pi R^2 L}{g_0} \dot{U}_c$$

Since $U_c = dx_c/dt$, the work done by F_{ex} increases the fluid kinetic energy at the rate

$$F_{ex} \frac{dx_c}{dt} = (\rho\pi R^2 L) \frac{d}{dt} \left(\frac{U_c^2}{2g_0} \right)$$

where $\rho\pi R^2 L$ is the virtual mass introduced in Eq. (9.76) from momentum considerations. Although both the momentum and energy principles gave the same virtual mass for a cylinder, the kinetic energy of a fluid generally is used to obtain the virtual mass of other simple geometric objects.

The kinetic energy of fluid surrounding a submerged object moving at velocity $\mathbf{V}_s$ is obtained from Fig. 9.29. The rectangular fluid element has kinetic energy

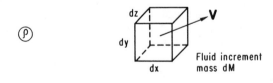

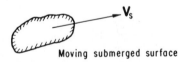

Moving submerged surface

Figure 9.29   Diagram for Fluid Kinetic Energy

TABLE 9.2   Virtual Mass for Two-Dimensional Structural Components (length L for all structures)

Body	Section through Body and Uniform Flow Direction	Virtual Mass (Patton, 1965)
Circle		$\rho \pi R^2 L$
		$\rho \pi a^2 L$
Ellipse		$\rho \pi b^2 L$
		$\rho \pi a^2 L$
Ellipse		
Plate		
Rectangle		

	a/b	
	∞	$\rho \pi a^2 L$
	10	1.14 $\rho \pi a^2 L$
	5	1.21 $\rho \pi a^2 L$
Diamond	2	1.36 $\rho \pi a^2 L$
	1	1.51 $\rho \pi a^2 L$
	1/2	1.70 $\rho \pi a^2 L$
	1/5	1.98 $\rho \pi a^2 L$
	1/10	2.23 $\rho \pi a^2 L$

	a/b	
	2	0.85 $\rho \pi a^2 L$
	1	0.76 $\rho \pi a^2 L$
	1/2	0.67 $\rho \pi a^2 L$
I-Beam	1/5	0.61 $\rho \pi a^2 L$
	$a/c = 2.6, b/c = 3.6$	
		$2.11 \rho \pi a^2 L$

TABLE 9.3 Virtual Mass for Three-Dimensional Structures

Description	Body and Flow Direction	Virtual Mass (Patton, 1965)
Circular Disk		$\dfrac{8}{3}\rho R^3$ $= 0.636\,\rho\left(\dfrac{4}{3}\,\pi R^3\right)$
Elliptical Disk		b/a $\infty\quad\quad\ \rho\,\dfrac{\pi}{6}\,ba^2$ $3\quad 0.900\,\rho\,\dfrac{\pi}{6}\,ba^2$ $2\quad 0.826\,\rho\,\dfrac{\pi}{6}\,ba^2$ $1.5\ \ 0.748\,\rho\,\dfrac{\pi}{6}\,ba^2$ $1.0\ \ 0.637\,\rho\,\dfrac{\pi}{6}\,ba^2$
Rectangular plate		b/a $1\quad 0.478\,\rho\,\dfrac{\pi}{4}\,a^2b$ $1.5\quad 0.680\,\rho\,\dfrac{\pi}{4}\,a^2b$ $2\quad 0.840\,\rho\,\dfrac{\pi}{4}\,a^2b$ $2.5\quad 0.953\,\rho\,\dfrac{\pi}{4}\,a^2b$ $3\quad\quad\ \ \rho\,\dfrac{\pi}{4}\,a^2b$ $\infty\quad\quad\ \ \rho\,\dfrac{\pi}{4}\,a^2b$
Triangular plate		$\rho a^3\,\dfrac{(\tan\theta)^{3/2}}{\pi}$
Sphere		$\rho\,\dfrac{2}{3}\,\pi R^3$

$$d(\text{KE}) = \frac{\mathbf{V} \cdot \mathbf{V}}{2g_0} \, dM = \frac{V^2}{2g_0} \, dM$$

The mass $dM = \rho \, dx \, dy \, dz$, with $V^2 = (\nabla \phi)^2$ and $\nabla^2 \phi = 0$, is used with Green's theorem

$$\int_{cv} [\phi \nabla^2 \phi + (\nabla \phi)^2] \, dx \, dy \, dz = \int_{cs} \phi (\nabla \phi) \cdot d\mathbf{A}$$

to obtain the fluid kinetic energy

$$\text{KE} = \frac{\rho}{2g_0} \int_{cs} \phi (\nabla \phi) \cdot d\mathbf{A} \tag{9.77}$$

If KE is expressed as $M_v \mathbf{V}_s \cdot \mathbf{V}_s / 2g_0$ where M_v is the virtual mass moving at velocity $\mathbf{V}_s$, we obtain

$$M_v = \frac{1}{V_s^2} \int_{cs} \phi (\nabla \phi) \cdot d\mathbf{A} \tag{9.78}$$

Equation (9.78) implies that M_v depends on the velocity amplitude V_s. However, ϕ generally is proportional to V_s and cancels from Eq. (9.78) so that M_v is independent of the moving object velocity for motion in a given direction. It should be recognized that the direction of $\mathbf{V}_s$ is included in the velocity potential ϕ. Thus, the virtual mass of a submerged object depends on its direction of motion. Exceptions include a sphere moving in any direction, and a circular cylinder moving in the x, y plane.

Tables 9.2 and 9.3 give the virtual mass for submerged two- and three-dimensional objects.

9.19 Waterhammer Formulation

When a disturbance time is of the order of L/C, propagation effects dominate and an appropriate response time is $\Delta t = L/C$. Propagation disturbances of Eqs. (6.78)–(6.80) were incorporated into the model coefficients of Table 6.3 for liquid flows. It was found that

$$\pi_1 = \pi_2 = \pi_3 = \pi_6 = \frac{\Delta V}{C}, \qquad \pi_2^0 = \frac{V_0}{C}$$

and π_4, π_5, π_7, and π_8 were approximately zero. All surviving terms in the mass conservation and momentum formulations, Eqs. (6.69) and (6.70), contain the term $\Delta V/C$, which can be cancelled to yield the multidimensional waterhammer equations

$$\frac{\partial P}{\partial t} + \frac{\rho C^2}{g_0} \nabla \cdot \mathbf{V} = 0 \tag{9.79}$$

$$\frac{\partial \mathbf{V}}{\partial t} + \frac{g_0}{\rho} \nabla P = 0 \tag{9.80}$$

If velocity is eliminated, we obtain

$$\frac{\partial^2 P}{\partial t^2} - C^2 \nabla^2 P = 0 \tag{9.81}$$

and if pressure is eliminated

$$\frac{\partial^2 \mathbf{V}}{\partial t^2} - C^2 \nabla(\nabla \cdot \mathbf{V}) = 0 \tag{9.82}$$

The velocity potential of Eq. (9.9) for inviscid, irrotational flow can be employed in Eqs. (9.82) and (9.80) to obtain

$$\frac{\partial^2 \phi}{\partial t^2} - C^2 \nabla^2 \phi = 0 \tag{9.83}$$

and

$$P = -\frac{\rho}{g_0} \frac{\partial \phi}{\partial t} \tag{9.84}$$

The ∇^2 operator is given in Eqs. (9.16)–(9.18) for rectangular, cylindrical polar, and spherical coordinates. Additional operations or terms appearing in the waterhammer equations are (see Fig. 9.4)

Rectangular

$$\nabla \cdot \mathbf{V} = \frac{\partial u}{\partial x} + \frac{\partial v}{\partial y} + \frac{\partial w}{\partial z}$$

$$\nabla \phi = \frac{\partial \phi}{\partial x} \mathbf{n}_x + \frac{\partial \phi}{\partial y} \mathbf{n}_y + \frac{\partial \phi}{\partial z} \mathbf{n}_z \tag{9.85}$$

Cylindrical polar

$$\nabla \cdot \mathbf{V} = \frac{1}{r} \frac{\partial}{\partial r} (rV_r) + \frac{1}{r} \frac{\partial V_\theta}{\partial \theta} + \frac{\partial V_z}{\partial z}$$

$$\nabla \phi = \frac{\partial \phi}{\partial r} \mathbf{n}_r + \frac{1}{r} \frac{\partial \phi}{\partial \theta} \mathbf{n}_\theta + \frac{\partial \phi}{\partial z} \mathbf{n}_z \tag{9.86}$$

Spherical

$$\nabla \cdot \mathbf{V} = \frac{1}{r^2} \frac{\partial}{\partial r} (r^2 V_r) + \frac{1}{r \sin \theta} \frac{\partial}{\partial \theta} (V_\theta \sin \theta) + \frac{1}{r \sin \theta} \frac{\partial V_\alpha}{\partial \alpha}$$

$$\nabla \phi = \frac{\partial \phi}{\partial r} \mathbf{n}_r + \frac{1}{r} \frac{\partial \phi}{\partial \theta} \mathbf{n}_\theta + \frac{1}{r \sin \theta} \frac{\partial \phi}{\partial \alpha} \mathbf{n}_\alpha \tag{9.87}$$

Equations (9.79) through (9.84) are used with appropriate boundary and initial conditions to predict propagative disturbances in regions of liquid, or slightly compressed gases, provided that $\Delta V/C \leq 0.2$. Since the equations are linear, superposition of solutions can be used with Eq. (9.89) to obtain an abundance of flow patterns.

9.20 Spherical Waterhammer Propagation

The disturbance field from a point source is spherically symmetric, for which Eqs. (9.81) and (9.83) take the form

$$\frac{\partial^2 \psi}{\partial t^2} - \frac{C^2}{r^2} \frac{\partial}{\partial r} \left(r^2 \frac{\partial \psi}{\partial r} \right) = 0, \qquad \psi = P, \phi \tag{9.88}$$

When the fluid is initially at rest,

$$V_r(r, 0) = \frac{\partial \phi}{\partial r} = 0, \qquad P(r, 0) = P_\infty \tag{9.89}$$

If the source is generating fluid at a volume rate $\dot{V}(t)$, boundary conditions are

$$4 \pi r^2 V_r(r, t)|_{r \to 0} = \dot{V}(t), \qquad V_r(\infty, t) = 0 \tag{9.90}$$

Solutions for outgoing propagation are given by

$$\phi(r, t) = \frac{1}{r} G\left(t - \frac{r}{C} \right) H_s \left(t - \frac{r}{C} \right)$$

$$P(r, t) = \frac{1}{r} F\left(t - \frac{r}{C} \right) H_s \left(t - \frac{r}{C} \right)$$

where F and G are arbitrary, twice-differentiable functions and H_s is the Heaviside step function. Substitution into Eqs. (9.79) and (9.80) with $\mathbf{V} = \nabla \phi$ yields

$$F' + \frac{\rho}{g_0} G'' = 0$$

$$G'' + \frac{C}{r} G' + \frac{g_0}{\rho} \left(F' + \frac{C}{r} F \right) = 0$$

where primes indicate differentiation with respect to the argument $t - r/C$. The source boundary condition of Eq. (9.90) is employed with Eq. (9.89) to obtain

$$G\left(t - \frac{r}{C} \right) = -\frac{\dot{V}(t - r/C)}{4\pi}$$

$$F\left(t - \frac{r}{C} \right) = \frac{\rho}{4\pi g_0} \dot{V}'\left(t - \frac{r}{C} \right)$$

It follows that the outgoing waterhammer solutions for an arbitrary point source are

Point Source, Arbitrary Volume Rate $\dot{V}(t)$

$$\phi(r, t) = -\frac{1}{4\pi r} \dot{V}\left(t - \frac{r}{C} \right) H_s\left(t - \frac{r}{C} \right)$$

$$V_r(r, t) = \frac{1}{4\pi r} \left[\frac{1}{C} \dot{V}'\left(t - \frac{r}{C} \right) + \frac{1}{r} \dot{V}\left(t - \frac{r}{C} \right) \right] H_s\left(t - \frac{r}{C} \right) \quad (9.91)$$

$$P(r, t) - P_\infty = \frac{1}{4\pi r} \frac{\rho}{g_0} \dot{V}'\left(t - \frac{r}{C} \right) H_s\left(t - \frac{r}{C} \right)$$

Additional solutions for incoming propagation would apply only to rare cases where a point source was at the exact center of a spherical boundary. The outgoing solutions are sufficient for most problems.

A ramp volume rate from a point source gives the velocity and pressure fields

Point Source, Ramp Volume Rate $\dot{V}(t) = Kt$

$$\frac{V_r(r, t)}{C} = \frac{K}{4\pi rC^2} \left(\frac{Ct}{r} \right) H\left(\frac{Ct}{r} - 1 \right) \quad (9.92)$$

$$\frac{P(r, t) - P_\infty}{\rho C^2/g_0} = \frac{K}{4\pi rC^2} H\left(\frac{Ct}{r} - 1 \right)$$

which are plotted in Fig. 9.30. It is seen that a pressure front decreases as it propagates, leaving a constant-pressure profile behind.

EXAMPLE 9.12: A CONSTANT ADVANCING PRESSURE FRONT
Determine the required volume rate $\dot{V}(t)$ from a point source which would result in

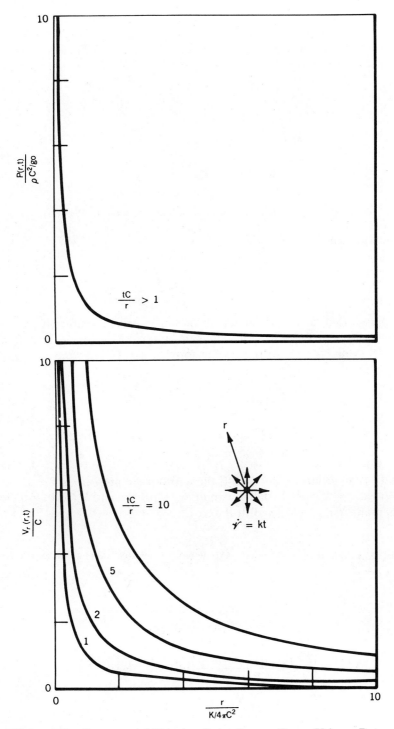

Figure 9.30 Pressure and Velocity, Point Source, Ramp Volume Rate

the propagation of a constant pressure front of magnitude $P - P_\infty = P_0$. Also determine the nature of resulting pressure profiles behind the front.

The pressure formulation of Eq. (9.91) is written with $r = Ct$ to follow the front. The integral of $\dot{\mathcal{V}}'$ gives

$$\dot{\mathcal{V}} = \frac{2\pi C g_0 P_0}{\rho}\, t^2$$

with corresponding pressure profiles

$$P(r, t) - P_\infty = P_0\, \frac{Ct}{r}\, H_s\!\left(t - \frac{r}{C}\right)$$

That is, a volume rate that increases as t^2 will cause a constant pressure front to propagate outward, with trailing pressure profiles which vary as $1/r$, but increase linearly with time.

Suppose that a waterhammer disturbance occurs uniformly over an imaginary spherical surface of radius R. Expressions for V_r and P with spherical symmetry are written in the form of Eq. (9.91) as

$$V_r(r, t) = -\frac{1}{r}\left(\frac{1}{C}\frac{dZ}{dt} + \frac{1}{r}Z\right)$$

$$P(r, t) = -\frac{\rho}{g_0}\frac{1}{r}\frac{dZ}{dt}, \qquad Z = Z\!\left(t - \frac{r - R}{C}\right) \tag{9.93}$$

where Z is an arbitrary function. If the undisturbed pressure is P_∞, solutions for $Z(t)$ at $r = R$ with the initial conditions $Z(0) = 0$ and $V_r(R, 0) = 0$ yield, for step and ramp velocity disturbances,

SPHERICALLY SYMMETRIC VELOCITY AND PRESSURE FIELDS, DISTURBANCES ON FIXED RADIUS $r = R$

Step

$$V_r(R, t) = V_0$$

$$\frac{V_r(r, t)}{V_0} = \left[\frac{R}{r}\left(1 - \frac{R}{r}\right)e^{-(Ct/r - r/R + 1)} + \left(\frac{R}{r}\right)^2\right]H_s\!\left(t - \frac{r - R}{C}\right)$$

$$\frac{P(r, t) - P_\infty}{\rho V_0 C/g_0} = \frac{R}{r}\, e^{-(Ct/r - r/R + 1)}H_s\!\left(t - \frac{r - R}{C}\right) \tag{9.94}$$

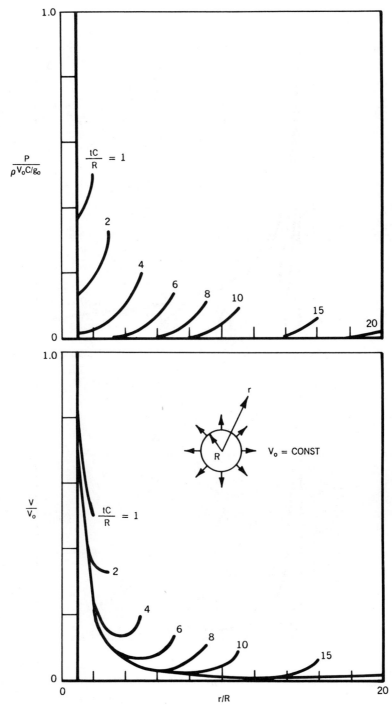

Figure 9.31 Pressure and Velocity, Point Source, Step Velocity, at $r = R$

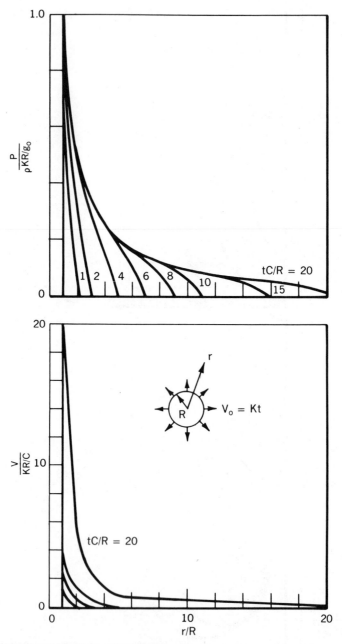

Figure 9.32 Pressure and Velocity, Point Source, Ramp Velocity at $r = R$

Ramp

$V_r(R, t) = Kt$

$$\frac{V_r(r, t)}{RK/C} = \frac{R}{r}\left[\left(\frac{R}{r} - 1\right)e^{-(Ct/r - r/R + 1)} + \frac{Ct}{r}\right]H_s\left(t - \frac{r - R}{C}\right)$$

$$\frac{P(r, t) - P_\infty}{\rho KR/g_0} = \frac{R}{r}\left[1 - e^{-(Ct/r - r/R + 1)}\right]H_s\left(t - \frac{r - R}{C}\right)$$

(9.95)

The velocity and pressure fields are graphed in Figs. 9.31 and 9.32

Spherical propagation from a fixed radius can be employed in large reservoirs where a waterhammer disturbance arrives from an attached pipe. Figure 9.33 shows three common geometries where a pipe enters a re-

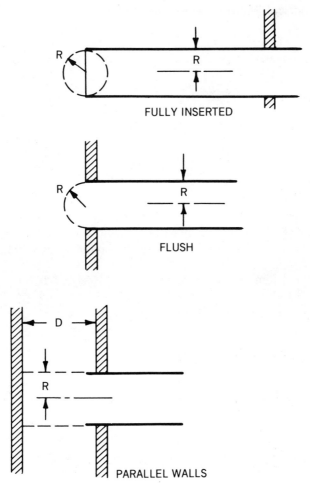

Figure 9.33 Plane Wave Transition Geometries

servoir. If a plane wave velocity disturbance $V_{\text{pipe}}(t)$ occurs in the pipe end where it enters the reservoir, it will be approximately spread over a hemisphere for the flush-mounted case, a sphere for the fully inserted case, and a cylinder for a fluid region between parallel walls. It follows from continuity that for each of these cases the disturbance velocity $V(R, t)$ is

$$V(R, t) \approx \begin{cases} \dfrac{1}{4} V_{\text{pipe}}(t) & \text{fully inserted pipe} \\[2mm] \dfrac{1}{2} V_{\text{pipe}}(t) & \text{flush-mounted} \\[2mm] \dfrac{1}{2} \dfrac{R}{D} V_{\text{pipe}}(t) & \text{parallel walls} \end{cases} \qquad (9.96)$$

9.21 Cylindrical Waterhammer Propagation

A uniform disturbance between parallel plane walls tends to propagate with cylindrical symmetry. Pressure and velocity responses are described by Eqs. (9.79) and (9.80) in the forms

$$\frac{\partial P}{\partial t} + \frac{\rho C^2}{g_0} \frac{1}{r} \frac{\partial}{\partial r} (rV_r) = 0 \qquad (9.97)$$

and

$$\frac{\partial V_r}{\partial t} + \frac{g_0}{\rho} \frac{\partial P}{\partial r} = 0 \qquad (9.98)$$

If the fluid starts from rest, the initial conditions of Eq. (9.89) apply. If the parallel walls are a distance D apart, a line source extending between them with volume rate $\dot{\mathcal{V}}(t)$ is consistent with the boundary conditions

$$\lim_{r \to 0} 2\pi r D V_r(r, t) = \dot{\mathcal{V}}(t), \qquad V_r(\infty, t) = 0 \qquad (9.99)$$

General solutions for V_r and P are given by

LINE SOURCE, ARBITRARY VOLUME RATE $\dot{\mathcal{V}}(t)$

$$V_r(r, t) = -\frac{1}{2\pi D C} \int_0^\infty \frac{d}{dt}\left[\dot{\mathcal{V}}\left(t - \frac{r}{C}\cosh\eta\right)\right] \cosh\eta\, H_s\left(\frac{tC}{r} - 1\right) d\eta \qquad (9.100)$$

$$P(r, t) - P_\infty = \frac{\rho}{g_0} \frac{1}{2\pi D} \int_0^\infty \frac{d}{dt}\left[\dot{\mathcal{V}}\left(t - \frac{r}{C}\cosh\eta\right)\right] H_s\left(\frac{tC}{r} - 1\right) d\eta \qquad (9.101)$$

A ramp volume rate yields the solutions

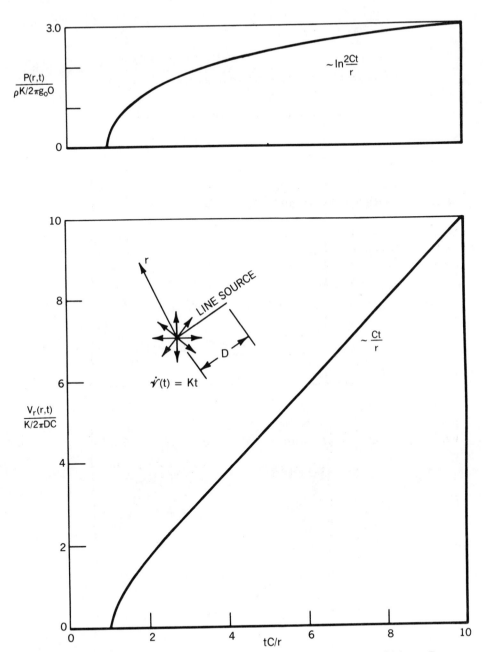

Figure 9.34 Pressure and Velocity, Cylindrical Geometry, Ramp Volume Rate

LINE SOURCE, RAMP VOLUME RATE, $\dot{V}(t) = Kt$

$$\frac{V_r(r, t)}{K/2\pi DC} = \left[\frac{1}{Ct/r - \sqrt{(Ct/r)^2 - 1}} - \frac{Ct}{r} \right] H_s\left(\frac{Ct}{r} - 1 \right) \quad (9.102a)$$

$$\frac{P(r, t) - P_\infty}{\rho K/g_0 2\pi D} = \ln \left[\frac{1}{Ct/r - \sqrt{(Ct/r)^2 - 1}} \right] H_s\left(\frac{Ct}{r} - 1 \right) \quad (9.102b)$$

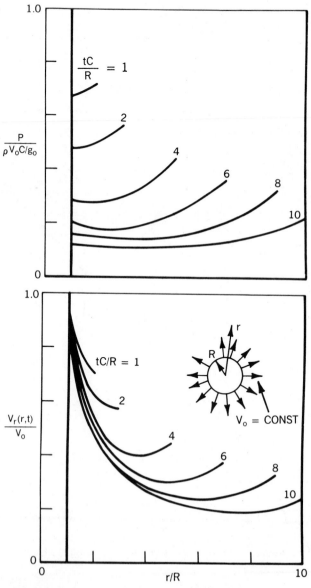

Figure 9.35 Pressure and Velocity, Cylindrical, Velocity Step at $r = R$

which are shown in Fig. 9.34. Cylindrical propagation solutions for step and ramp volume rates at a fixed boundary $r = R$ were obtained from Eqs. (9.97) and (9.98) by the method of characteristics (see Section 7.21). Results are given in Figs. 9.35 and 9.36.

EXAMPLE 9.13: SPHERICAL AND CYLINDRICAL ATTENUATION

Step velocity disturbances to V_0 occur at a radius R. Compare the resulting pressure values at the surface of origination $r/R = 1.0$ and also at $r/R = 5.0$ for cases of

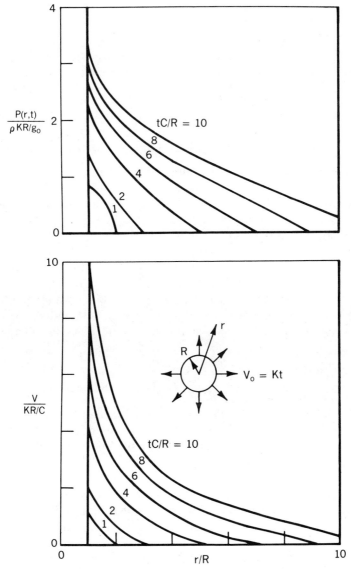

Figure 9.36   Pressure and Velocity, Cylindrical, Ramp Velocity at $r = R$

TABLE 9.4 Pressure $P^* = P/(\rho V_0 C/g_0)$ from Step Velocity Disturbance

Time $\dfrac{tC}{R}$	P^* at $r/R = 1.0$		P^* at $r/R = 5.0$	
	Spherical	Cylindrical	Spherical	Cylindrical
0	1.00	1.00	0	0
1	0.38	0.68	0	0
2	0.14	0.48	0	0
4	0.02	0.28	0.20	0.43
6	0	0.20	0.04	0.24
8	0	0.15	0	0.15
10	0	0.10	0	0.10

spherical and cylindrical propagation. Let P_∞ be arbitrarily set equal to zero.

Figures 9.31 and 9.35 were employed to obtain the normalized pressure values in Table 9.4. It is seen that cylindrical pressure propagation does not attenuate with distance as strongly as spherical propagation.

9.22 Superposition of Waterhammer Flow Patterns

Velocity potentials for incompressible bulk flow patterns satisfy the linear equation $\nabla^2 \phi = 0$. Therefore, Sections 9.10–9.12 employed the superposition of more than one ϕ_n to obtain various flow patterns. The velocity and pressure fields then were obtained from $\mathbf{V} = \nabla \phi$ and Bernoulli's equation.

The velocity potential in waterhammer flows satisfies Eq. (9.83), which also is linear and permits superposition of ϕ. However, pressure and velocity fields also can be superimposed directly in waterhammer flows because Eqs. (9.81) and (9.82) are linear. That is, if V_n and P_n are solutions to the waterhammer equations, additional solutions can be obtained from

$$V = \sum_n V_n, \qquad P = \sum_n P_n, \qquad n = 1, 2, \ldots \qquad (9.103)$$

Furthermore, incompressible flows often require a large number of images to obtain solutions because they all act simultaneously. However, at any time during a waterhammer transient, only those images within a distance $r = Ct$ influence the solution. Waterhammer superposition for several useful geometries is discussed in the next sections.

9.23 Spherical Waterhammer Source Near a Flat Wall

Figure 9.37(a) shows a large region of fluid containing a spherical source of radius R on which a flow disturbance occurs. A flat wall is shown at distance

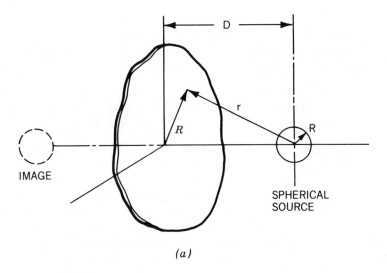

(a)

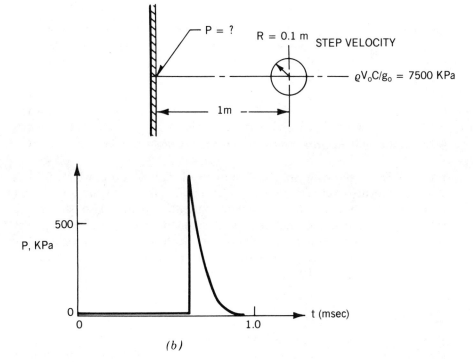

(b)

Figure 9.37 *(a)* Flat Plate by Superposition, *(b)* Pressure on Flat Plate

D. The velocity field is obtained by the superposition of an identical image as shown. If the velocity disturbance of Eq. (9.93) is employed, horizontal velocity components cancel on the plate. However, when the corresponding pressure fields are added according to Eq. (9.103), it is found that

$$P(\mathcal{R}, t) = 2P(r, t) = -\frac{2\rho}{g_0} \frac{1}{r} \frac{dZ}{dt} H_s\left(t - \frac{r-R}{C}\right)$$

$$Z = Z\left(t - \frac{r-R}{C}\right), \quad r = \sqrt{D^2 + \mathcal{R}^2} \tag{9.104}$$

EXAMPLE 9.14: TIME-DEPENDENT WALL PRESSURE
Water of density $1.0 \, \text{g/cc}$ ($62.4 \, \text{lbm/ft}^3$) and sonic speed $1500 \, \text{m/s}$ ($4920 \, \text{ft/s}$) flows in a pipe of radius $R = 0.1 \, \text{m}$ ($0.33 \, \text{ft}$). A valve creates a step velocity disturbance of $V_0 = 20 \, \text{m/s}$ ($65.6 \, \text{ft/s}$). If the pipe is fully inserted into a large reservoir from which water is discharging, determine the time-dependent pressure on a large flat wall at its closest point $D = 1.0 \, \text{m}$ ($3.28 \, \text{ft}$) from the pipe opening.

Equation (9.96) gives the velocity on an imaginary spherical surface of radius R as

$$V(R, t) = \tfrac{1}{4}V_0 = 5 \, \text{m/s} \ (16.4 \, \text{ft/s})$$

for which

$$\frac{\rho V_0 C}{g_0} = 7500 \, \text{kPa} \ (1102 \, \text{psia})$$

The graph of disturbance pressure P on the plate shown in Fig. 9.37(b) was obtained from Eq. (9.94) for superposition of the source with its image, according to Eq. (9.103). It is seen that although the velocity disturbance step is sustained, a wall pressure spike occurs only for a brief period.

9.24 Spherical Source and Parallel Flat Walls

Figure 9.38(a) shows a spherical surface of radius R on which flow is disturbed, located between two extended parallel flat plates. It is necessary to determine the pressure distibution on either plate. Also shown are spherical source images symmetrically spaced about each plate so that every time a wave reaches either wall, it meets an identical wave from a symmetric image. The corresponding pressure at location $\mathcal{R}$ on the left plate can be obtained from Eq. (9.103) as

$$P(\mathcal{R}, t) - P_\infty = 2\sum_n P(r_n, t), \quad r_n = \sqrt{x_n^2 + \mathcal{R}^2} \tag{9.105}$$

EXAMPLE 9.15: REFLECTIONS BETWEEN PARALLEL PLATES
Two extended parallel flat plates like those of Fig. 9.38(a) are separated by a distance $D = 2.0 \, \text{m}$ in liquid of density $1.0 \, \text{g/cc}$ and sonic speed $1500 \, \text{m/s}$. It has been

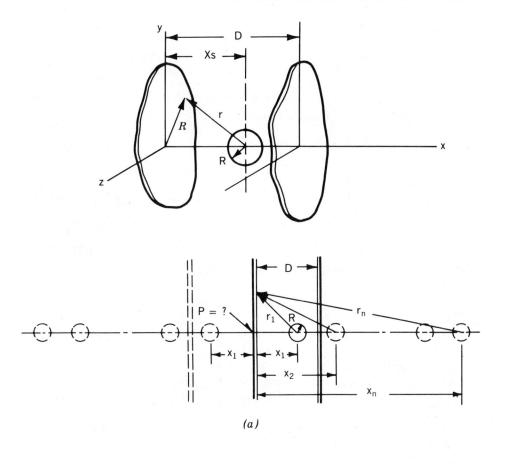

(a)

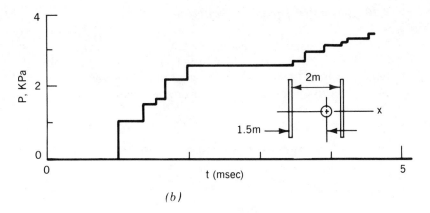

(b)

Figure 9.38 (*a*) Spherical Source between Parallel Plates by Superposition (*b*) Pressure on Parallel Flat Plates

found that flow in a fully inserted pipe produce a ramp velocity disturbance of the form $V(R, t) = Kt$ where $K = 20 \, \text{m/s}^2$, on a spherical surface of radius 0.25 m, which is a distance $x_s = 1.5 \, \text{m}$ from the left plate. Determine pressure on the left plate where it is intersected by the x axis for several reflections.

The pressure solution implied by Eq. (9.105) is obtained from Eq. (9.95) for a ramp. The calculated pressure on the specified surface is $\rho KR/g_0 = 5000 \, \text{Pa}$, which gave the solution for several reflections shown in Fig. 9.38(b).

Waterhammer superposition of sources for two- and three-sided corners is accomplished by the same pattern shown for incompressible bulk flow in Fig. 9.13.

9.25 Superposition of Cylindrical Sources

Suppose that fluid in an annular region between two closely spaced cylinders is disturbed locally. The resulting pressure field is approached by cylindrical propagation from an imaginary tubular surface extending between and normal to the inner and outer walls of Fig. 9.39. Pressure in the unfolded annular region can be obtained from the source/sink image pattern shown, employing sources (a) and (b) and horizontally repeating images. Since the pressure is constant at the top and bottom ends of the annular region, symmetric image sinks (negative sources) are required as shown. Equation (9.103) again is employed to determine pressure from all applicable sources and sinks at any instant.

9.26 Pressure Force on Structural Surfaces

The total pressure force acting on a surface is obtained from the integral of $(P - P_\infty) \, dA$. Pressure profiles from spherical step and ramp velocity disturbances of Eqs. (9.94) and (9.95) were integrated over an infinite flat wall to obtain the following forces where time t is zero at the time of disturbance arrival:

FORCES ON AN INFINITE FLAT WALL; FLUID DISTURBANCE SPECIFIED ON SPHERE $r = R$
WITH CENTER DISTANCE D FROM WALL (FIG. **9.40**)

Step Velocity $V(R, t) = V_0$

$$\frac{F}{(\rho V_0 C/g_0)\pi R^2} = 4[1 - e^{-ct/R}] \qquad (9.106)$$

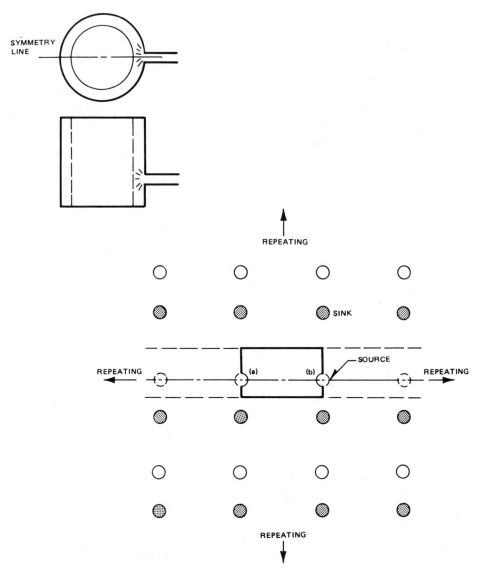

Figure 9.39   Superposition for Finite Cylinders

Ramp Velocity, $V(R, t) = Kt$

$$\frac{F}{(\rho KR/g_0)\pi R^2} = 4\,\frac{Ct}{R} \tag{9.107}$$

These forces are graphed in Fig. 9.40. It is interesting that the asymptotic total force on an infinite wall, caused by a step velocity, reaches a value of

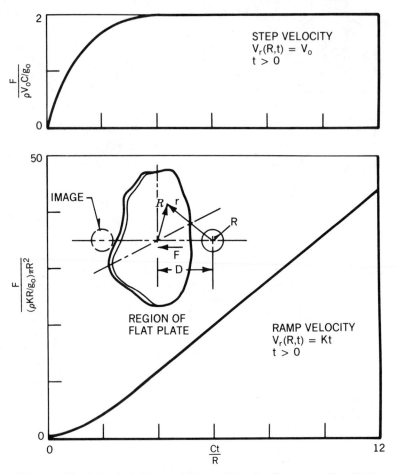

Figure 9.40 Spherical Step and Ramp Velocity Forces on Flat Wall

twice the waterhammer pressure $\rho V_0 C/g_0$, acting over area πR^2 regardless of distance D! Disturbance forces also were obtained for the following:

Pressure Step from Zero to P_0 on $r = R$

$$\frac{F}{2\pi R^2 P_0} = \frac{Ct}{R} \tag{9.108}$$

Velocity Impulse $V_0 \tau = \mathcal{V}/4\pi R^2$ on $r = R$

$$\frac{F}{4\pi R^2(\rho C V_0/g_0)(C\tau/R)} = e^{-Ct/R} - 1 \tag{9.109}$$

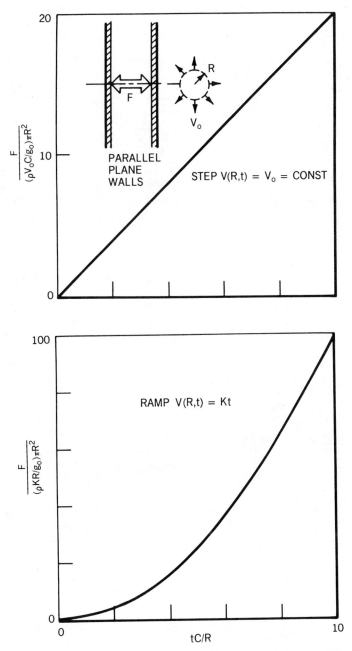

Figure 9.41 Total Force between Parallel Planes, Cylindrical Step and Ramp at R

Oscillating Pressure $P_0 \sin \omega t$ on $r = R$

$$\frac{F}{2\pi R^2 P_0 (C/\omega R)} = 1 - \cos \omega t \qquad (9.110)$$

Sudden Bubble Collapse, Eqs. (9.33) *and* (9.36)

$$\frac{F}{(2.4)4\pi R^2 [(\rho C/g_0)\sqrt{g_0 P_\infty/\rho}]} \approx t^{*5.8}[1 - H_s(t^* - 0.915)]$$

$$t^* = \frac{t}{R}\sqrt{\frac{g_0 P_\infty}{\rho}} \qquad (9.111)$$

Note that the positive velocity impulse of Eq. (9.109) causes a negative force on the wall! The opposite is true for a negative velocity impulse, as in the case of an instantaneous bubble collapse, which causes a positive outward force on the wall.

EXAMPLE 9.16: BUBBLE COLLAPSE FORCE
A steam bubble of radius $R = 2.5$ cm (1.0 in) is discharged into water and undergoes rapid collapse. Determine the maximum force exerted on a nearby flat wall of very large extent, whose closest point to the bubble is 1.0 m (3.28 ft). Take $\rho = 1000$ kg/m³ (62.4 lbm/ft³), $C = 1500$ m/s (4920 ft/s), and $P = 101$ kPa (14.7 psia).
Equation (9.111) shows that the maximum force occurs when $t^* = 0.915$. Therefore,

$$\frac{\rho C}{g_0} = 1500 \text{ kN-s/m}^3, \qquad \sqrt{\frac{g_0 P_\infty}{\rho}} = 10 \text{ m/s (32.8 ft/s)}$$

and $F = 169$ kN (37856 lbf). Note that for a large wall the bubble distance does not influence the force magnitude.

The cylindrical pressure fields of Figs. 9.34 and 9.35 were integrated numerically to determine the total force on either of the two parallel flat plates. Results are given in Fig. 9.41 for step and ramp disturbances.

References

9.1   Marble, W. J., et al., "Preliminary Report on the Fission Product Scrubbing Program," GE Report No. NEDO-30017, 1983.

9.2   Moody, F. J., W. C. Reynolds, "Liquid Surface Motion Induced by Acceleration and External Pressure," *ASME JBE*, **94D**, No. 3 (1972).

9.3   Taylor, G. I., "The Instability of Liquid Surfaces When Accelerated in a Direction Perpendicular to Their Planes," *Proc. Roy. Soc. A*, (1950).

9.4 Moody, F. J., L. C. Chow, and L. E. Lasher, "Analytical Model for Estimating Drag Forces on Rigid Submerged Structures Caused by LOCA and Safety Relief Valve Ramshead Air Discharges," GE Report NEDO-21479, September, 1977.

9.5 Florsheutz, L. W., and B. T. Chao, "On the Mechanisms of Vapor Bubble Collapse," *ASME JHT*, **87** No. 2 (1965).

9.6 McCauley, E. W., and J. H. Pitts, "Trial Air Test Results for the 1/5 Scale Mk I BWR Pressure Suppression Experiment," Lawrence Livermore Laboratory, UCRL-52371, 1977.

9.7 Streeter, V. L., *Fluid Dynamics*, McGraw-Hill, New York, 1948.

9.8 Lamb, H., *Hydrodynamics*, Dover, New York.

9.9 "The Dynamics of Liquids in Moving Containers," NASA SP-106, H. N. Abramson, Ed., 1966.

9.10 Hirt, C. W., et al. "Multidimensional Analysis for Pressure Suppression Systems," Los Alamos Scientific Laboratory Report, U. S. Department of Energy, Contract W-7405-ENG, 36, April 1979.

9.11 Moody, F. J., "Vessel Internal and External Pressure Loads Caused by Flow Disturbances," *Dynamics of Fluid-Structure Systems in the Energy Industry*, ASME Special Publication PVP-39, M. K. Au-Yang, S. Brown, eds. 1979.

9.12 McCready, J. L., et. al., "Steam Vent Clearing Phenomena and Structural Response of the BWR Torus (Mark I Containment)," General Electric Report NEDO-10859, April, 1973.

9.13 Cole, R. H., *Underwater Explosions*, Princeton University Press, Princeton, N.J., 1948.

9.14 Carre, B. A., "The Determination of the Optimum Accelerating Factor for Successive Over-Relaxation," *Computer J.*, **4**(1) (1961), 73.

9.15 Chan, R. K. C., and R. L. Street, "Computer Studies of Finite Amplitude Water Waves," Technical Report No. 104, Dept. of Civil Engineering, Stanford University, 1969.

9.16 Street, R. L., and F. E. Camfield, "Observations and Experiments on Solitary Wave Deformation," *Proc. 10th Conference on Coastal Engineering*, Tokyo, Japan, Published by ASCE, 1966.

9.17 Welch, J. E., et al., "The MAC Method—A Computing Technique for Solving Viscous, Incompressible, Transient Fluid Flow Problems Involving Free Surfaces," Los Alamos Scientific Laboratory Report No. 3425, 1966.

9.18 Patton, K. T., "Tables of Hydrodynamic Mass Factors for Translational Motion," ASME Paper 65-WA/UNT-2.

Problems

9.1 Inviscid liquid fills the space between two long, concentric cylinders of inner radius R_i and outer radius R_0. The cylinders are aligned with the z axis. Consider the case of motion in the x direction where the inner is moving with velocity $u_i(t)$ and the outer with velocity $u_0(t)$. Show that the velocity potential for fluid between the two cylinders is given by

$$\phi = \left(Ar + \frac{B}{r} \right) \cos \theta$$

$$A = \frac{R_0^2 u_0 - R_i^2 u_i}{R_0^2 - R_i^2}, \qquad B = \frac{(u_0 - u_i) R_i^2 R_0^2}{R_0^2 - R_i^2}$$

9.2   Consider a vessel of liquid with density ρ, pressurized to P_0. If an attached pipe should rupture suddenly, the liquid discharge would undergo a period of acceleration to its quasi-steady flow rate. Acceleration of liquid in the pipe controls the unsteady discharge as discussed in Chapter 7. However, if the pipe is short, acceleration of vessel liquid into the pipe can dominate the discharge response. An estimate of the vessel liquid flow inertial response can be obtained if the discharge is simulated by a point sink on the vessel wall and an imaginary discharge pressure acting on a hemisphere of pipe radius R, which covers the short pipe entrance as shown in Fig. P9.2. Employ

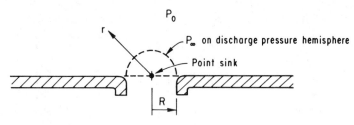

Figure P9.2

the velocity potential for a point sink [the sign of $\dot{V}(t)$ is changed to $+$ in Eq. (9.13)] and Bernoulli's equation without gravity to show that the unsteady discharge volume flow rate is described by

$$\dot{V}(t) = 2\pi R^2 \sqrt{\frac{2g_0}{\rho} (P_0 - P_\infty)} \tanh \frac{t}{2\tau}$$

for which the response time is

$$\tau = \frac{R}{\sqrt{\dfrac{2g_0}{\rho} (P_0 - P_\infty)}}$$

9.3   The velocity potential in bulk flow satisfies Laplace's equation, $\nabla^2 \phi = 0$. Show with the help of Bernoulli's equation (9.25) that at times when the velocity everywhere in a flow field is zero, the pressure P also satisfies Laplace's equation.

9.4 An idealized gas bubble in the shape of a sphere rises at velocity U_b in liquid due to buoyancy and simultaneously expands as its pressure decreases. Show that the velocity potential for this case is given by

$$\phi = \frac{U_b}{2}\frac{R^3}{r^2}\cos\theta - \frac{R^2}{r}\dot{R}$$

(Hint: Superimpose solutions for a translating rigid sphere, and a stationary expanding sphere.)

9.5 Use the procedure of Section 9.7 to show that the small amplitude bubble oscillation frequency of an *isothermal* gas bubble is given by

$$f = \frac{1}{2\pi R_\infty}\sqrt{\frac{3g_0 P_\infty}{\rho}}$$

9.6 A depth charge explodes in water where the pressure is 20 atm (294 psia), instantly creating a hot gas bubble of 1.0 m (3.28 ft) initial radius at 200 atm (2940 psia). Assume bulk water motion.

(a) Use Fig. 9.8 to show that the oscillation frequency is about 20 Hz.

(b) Show that the maximum pressure increase at a distance of 20 m (65.6 ft) from the bubble is 9 atm (132 psia).

9.7 A small gas bubble is submerged in liquid of density ρ where the local pressure is P_∞. Let the bubble be small enough that surface tension influences its oscillation frequency. The Rayleigh bubble equation of Section 9.5 is based on zero surface tension so that P_b is pressure exerted by gas in the bubble on the liquid. However, if surface tension is included, the bubble gas pressure P becomes $P = P_b + 2\sigma/R$. The equilibrium gas pressure is $P_{eq} = P_\infty + 2\sigma/R_\infty$ where R_∞ is the equilibrium radius. Thus, for adiabatic bubble volume oscillations, $P = P_{eq}(R_\infty/R)^{3k}$.

(a) Show that the Rayleigh equation becomes

$$R\ddot{R} + \frac{3}{2}\dot{R}^2 + \frac{2g_0\sigma}{\rho}\frac{1}{R} = \frac{g_0}{\rho}\left[\left(P_\infty + \frac{2\sigma}{R_\infty}\right)\left(\frac{R_\infty}{R}\right)^{3k} - P_\infty\right]$$

(b) Let the initial conditions be given by

$$t = 0, \qquad R = R_\infty + \epsilon, \qquad \dot{R} = 0$$

and introduce a regular perturbation solution (section C.2 of Appendix C) $R = R_\infty + \epsilon R_1 + \cdots$ to obtain a differential equation for function R_1 in the form

$$\ddot{R}_1 + \left(\frac{3kg_0P_\infty}{\rho R_\infty^2}\right)\left[1 + \frac{2(3k-1)}{3k}\frac{\sigma}{R_\infty P_\infty}\right]R_1 = 0$$

Note that if surface tension is neglected at this point, the frequency reduces to that of Eq. (9.39). Does surface tension increase or decrease the oscillation frequency?

(c) Consider an air bubble in water at one atmosphere with $\sigma = 70$ dynes/cm (0.0048 lbf/ft). Show that the bubble radius below which surface tension has at least a 10 percent effect on frequency is only 1.06 μm (4.17×10^{-5} in).

9.8 The internal vapor pressure is suddenly reduced to zero in a bubble of 20 cm (7.87 in) radius, submerged in water at 1 atm (14.7 psia). At what radius during its collapse will the bubble wall velocity reach the water sonic speed of 1370 m/s (4500 ft/s)?

Answer:

0.66 cm (0.26 in)

9.9 A spherical bubble of radius R_i submerged in water at pressure P_i is suddenly pressurized to $P_\infty > P_i$.

(a) Employ Eq. (9.32) with $P_\infty - P_{sat}$ replaced by $P_\infty - P_i$ to show that the bubble wall acceleration is

$$\ddot{R} = -\frac{g_0}{\rho}(P_\infty - P_i)\frac{R_i^3}{R^4}$$

(b) Show that the initial acceleration of a 1.0 cm (0.39 in) radius bubble in water subjected to $P_\infty - P_i = 1.0$ atm (14.7 psi) is -1010 m/s^2 (-3313 ft/s^2).

(c) Show that the collapse time of the bubble described in part (b) is 0.91 ms.

9.10 A gas bubble whose center is at distance D from a rigid flat wall undergoes a sinusoidal oscillation. Pressure on the wall is desired. The radii of both the bubble and its image vary as $R = R_0 + \epsilon \sin \omega t$. Neglect gravity.

(a) Use Eq. (9.28) and superimpose the *velocity potentials* of the bubble and its image at $x = \pm D$ respectively. Show that pressure anywhere on the wall is given by

$$P - P_\infty = \frac{2\rho R}{g_0\sqrt{D^2 + \mathcal{R}^2}}\left\{R\ddot{R} + \dot{R}^2\left[2 - \frac{R^3\mathcal{R}^2}{(D^2 + \mathcal{R}^2)^{5/2}}\right]\right\}$$

(b) Equation (9.29) gives the pressure field from a bubble source. Assume that the *pressure fields* for a bubble and its image at $x = \pm D$ can be superimposed as in part (a) to show that on the simulated wall,

$$P - P_\infty = \frac{2\rho R}{g_0\sqrt{D^2 + \mathcal{R}^2}}\left\{R\ddot{R} + \dot{R}^2\left[2 - \frac{R^3}{2(D^2 + \mathcal{R}^2)^{3/2}}\right]\right\}$$

(c) Compare results of parts (a) and (b), and show that, at instants when $\dot{R} = 0$, the superposition of velocity potentials or pressure fields give identical results, which is consistent with the conclusion of Problem 9.3.

9.11 Estimate the stable droplet radius r formed when water condenses on an overhead structure and falls. (Hint: The weight of a hemispherical half-drop just overcomes its surface tension force.)

Answer:

$$r \cong \frac{1}{2^{1/3}}\sqrt{\frac{3g_0\sigma}{\rho_g}}$$

9.12 The droplet size formed when liquid is discharged from a spray nozzle depends on the liquid velocity relative to the surrounding gas and on the surface tension. Consider a model for estimating the droplet size based on a balance of the aerodynamic (lift) force, which tends to pull a spherical droplet apart, and the surface tension force, which tends to hold it together.

(a) Use Fig. 9.5(a) and Eqs. (9.19) and (9.25) to show that pressure on the surface of a sphere of radius R traveling in still gas of density ρ_g at spray velocity V_s is

$$P(r, \theta) = P_\infty - \frac{9}{4}\frac{\rho_g}{2g_0}V_s^2\sin^2\theta$$

(b) Let the sphere of Fig. 9.5(a) be treated as two hemispheres held together at the x, z plane by surface tension. Integrate the pressure of part (a) over the upper hemisphere to show that the total downward vertical force component is

$$F_y = \pi R^2\left(P_\infty - \frac{27}{16}\rho_g\frac{V_s^2}{2g_0}\right)$$

(Hint: Consider surface area increments $dA = (R\,d\theta)(R\sin\theta\,d\alpha)$ where α is the angle measured in the y, z plane of Fig. 9.5(a) from the negative z axis. The normal force on dA is $dF = P\,dA$, and the vertical component is $dF_y = dF\sin\theta\sin\alpha$.)

(c) If the liquid drop is stationary in gas at P_∞, its internal pressure is $P_\infty + 2\sigma/R$. If it is moving, the external force applied to a hemisphere by surrounding gas is F_y from part (b). However, if R is greater than the stable size, the droplet will divide. If the two hemispheres are being pulled apart, the internal pressure is relieved to P_∞, but surface tension still resists division. Show from a force balance on the hemisphere that the droplet will no longer divide when

$$\text{We} = \text{Weber Number} = \frac{\rho_g R V_s^2}{g_0 \sigma} \le \frac{64}{27}$$

Experimental results indicate that droplet breakup will continue until the Weber number is less than about 6. How can this model be improved?

9.13 A spherical bubble of maximum radius R_{max} is oscillating in a corner obtained by the intersection of two perpendicular rigid plane walls. Since this geometry can be simulated by the superposition of four equal bubbles, determine the closest distance D from both walls for a symmetrically located bubble, based on a criterion similar to that used in obtaining Eq. (9.46).

Answer:

$D/R \ge 4$

9.14 Wind blows across a deep pool of stagnant water. It is necessary to determine the maximum air velocity V_g that will not entrain droplets of any size (or cause unstable wave growth of any wavelength at the interface). Take air and water densities of $1.0\,\text{kg/m}^3$ ($0.062\,\text{lbm/ft}^3$) and $1000\,\text{kg/m}^3$ ($62.4\,\text{lbm/ft}^3$) respectively, and surface tension at the air/water interface of $0.07\,\text{N/m}$ ($0.0048\,\text{lbf/ft}$), and $a_y = g = 9.8\,\text{m/s}^2$ ($32.2\,\text{ft/s}^2$).

(a) Use the condition of Eq. (9.70a) to show that interface waves of wavelength λ will be stable when $(V_g - V_L)^2 \le f(\lambda)$ is satisfied. Determine $f(\lambda)$.

(b) Show that $(V_g - V_L)$ has a minimum value at

$$\lambda = 2\pi \sqrt{\frac{g_0 \sigma}{g(\rho_L - \rho_g)}} = 1.6\,\text{cm} \ (0.63\,\text{in})$$

which, for stagnant water, corresponds to

$$V_g \le \left(\frac{\rho_L + \rho_g}{\rho_L \rho_g} \right)^{1/2} \sqrt{2} [g g_0 \sigma(\rho_L - \rho_g)]^{1/4}$$

$$\le 7.23\,\text{m/s} \ (23.7\,\text{ft/s})$$

9.15 Wind blows over stagnant water, similar to Problem 9.14, except that the water depth is only $H_L = 1.0$ cm (0.39 in).

(a) Show that as the wind velocity V_g is increased, the first wavelength to go unstable is obtained from a solution to

$$\frac{F_\sigma + (2\pi H_L/\lambda)^2}{F_\sigma - (2\pi H_L/\lambda)^2}$$

$$= \frac{(\rho_L/\rho_g)\cosh^2(2\pi H_L/\lambda) + \sinh(2\pi H_L/\lambda)\cosh(2\pi H_L/\lambda)}{2\pi H_L/\lambda}$$

where

$$F_\sigma = \frac{gH_L^2(\rho_L - \rho_g)}{g_0\sigma}$$

(b) Show that the first wavelength to go unstable is 1.7 cm, which would be excited at a gas velocity of 7.2 m/s.

9.16 Mercury is introduced at the top of two vertical, parallel plates separated by a distance D as shown in Fig. P9.16. At what spacing D will the first antisymmetric interface mode become unstable for $\rho_g \ll \rho_L$, $\rho_L = 13550$ kg/m^3, and $\sigma = 480$ dynes/cm?

Answer:

3 mm

Figure P9.16

9.17 A two-dimensional curve is expressed by the equation $y = f(x)$.

(a) Refer to Fig. P9.17 to show that the straight lines forming angle $d\theta$ are described by the equations

$$y - y_0 = -\frac{1}{f'(x_0)}(x - x_0)$$

$$y - [y_0 + f'(x_0)\,dx] = -\frac{1}{f'(x_0 + dx)}(x - x_0 - dx)$$

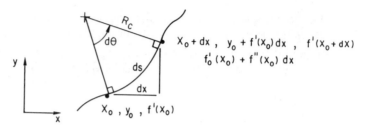

Figure P9.17

(b) Both of these lines intersect when the y values are equal. Show that this condition yields

$$x - x_0 = -\frac{f'(x_0)}{f''(x_0)}[1 + f'(x_0)^2]$$

(c) Finally show that the radius of curvature is

$$R_c = \sqrt{(x - x_0)^2 + (y - y_0)^2} = \frac{[1 + f'(x_0)^2]^{3/2}}{f''(x_0)}$$

9.18 Use the results of Problem 9.17 to show that the pressure difference across a two-dimensional concave-upward interface $y = f(x)$ with pressure P above and P_∞ below is given by

$$P - P_\infty = \frac{\sigma}{R_c} = \frac{\sigma f''(x)}{[1 + f'(x)^2]^{3/2}}$$

Let $y = 0$ represent a horizontal, flat interface, and $y = f(x) = \epsilon g(x)$ describe a distorted interface of small amplitude ϵ. If the pressure distortion caused by a small interface curvature is $P - P_\infty = \epsilon \Delta P$, verify that since ϵ is small, $\Delta P = \sigma g''(x)$.

9.19 Consider a cylindrical coordinate system with r, θ in the horizontal plane and y vertical. The total radius of curvature for the interface $y = \eta(r, \theta, t)$ between two fluids is given by [see Eq. (9.42) and Ref. 9.9]

$$\frac{1}{R_{c1}} + \frac{1}{R_{c2}} = \frac{1}{r}\frac{\partial}{\partial r}\left(\frac{r\,\partial\eta/\partial r}{\sqrt{1 + (\partial\eta/\partial r)^2 + (1/r^2)(\partial\eta/\partial\theta)^2}}\right)$$

$$+ \frac{1}{r^2}\frac{\partial}{\partial\theta}\left(\frac{\partial\eta/\partial\theta}{\sqrt{1 + (\partial\eta/\partial r)^2 + (1/r^2)(\partial\eta/\partial\theta)}}\right)$$

Let water enter a vertical tube at the top, similar to the case in Problem 9.16, except that the tube is tapered, its radius R increasing in the downward direction. The water column slowly lengthens as its interface radius increases until it no longer is stabilized by surface tension. Then small amplitude waves grow, and the water runs out of the tube. Ambient air pressure is P_∞.

(a) Verify that for small amplitude interface perturbations about the horizontal, that is, when η is $O(\epsilon)$ in magnitude, the linearized pressure difference across the interface is

$$\Delta P = \sigma \left(\frac{\partial^2 \eta}{\partial r^2} + \frac{1}{r}\frac{\partial \eta}{\partial r} + \frac{1}{r^2}\frac{\partial^2 \eta}{\partial \theta^2} \right)$$

(b) Show that for small amplitude velocity and interface motion, the linearized Bernoulli equation in water adjacent to the interface can be written as

$$\frac{\partial \phi}{\partial t} + g\eta = -\frac{g_0 \sigma}{\rho}\left(\frac{\partial^2 \eta}{\partial r^2} + \frac{1}{r}\frac{\partial \eta}{\partial r} + \frac{1}{r^2}\frac{\partial^2 \eta}{\partial \theta^2} \right)$$

on $y = 0$ where the arbitrary function $F(t)$ has been absorbed in ϕ.

(c) Let the interface between the air and water in cylindrical geometry be expressed as $y = \eta(r, \theta, t)$. Follow the procedure of Section 9.13 to obtain the kinematic condition,

$$\frac{\partial \phi}{\partial y} = \frac{\partial \eta}{\partial t} + \frac{\partial \eta}{\partial r}\frac{\partial \phi}{\partial r} + \frac{1}{r^2}\frac{\partial \eta}{\partial \theta}\frac{\partial \phi}{\partial \theta}$$

(d) Employ the conditions of zero radial velocity at the tube wall $r = R$, periodicity of the interface in θ, and stagnant water at large y to show that a solution of Eq. (9.17) is

$$\phi(r, \theta, y, t) = \sum_{m=1}^{\infty}\sum_{n=1}^{\infty} T_{n,m}(t)J_m(\lambda_{n,m}r)e^{-\lambda_{n,m}y}\cos m\theta$$

where $\lambda_{n,m}$ satisfies the condition,

$$\frac{dJ_m(\lambda_{n,m}r)}{dr}\bigg|_{r=R} = -(\lambda_{n,m}R)J_{m+1}(\lambda_{n,m}R) + mJ_m(\lambda_{n,m}R) = 0$$

(e) Employ ϕ of part (d) with the linearized dynamic and kinematic interface conditions of parts (b) and (c) to obtain the time-dependent function $T_{n,m}(t)$ in the form

$$\frac{d^2 T_{n,m}}{dt^2} + \left(\frac{g_0 \sigma}{\rho}\lambda_{n,m}^3 - g\lambda_{n,m} \right)T_{n,m} = 0$$

(f) Show that the interface can undergo small amplitude stable oscillations in the first symmetric mode, $m = 0$, when

$$\frac{\rho g R^2}{g_0 \sigma} < (3.832)^2$$

and that the first antisymmetric mode, $m = 1$, is stable for

$$\frac{\rho g R^2}{g_0 \sigma} < (1.841)^2$$

(g) Show that with $\rho = 1.0\,\text{g/cm}^3$ (62.4 lbm/ft^3), $\sigma = 70\,\text{dynes/cm}$ (0.0048 lbf/ft), the largest tube radius for a stable interface is 0.49 cm (0.19 in).

9.20 A swimmer "cannonballs" into a swimming pool, hitting the water surface at a perpendicular velocity of $V_\infty = 5\,\text{m/s}$ (16.4 ft/s). He is curled up with his knees on his chest and his arms locked about his ankles so that he forms a crude sphere of radius $R = 0.5\,\text{m}$ (1.64 ft).

(a) Estimate the decelerating force acting on him when he enters the water. Assume that his density is approximately equal to that of water. Hint: Estimate his sudden change in velocity from the assumption that his momentum before impact is equal to the momentum of himself plus that of a virtual water mass moving with him after impact. Then employ Newton's law in the form $F\,\Delta t = M\,\Delta V/g_0$ to the person, with Δt equal to the time it takes for a sphere to fully enter the water at velocity V_∞.

Answer:

$F = 2\pi R^2 V_\infty^2 \rho/9 g_0 = 4363\,\text{N}$ (981 lbf)

(b) Compare this result to a drag force calculation for a fully submerged sphere moving at the entrance velocity with a drag coefficient of 1.0.

Answer:

9817 N (2210 lbf)

(c) Do you think the analysis of part (a) is reasonable? If not, how might it be improved? What about the analysis of part (b)?

9.21 The natural frequency of a spring-mass system in a vacuum is $f = \omega/2\pi = \sqrt{Kg_0/M}/2\pi$ where K is the spring constant, $M = \rho \mathcal{V}$ is the mass, $\mathcal{V}$ is its volume, and ρ is its density. If the system is submerged in liquid of density ρ_L, determine the ratios of submerged frequency f_s to f for masses in the shape of (a) a sphere and (b) a circular cylinder.

Answer:

$$\frac{f_s}{f} = \begin{cases} \left[\left(1 + \frac{\pi\rho}{2\rho_L}\right)^{-1/2}\right] & \text{sphere} \\[2ex] \left(1 + \frac{\rho}{\rho_L}\right)^{-1/2} & \text{cylinder} \end{cases}$$

9.22 Show that Eq. (9.88) reduces to the form of Eq. (9.53) for one-dimensional propagation of a plane wave if one makes the substitution $\zeta = r\psi$.

9.23 Employ nondimensional variables $r^* = r/r_{ref}$, $t^* = t/t_{ref}$, and $P^* = P/P_{ref}$ in Eq. (9.88) for waterhammer pressure, and show that the resulting nondimensional group is Ct_{ref}/r_{ref}. This group implies that the problem could be formulated in terms of a similarity variable $\eta(r, t) = Ct/r$. Show that if $P(r, t) = P(\eta)$ is substituted, Eq. (9.88) is transformed to

$$(1 - \eta^2)\frac{d^2P}{d\eta^2} = 0$$

for which a solution is $P = a + b(ct/r)$ where a and b are constants. What kind of behavior is described by this solution?

9.24 A spherical boundary is submerged in a large expanse of liquid, and its radius grows in time according to $R(t)$, on which the pressure is described by $P_R(t)$. Show that the corresponding solution of Eq. (9.88) for outgoing pressure waves is

$$P(r, t) = \frac{1}{r} R\left(t - \frac{r}{C}\right) P_R\left(t - \frac{r}{C}\right), \qquad t > \frac{r}{C}$$

9.25 A submerged explosion creates an instant spherical volume of high pressure gas, which undergoes an adiabatic expansion. If the initial pressure and radius are P_i and R_i, employ Eq. (9.91) to show that the expanding radius is determined by the differential equation,

$$R\frac{d^2R}{dt^2} + 2\left(\frac{dR}{dt}\right)^2 = \frac{g_0}{\rho} P_\infty\left[\frac{P_i}{P_\infty}\left(\frac{R_i}{R}\right)^{3k} - 1\right]$$

where P_∞ is the distant pressure (neglecting gravity) and k is the adiabatic exponent. Compare this result with the Rayleigh equation (9.30). Which bubble do you think will expand faster?

9.26 Use Eq. (9.93) to show that if a constant pressure disturbance P_0 is to be maintained at a spherical boundary of fixed radius R, the required liquid flow velocity crossing the boundary is

$$V_r(R, t) = \frac{g_0 P_0}{R\rho} t$$

Also show that the propagating pressure is given by $P(r, t) = RP_0/r$ where $t > (r - R)/C$.

9.27 If the liquid velocity at $r = R$ on a spherical surface increases according to $V_r(R, t) = Kt^2$, where K is a constant, show that the function Z of Eq. (9.93) is

$$Z = -\frac{KR^4}{C^2}\left[\left(\frac{Ct}{R}\right)^2 - 2\left(\frac{Ct}{R}\right) + 2 - 2e^{-Ct/R}\right]$$

Also show that if terms of $O(1/C)$ are neglected, the pressure field is given by

$$P(r, t) = \frac{\rho}{g_0}\frac{2}{r} KR^2\left(t - \frac{r - R}{C}\right), \qquad t > \frac{r - R}{C}$$

9.28 Consider symmetric waterhammer propagation in cylindrical geometry.

(a) Eliminate V_r from Eqs. (9.97) and (9.98) to obtain the corresponding cylindrical waterhammer equation,

$$\frac{\partial^2 P}{\partial t^2} - C^2\left(\frac{\partial^2 P}{\partial r^2} + \frac{1}{r}\frac{\partial P}{\partial r}\right) = 0$$

(b) Nondimensionalize the equation of part (a) with $P^* = P/P_{ref}$, $t^* = t/t_{ref}$, and $r^* = r/r_{ref}$, and show that the surviving nondimensional group is Ct_{ref}/r_{ref}. This implies a possible similarity variable of the form $\eta = Ct/r$, as in Problem 9.23.

(c) Assume that $P = P(\eta)$ and show that the partial differential equation of part (a) reduces to the ordinary differential equation

$$(1 - \eta^2)\frac{d^2 P}{d\eta^2} - \eta\frac{dP}{d\eta} = 0$$

(d) Show that the pressure solution of Eq. (9.102b), written in terms of η, indeed satisfies the differential equation of part (c).

9.29 A step pressure disturbance $P_0 - P_\infty$, propagating from a spherical surface $r = R$, corresponds to

$$P(r, t) = \frac{R}{r}(P_0 - P_\infty), \qquad t > \frac{r - R}{C}$$

How far away must a rigid flat wall be located for pressure at the closest point to double, as is the case of plane waves?

Answer:

Only if the surface $r = R$ touches the wall will the pressure double. Otherwise, it is less.

9.30 Determine the wall pressure closest to the source of Problem 9.29 if its center is located a distance D (a) from a plane wall, and (b) from both perpendicular walls in a two-sided corner.

Answer:

$$P = \frac{2R}{D} P_0, \qquad\qquad \text{plane wall}$$

$$P = \frac{2R}{D} P_0\left(1 + \frac{1}{\sqrt{5}}\right), \qquad \text{two-sided corner}$$

9.31 A sudden pressure step $P_0 = 10\,\text{MPa}$ (1450 psia) occurs at a fixed spherical boundary $r = R = 0.1\,\text{m}$ (0.328 ft). Determine the maximum force exerted on an infinite plane wall a distance $D = 10\,\text{m}$ (32.8 ft) away.

Answer:

$314,000\,\text{N}$ $(70,587\,\text{lbf})$

9.32 A waitress carries a 3 cm radius glass of water with a depth of 6 cm. Show that the lowest sloshing frequency is about 3.8 Hz.

9.33 Consider a point source midway between parallel plates, as shown in Fig. 9.14. Use Eq. (9.52) to show that the velocity on the wall is given by

$$V_r = \frac{\dot{V}\mathscr{R}}{2\pi} \sum_{n=0}^{\infty} \left\{ \frac{1}{[4n^2 b^2 + \mathscr{R}^2]^{3/2}} + \frac{1}{[4(n+1)^2 b^2 + \mathscr{R}^2]^{3/2}} \right\}$$

9.34 Show for the oscillating bubble of Example 9.4 that if $\Delta R/R_\infty = 1/\sqrt{2}$, the maximum pressure increase on the wall corresponds to 59 Pa.

9.35 Consider a rectangular tank of length L in the x direction which contains inviscid liquid with undisturbed depth H. Suppose that horizontal ground acceleration from an earthquake occurs in x so that $\Omega = gy + a_x(t)x$ in Eq. (9.25). Follow the procedure described in Section 9.13 to show that for an initially stationary pool and two-dimensional liquid motion in the x, y plane, the linearized free surface disturbance elevation $\eta(x, t)$ is described by

$$\eta(x, t) = \frac{2}{\pi} \sum_{n=0}^{\infty} T_n(t) \frac{\tanh(n\pi H/L)}{n\omega_n} [1 - (-1)^n] \cos \frac{n\pi x}{L}$$

$$T_n(t) = \int_0^t \left[\int_0^t \dot{a}_x(t - \tau) \sin \omega_n \tau \, d\tau \right] dt$$

$$\omega_n = \left(\frac{gn\pi}{L} \tanh \frac{n\pi H}{L} \right)^{1/2}$$

9.36 Suppose that in Problem 9.35 the horizontal acceleration is given by $a_x(t) = A \sin \omega t$, where A is a constant. Show that

$$T_n(t) = \frac{A\omega\omega_n}{\omega_n^2 - \omega^2} \left(\frac{1}{\omega} \sin \omega t - \frac{1}{\omega_n} \sin \omega_n t \right)$$

Also show that in the limiting case when ω approaches one of the natural frequencies ω_n, the wave amplitude grows as $t \cos \omega_n t$.

9.37 A rectangular tank of length L in the x direction contains inviscid liquid with undisturbed depth H. It rests on a platform which begins vertical oscillatory motion with displacement given by

$$Y = -Y_0(1 - \cos \omega t)$$

where Y_0 is the amplitude and ω is the circular frequency. The corresponding body force acceleration potential of Eq. (9.25) is

$$\Omega = (g + \ddot{Y})y = (g - Y_0\omega^2 \cos \omega t)y$$

Consider two-dimensional motion in the x, y plane. Follow the procedure of section 9.13 to show that the solution for $\phi(x, y, t)$ is

$$\phi = \sum_{n=0}^{\infty} \left[e^{n\pi y/L} + e^{-n\pi(2H+y)/L} \right] \cos \frac{n\pi x}{L} T_n(t)$$

where $T_n(t)$ must satisfy

$$\ddot{T}_n + \left(\frac{n\pi}{L} \tanh \frac{n\pi H}{L} \right)(g + Y_0\omega^2 \cos \omega t) T_n = 0$$

Show that this can be put into the form of Mathiu's differential equation,

$$\frac{d^2 T_n^*}{dt^{*2}} + (\mathscr{P} - 2Q \cos 2t^*) T_n^* = 0$$

where

$$T_n^* = \frac{2T_n}{L\omega}, \qquad t^* = \frac{\omega t}{2}$$

$$\mathscr{P} = \frac{4n\pi}{L\omega^2} g \tanh \frac{n\pi H}{L}, \qquad Q = \frac{2n\pi}{L} Y_0 \tanh \frac{n\pi H}{L}$$

which has the stability properties shown in Fig. P9.37. The clear regions represent stable behavior with limited amplitude growth. The shaded regions correspond to unstable surface amplitude growth with time.

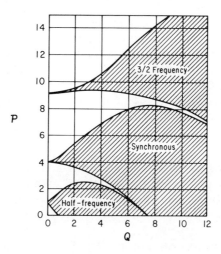

Figure P9.37

9.38 The tank of liquid in Problem 9.37 undergoes vertical oscillatory motion with the properties,

$$g = 9.8 \text{ m/s}^2 \ (32.2 \text{ ft/s}^2), \qquad \omega = 3.0 \text{ rad/s}$$

$$L = 2 \text{ m} \ (6.56 \text{ ft}), \qquad H = 1.0 \text{ m} \ (3.28 \text{ ft})$$

Use the stability map of Problem 9.37 to show that the first surface harmonic $(n = 1)$ is unstable if the vertical amplitude $Y_0 > 0.955$ m (3.13 ft). Also, show that the second harmonic $(n = 2)$ will be unstable if $Y_0 > 1.11$ m (3.64 ft).

10 MISCELLANEOUS TOPICS IN UNSTEADY THERMOFLUID MECHANICS

The purpose of this chapter is to introduce several more unsteady thermofluid topics, formulations, and solution methods which were not included in earlier chapters. Moving reference frames, acoustic thermal damping, and the effect of acoustic wave arrival at submerged structures are discussed. A normalization procedure is described which is useful in formulating coupled fluid-structure interaction problems. Finally, the energy method of Lagrangian mechanics is introduced and applied to example problems involving fluid-structure systems.

Although dozens of other unsteady thermofluid applications have not been addressed in this book, the additional applications discussed in this chapter should help to illustrate the broadness of this fascinating subject.

10.1 The Momentum Equation for Moving Reference Frames

The momentum equation in Chapter 1 was formulated for a reference frame attached to the earth. One form was expressed by Eq. (1.37) as

$$\sum_{\text{cs}} (\mathbf{V}\,\delta\dot{m})_{\text{out}} - \sum_{\text{cs}} (\mathbf{V}\,\delta\dot{m})_{\text{in}} + \frac{d}{dt}\sum_{\text{cv}} (\mathbf{V}\,\delta M) = g_0\mathbf{F} \qquad (10.1)$$

where mass flow rates $\delta\dot{m}_{\text{out}}$ and $\delta\dot{m}_{\text{in}}$ are relative to the CS, and the velocity vector $\mathbf{V}$ is relative to the earth. It is sometimes desirable to express Eq. (10.1) in terms of the coordinates of a moving reference frame which can be translating and rotating.

586

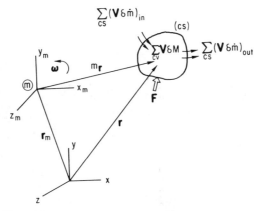

Figure 10.1 Momentum Diagram for Moving Reference Frame

Consider the diagram of Fig. 10.1, which shows the CS, a position vector **r** relative to earth, a moving reference frame ⓜ with origin at position $\mathbf{r}_m$ and angular velocity **ω** relative to earth, and position vector $^m\mathbf{r}$ of the CS relative to ⓜ.

The position of any fluid particle within the CS can be written as

$$\mathbf{r} = \mathbf{r}_m + {}^m\mathbf{r}$$

and its velocity relative to earth is

$$\mathbf{V} = \frac{d\mathbf{r}_m}{dt} + \frac{d{}^m\mathbf{r}}{dt}$$

But from vector differentiation (section C.3 of Appendix C)

$$\frac{d\,{}^m\mathbf{r}}{dt} = {}^m\mathbf{V} + \boldsymbol{\omega} \times {}^m\mathbf{r} \qquad \text{and} \qquad \frac{d\mathbf{r}_m}{dt} = \mathbf{V}_m$$

so that

$$\mathbf{V} = \mathbf{V}_m + {}^m\mathbf{V} + \boldsymbol{\omega} \times {}^m\mathbf{r} \qquad (10.2)$$

The acceleration of a fluid particle relative to earth is $d\mathbf{V}/dt$, which introduces other derivatives from terms in Eq. (10.2). These include the acceleration of frame ⓜ relative to earth, $\mathbf{a}_m = d\mathbf{V}_m/dt$, and the time rate of change of $^m\mathbf{V}$ relative to earth, which is written as

$$\frac{d\,{}^m\mathbf{V}}{dt} = \frac{{}^m d\,{}^m\mathbf{V}}{dt} + \boldsymbol{\omega} \times {}^m\mathbf{V}$$

It follows that differentiation of Eq. (10.2) yields

$$\frac{d\mathbf{V}}{dt} = \frac{d\mathbf{V}_m}{dt} + \frac{{}^m d\,{}^m\mathbf{V}}{dt} + 2\boldsymbol{\omega} \times {}^m\mathbf{V} + \boldsymbol{\alpha} \times {}^m\mathbf{r} + \boldsymbol{\omega} \times \boldsymbol{\omega} \times {}^m\mathbf{r} \qquad (10.3)$$

Therefore, Eq. (10.1) can be written as

$$\sum_{cs} ({}^m\mathbf{V}\,\delta\dot{m})_{out} - \sum_{cs} ({}^m\mathbf{V}\,\delta\dot{m})_{in} + \frac{{}^m d}{dt} \sum_{cv} {}^m\mathbf{V}\,\delta M$$

$$= g_0\left(\mathbf{F} - \frac{M}{g_0}\frac{d\mathbf{V}_m}{dt}\right) - \mathbf{F}_\omega \qquad (10.4)$$

where

$$\mathbf{F}_\omega = \sum_{cs} (\boldsymbol{\omega} \times {}^m\mathbf{r}\,\delta\dot{m})_{out} - \sum_{cs} (\boldsymbol{\omega} \times {}^m\mathbf{r}\,\delta\dot{m})_{in} + \frac{{}^m d}{dt} \sum_{cv} \boldsymbol{\omega} \times {}^m\mathbf{r}\,\delta M$$

$$+ \boldsymbol{\omega} \times \sum_{cv} {}^m\mathbf{V}\,\delta M + \boldsymbol{\omega} \times \boldsymbol{\omega} \times \sum_{cv} {}^m\mathbf{r}\,\delta M \qquad (10.5)$$

Equation (10.4) expresses the momentum principle relative to an observer on moving reference frame ⓜ. The time derivative operator ${}^m d/dt$ is relative to an observer moving with ⓜ, and also the velocity components and poition vectors are relative to ⓜ. If ⓜ is translating without rotation, $\mathbf{F}_\omega = 0$, and the only difference from the momentum formulation of Eq. (10.1) is the acceleration term $(M/g_0)(d\mathbf{V}_m/dt)$. Moreover, if ⓜ translates at constant speed, $d\mathbf{V}_m/dt = 0$ and the formulation reduces to Eq. (10.1). If ⓜ is undergoing free-fall, we have $d\mathbf{V}_m/dt = -g\mathbf{n}_y$.

EXAMPLE 10.1: MANOMETER MOTION, TRANSLATING FRAME
A frictionless U-tube of uniform area A contains liquid as shown in Fig. 10.2, and is attached to a structure ⓜ which undergoes vertical and horizontal translation without rotation. Obtain the equation of motion for an arbitrary structural acceleration

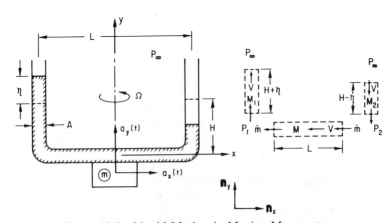

Figure 10.2 Liquid Motion in Moving Manometer

$$\frac{d\mathbf{V}_m}{dt} = a_x(t)\mathbf{n}_x + a_y(t)\mathbf{n}_y$$

A CS which contains all the liquid has no inflow terms and would require a specification of the force $\mathbf{F}$ in Eq. (10.4), which is not known. However, if Eq. (10.4) is applied to one CV for the horizontal section and two CVs for the vertical sections, we obtain

MASS CONSERVATION

$$-\dot{m} + \frac{dM_1}{dt} = 0, \qquad \text{left}$$

$$\frac{dM}{dt} = 0, \qquad \text{horizontal}$$

$$\dot{m} + \frac{dM_2}{dt} = 0, \qquad \text{right}$$

MOMENTUM, BASED ON EQ. (10.4) WITH $F_\omega = 0$:

$$-\dot{m}V + \frac{d}{dt}(M_1 V) = g_0 \left[(P_1 - P_\infty)A - W_1 - \frac{M_1}{g_0} a_y(t) \right] \qquad \text{left}$$

$$\frac{d}{dt}(MV) = g_0 \left[(P_1 - P_2)A - \frac{M}{g_0} a_x(t) \right] \qquad \text{horizontal}$$

$$-\dot{m}V + \frac{d}{dt}(-M_2 V) = g_0 \left[(P_2 - P_\infty)A - W_2 - \frac{M_2}{g_0} a_y(t) \right] \qquad \text{right}$$

If we put $\dot{m} = \rho A V$, $M_1 = \rho A(H + \eta)$, $M_2 = \rho A(H - \eta)$, $M = \rho AL$, $W_1 = M_1 g/g_0$, $W_2 = M_2 g/g_0$, $V = d\eta/dt$ and eliminate P_1 and P_2, we obtain

$$\frac{d^2\eta}{dt^2} + \frac{2[g + a_y(t)]}{2H + L}\eta = \frac{L}{2H + L} a_x(t) \qquad (10.6)$$

It is seen that the acceleration component $a_x(t)$ plays the role of a forcing function, and $a_y(t)$ is like a time-dependent gravity.

The rotational term, Eq. (10.5), would play a role in any system which rotates relative to earth. Most frequently F_ω occurs in analysis involving spacecraft, aircraft, marine, and transportation applications.

EXAMPLE 10.2: LIQUID OSCILLATION, SPINNING FRAME
Let the manometer of Example 10.1 spin at constant angular velocity $\boldsymbol{\omega} = \Omega\mathbf{n}_y$ about the y axis, with the x, y origin remaining fixed. Determine the effect of angular velocity on motion of the contained liquid.

Each section of the manometer liquid is considered separately, as in Example 10.1. Only those momentum terms, forces, and terms from Eq. (10.5) with $\mathbf{n}_y$

components are employed in Eq. (10.4) for the two vertical sections. Furthermore, terms with $\mathbf{n}_x$ components are employed for the horizontal section. The left side of Eq. (10.4) is the same for each section as in Example 10.1, and so are the mass conservation formulations. The acceleration term $d\mathbf{V}_m/dt = 0$. It is found that all $\mathbf{n}_y$ components of Eq. (10.5) vanish in the vertical sections, and the $\mathbf{n}_x$ components vanish in the horizontal section. Combination of the momentum and mass conservation equations and elimination of pressures P_1 and P_2 as before yield the dynamic equation

$$\frac{d^2\eta}{dt^2} + \frac{2g}{2H + L}\eta = 0 \tag{10.7}$$

It may seem surprising that spinning does not affect the liquid motion. If the device was made to spin about a nonsymmetric axis, motion would occur about a different equilibrium state with one vertical column longer than the other.

10.2   Thermal Damping in Bubbly Mixtures

The phenomenon of thermal damping was introduced in Chapter 2 for the oscillation of a gas bubble in surrounding liquid. Idealized adiabatic and isothermal oscillations were undamped, but any process between these two limiting cases resulted in a damped oscillation. The same phenomenon can play a role in damping waterhammer or acoustic waves in a mixture of gas bubbles in liquid, even when the bubbles are so small and finely dispersed that the liquid has a milky appearance. Propagation in bubbly mixtures has some interesting and important features.

The fluid in this model consists of a frictionless liquid with a homogeneous distribution of small gas bubbles. Gravity is considered negligible, and no phase change occurs. Compression or expansion alters the gas temperature, which causes heat transfer between the bubbles and liquid. Interphase heat exchange does not alter the fluid mass conservation and momentum laws, which are rewritten from Eq. (6.20) and the inviscid form of Eq. (6.29) as

$$\frac{\partial\rho}{\partial t} + \mathbf{V}\cdot\nabla\rho + \rho\nabla\cdot\mathbf{V} = 0 \tag{10.8}$$

and

$$\frac{\partial\mathbf{V}}{\partial t} + \mathbf{V}\cdot\nabla\mathbf{V} = -\frac{g_0}{\rho}\nabla P \tag{10.9}$$

Heat transfer causes nonisentropic state changes. Consequently, the mixture density cannot be expressed as a function of pressure only, which is customary in classical waterhammer analysis. Instead, density of the bubbly mixture is expressed as

$$\frac{1}{\rho} = \frac{x}{\rho_g} + \frac{1-x}{\rho_L} \qquad (10.10)$$

where x is M_g/M, or quality, which is the mass ratio of gas to total liquid and gas in a small sample. Since no phase change occurs, quality remains constant during compression or expansion, provided that the gas is relatively insoluble. Therefore, the total differential of density is

$$d\rho = x\left(\frac{\rho}{\rho_g}\right)^2 d\rho_g + (1-x)\left(\frac{\rho}{\rho_L}\right)^2 d\rho_L \qquad (10.11)$$

If Eq. (10.11) is employed in Eq. (2.45), we obtain the mixture sonic speed in the form

$$C = \left[x\left(\frac{\rho}{\rho_g C_g}\right)^2 + (1-x)\left(\frac{\rho}{\rho_L C_L}\right)^2 \right]^{-1/2} \qquad (10.12)$$

Equation (10.11) gives $d\rho$ in terms of gas and liquid density differentials, which depend on heat transfer between the two phases. Figure 10.3 shows an incremental region of fluid, for which energy conservation yields

$$P\frac{d\mathcal{V}}{dt} + q + \frac{dU}{dt} = 0 \qquad (10.13)$$

If $\mathcal{V} = M/\rho$ and $U = Me$ are substituted, Eq. (10.13) becomes

$$-\frac{P}{\rho^2}\frac{d\rho}{dt} + \frac{q}{M} + \frac{de}{dt} = 0 \qquad (10.14)$$

Equation (2.117) is used to write

$$\frac{de}{dt} = \left(\frac{g_0 c_p}{\beta\rho C^2}\right)\frac{dP}{dt} + \left(\frac{P}{\rho^2} - \frac{c_p}{\beta\rho}\right)\frac{d\rho}{dt} \qquad (10.15)$$

from which Eq. (10.14) can be written for the gas phase as

$$\frac{d\rho_g}{dt} = \left(\frac{\beta\rho}{c_p}\right)_g \frac{q}{M_g} + \frac{g_0}{C_g^2}\frac{dP}{dt} \qquad (10.16)$$

Figure 10.3   Fluid Region for Energy Conservation

A similar procedure is employed for liquid adjacent to a gas bubble, where heat transfer q from the gas has the opposite sign when entering the liquid. It follows that

$$\frac{d\rho_L}{dt} = -\left(\frac{\beta\rho}{c_p}\right)_L \frac{q}{M_L} + \frac{g_0}{C_L^2}\frac{dP}{dt} \qquad (10.17)$$

Every gas bubble of mass M_g is surrounded by a discrete liquid mass M_L with which it exchanges energy. Each bubble and associated liquid mass in a homogeneous mixture should have, on the average, the same ratio M_L/M_g. Therefore, quality x can be expressed as

$$x = \frac{M_g}{M} = \frac{M_g}{M_g + M_L} \qquad (10.18)$$

Since $M_g = xM$ and $M_L = (1-x)M$, Eqs. (10.14) through (10.17) can be employed in (10.8) to give

$$\frac{dP}{dt} + \frac{C^2}{g_0}F_b\frac{q}{M} + \frac{\rho C^2}{g_0}\nabla\cdot\mathbf{V} = 0 \qquad (10.19)$$

where

$$F_b = \left(\frac{\rho}{\rho_g}\right)^2\left(\frac{\beta\rho}{c_p}\right)_g - \left(\frac{\rho}{\rho_L}\right)^2\left(\frac{\beta\rho}{c_p}\right)_L \qquad (10.20)$$

Gas bubbles are expected to undergo slight motion relative to the surrounding liquid. Consequently, the interface heat transfer probably is dominated by convection with

$$q = \mathcal{H}A_q(T_q - T_L) \qquad (10.21)$$

where A_q is the total bubble-liquid interface area in a given region of mixture. Equation (10.19) becomes

$$\frac{\partial P}{\partial t} + \mathbf{V}\cdot\nabla P + \frac{C^2}{g_0}F_b\frac{\mathcal{H}A_g}{M}(T_g - T_L) + \frac{\rho C^2}{g_0}\nabla\cdot\mathbf{V} = 0 \qquad (10.22)$$

Equations (10.9) and (10.22) describe the unsteady pressure and velocity fields in a bubbly gas-liquid mixture with convection-limited heat transfer between the two phases. Several more idealizations lead to a modified waterhammer formulation.

Since liquid has large heat capacity relative to gas, liquid temperature T_L is assumed to remain constant. The perfect gas idealization gives

$$T_g = \frac{P}{R_g\rho_g}, \qquad T_L = \frac{P_\infty}{R_g\rho_g} \qquad (10.23)$$

for use in Eq. (10.22). The convective terms $\mathbf{V}\cdot\nabla P$ and $\mathbf{V}\cdot\nabla\mathbf{V}$ are negligible in waterhammer or acoustic problems, and all fluid properties except P and $\mathbf{V}$ are regarded as a constant. Thus, Eqs. (10.22) and (10.9) become

$$\frac{\partial P}{\partial t} + \frac{C^2}{g_0} F_b \frac{\mathcal{H}A_q}{MR_g\rho_g}(P - P_\infty) + \frac{\rho C^2}{g_0}\nabla\cdot\mathbf{V} = 0 \qquad (10.24)$$

and

$$\frac{\partial \mathbf{V}}{\partial t} + \frac{g_0}{\rho}\nabla P = 0 \qquad (10.25)$$

Either $\mathbf{V}$ or P can be eliminated from Eqs. (10.24) and (10.25) to give the linear wave equations

$$\frac{\partial^2 P}{\partial t^2} + D\frac{\partial P}{\partial t} - C^2\nabla^2 P = 0 \qquad (10.26)$$

and

$$\frac{\partial^2 \mathbf{V}}{\partial t^2} + D\frac{\partial \mathbf{V}}{\partial t} - C^2\nabla(\nabla\cdot\mathbf{V}) = 0 \qquad (10.27)$$

where the damping coefficient D is given by

$$D = \frac{C^2}{g_0} F_b \frac{\mathcal{H}A_q}{MR_g\rho_g} \qquad (10.28)$$

The heat transfer area per unit mass A_q/M depends on the number and size of bubbles present in a region of liquid. If the interphase heat transfer is made zero by setting $\mathcal{H} = 0$, then D becomes zero and Eqs. (10.26) and (10.27) reduce to the classical wave equations.

Important fluid pressure response characteristics can be obtained from a simplified one-dimensional version of Eqs. (10.26) and (10.27), written as

$$\frac{\partial^2 P}{\partial t^2} + D\frac{\partial P}{\partial t} - C^2\frac{\partial^2 P}{\partial z^2} = 0 \qquad (10.29)$$

for pressure and

$$\frac{\partial^2 u}{\partial t^2} + D\frac{\partial u}{\partial t} - C^2\frac{\partial^2 u}{\partial z^2} = 0 \qquad (10.30)$$

for velocity. If a solution for either $P(z, t)$ or $u(z, t)$ is known, it is often easier to obtain the other from the one-dimensional form of Eq. (10.25), given by

$$\frac{\partial u}{\partial t} + \frac{g_0}{\rho} \frac{\partial P}{\partial z} = 0 \tag{10.31}$$

than it is to solve the other second-order equation.

10.3   One-Dimensional Bubbly Mixture Response

The rigid cylinder in Fig. 10.4 is closed at the left end by a stationary wall. The fluid column of length L is exposed to ambient pressure P_∞ at the right end. An arbitrary initial pressure distribution $f(z)$ with zero pressure rate is specified. Initial and boundary conditions are given formally as

$$u(0, t) = \frac{\partial u(0, t)}{\partial t} = \frac{\partial P(0, t)}{\partial z} = 0$$

$$P(L, t) - P_\infty = 0 \tag{10.32}$$

$$P(z, 0) - P_\infty = f(z)$$

The corresponding solution of Eq. (10.29) for an underdamped system is

$$P(z, t) - P_\infty = e^{-Dt/2} \sum_{n=0}^{\infty} A_n \cos \omega_n t \cos \gamma_n z \tag{10.33}$$

where

$$\omega_n = \frac{D}{2} \sqrt{\left(\frac{2}{D} \gamma_n C\right)^2 - 1}$$

$$\gamma_n = \left(\frac{2n + 1}{2}\right) \frac{\pi}{L} \tag{10.34}$$

$$A_n = \frac{2}{L} \int_0^L f(z) \cos\left(\frac{2n + 1}{2} \frac{\pi z}{L}\right) dz$$

Equation (10.33) shows that any arbitrary initial disturbance is damped exponentially according to $\exp(-Dt/2)$. Moreover, the frequency of each

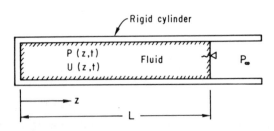

Figure 10.4   Fluid Response in Rigid Cylinder

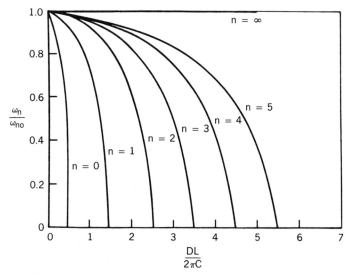

Figure 10.5   Ratio of Damped to Undamped Frequency

harmonic is affected by the damping coefficient D in much the same way as a spring-mass system with linear damping. The nth undamped frequency ω_{n0} is obtained from Eq. (10.34) with $D = 0$, which gives

$$\omega_{n0} = \gamma_n C = \frac{2n + 1}{2} \frac{\pi C}{L}$$ (10.35)

Therefore, the ratio of damped to undamped frequencies is

$$\frac{\omega_n}{\omega_{n0}} = \frac{DL}{2\pi C} \left(\frac{2}{2n + 1}\right) \sqrt{\left[\frac{2\pi C}{DL}\left(\frac{2n + 1}{2}\right)\right]^2 - 1}$$ (10.36)

which is plotted in Fig. 10.5. It is seen that higher harmonics are less affected by damping than lower harmonics. Thermal damping reduces the frequency of lower harmonics. If $DL/2\pi C = 2.0$ in Fig. 10.5, frequency of the $n = 2$ harmonic would be 0.6 of the undamped frequency. Moreover, if $DL/2\pi C = 2.5$, the $n = 2$ harmonic would be critically damped, lower harmonics would be overdamped, and higher harmonics still would be underdamped. High frequencies (as $n \to \infty$) are unaffected by thermal damping.

10.4 Acoustic Penetration

A solution of Eq. (10.29) is obtained from the general form $\exp(az + bt)$. Coefficient a is obtained as a function of b when the solution form is substituted into Eq. (10.29). Then coefficient b must be determined from boundary conditions. If pressure at the boundary $z = 0$ varies sinusoidally as $P(0, t) = P_0 \sin \omega t$, the pressure at any other location z corresponds to

$$P(z, t) = P_0 e^{-\sigma z} \sin \omega \left(t - \frac{z}{C_\omega} \right) \tag{10.37}$$

where for a bubbly mixture

$$\sigma = \frac{D}{C} \left\{ \frac{1}{2} \left[\frac{\omega}{D} \sqrt{\left(\frac{\omega}{D} \right)^2 + 1} - \left(\frac{\omega}{D} \right)^2 \right] \right\}^{1/2} \tag{10.38a}$$

$$C_\omega = C\sqrt{2} \left(1 + \sqrt{1 + \left(\frac{D}{\omega} \right)^2} \right)^{-1/2} \tag{10.38b}$$

and C is from Eq. (10.12). The damping envelope $\exp(-\sigma z)$ is given in Fig. 10.6. Low- and high-frequency limits correspond to

$$\exp(-\sigma z)|_{\omega \to 0} = 1.0$$

and $$\tag{10.39}$$

$$\exp(-\sigma z)|_{\omega \to \infty} = \exp\left(-\frac{1}{2} \frac{Dz}{C} \right)$$

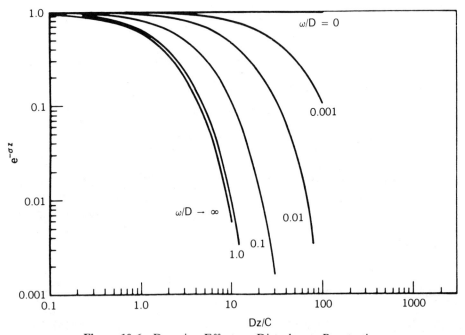

Figure 10.6 Damping Effect on Disturbance Penetration

It is seen that lower frequencies tend to penetrate further without attenuation, whereas higher frequencies penetrate shorter distances. When ω/D exceeds 1.0, further reduction in penetration depth is slight. A bubbly mixture acts as a selective depth filter for various frequencies.

EXAMPLE 10.3: OCEAN FLOOR MAPPING BY SONAR

Sonar involves the transmission and reflection of sound waves to record the topology of the ocean floor and to locate sunken ships. It even has been used in attempts to locate the legendary Loch Ness monster!

A crew is preparing to map the ocean floor with sonar in a region of mild subterranean volcanic activity, which has an average depth of $H = 6\,\text{km}$. The continuous release of gas causes an almost constant suspension of small bubles throughout the water depth with an average gas volume fraction of $\alpha = 0.01$. The bubbles have an average diameter of $D_b = 0.5\,\text{cm}$. One engineer recalls that only foghorns, not whistles, penetrate a London fog because the suspended water drops filter higher frequencies. This has led him to worry that sonar signals might be lost in a bubble-filled liquid. Are his worries well-founded?

The gas is approximated by air properties at 10°C, and an average heat transfer coefficient across a bubble boundary is $30\,\text{W/m}^2\text{-K}$. Other data are listed below:

Liquid (Water)	Gas (Air)
$\rho_L = 1000\,\text{kg/m}^3$	$P_{avg} = 29500\,\text{kPa abs}$
$C_L = 1400\,\text{m/s}$	$T = 10°C = 283\,\text{K}$
$\beta_L = 1.8 \times 10^{-4}/\text{K}$	$\rho_g = 360\,\text{kg/m}^3$
$c_{pL} = 4.17\,\text{kJ/kg-K}$	$\beta_g = 1/283\,\text{K}$
	$c_{pg} = 1000\,\text{J/kg-K}$
	$R_g = 287\,\text{J/kg-K}$

Mixture density is obtained from the gas volume fraction α as

$$\rho = \rho_g \alpha + \rho_L(1 - \alpha) = 993.6\,\text{kg/m}^3$$

Also, quality x, or vapor mass fraction, is obtained from the specific volume

$$v = \frac{1}{\rho} = \frac{x}{\rho_g} + \frac{1-x}{\rho_L}$$

which yields

$$x = \left(1 + \frac{1-\alpha}{\alpha}\frac{\rho_L}{\rho_g}\right)^{-1} = 0.000036$$

It follows from Eq. (10.20) that

$$F_b = 0.0096\,\text{kg}^2/\text{m}^3\text{-J}$$

The bubble area per unit mass is obtained from

$$\frac{A}{M} = \frac{A}{\text{bubble}} \cdot \frac{\text{bubble}}{\text{gas vol}} \cdot \frac{\text{gas vol}}{\text{total vol}} \cdot \frac{\text{total vol}}{\text{liquid vol}} \cdot \frac{\text{liquid vol}}{\text{liquid mass}}$$

$$= 4\pi R^2 \left(\frac{1}{(4/3)\pi R^3}\right) \alpha \frac{1}{1-\alpha} \frac{1}{\rho_L} = \frac{6\alpha}{(1-\alpha)D_b\rho_L}$$

$$= 0.012 \text{ m}^2/\text{kg}$$

The bubbly mixture sound speed, obtained from Eq. (10.12), is

$$C = 1400 \text{ m/s}$$

and, furthermore,

$$D = \frac{C^2 F_b \mathcal{H}}{g_0 R_g \rho_g} \frac{A}{M} = 0.0655 \text{ s}^{-1}$$

A sound wave must travel $z = 2H = 12$ km round trip so that

$$\frac{Dz}{C} = 0.56$$

It is seen from Fig. 10.6 that the bubbles will not cause significant filtering of any frequency. Therefore, the engineer does not need to worry that sonar signals might be lost in this bubbly mixture.

10.5 Ringout Damping

A phenomenon which displays obvious damping is called *chug ringout*. Periodic steam discharge occurs from a submerged vent into cool water, and sudden condensation void collapse creates a sharp acoustic disturbance. Figure 10.7 gives an experimental chug pressure trace for a 43-cm-diameter cylindrical tank with a water depth $H = 0.96$ m. The decaying oscillation, or ringout, is not explained by structural damping or fluid viscosity, both of which are comparatively weak effects. However, thermal damping easily provides the strong dissipation mechanism observed in Fig. 10.7.

Most available chugging experiments in which ringout occurs begin with the purge of an air-filled vent. When steam discharge from the source begins, the air is blown into the water where it breaks into small bubbles. Even when the larger bubbles have risen, it has been noted that the water retains many small bubbles down to a size which may give a cloudy appearance. Even though bubble sizes and mixture void fraction were not measured specifically, the role of thermal damping in ringout can be assessed from Eq. (10.34). A pressure calculation at arbitrary z yields, from Eq. (10.33),

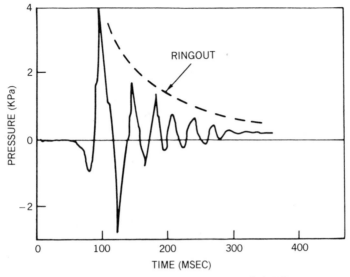

Figure 10.7   Chug Pressure Trace [10.11]

$$P(z, t) - P_\infty = e^{-Dt/2} \cos \frac{Dt}{2} \sqrt{\left[\frac{2}{D}\left(\frac{2n+1}{2}\right)\frac{\pi}{L} C\right]^2 - 1}$$
$$\times \cos \frac{2n+1}{2} \frac{\pi z}{L} \qquad\qquad (10.40)$$

The thermal damping coefficient D and bubble mixture sound speed C are graphed for air in water at standard conditions in Fig. 10.8. A thermal damping coefficient $D = 24\,\mathrm{s}^{-1}$ gives the dashed envelope in Fig. 10.7 and corresponds to a void fraction greater than 10^{-5}, bubble diameters of 0.2 mm and larger, and an average convection heat transfer coefficient of $30\,\mathrm{W/m^2}$-K, all of which are physically reasonable. Furthermore, a small amount of undissolved air can reduce the sound speed in water by an order of magnitude or more. The 25-Hz ringout frequency of Fig. 10.7 is consistent with a bubbly mixture sound speed of 100 m/s, for which Fig. 10.8 requires a void fraction of 0.02. Therefore, it appears that thermal dissipation in a fine bubbly mixture provides the strong damping associated with ringout. Mixture properties usually are known in two-phase flow applications, which would permit a ready estimate of the thermal damping.

10.6   Fluid Structure Interaction (FSI)

If either a submerged elastic structure or the fluid surrounding it is disturbed, the structure and fluid may undergo coupled motion, referred to as

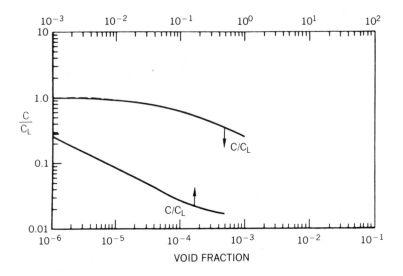

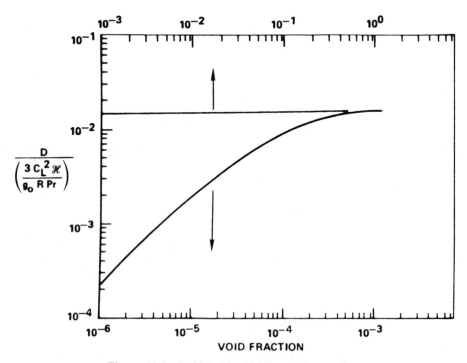

Figure 10.8   Bubble Liquid Mixture Properties

fluid structure interaction (FSI). Bulk flow occurs with incompressible fluid, whereas propagative flow occurs in compressible fluid. Differences in FSI behavior arising from incompressible and compressible fluid are discussed for the specific example of Fig. 10.9, which shows a sphere of mass M_s with radius R attached to a spring of constant K submerged in an infinite, stationary fluid of density ρ. Its motion is along the x axis only, with sphere velocity $U_s(t)$.

The velocity potential for incompressible, irrotational liquid response with axial symmetry about x is described by Eq. (9.18) in the form

$$\frac{1}{r^2}\frac{\partial}{\partial r}\left(r^2 \frac{\partial\phi}{\partial r}\right) + \frac{1}{r^2 \sin\theta}\frac{\partial}{\partial\theta}\left(\sin\theta\,\frac{\partial\phi}{\partial\theta}\right) = 0$$

for which a solution is

$$\phi(r,\theta,t) = -U_s(t)\frac{R^3}{2r^2}\cos\theta \tag{10.41}$$

If gravity is neglected, $\Omega = gy = 0$ and the local fluid pressure of Eq. (9.25) with $V^2 = (\nabla\phi)^2$ becomes

$$P = \frac{\rho}{g_0}\left(f(t) - \frac{\partial\phi}{\partial t} - \frac{V^2}{2}\right) \tag{10.42}$$

The differential x-directional fluid force on the sphere is expressed as

$$dF_x = 2\pi R^2 P \cos\theta \sin\theta\, d\theta \tag{10.43}$$

where P is from Eq. (10.42). Integration over the entire surface from $\theta = 0$ to π yields

$$F_x = \frac{M_v}{g_0}\dot{U}_s(t), \qquad M_v = \frac{2}{3}\pi R^3\rho \tag{10.44}$$

where F_x is positive to the left, as indicated in Fig. 10.9. The virtual mass M_v corresponds to that of a sphere in Table 9.3. The function $f(t)$ adds nothing

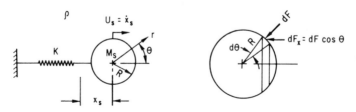

Figure 10.9 Fluid Structure Interaction Model

to the pressure integration, and the integral of $V^2 = (\nabla\phi)^2$ over the surface also is zero (d'Alembert's principle). Thus, the only pressure term surviving the integration of dF_x is $-(\rho/g_0)(\partial\phi/\partial t)$ for incompressible, inviscid flow.

Newton's law for motion of the sphere is written as

$$-F_x - Kx_s = \frac{M_s}{g_0}\frac{d^2x_s}{dt^2} \tag{10.45}$$

If we employ F_x from Eq. (10.44), then

$$\frac{d^2x_s}{dt^2} + \left(\frac{Kg_0}{M_s + M_v}\right)x_s = 0, \qquad M_v = \frac{2}{3}\pi R^3\rho \tag{10.46}$$

Note that response of the sphere corresponds to a harmonic oscillator with the sphere mass M_s increased by the virtual mass M_v. No damping occurs in this inviscid, incompressible system.

Behavior of the sphere is different when fluid compressibility is important. The describing equation for the compressible case is obtained from Eq. (9.83) and the ∇^2 operator of (9.18) as

$$\frac{\partial^2\phi}{\partial t^2} - C^2\left[\frac{1}{r^2}\frac{\partial}{\partial r}\left(r^2\frac{\partial\phi}{\partial r}\right) + \frac{1}{r^2\sin\theta}\frac{\partial}{\partial\theta}\left(\sin\theta\frac{\partial\phi}{\partial\theta}\right)\right] \tag{10.47}$$

A general solution of Eq. (10.47) for small-amplitude motion of the sphere and outward propagation into an infinite, stationary fluid can be written as

$$\phi = \left[\frac{R}{Cr}F'\left(t - \frac{r-R}{C}\right) + \frac{R}{r^2}F\left(t - \frac{r-R}{C}\right)\right]\cos\theta \tag{10.48}$$

where the function F is to be determined, and F' is its derivative with respect to the argument $t - (r - R)/C$. Equation (10.48) shows that as $r \to \infty$, ϕ approaches zero. The $\cos\theta$ function satisfies symmetry about the x axis with periodicity over an interval $0 \le \theta \le 2\pi$.

The local fluid pressure of Eq. (9.84) is employed with (10.48) in (10.43), which is integrated over the spherical surface ($r = R$, θ, t) to give

$$F_x = -M_L\left[\frac{1}{CR}F''(t) + \frac{1}{R^2}F'(t)\right], \qquad M_L = \frac{4}{3}\pi R^3\rho \tag{10.49}$$

where M_L corresponds to the liquid mass displaced by the sphere. The force F_x is considered positive to the left, as in the incompressible case. Equation (10.45) yields

$$\frac{d^2x_s}{dt^2} + \frac{Kg_0}{M_s}x_s = \frac{M_L}{M_s}\left[\frac{1}{RC}F''(t) + \frac{1}{R^2}F'(t)\right] \tag{10.50}$$

where $F(t)$ must yet be determined.

Every surface point on the sphere moves with velocity $U_s(t)$. The normal component of velocity $U_s(t) \cos \theta$ is equated to adjacent fluid velocity normal to the sphere, $\partial \phi / \partial r$ at $r = R$. This boundary condition with Eq. (10.48) yields

$$F'' + 2\left(\frac{C}{R}\right)F' + 2\left(\frac{C}{R}\right)^2 F = -U_s(t)C^2 = -\frac{dx_s}{dt}C^2 \qquad (10.51)$$

If x_s is eliminated from Eqs. (10.50) and (10.51), a differential equation for $F(t)$ is obtained in the form

$$\left(\frac{R}{C}\right)^2 F^{\mathrm{iv}} + \left(\frac{R}{C}\right)\left(2 + \frac{M_L}{M_s}\right)F''' + \left[2 + \frac{M_L}{M_s} + \omega^2\left(\frac{R}{C}\right)^2\right]F''$$

$$+ 2\omega^2\left(\frac{R}{C}\right)r' + 2\omega^2 F = 0 , \qquad \omega^2 = \frac{Kg_0}{M_s} \qquad (10.52)$$

where ω is the spring-mass circular frequency of a dry (non-submerged) system. A solution for $F(t)$ can be used in Eq. (10.50) or (10.51) to obtain the equation of motion of the submerged sphere. It is often permissible to neglect terms with coefficients $(R/C)^2$ and R/C. However, terms with $\omega R/C$ can be important for high ω. We can write

$$\frac{\omega R}{C} = 2\pi \frac{R}{\lambda_a} , \qquad \lambda_a = \frac{C}{f} \qquad (10.53)$$

where λ_a is the *acoustic wavelength*. Retention of the $\omega R/C$ term is appropriate whenever the acoustic wavelength is within an order of magnitude of the sphere radius. It follows that for many cases, Eq. (10.52) can be reduced to

$$F'' + DF' + \Omega^2 F = 0$$

$$\Omega^2 = \frac{2\omega^2}{2 + M_L/M_s + (\omega R/C)^2} , \qquad D = \frac{R}{C}\Omega^2 , \qquad M_L = \frac{4}{3}\pi R^3 \rho \qquad (10.54)$$

which describes an oscillator with linear damping and submerged circular frequency Ω. The displaced liquid mass M_L and propagation effects result in a submerged frequency Ω which is lower than the nonsubmerged frequency ω. The limiting submerged oscillation frequency for a stiff spring becomes

$$\Omega\big|_{\omega \to \infty} = \frac{C}{R}\sqrt{2} \qquad (10.55)$$

Damping is exponential according to $e^{-Dt/2}$. Small values of D correspond to low ω and slow damping. Higher values of ω damp rapidly according to $e^{-Ct/R}$.

This analysis is based on inviscid fluid with no frictional force, and yet system damping occurs. The reason is that the sphere transfers energy to the surrounding fluid by performing work on it. Energy leaves the elastic system by acoustic propagation outward without returning.

EXAMPLE 10.4: ACOUSTIC ATTENUATION, WATER AND AIR
A spherical mass-spring system has properties $M_s = 2500$ g, $K = 20$ kN/cm, and $R = 5$ cm. An impulsive force puts the system in a small-amplitude vibrational mode. Compare the acoustic damping for oscillations in a large expanse of water with $\rho_L = 1.0$ g/cc, $C_L = 1400$ m/s, and air with $\rho_a = 0.0012$ g/cc, $C_a = 345$ m/s.
  The natural frequency of the nonsubmerged system is

$$\omega = \sqrt{\frac{Kg_0}{M_s}} = 890 \text{ rad/s}$$

and the displaced fluid volume is

$$\frac{4}{3} \pi R^3 = 524 \text{ cc}$$

It follows from Eq. (10.54) that for water and air,

Quantity	Air	Water
$\dfrac{\omega R}{C}$	0.129	0.0318
Ω rad/s	890	847
$D \text{ s}^{-1}$	115	25.6

Since the damping proceeds according to $e^{-Dt/2}$, this example shows that the motion damps faster in air than in water! The result is based on negligible viscous effects and fluid drag, which is approachable only in small amplitude motions.

10.7 Plane Wave Arrival at a Submerged Structure

A plane wave will undergo local scattering upon arrival at a submerged object while exerting an unsteady force. This fact has proved useful in such procedures as the disintegration of kidney stones with sound waves.
  Consider the submerged spherical mass M of radius R attached to a spring of constant K in Fig. 10.10. Initial pressure is assigned the value of zero. A plane wave with fluid velocity U_0 and pressure P_0 arrives from the left with the velocity potential

$$\phi_w = -U_0 C \left(t - \frac{x+R}{C} \right) H_s \left(t - \frac{x+R}{C} \right), \qquad x = r \cos \theta \qquad (10.56)$$

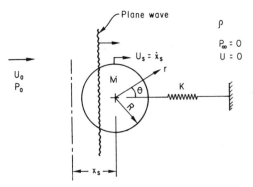

Figure 10.10 Plane Wave Arrival at Sphere

Time $t = 0$ corresponds to arrival at the point $x = -R$. We can superimpose ϕ_w and the general velocity potential of Eq. (10.48) for spherical acoustic flows in the form

$$\phi = \phi_w + \left[\frac{R}{Cr} F'\left(t - \frac{r - R}{C}\right) + \frac{R}{r^2} F\left(t - \frac{r - R}{C}\right) \right] \cos \theta \qquad (10.57)$$

subject to the boundary condition

$$\left. \frac{\partial \phi}{\partial r} \right|_{r=R} = U_s \cos \theta$$

to obtain

$$U_0 - \frac{1}{C^2} F''(t) - \frac{2}{CR} F'(t) - \frac{2}{R^2} F(t) = U_s(t) = \dot{x}_s \qquad (10.58)$$

The function F depends on $\dot{x}_s$, which is yet unknown, but is governed by Eq. (10.45). The differential fluid force of Eq. (10.43) is integrated with pressure P of Eq. (9.84) in terms of the velocity potential of Eq. (10.57) on $r = R$, $t > 0$. Surface pressure does not include the plane wave from $\theta = 0$ to $\cos^{-1}(Ct/R - 1)$, which corresponds to that part of the sphere to the right of the advancing plane wave. That part of the integral from $\theta = \cos^{-1}(Ct/R - 1)$ to π exerts force F_x, given by

$$\frac{F_x}{2\pi R^2 \rho U_0 C / g_0} = -\frac{1}{2}\left[1 - \left(\frac{Ct}{R} - 1\right)^2 \right]$$

$$-\frac{1}{3U_0 C}\left(\frac{1}{C} F'' + \frac{1}{R} F'\right)\left[1 + \left(\frac{Ct}{R} - 1\right)^3 \right] H_s\left(1 - \frac{Ct}{2R}\right)$$

$$(10.59)$$

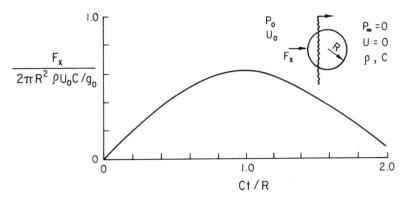

Figure 10.11   Plane Wave Force on Rigid Sphere

If the sphere is rigid with $\dot{x}_s = 0$ in Eq. (10.58), pressure to the right of the wave is zero. The function $F(t)$ is integrated directly for initial conditions $t = 0$, $\phi = \partial\phi/\partial t = 0$ on $r = R$ to give

$$F(t) = \frac{U_0 R^2}{2} \left[e^{-Ct/R}\left(\cos \frac{C}{R}t - \sin \frac{C}{R}t\right) + 1\right]$$ (10.60)

for which Eq. (10.59) gives the force

$$\frac{F_x}{2\pi R^2 \rho U_0 C/g_0} = -\frac{1}{2}\left[1 - \left(\frac{Ct}{R} - 1\right)^2\right]$$

$$-\left(\frac{1}{3}\, e^{-Ct/R} \sin \frac{Ct}{R}\right)\left[1 + \left(\frac{Ct}{R} - 1\right)^3\right]H_s\left(t - \frac{2R}{C}\right)$$ (10.61)

This result is shown in Fig. 10.11. Note that the force exerted on the rigid sphere reaches a maximum value of 60 percent of the full waterhammer pressure $\rho U_0 C/g_0$ applied over a hemispherical area $2\pi R^2$.

If the structure is flexible, $\dot{x}_s(t)$ can be eliminated from Eqs. (10.45), (10.58), and (10.59) to first obtain a fourth-order differential equation in the function $F(t)$. A solution for $F(t)$ then can be employed in Eq. (10.58) to give the velocity $\dot{x}_s$.

10.8   Fluid-Structure Interaction Normalization

Sometimes a coupled fluid-structure system cannot be readily formulated for either a bulk (incompressible) or propagative (acoustic) flow solution. When it is desirable to determine the natural modes of vibration without specifying a particular disturbance, an entire range of eigenfrequencies can be over-looked by an improper formulation. A procedure which helps in the formulation of problems for such cases is called a *full spectrum normali-zation*.

Generally, the procedure begins with a solution to the incompressible problem, which then is employed to determine the significance of acoustic effects. It is sometimes found that structural response modes in one range are dominated by incompressible flow, and by acoustic flow in another range. Surprisingly, the higher modes with higher frequencies often are dominated by incompressible liquid behavior, even though intuition tells us that acoustic effects should dominate. A general formulation for a class of fluid-structure problems is described next to illustrate the full spectrum normalization procedure.

The mass conservation and momentum principles for inviscid, thermally nonconducting fluids and negligible body force acceleration are expressed by Eqs. (6.69) and (6.70) in the forms

$$\frac{\partial P}{\partial t} + \mathbf{V}\cdot\nabla P + \frac{\rho C^2}{g_0}\nabla\cdot\mathbf{V} = 0 \qquad (10.62a)$$

$$\frac{\partial \mathbf{V}}{\partial t} + \mathbf{V}\cdot\nabla\mathbf{V} + \frac{g_0}{\rho}\nabla P = 0 \qquad (10.62b)$$

The fluid region can be bounded by a combination of rigid walls, flexible walls, or free surfaces. Unspecified pressure or velocity disturbances occur at the fluid boundaries. Free surfaces bounding the fluid are characterized by uniform, steady pressure. Flexible walls have pressure and normal velocity common to that of the adjacent fluid. An unsteady pressure $P(x, y, z, t)$ on a flexible wall boundary will excite a dynamic response. If the boundary is a flat plate, its normal displacement from an equilibrium configuration, $w = w(x, y, z, t)$, is described by

$$D\nabla^4 w + \frac{\rho''}{g_0}\frac{\partial^2 w}{\partial t^2} = P \qquad (10.63)$$

where ρ'' is the plate mass per unit area, and $D = E_m\delta^3/12(1-v^2)$ is the bending stiffness.

The response of fluid-structure systems is between incompressible and acoustic extremes, although it is usually difficult to determine a priori which extreme, if either dominates the behavior. The full spectrum normalization provides a systematic procedure for obtaining the correct formulation. The pressure is normalized with the disturbance magnitude, P_0. If the fluid response was acoustic, velocity would be normalized with $g_0 P_0/\rho C$ from waterhammer theory. If the fluid response was closer to incompressible, it would be appropriate to normalize velocity with $w_0\omega_i$, which is the maximum velocity of a flexible boundary oscillating at frequency ω_i whose displacement amplitude w_0 corresponds to a steadily applied pressure P_0. Time would be normalized with L/C for acoustic response where L is a characteristic fluid dimension, and with a natural frequency corresponding to bulk flow, ω_i, for incompressible response. Finally, space coordinates x, y, z are normalized with L. Therefore, in order to include the full spectrum between incompressible and acoustic responses, normalized variables are defined by

$$P^* = \frac{P}{P_0}, \qquad V^* = \frac{V}{(g_0 P_0/\rho C) + w_0 \omega_i}$$

$$(x^*, y^*, z^*) = \frac{(x, y, z)}{L}, \qquad \nabla^* = L\nabla \qquad (10.64)$$

$$t^* = \frac{t}{(L/C) + (1/\omega_i)}, \qquad w^* = \frac{w}{w_0}$$

It follows that the normalized forms of Eqs. (10.62a), (10.62b), and (10.63) are

$$a_1 \frac{\partial P^*}{\partial t^*} + a_2 V^* \cdot \nabla^* P^* + \nabla^* \cdot V^* = 0 \qquad (10.65a)$$

$$a_3 \frac{\partial V^*}{\partial t^*} + a_4 V^* \cdot \nabla^* V^* + \nabla^* P^* = 0 \qquad (10.65b)$$

and

$$b_1 \nabla^{*4} w^* + b_2 \frac{\partial^2 w^*}{\partial t^{*2}} = P^* \qquad (10.66)$$

in which

$$a_1 = \left[\left(1 + \frac{C}{L\omega_i}\right)\left(1 + \frac{w_0 \omega_i \rho C}{g_0 P_0}\right) \right]^{-1}$$

$$a_2 = \frac{g_0 P_0}{\rho C^2}$$

$$a_3 = \frac{1 + w_0 \omega_i \rho C/g_0 P_0}{1 + C/L\omega_i}$$

$$a_4 = \frac{g_0 P_0}{\rho C^2}\left(1 + \frac{w_0 \omega_i \rho C}{g_0 P_0}\right)^2 \qquad (10.67)$$

$$b_1 = \frac{Dw_0}{L^4 P_0}$$

$$b_2 = \frac{\rho'' w_0 C^2}{L^2 g_0 P_0 (1 + C/L\omega_i)^2}$$

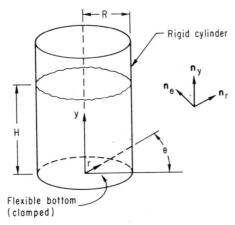

Figure 10.12 Fluid-Structure Coupled System

Both the normalized properties and derivatives in Eqs. (10.65) and (10.66) are $O(1)$. Therefore, the magnitudes of coefficients a_1, a_2, a_3, and a_4 determine which physical effects in the fluid are significant. A coefficient is considered negligible if its magnitude is about 0.1 or less. If a_2 and a_4 are negligible , the fluid system responds acoustically, whereas if a_1 and a_2 are negligible, the fluid response is incompressible.

The full spectrum normalization is applied to the cylindrical container of liquid in Fig. 10.12, with a rigid vertical wall and a clamped flexible bottom. Data are listed in Table 10.1 for two tanks of 1.0 and 0.5 m radii. It is necessary to determine the fluid-structure frequencies which could be excited by an arbitrary disturbance.

The system with incompressible liquid is described by Eqs. (10.65a), (10.65b), and (10.66) with $C \to \infty$ and small w_0, giving

TABLE 10.1 Cylindrical Tank Data

Cylinder radius	$R = 1\,\text{m}$ (Tank 1), $R = 0.5\,\text{m}$ (Tank 2)
Water depth	$H = 3\,\text{m}$
Water density	$\rho = 1000\,\text{kg/m}^3$
Water sonic speed	$C = 1500\,\text{m/s}$
Thickness, flexible bottom	$\delta = 4\,\text{cm}$
Plate mass/area	$\rho'' = 80\,\text{kg/m}^2$
Poisson's ratio	$\nu = 0.3$
Elastic modulus	$E_m = 2 \times 10^{11}\,\text{N/m}^2$
Bending stiffness	$D = 1.15 \times 10^6\,\text{N-m}$
Pressure disturbance	$P_0 = 100\,\text{kPa}$

DEs: $\nabla \cdot V_0$, $\dfrac{\partial V}{\partial t} + \dfrac{g_0}{\rho} \nabla P = 0$

$$\text{(10.68)}$$

$\quad\quad y = H$, $P = 0$

BCs: $r = R$, $V \cdot n_r = 0$

$$\text{(10.69)}$$

$\quad\quad y = 0$, $D\nabla^4 w + \dfrac{\rho''}{g_0} \dfrac{\partial^2 w}{\partial t^2} = P$

where, for cylindrical coordinates,

$$\nabla = n_r \frac{\partial}{\partial r} + n_\theta \frac{1}{r}\frac{\partial}{\partial \theta} + n_y \frac{\partial}{\partial y}$$

$$\nabla \cdot \nabla = \nabla^2 = \frac{\partial^2}{\partial r^2} + \frac{1}{r}\frac{\partial}{\partial r} + \frac{1}{r^2}\frac{\partial^2}{\partial \theta^2} + \frac{\partial^2}{\partial y^2}$$

$$\text{(10.70)}$$

Since only eigenfrequencies are required, initial conditions are not specified.

The flexible bottom vertical displacement is $w = w(r, \theta, t)$, and its vertical velocity $\partial w/\partial t$ must be equal to the adjacent liquid velocity $V \cdot n_y$ at $y = 0$. Eliminating V from the differential equations, and differentiating the bottom boundary condition of Eq. (10.69) so that it can be written in terms of P, the incompressible formulation becomes

DE: $\nabla^2 P = 0$

BCs: $y = H$, $P = 0$

$$\text{(10.71)}$$

$\quad\quad r = R$, $\dfrac{\partial P}{\partial r} = 0$

$\quad\quad y = 0$, $D\dfrac{g_0}{\rho}\nabla^4\left(\dfrac{\partial P}{\partial y}\right) + \dfrac{\rho''}{\rho}\left(\dfrac{\partial^3 P}{\partial y\,\partial t^2}\right) + \dfrac{\partial^2 P}{\partial t^2} = 0$

The differential equation, and conditions at the top and cylinder wall are satisfied by the product solution,

$$P_{n,m}(r, \theta, y, t) = T_{n,m}(t)(e^{\lambda_{n,m}y} - e^{-\lambda_{n,m}(y-2H)})J_m(\lambda_{n,m}r)\cos m\theta$$

$$\text{(10.72a)}$$

$$m = 0, 1, 2, \ldots$$

where the $\lambda_{n,m}$ are roots of

$$J_{m+1}(\lambda_{n,m}R) = \frac{m}{R}J_m(\lambda_{n,m}R)$$

$$\text{(10.72b)}$$

If we consider only the axisymmetric modes with $m = 0$, $\lambda_{n,m} = \lambda_{n,o} = \lambda_n$, the flexible wall condition at $y = 0$ yields

$$\ddot{T}_n + \omega_{i,n}^2 T_n = 0 \tag{10.73a}$$

for which the incompressible natural frequencies are given by

$$\omega_{i,n} = 2\pi f_{i,n} = \left[\left(\frac{\lambda_n^5 D g_0}{\rho} \right) \Big/ \left(\lambda_n \frac{\rho''}{\rho} + \tanh 2H\lambda_n \right) \right]^{1/2} \tag{10.73b}$$

The steady maximum displacement w_0 for a uniform pressure P_0 applied to the flexible bottom with a clamped edge is obtained from a solution to the steady form of Eq. (10.63), giving

$$\frac{w_0}{P_0} = \frac{R^4}{64 D} \tag{10.74}$$

Constants, a_1, a_2, a_3, and a_4 are obtained from Eqs. (10.67) with $L = H$. Computations are summarized in Table 10.2.

Consider tank 1 with a 1.0 m radius. Coefficients a_1 and a_2 for both the first and second (and higher) eigenfrequencies are relatively small, showing that the dominant fluid response is incompressible. It follows that the first two axisymmetric eigenfrequencies are 135 and 564 Hz. The fact that higher frequencies remain dominated by incompressible liquid can be explained by the fact that at high frequencies, fluid-structure coupling is achieved by virtual mass effects. Only that fluid close to the structural surface has

TABLE 10.2   Coefficients for Example Fluid-Structure Systems

Eigenvalue	Tank 1, $R = 1.0$ m	Tank 2, $R = 0.5$ m
First		
λ_1 (1/m)	3.83	7.66
$\omega_{i,1}$ (rad/s)	851 (135 Hz)	4336 (690 Hz)
w_0/P_0 (m^3/N)	1.4×10^{-8}	8.5×10^{-10}
a_1	0.03	0.14
a_2	4.4×10^{-5}	4.4×10^{-5}
a_3	11.6	5.84
a_4	0.02	0.002
Second		
λ_2 (1/m)	7.02	14.0
$\omega_{i,2}$ (rad/s)	3543 (564 Hz)	17162 (2731 Hz)
a_1	0.01	0.04
a_2	4.4×10^{-5}	4.4×10^{-5}
a_3	65.8	22.0
a_4	0.25	0.02

significant motion at higher eigenmodes, leaving the more distant fluid relatively undisturbed. That is, a higher mode of vibration, consisting of a sinusoidal profile with many small amplitude waves extending over the structural surface, simply transfers small amounts of fluid back and forth between adjacent peaks and valleys as they oscillate. The small amount of fluid involved is dominated by incompressible flow rather than acoustic effects.

Next, consider tank 2 with a 0.5 m radius. Coefficient $a_1 = 0.14$ for the first eigenfrequency is large enough to show that acoustic effects play a role. The acoustic problem is formulated with the differential equation,

$$\frac{\partial^2 P}{\partial t^2} - C^2 \nabla^2 P = 0 \tag{10.75}$$

and the boundary conditions previously used are unchanged. A product solution which satisfies the differential equation, top, and cylindrical wall conditions is given by

$$P_{n.m}(r, \theta, y, t) = \beta_{n,m} \cos \lambda_{n,m} Ct(e^{\beta_{n,m}y}$$
$$- e^{-\beta_{n,m}(y-2H)}) J_m(\sqrt{\lambda_{n,m}^2 + \beta_{n,m}^2}\, r) \cos m\theta \tag{10.76}$$

where

$$J_{m+1}(\sqrt{\lambda_{n,m}^2 + \beta_{n,m}^2}\, R) = \frac{m}{R} J_m\left(\sqrt{\lambda_{n,m}^2 + \beta_{n,m}^2}R\right) \tag{10.77}$$

Again, considering only axisymmetric modes with $m = 0$, $\lambda_{n,m} = \lambda_{n,o} = \lambda_n$, and $\beta_{n,m} = \beta_{n,o} = \beta_n$, the flexible bottom condition yields

$$\frac{g_0}{\rho} D \frac{(\lambda_n^2 + \beta_n^2)^2}{\lambda_n^2 C^2} \beta_n - \frac{\rho''}{\rho} \beta_n + \tanh H\beta_n = 0 \tag{10.78}$$

A solution of the last two equations yields $\lambda_1 = 2.75\,\text{m}^{-1}$ and $\beta_1 = 7.15\,\text{m}^{-1}$. Since the first eigenfrequency is $\omega_1 = \lambda_1 C$, it follows that $\omega_1 = 4125\,\text{rad/s}$ (656 Hz) for acoustic fluid response instead of 4336 rad/s (690 Hz) obtained for incompressible fluid response in tank 2.

The second incompressible, axisymmetric eigenfrequency for tank 2 yielded $a_1 = 0.04$, which with negligible a_2, implies incompressible fluid response. It may seem strange that the lower frequency response can be acoustic and higher frequency response incompressible, although this trend again is explained in terms of hydrodynamic mass, which decreases at higher modes, eventually approaching dry structure response.

10.9   The Method of Lagrangian Mechanics

Many problems in unsteady thermofluid mechanics, which involve coupled dynamic systems and multiple degrees of freedom, are conveniently formulated by the method of Lagrangian mechanics. The generalized formulation is first discussed, followed by several examples of its use.

Figure 10.13 shows a particular mass M at position $\mathbf{r}$, moving with acceleration $\mathbf{a}$ relative to the earth. The resultant of all surface forces exerted by pressure, shear stress, and direct contact with adjacent mass, is $\mathbf{F}_s$. The body force is $\mathbf{F}_B$. These forces are kept separate in Newton's second law

$$\mathbf{F} = \mathbf{F}_s + \mathbf{F}_B = \frac{M\mathbf{a}}{g_0} \tag{10.79}$$

The differential work done by the resultant surface force in moving M an amount $d\mathbf{r}$ is

$$dW = \mathbf{F}_s \cdot d\mathbf{r} \tag{10.80}$$

Let the position vector of mass M be specified as

$$\mathbf{r} = \mathbf{r}(q_1, q_2, \ldots, q_r, \ldots, q_N) \tag{10.81}$$

where $q_1, q_2, \ldots, q_r, \ldots,$ are generalized coordinates which designate translational or angular displacements. The use of generalized coordinates makes it possible to have many systems, even with their own coordinates, connected together in various ways. It follows that

$$d\mathbf{r} = \frac{\partial \mathbf{r}}{\partial q_1} dq_1 + \frac{\partial \mathbf{r}}{\partial q_2} dq_2 + \cdots = \sum_{r=1}^{N} \frac{\partial \mathbf{r}}{\partial q_r} dq_r \tag{10.82}$$

Equations (10.79) through (10.82) are combined to express

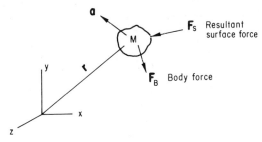

Figure 10.13   Diagram for Lagrangian Mechanics. F_s = Resultant Surface Force; F_B = Body Force

$$dW = \frac{M}{g_0} \sum_{r=1}^{N} \mathbf{a} \cdot \frac{\partial \mathbf{r}}{\partial q_r} dq_r - \sum_{r=1}^{N} \mathbf{F}_B \cdot \frac{\partial \mathbf{r}}{\partial q_r} dq_r \tag{10.83}$$

Suppose that only one generalized coordinate in Eq. (10.83), say q_r, is allowed to move while all the others remain stationary. Then only the rth term is nonzero, which gives

$$\frac{\partial W}{\partial q_r} = \mathbf{F}_s \cdot \frac{\partial \mathbf{r}}{\partial q_r} = \frac{M}{g_0} \mathbf{a} \cdot \frac{\partial \mathbf{r}}{\partial q_r} - \mathbf{F}_B \cdot \frac{\partial \mathbf{r}}{\partial q_r}, \qquad r = 1, 2, \dots, N \tag{10.84}$$

The term $\partial W / \partial q_r$ is called the virtual work per unit displacement of the rth generalized coordinate, with all the others held stationary.

The last two terms of Eq. (10.84) can be written in terms of kinetic and potential energies. First, note that

$$\mathbf{a} \cdot \frac{\partial \mathbf{r}}{\partial q_r} = \ddot{\mathbf{r}} \cdot \frac{\partial \mathbf{r}}{\partial q_r} \tag{10.85}$$

and

$$\frac{d}{dt} \left(\dot{\mathbf{r}} \cdot \frac{\partial \mathbf{r}}{\partial q_r} \right) = \ddot{\mathbf{r}} \cdot \frac{\partial \mathbf{r}}{\partial q_r} + \dot{\mathbf{r}} \cdot \frac{\partial \dot{\mathbf{r}}}{\partial q_r} \tag{10.86}$$

Equation (10.82) can be written as

$$\dot{\mathbf{r}} = \frac{d\mathbf{r}}{dt} = \sum_{r=1}^{N} \frac{\partial \mathbf{r}}{\partial q_r} \dot{q}_r = \dot{\mathbf{r}}(q_1, q_2, \dots ; \dot{q}_1, \dot{q}_2, \dots) \tag{10.87}$$

That is, $\dot{\mathbf{r}}$ can be expressed as a function of the generalized displacements q_r and their time derivatives $\dot{q}_r$. Partial differentiation of $\dot{\mathbf{r}}$ with respect to one of the $\dot{q}_r$ in Eq. (10.87) yields

$$\frac{\partial \dot{\mathbf{r}}}{\partial \dot{q}_r} = \frac{\partial \mathbf{r}}{\partial q_r} \tag{10.88}$$

so that Eq. (10.86) becomes

$$\ddot{\mathbf{r}} \cdot \frac{\partial \mathbf{r}}{\partial q_r} = \frac{d}{dt} \left(\frac{1}{2} \frac{\partial \dot{\mathbf{r}}^2}{\partial \dot{q}_r} \right) - \frac{1}{2} \frac{\partial \dot{\mathbf{r}}^2}{\partial q_r} \tag{10.89}$$

If we employ the kinetic energy

$$KE = \frac{M}{2g_0} \dot{\mathbf{r}}^2 \tag{10.90}$$

then Eq. (10.89) with $\ddot{\mathbf{r}} = \mathbf{a}$ gives

$$\frac{M}{g_0} \mathbf{a} \cdot \frac{\partial \mathbf{r}}{\partial q_r} = \frac{d}{dt}\left(\frac{\partial(\text{KE})}{\partial \dot{q}_r}\right) - \frac{\partial(\text{KE})}{\partial q_r} \tag{10.91}$$

Next, if a body force is due to gravity, we have

$$\mathbf{F}_B = -\frac{Mg}{g_0}\,\mathbf{n}_y \tag{10.92}$$

The position vector in x, y, z coordinates is

$$\mathbf{r} = x\mathbf{n}_x + y\mathbf{n}_y + z\mathbf{n}_z \tag{10.93}$$

and the gravitational potential energy is

$$\text{PE} = M\frac{g}{g_0}\,y \tag{10.94}$$

Therefore,

$$\mathbf{F}_B \cdot \frac{\partial \mathbf{r}}{\partial q_r} = -\frac{Mg}{g_0}\frac{\partial y}{\partial q_r} = -\frac{\partial(\text{PE})}{\partial q_r} \tag{10.95}$$

Since PE is not a function of $\dot{q}_r$, the Lagrangian

$$\ell = \text{KE} - \text{PE} \tag{10.96}$$

can be employed in Eqs. (10.95), (10.91), and (10.84) to give the equation of motion for mass M as

$$\frac{d}{dt}\left(\frac{\partial \ell}{\partial \dot{q}_r}\right) - \frac{\partial \ell}{\partial q_r} = \mathbf{F}_s \cdot \frac{\partial \mathbf{r}}{\partial q_r} \tag{10.97}$$

Equation (10.97) was obtained for one mass increment. If connected systems are indexed by $1, 2, \ldots, i, \ldots, I$, the motion of each system is described by Eq. (10.97) with $\ell = \ell_i$, $\mathbf{F}_s = \mathbf{F}_{si}$, and $\mathbf{r} = \mathbf{r}_i$. these can be added to give

$$\frac{d}{dt}\left(\frac{\partial \mathscr{L}}{\partial \dot{q}_r}\right) - \frac{\partial \mathscr{L}}{\partial q_r} = \sum_{i=1}^{I}\mathbf{F}_{si} \cdot \frac{\partial \mathbf{r}_i}{\partial q_r} \tag{10.98}$$

$$r = 1, 2, \ldots, N \qquad \text{generalized coordinates}$$
$$i = 1, 2, \ldots, I \qquad \text{index of connected systems}$$

where

$$\mathscr{L} = \sum_{i=1}^{I} \ell_i = (\text{KE} - \text{PE})_{\text{all systems}} \tag{10.99}$$

Equation (10.98) generally gives N equations, which collectively are called Lagrange's equations of motion.

Usually KE and PE can be obtained for each system in terms of various generalized coordinates. It can be shown that PE also can include spring energy $Kx^2/2$, or other forms of energy stored from reversible processes.

Each index r provides one equation of motion, which may contain more than one generalized coordinate. The right side of Eq. (10.98) is summed over the i systems for each r. That is, the rth equation of motion requires a summation over I systems, as does the $r+1^{th}$ system, and so on. Interior forces of adjacent systems are equal, but of opposite direction. If the point of contact is described by a nonslipping condition, there is no contribution to the summation of Eq. (10.98).

EXAMPLE 10.5: CONCENTRIC CYLINDER COUPLING
Consider two concentric cylinders of length L, which can move horizontally as shown in Fig. 10.14. The annular space between them is filled with liquid of density ρ. The inner and outer cylinders of radii R_i and R_o and masses M_i and M_o are connected by a spring of elastic constant K_i. The outer is connected to a stationary anchor by a spring of constant K_o. The differential equations of motion for small displacements x_i and x_o are to be obtained.

The liquid can be treated as inviscid, two-dimensional flow with negligible end effects and gravity.

The two generalized coordinates for this problem are $q_1 = x_i$ and $q_2 = x_o$. Kinetic energies of the cylinders are

$$\text{KE}_o = \frac{M_o}{2g_o} \dot{x}_o^2 \quad \text{and} \quad \text{KE}_i = \frac{M_i}{2g_o} \dot{x}_i^2$$

A solution of Eq. (9.17) with boundary conditions corresponding to the moving inner and outer cylinders, namely,

$$\left.\frac{\partial \phi}{\partial r}\right|_{r=R_o} = \dot{x}_o \cos \theta , \qquad \left.\frac{\partial \phi}{\partial r}\right|_{r=R_i} = \dot{x}_i \cos \theta$$

yields the potential function for liquid motion in the form,

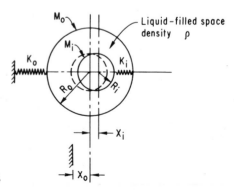

Figure 10.14   Liquid-Coupled Cylinders

$$\phi = \left[Ar + \frac{B}{r} \right] \cos \theta$$

where

$$A = \frac{R_o^2 \dot{x}_o - R_i^2 \dot{x}_i}{R_o^2 - R_i^2}, \qquad B = \frac{(\dot{x}_o - \dot{x}_i)R_i^2 R_o^2}{R_o^2 - R_i^2}$$

The liquid KE is obtained from Section 9.18 as

$$KE_{liq} = \frac{\pi \rho L}{2g_0} \frac{(\dot{x}_o R_o^2 + \dot{x}_i R_i^2)^2 + R_i^2 R_o^2 (\dot{x}_o + \dot{x}_i)^2}{R_o^2 - R_i^2}$$

Also, the potential energies of the inner and outer springs are

$$PE_i = (1/2)K_i x_i^2, \qquad PE_o = (1/2)K_o x_o^2$$

The point of contact of the external force applied by the anchor to the outer spring does not move and contributes nothing to the summation of Eq. (10.98). First putting $q_1 = x_i$ and then $q_2 = x_0$, direct substitution into Eqs. (10.98) and (10.99) yields the two equations of motion

$$a_1 \ddot{x}_i + a_2 x_i = -a_3 \ddot{x}_o$$

$$b_1 \ddot{x}_o + b_2 x_o = -b_3 \ddot{x}_i$$

where

$$a_1 = 1 + \left(\frac{\pi \rho L R_i^2}{M_i} \right) \left(\frac{R_0^2 + R_i^2}{R_0^2 - R_i^2} \right)$$

$$a_2 = \frac{K_i g_0}{M_i}, \qquad a_3 = 2 \left(\frac{\pi \rho L R_i^2}{M_i} \right) \left(\frac{R_0^2}{R_0^2 - R_i^2} \right)$$

$$b_1 = 1 + \left(\frac{\pi \rho L R_o^2}{M_o} \right) \left(\frac{R_o^2 + R_i^2}{R_o^2 - R_i^2} \right)$$

$$b_2 = \frac{K_o g_o}{M_o}, \qquad b_3 = 2 \left(\frac{\pi \rho L R_o^2}{M_o} \right) \left(\frac{R_i^2}{R_o^2 - R_i^2} \right)$$

This problem gives two coupled differential equations in each of the generalized coordinates.

The method of Lagrangian mechanics also can be applied to a fluid system which is on a moving reference frame, as shown in the next example.

EXAMPLE 10.6: MANOMETER MOTION, LAGRANGE'S METHOD
Consider the manometer problem of Example 10.1 in which it was necessary to write mass and momentum equations for each of the three liquid columns. This same problem can be formulated easier with the method of Lagrangian mechanics.

Figure 10.2 shows the known accelerations a_x and a_y in x and y. It follows that manometer velocity components are

$$u = \int_0^t a_x(t)\, dt\,, \qquad v = \int_0^t a_y(t)\, dt$$

Elevation of the left liquid column is η, and its vertical velocity relative to the manometer is $\dot{\eta}$. If we use the horizontal section as a datum for potential energy, we have

$$\mathrm{PE} = \mathrm{PE}_{\mathrm{left}} + \mathrm{PE}_{\mathrm{right}}$$

$$= \frac{\rho A g}{2 g_0}\left[(H + \eta)^2 + (H - \eta)^2\right]$$

The kinetic energy of the liquid relative to the earth is

$$\mathrm{KE}_{\mathrm{left}} = \frac{\rho A(H + \eta)}{2 g_0}\left[u^2 + (v + \dot{\eta})^2\right]$$

$$\mathrm{KE}_{\mathrm{horiz}} = \frac{\rho A L}{2 g_0}\, (u - \dot{\eta})^2$$

$$\mathrm{KE}_{\mathrm{right}} = \frac{\rho A(H - \eta)}{2 g_0}\left[u^2 + (v - \dot{\eta})^2\right]$$

These energy components are employed in Eqs. (10.98) and (10.99) for the single generalized coordinate η to obtain the result given in Example 10.1.

Dissipative terms are readily handled by the right side of Eq. (10.98) in problems formulated by the method of Lagrangian mechanics.

EXAMPLE 10.7: SPRING-MASS SYSTEM WITH DRAG
Consider the submerged spring-mass system of Fig. 10.15 on which the surrounding liquid of density ρ exerts a quadratic drag force, $D|\dot{x}|\dot{x}$. The spring potential energy is

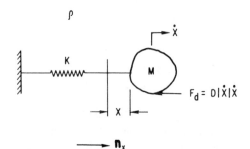

Figure 10.15   Spring-Mass System with Quadratic Drag

$$PE_{spring} = (1/2)Kx^2$$

Furthermore, the kinetic energy of the mass M is

$$KE = \frac{M}{2g_0} \dot{x}^2$$

where M includes the virtual mass of the liquid (Table 9.3). The right side of Eq. (10.98) is obtained by writing the drag force acting on the object as

$$\mathbf{F}_d = -D|\dot{x}|\dot{x}\mathbf{n}_x$$

If $\mathbf{F}_d$ acted at one point on the object, the position vector of its point of application could be written as

$$\mathbf{r} = x\mathbf{n}_x$$

so that we have, for $q_r = x$,

$$\frac{\partial \mathbf{r}}{\partial x} = \mathbf{n}_x$$

from which

$$\sum_{i=1}^{I} \mathbf{F}_{si} \cdot \frac{\partial \mathbf{r}_i}{\partial q_r}\bigg|_{q_r=x} = \mathbf{F}_d \cdot \frac{\partial \mathbf{r}}{\partial x} = -D|\dot{x}|\dot{x}$$

The final result of employing Eqs. (10.98) and (10.99) is

$$\ddot{x} + \frac{Dg_0}{M}|\dot{x}|\dot{x} + \frac{Kg_0}{M}x = 0$$

which is the classical equation of a spring-mass oscillator with quadratic damping.

The right side of Eq. (10.98) can be evaluated for other fluid structure systems where the fluid and structure contact points slip or move relative to each other. Consider a horizontal section of pipe which is moving at velocity $\dot{x}_p$, as shown in Fig. 10.16. Contained liquid is moving at velocity $\dot{x}_L$. Both velocities are relative to the earth. The pressure loss across the orifice is

$$P_1 - P_2 = \frac{K\rho}{2g_0}|(\dot{x}_L - \dot{x}_p)|(\dot{x}_L - \dot{x}_p) \tag{10.100}$$

The momentum law can be applied to liquid in the vicinity of the orifice to show that the liquid exerts a force on the pipe via the orifice, given by

$$\mathbf{F}_p = (P_1 - P_2)A\mathbf{n}_x \tag{10.101}$$

whereas the pipe exerts an equal and opposite force on the liquid

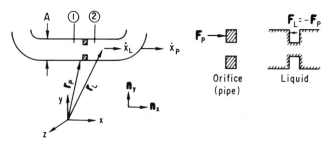

Figure 10.16   Motion of Pipe and Contained Liquid

$$\mathbf{F}_L = -\mathbf{F}_p = -(P_1 - P_2)A\mathbf{n}_x \tag{10.102}$$

The position vector of some point on the pipe, say, at the orifice, is written as

$$\mathbf{r}_p = x_p\mathbf{n}_x + y_p\mathbf{n}_y \tag{10.103}$$

so that we have for the partial derivatives

$$\frac{\partial \mathbf{r}_p}{\partial x_p} = \mathbf{n}_x \quad \text{and} \quad \frac{\partial \mathbf{r}_p}{\partial x_L} = 0 \tag{10.104}$$

Moreover, since force $\mathbf{F}_L$ acts on liquid which is moving relative to the pipe, we consider a liquid particle with the position vector

$$\mathbf{r}_L = x_L\mathbf{n}_x + y_L\mathbf{n}_y \tag{10.105}$$

which is differentiated to give

$$\frac{\partial \mathbf{r}_L}{\partial x_L} = \mathbf{n}_x \quad \text{and} \quad \frac{\partial \mathbf{r}_L}{\partial x_p} = 0 \tag{10.106}$$

It follows that the right side of Eq. (10.98) becomes, for $q_r = x_p$,

$$\mathbf{F}_p \frac{\partial \mathbf{r}_p}{\partial x_p} + \mathbf{F}_L \cdot \frac{\partial \mathbf{r}_L}{\partial x_p} = (P_1 - P_2)A + 0 \tag{10.107}$$

whereas, for $q_r = x_L$,

$$\mathbf{F}_p \cdot \frac{\partial \mathbf{r}_p}{\partial x_L} + \mathbf{F}_L \cdot \frac{\partial \mathbf{r}_L}{\partial x_L} = 0 - (P_1 - P_2)A \tag{10.108}$$

The same overall procedure is readily extended to systems with additional i components and r generalized coordinates.

EXAMPLE 10.8: MOTION DAMPING DEVICE

Consider the spring-mass system of Fig. 10.17 with an orificed liquid manometer attached. The orifice damps liquid motion by a pressure drop given by Eq. (10.100). The equations of motion are to be obtained for both the x displacement of the device and the liquid displacement y in the manometer.

The potential energies needed are given by

$$PE_{left} = \frac{\rho A(H+y)^2}{2g_0}$$

$$PE_{right} = \frac{\rho A(H-y)^2}{2g_0}$$

$$PE_{spring} = (1/2)K_s x^2$$

The kinetic energies must include the total velocity of each liquid increment. Thus,

$$KE_{left} = \frac{\rho A(H+y)}{2g_0}(\dot{x}^2 + \dot{y}^2)$$

$$KE_{horiz} = \frac{\rho AL}{2g_0}(\dot{x} - \dot{y})^2$$

$$KE_{right} = \frac{\rho A(H-y)}{2g_0}(\dot{x}^2 + \dot{y}^2)$$

$$KE_M = \frac{M\dot{x}^2}{2g_0}$$

The liquid force on the pipe (orifice) is

$$\mathbf{F}_p = (P_1 - P_2)A\mathbf{n}_x$$

where

$$P_1 - P_2 = \frac{-K\rho}{2g_0}|\dot{y}|\dot{y}$$

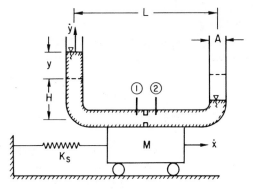

Figure 10.17 Motion Damping Device

The pipe force on liquid is equal and opposite; that is, $\mathbf{F}_L = -\mathbf{F}_p$. The pipe position vector is

$$\mathbf{r}_p = x\mathbf{n}_x$$

and the position vector of a liquid particle in the pipe is

$$\mathbf{r}_L = (x - y)\mathbf{n}_x$$

All of the necessary quantities are ready to be employed in Eqs. (10.98) and (10.99) to obtain the equations of motion

$$\left(1 + \frac{2\rho HA}{M} + \frac{\rho AL}{M}\right)\ddot{x} + \frac{K_s g_0}{M}x = \frac{\rho AL}{M}\ddot{y}$$

and

$$\ddot{y} + \left(\frac{2g}{L + 2H}\right)y + \frac{K}{2}\frac{1}{L + 2H}|\dot{y}|\dot{y} = \left(\frac{L}{L + 2H}\right)\ddot{x}$$

References

10.1 Lamb, H., *Hydrodynamics*, Dover, New York.

10.2 Davies, R. M., and G. I. Taylor, "The Mechanics of Large Bubbles Rising through Extended Liquids and through Liquids in Tubes," *Proc. Roy. Soc.*, **200** (1950) 375.

10.3 Moody, F. J., "Dynamic and Thermal Behavior of Hot Gas Bubbles Discharged into Water," *Nuclear Engineering and Design*, **95** (1986) 47–54.

10.4 Chu, H. Y., et al., "Analysis of the Fluid-Structure Interaction in 3D Space using the PISCES 3 DELK Code," ASME Special Publication PVP-Vol. 75, *Advances in Fluid-Structure Interaction Dynamics*, 1983.

10.5 Riddell, R. A., and R. E. Schwirian, "Simulation of Fluid-Structure Interaction during a LOCA using Existing Finite Element Formulations," ASME Special Publication PVP-Vol. 75, *Advances in Fluid-Structure Interaction Dynamics*, 1983.

10.6 Takeuchi, K., et al., "Multiflex-A Fortran IV Computer Program for Analyzing Thermal-Hydraulic Structure System Dynamics," WCAP-8709, Feb. 1976.

10.7 Moody, F. J., "Unsteady Condensation and Fluid-Structure Frequency Dependence on Parameters of Vapor Quench Systems," ASME Special Publication PVP-Vol. 46, *Ineractive Fluid Structural Dynamic Problems in Power Engineering*, 1981.

10.8 Au-Yang, M. K., and J. E. Galford, "Fluid-Structure Interaction—A Survey with Emphasis on Its Application to Nuclear Steam System Design," ASME Special Publication PVP-Vol. 46, *Interactive Fluid Structural Dynamic Problems in Power Engineering*, 1981.

10.9 Moody, F. J., "Vessel Internal and External Pressure Loads Caused by Flow Disturbance," ASME Special Publication PVP-39, *Dynamics of Fluid-Structure Systems in the Energy Industry*, 1979.

10.10 Shin, Y. W., F. J. Moody, and M. K. Au-Yang, Eds, *Fluid Transients and Fluid-Structure Interaction*, ASME Special Publication PVP-64, 1982.

10.11 Marks, J. S., and G. B. Andeen, "Chugging and Condensation Oscillation Tests," EPRI NP-1167, Sept. 1979.

Problems

10.1 Consider a manometer like the one shown in Fig. 10.2. It rests on a platform that is undergoing vertical oscillatory motion with acceleration $a_y(t) = \epsilon \cos \omega t$. Show that the differential equation which describes motion of the liquid can be put into the form of Mathiu's equation (see Problem 9.37).

10.2 Show from Eq. (10.6) that liquid motion in the manometer of Fig. 10.2 with zero horizontal acceleration can be made to undergo stable, neutrally stable, and unstable motions for constant vertical accelerations of $a_y \geq 0$, $a_y = -g$ (free fall), and $a_y < -g$ respectively.

10.3 Show that the Bernoulli equation, Eq. (9.25), for a translating reference frame, with acceleration components relative to earth of $a_x(t)$, $a_y(t)$, and $a_z(t)$, can be derived in the form

$$\frac{\partial \phi}{\partial t} + \frac{{}^m V^2}{2} + \frac{g_0}{\rho} P + \Omega(t) = F(t)$$

where

$$\Omega(t) = a_x(t)x + a_y(t)y + a_z(t)z$$

10.4 Suppose that an upright cylindrical container of liquid is rotated about its vertical axis at constant angular velocity $\omega = \omega n_y$ for a long time so that the liquid rotates with the container.

(a) Apply Eqs. (10.4) and (10.5) to an element of liquid mass $dM = \rho r \, d\theta \, dr \, dy$ where r, θ are in the horizontal plane, y is in the vertical direction, and only gravity and pressure forces act on it. Employ $\mathbf{r} = r\mathbf{n}_r$ to obtain

$$\left(-\frac{\partial P}{\partial r} + \frac{r\Omega^2 \rho}{g_0}\right)\mathbf{n}_r - \left(\frac{\partial P}{\partial y} + \rho\frac{g}{g_0}\right)\mathbf{n}_y = 0$$

(b) Show that a solution for pressure in the liquid is

$$P = \rho \frac{g}{g_0} \left(\frac{\Omega^2 r^2}{2g} - y \right) + \text{constant}$$

Note that on the interface where P is equal to ambient pressure P_∞ the interface has a parabolic shape.

10.5   It has been determined that the void fraction in a bubbly mixture of gas and liquid is 0.01 in a long pipe.

(a) Show that for bubbles of $r = 1.0$ mm (0.039 in) radius, a bubble/liquid heat transfer coefficient of $\mathscr{H} = 50$ W/m^2-K (8.9 Btu/h-ft^2-F), liquid sound speed of $C_L = 1500$ m/s (4920 ft/s), a gas constant $R_g = 287$ J/kg-K (53.3 ft-lbf/lbm-R), and pressure $P = 101$ kPa, (14.7 psia), the acoustic damping coefficient is $D = 163$ s^{-1}.

(b) Show from Fig. 10.6 that for the bubbly mixture of part (a), the amplitude of a high frequency disturbance will be attenuated to 0.1 of its originating value in a distance of $z = 36.8$ m (121 ft), whereas a frequency of $\omega = 16.3$ rad/s will only be attenuated to 0.4 of its originating amplitude.

10.6   Show by direct substitution of the solution form

$$P(z, t) = P_0 e^{-\sigma z} \sin 2\pi f \left(t - \frac{z}{C_\omega} \right)$$

that σ and C_ω are given by Eq. (10.38). Compare with the results of Section 7.15.

10.7   Verify that the wavelength in Problem 10.6 or Section 10.4 depends on the frequency according to

$$\lambda = \frac{2\pi(C/\omega)\sqrt{2}}{[1 + \sqrt{1 + (D/\omega)^2}]^{1/2}}$$

Note that $2\pi C/\omega$ is the wavelength without thermal damping ($D = 0$). Discuss how the wavelength changes with: increasing D, increasing and decreasing ω.

10.8   Consider Eq. (10.29) for thermally damped time- and space-dependent pressure $P(z, t)$ in a uniform flow passage that contains gas bubbles in liquid. Suppose that an oscillating pressure disturbance $P(0, t) = P_0 \sin \omega t$ occurs at the left boundary and propagates into the mixture.

(a) Start with the general solution form $P(z, t) = A e^{i(\omega t + 2\pi z/\lambda)}$ where A is a constant and λ is the wavelength, and show that λ and ω must satisfy

$$\lambda^2 = \frac{C^2}{\omega^2 - iD\omega}$$

(b) Show that the solution of part (a) will satisfy the oscillatory boundary condition if

$$P(0, t) = P_0 \, \text{Imag}(e^{i\omega t})$$

10.9 Consider a solid cylinder of length L, cross-sectional area a, and mass M floating vertically in liquid of density ρ, which is contained in a tank of cross-sectional area A as shown in Fig. P10.9. The liquid elevation is h, and y is elevation to the cylinder. Equilibrium elevations are h_e and y_e. The cylinder is lifted to elevation y_i and released, and its bobbing motion is to be determined by Lagrange's method. Viscous effects are negligible, and vertical liquid velocity between the cylinder and tank walls is uniform, although time-dependent.

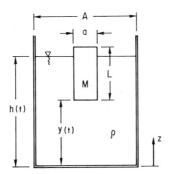

Figure P10.9

(a) Show from buoyancy considerations that at the equilibrium state,

$$h_e = y_e + \frac{M}{\rho a}$$

(b) Show that at any time while bobbing continues, h and y are related by

$$h = \frac{M}{\rho a} + \left(\frac{A}{A - a}\right) y_e - \left(\frac{a}{A - a}\right) y$$

(c) Verify that the gravitational potential energy of the system relative to $y = 0$ is

$$\text{PE} = \frac{\rho g}{2g_0} [(A - a)h^2 + ay^2] + M \frac{g}{g_0} \left(y + \frac{L}{2}\right)$$

(d) Assume that for vertical motion of the cylinder, liquid velocity is zero in $0 \le z \le y(t)$, and vertically uniform in $y(t) < z < h(t)$. Verify that $\dot{h} = -a\dot{y}/(A - a)$, and that the system kinetic energy is

$$KE = \left[M + \frac{\rho a^2}{A - a} (y_e - y) \right] \left(\frac{A}{A - a} \right) \frac{\dot{y}^2}{2g_0}$$

(e) Obtain the differential equation of bobbing motion in terms of $y(t)$ using Lagrange's method. Show that if the initial conditions are $y(0) = y_i$ and $\dot{y}(0) = 0$, the motion is described by

$$\text{DE:} \quad (1 - B\eta^*) \frac{d^2\eta^*}{dt^{*2}} + \frac{B}{2} \left(\frac{d\eta^*}{dt^*} \right)^2 + \eta^* = 0$$

$$\text{IC:} \quad t^* = 0, \quad \eta^* = 1, \quad \frac{d\eta^*}{dt^*} = 0$$

where

$$\eta^* = \frac{y - y_e}{y_i - y_e}, \quad t^* = t\sqrt{\frac{\rho g a}{M}}, \quad B = \frac{\rho a^2(y_i - y_e)}{M(A - a)}$$

(f) Show that a sinusoidal oscillation occurs with a circular frequency $\omega = \sqrt{\rho g a / M}$ if $A \to \infty$.

(g) Show that the first integral of the large amplitude solution is given by

$$\frac{d\eta^*}{dt^*} = \pm \frac{\sqrt{2}}{B} \sqrt{\frac{(1 - B\eta^*)}{(1 - B)} - 1 - \ln \frac{(1 - B\eta^*)}{(1 - B)}}$$

10.10   Consider the two submerged square channels enclosed by a rigid rectangular container, shown in the top view of Fig. P10.10. The square channels have sides of width $2H$ and mass M. The dotted

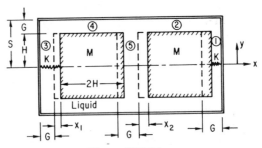

Figure P10.10

lines show the channels in their equilibrium positions. Length S is $H + G$ where G is the normal gap width. The channels extend to a depth D into the page. The elastic property for sideways motion is simulated by linear springs of constant K, which attach the channels to the rigid frame. It is reasonable to assume that the channels translate uniformly throughout depth D. If the left channel moves a distance x_1 from the equilibrium dotted position, the attached spring exerts a leftward force Kx_1. Neglect flow velocity in the gaps perpendicular to the sketch. Water in the gaps will flow around the channels as they move sideways in the frame. That is, as x_2 increases, water is squeezed out of the rightmost gap ①, leftward past the channel into either of the other gaps ⑤ or ③ or both. Thus, there is a kinetic energy associated with water in the gap. The water kinetic energy can be approximated by one-dimensional flow in each of the gaps. Consider only sideways left-to-right motion of the channels

(a) Assume symmetric flow about the x axis and show that the total liquid kinetic energy is given by

$$\begin{array}{l} \text{KE}_L \\ \text{(gap)} \end{array} = \frac{\rho D H^3}{g_0}$$

$$\times \left[\underbrace{\frac{\dot{x}_1^2}{3(G+x_1)}}_{③} + \underbrace{\frac{\dot{x}_2^2}{3(G-x_2)}}_{①} + \underbrace{\frac{(\dot{x}_1-\dot{x}_2)^2}{3(G+x_2-x_1)}}_{⑤} + \underbrace{\frac{2}{G}(\dot{x}_2^2+\dot{x}_1^2)}_{②\quad④} \right]$$

Hint: Kinetic energy in gaps ①, ③, and ⑤ requires integration of $V^2/2g_0 \, dM$ where $V = V(y)$.

(b) Neglect liquid potential energy, but include spring energy $Kx^2/2$ and channel kinetic energy, and show that for x_1 and x_2 small relative to G, the describing equations of motion are

$$\left(\frac{16}{3}\beta+1\right)\begin{Bmatrix}\ddot{x}_1\\\ddot{x}_2\end{Bmatrix} - \frac{2}{3}\beta\begin{Bmatrix}\ddot{x}_2\\\ddot{x}_1\end{Bmatrix} + \omega^2\begin{Bmatrix}x_1\\x_2\end{Bmatrix} = 0$$

where $\omega^2 = Kg_0/M$, and the liquid coupling coefficient is $\beta = \rho DH^3/GM$.

(c) Discuss how the coupling coefficient β affects the motion. How many frequencies are possible? What happens to the motion when $\beta = 0$ and when $\beta \to \infty$?

10.11 Figure P10.11 shows a fluid column of length L and mass ρAL with a prescribed pressure $P(0, t)$ at the left boundary. The right end is bounded by a frictionless, nonleaking piston of mass M and area A,

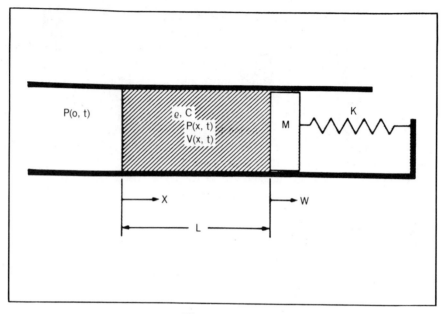

Figure P10.11

attached to a linear spring of constant K. The net force on the piston at any time is $P(L, t)A$, less the spring force Kw. Piston displacement w is measured from the unstressed position, and is small relative to the fluid length L. Hence, the right boundary condition, corresponding to Eq. (10.63) is

$$P^*(1, t^*) - b_1 w^* = b_2 \frac{d^2 w^*}{dt^{*2}}$$

$$b_1 = \frac{Kw_0}{AP_0}, \qquad b_2 = \frac{Mw_0 C^2}{L^2 P_0 A g_0 (1 + C/L\omega_i)^2}$$

The one-dimensional forms of Eqs. (10.65a) and (10.65b) describe fluid response in the cylinder. If the response is incompressible, the combined liquid and piston fundamental frequency is

$$\omega_i = \sqrt{\frac{Kg_0}{M + \rho AL}}$$

A sustained pressure P_0 would create the piston displacement,

$$w_0 = \frac{P_0 A}{K}$$

(a) Show that Eqs. (10.67) yield

$$a_1 = \left[\left(1 + \frac{C}{L\omega_i}\right)\left(1 + \frac{C}{L\omega_i} \frac{1}{1 + \frac{M}{\rho AL}}\right)\right]^{-1}$$

$$a_2 = \left(1 + \frac{\overset{\bullet}{M}}{\rho AL}\right)\left(\frac{L\omega_i}{C}\right)^2 \frac{w_0}{L}$$

$$a_3 = \left(1 + \frac{C}{L\omega_i} \frac{1}{1 + M/\rho AL}\right)\left(1 + \frac{C}{L\omega_i}\right)^{-1}$$

$$a_4 = \left[1 + \left(\frac{L\omega_i}{C}\right)\left(1 + \frac{M}{\rho AL}\right)\right]^2 \frac{w_0}{L}\left(1 + \frac{M}{\rho AL}\right)^{-1}$$

(b) If structural displacement w_0 is considered small so that $w_0/L \approx 0$, show that Eqs. (10.65a) and (10.65b) can be combined to give

$$\frac{\partial^2 P^*}{\partial t^{*2}} - C^{*2} \frac{\partial^2 P^*}{\partial x^{*2}} = 0$$

with

$$C^* = 1 + \frac{C}{L\omega_i}$$

(c) Employ the left boundary condition

$$P^*(0, t^*) = 0$$

to show that the differential equation of part (b) is satisfied by

$$P_n^*(x^*, t^*) = (B_1 \cos \omega_n^* t^* + B_2 \sin \omega_n^* t^*) \sin \frac{\omega_n^* x^*}{C^*}$$

where ω_n^* is a nondimensional frequency

$$\omega_n^* = \omega_n \left(\frac{L}{C} + \frac{1}{\omega_i}\right) = \frac{\omega_n L}{C} C^*$$

yet to be determined.

(d) Show that Eq. (10.65b) yields the boundary condition,

$$\frac{\partial V_n^*}{\partial t_n^*}(1, t^*) = -\frac{1}{a_3} \frac{\partial P_n^*}{\partial x^*}(1, t^*)$$

and that since the piston and adjacent fluid have the same velocity, the nondimensional coupling equation is

$$\frac{w_0}{L} \frac{dw_n^*}{dt^*} = \frac{a_2}{a_1} V_n^*(1, t^*)$$

(e) Show that the right boundary condition can be differentiated twice with respect to t^* in order to express it in terms of P^* as

$$\frac{\partial^3 P_n^*(1, t^*)}{\partial x^* \partial t^{*2}} + \left(\frac{\rho AL}{M}\right) \frac{\partial^2 P_n^*(1, t^*)}{\partial t^{*2}} + \left(1 + \frac{\rho AL}{C}\right)$$

$$\times \left(1 + \frac{\omega_i L}{C}\right)^2 \frac{\partial P_n^*(1, t^*)}{\partial x^*} = 0$$

(f) Show that the solution form of part (c) and the piston boundary condition of part (e) yield the eigenfrequency equation,

$$\tan\left(\frac{\omega_n L}{C}\right) = \left[\left(1 + \frac{\rho AL}{M}\right)\left(\frac{\omega_i L}{C}\right)^2 - \left(\frac{\omega_n L}{C}\right)^2\right]\left[\left(\frac{\rho AL}{M}\right)\left(\frac{\omega_n L}{C}\right)\right]^{-1}$$

(g) Several natural frequencies were calculated from the eigenfrequency equation of part (f), and are given in a table below for various ranges of $\omega_i L/C$ and $\rho AL/M$. Study the table and list any significant observations.

Liquid Column-Piston-Spring Natural Frequencies, $\omega_n L/C$

	0.2	0.2	0.2	0.2	0.2	0.2	0.2
	0.5	0.5	0.5	0.5	0.4	0.4	0.4
	1.0	0.9	0.9	0.9	0.9	0.9	0.9	First
	2.0	1.5	1.4	1.4	1.3	1.3	1.3	$n = 1$
	5.0	1.6	1.6	1.6	1.6	1.6	1.6
	10.0	1.6	1.6	1.6	1.6	1.6	1.6
	0.2	1.6	1.8	2.1	2.3	2.6	2.8
	0.5	1.6	1.9	2.1	2.3	2.6	2.8
$\frac{\omega_i L}{C} =$	1.0	1.7	2.0	2.3	2.5	2.8	3.2	Second
	2.0	2.3	2.6	3.0	3.4	3.7	3.9	$n = 2$
	5.0	4.6	4.6	4.6	4.5	4.5	4.5
	10.0	4.7	4.7	4.7	4.7	4.7	4.7
$\frac{L}{C}\sqrt{\frac{Kg_0}{M + \rho AL}}$	0.2	4.7	4.7	4.8	5.1	5.4	5.7
	0.5	4.7	4.7	4.8	5.1	5.5	5.7
	1.0	4.7	4.7	4.9	5.1	5.5	5.8	Third
	2.0	4.7	4.7	5.0	5.3	5.7	6.3	$n = 3$
	5.0	5.6	6.1	6.8	7.3	7.5	7.5
	10.0	7.8	7.8	7.8	7.8	7.8	7.8
	0.2	7.8	7.8	8.0	8.1	8.5	8.8
	0.5	7.8	7.8	8.0	8.1	8.5	8.8
	1.0	7.8	7.8	8.0	8.1	8.5	8.8	Fourth
	2.0	7.8	7.8	8.0	8.2	8.6	8.8	$n = 4$
	5.0	7.8	8.0	8.3	9.1	10.1	10.4
	10.0	10.1	10.9	10.9	10.9	10.9	10.9
$\frac{\rho AL}{M} =$		0.2	0.5	1.0	2.0	5.0	10.0

10.12 A rigid sphere of radius R, submerged in inviscid, slightly compressible fluid of density ρ, oscillates on the x axis according to $x_s(t) = \epsilon \sin \omega t$ where ϵ is a small amplitude.

(a) Use Eq. (10.48) and the velocity boundary condition normal to the spherical surface to show that the function F is determined from

$$\frac{1}{C^2} F'' + \frac{2}{RC} F' + \frac{2}{R^2} F = -\epsilon \omega \cos \omega t$$

(b) Show that for sustained motion only the particular solution survives and is given by

$$F(t) = -\frac{\epsilon R^2 \omega}{2} \left\{ \frac{[1 - \frac{1}{2}(\omega R/C)^2]}{[1 + \frac{1}{4}(\omega R/C)^4]} \cos \omega t + \frac{(\omega R/C)}{[1 + \frac{1}{4}(\omega R/C)^4]} \sin \omega t \right\}$$

and that if terms of $O(1/C)$ and smaller are neglected, a reduced function $F(t) = -(\epsilon R^2 \omega/2) \cos \omega t$ could be obtained directly from the differential equation without solving it.

(c) Neglect terms of $O(1/C)$ and show that the horizontal force F_x which must be exerted by the sphere on the surrounding fluid is

$$F_x = -\frac{M_L}{g_0} \frac{\epsilon \omega^2}{2} \sin \omega t, \qquad M_L = \frac{4}{3} \pi R^3 \rho$$

(d) If the mass of the sphere is M_s, the force which must be applied to sustain the motion is $F_{total} = F_x + M_s \ddot{x}_s/g_0$. Show that the power requirement is

$$\mathcal{P} = -\frac{(M_L + M_s)}{g_0} \frac{\epsilon^2 \omega^3}{2} \sin \omega t \cos \omega t$$

Note that power increases as the cube of the frequency.

10.13 Consider the sound wave process of disintegrating a stationary solid sphere, similar to Section 10.7. Let an oscillator generate an x-directional velocity disturbance $u_0 \cos \omega t$ that occurs for a long enough time to establish a wave train throughout the flow field with a velocity potential given by

$$\phi_w(x, t) = -\frac{u_0 C}{\omega} \sin \omega \left(t - \frac{x}{C} \right), \qquad x = r \cos \theta$$

Assume that a rigid sphere of radius R is placed at the origin with the general velocity potential ϕ_s of Eq. (10.48).

(a) Superimpose ϕ_w and ϕ_s, and for the boundary condition $\partial \phi / \partial r = 0$ at $r = R$, with $t \gg R \cos \theta / C$, show that the function $F(t)$ must satisfy

$$\frac{1}{C^2} F'' + \frac{2}{RC} F' + \frac{2}{R^2} F = \frac{u_0 C}{\omega} \cos \omega t$$

and that the integration of presure over the spherical surface gives the force of Eq. (10.49).

(c) Show that if terms of $O(1/C)$ and smaller are neglected, the force exerted by the pressure wave train on the stationary sphere is

$$F_x = \frac{2}{3} \pi R^3 \frac{\rho}{g_0} u_0 \omega \sin \omega t$$

(d) Show that the fully incompressible (bulk flow) problem yields

$$F_x = 2 \pi R^3 \frac{\rho}{g_0} u_0 \omega \sin \omega t$$

APPENDIX   A

A.1 Unit Conversion Factors

Length	1.0 m	= 1000 mm
		= 100 cm
		= 0.001 km
		= 3.28 ft
		= 39.36 in
		= 0.000621 mi
Area	$1.0 \ m^2$ = $10.758 \ ft^2$	
Time	1.0 s = 0.01667 min	
	= 0.0002778 h	
Force	$1.0 \ N = 10^5$ dynes	
	= 0.2248 lbf	
Mass	1.0 kg = 1000 g	
	= 2.20 lbm	
	= 0.0685 slugs	
Angle	$1.0 \ rad = (360/2\pi)°$	
Temperature	$(°F - 32)/1.8 = °C$	
	$1.8(°C) + 32 = °F$	
	$°R = °F + 460$ (absolute)	
	$°K = °C + 273$ (absolute)	
Mass Flux	$1.0 \ kg/s\text{-}m^2 = 0.2045 \ lbm/s\text{-}ft^2$	

Work, Energy	$1.0 \, J = 1.0 \, N\text{-}m$
	$= 10^7 \, ergs = 10^7 \, dyne\text{-}cm$
	$= 0.239 \, cal$
	$= 0.000948 \, B$
	$= 0.737 \, ft\text{-}lbf$
Power	$1.0 \, W = 1.0 \, J/s$
	$= 0.737 \, ft\text{-}lbf/s$
	$= 0.00134 \, hp$
	$= 0.000948 \, B/s$
Pressure, Shear Stress	$1.0 \, Pa = 1.0 \, N/m^2$
	$= 0.000145 \, lbf/in^2$
	$= 9.867 \times 10^{-6} \, atm$
	$101,325 \, Pa = 101.3 \, kPa = 1.0 \, atm$
Viscosity, Dynamic	$1.0 \, N\text{-}s/m^2 = 10 \, P \, (poise)$
	$= 10 \, dyne\text{-}s/cm^2$
	$= 0.0209 \, lbf\text{-}s/ft^2$
Viscosity, Kinematic	$1.0 \, m^2/s = 10.758 \, ft^2/s$
Surface Tension	$1.0 \, N/m = 0.0685 \, lbf/ft$
	$= 1000 \, dyne/cm$
Volume	$1.0 \, L = 1000 \, cm^3$
	$= 0.001 \, m^3$
	$= 0.0353 \, ft^3$
	$= 0.264 \, gal$
	$1.0 \, L/s = 15.84 \, gal/min$
Density	$1.0 \, kg/m^3 = 0.06232 \, lbm/ft^3$
Specific Heat	$1.0 \, J/g\text{-}K = 0.2396 \, B/lbm\text{-}°F$
Thermal Conductivity	$1.0 \, W/m\text{-}K = 0.578 \, B/h\text{-}ft\text{-}°F$
Gas Constant	$1.0 \, m\text{-}N/kg\text{-}K = 0.186 \, ft\text{-}lbf/lbm\text{-}°F$
Convection Coefficient	$1.0 \, W/m^2\text{-}K = 0.176 \, B/h\text{-}ft^2°F$
Heat Flux	$1.0 \, W/m^2 = 0.317 \, B/h\text{-}ft^2$
Specific Energy	$1.0 \, J/g = 0.431 \, B/lbm$

A.2 Approximate Propagation Speeds

Sonic Speed

$$\sqrt{g_0 \left(\frac{\partial p}{\partial \rho} \right)_s}$$

Perfect gas $\sqrt{k g_0 R_g T}$	Approximate Values
Standard air	335 m/s
Saturated steam	450 m/s
Cold water	1500 m/s
Steel	5000 m/s

Shock Wave in Still Gas

$$C_x \sqrt{\frac{k+1}{2k} \frac{P_y}{P_x} + \frac{k-1}{2k}}$$

C_x = Sonic speed in undisturbed gas
k = ratio of specific heats
P_x = undisturbed pressure
P_y = shock pressure

Speed of Hydrostatic Waves

$$\sqrt{gH}, \qquad H = \text{depth}$$

Surface Wave Speed in Deep Liquid

$$\sqrt{\frac{g\lambda}{2\pi} \tanh \frac{2\pi H}{\lambda}}$$

λ = wavelength
H = liquid depth
g = acceleration of gravity

Heat Conduction Front Propagation Speed in x Direction

$$2\alpha/x$$

Laminar Shear Stress Propagation Speed in y Direction (y normal to flow direction x)

$$2\nu/x$$

Turbulent Shear Stress in y Direction (y normal to flow direction x, U = stream velocity)

$$(0.238) \frac{\nu}{y} \left(\frac{U_\infty y}{\nu} \right)^{3/4}$$

A.3 Selected Transport Properties, 1.0 ATM

	Air	Sat. Steam	Water
Kinematic Viscosity ν			
cm^2/s	0.14	0.218	0.011
ft^2/s	1.5×10^{-4}	2.34×10^{-4}	1.22×10^{-5}
Thermal Conductivity κ			
W/m-K	0.026	0.023	0.60
B/h-ft-°F	0.015	0.014	0.34
Thermal Diffusivity α			
cm^2/s	0.186	0.223	0.0014
ft^2/h	0.72	0.864	5.47×10^{-3}
Dynamic Viscosity μ			
$N\text{-}s/m^2$	1.8×10^{-5}	1.3×10^{-5}	1.13×10^{-3}
lbm/ft-s	1.2×10^{-5}	0.87×10^{-5}	0.76×10^{-3}
Density ρ			
kg/m^3	1.206	0.597	1000
lbm/ft^3	0.0752	0.0372	62.4
Specific Heats $c_p(c_v)$			
J/g-K	1.0 (0.71)	1.88 (1.45)	4.17
B/lbm-F	0.24 (0.17)	0.45 (0.35)	1.00

APPENDIX B

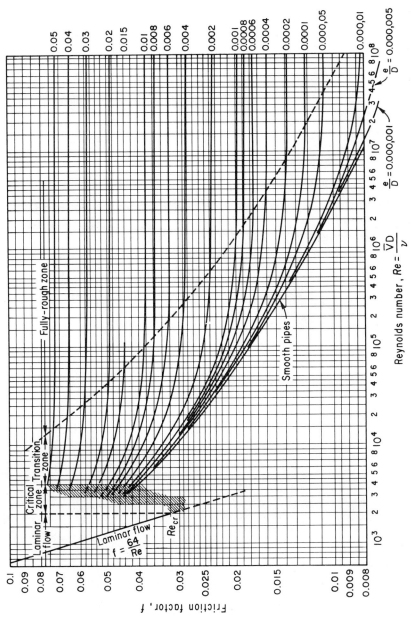

Fig. B.1 Pipe Friction Factor (Fox & McDonald, *Fluid Mechanics*, Wiley)

APPENDIX C

C.1 Second-Order Runge-Kutta Solution Method for Second-Order Differential Equations

Suppose it is necessary to solve a second-order ordinary differential equation of the form,

$$y'' = f(y', y, t), \qquad y(0) = y_i, \quad y'(0) = y'_i$$

First, rewrite as two first order differential equations,

$$y' = z, \qquad y(0) = y_i$$

$$y'' = z' = f(z, y, t), \qquad z(0) = y'(0) = y'_i$$

Input initial values of y, z, and also Δt.
First Coefficients
$\quad y1 = z$
$\quad z1 = f(z, y, t)$
$\quad ys = y + y1 \cdot \Delta t$
$\quad zs = z + z1 \cdot \Delta t$
Second Coefficients
$\quad y2 = zs$
$\quad z2 = f(zs, ys, t)$
Current values of y and z
$\quad y = y + 0.5(y1 + y2)\Delta t$
$\quad z = z + 0.5(z1 + z2)\Delta t$

Print t, y, $z = y'$
Advance time $t = t + \Delta t$
Continue calculations

The Δt step should be estimated. If it is not obvious, calculate with successively smaller values of Δt until minor changes in results are obtained.

C.2 Regular Perturbation Solution of Differential Equations with a Small Parameter or Small Disturbance

The so-called *regular perturbation method* provides an orderly procedure for obtaining approximate solutions of non-linear or otherwise complicated differential equations which contain a small parameter, or have a small disturbance imposed on an otherwise steady condition.

Differential Equation with a Small Parameter

Consider the problem

$$\text{DE:} \quad \frac{dy}{dt} = \frac{\epsilon}{2} \frac{1}{\sqrt{1 + \epsilon t}}, \quad \epsilon \text{ is small} \quad \text{(C.1)}$$

$$\text{IC:} \quad t = 0, \quad y = 1 \quad \text{(C.2)}$$

Suppose that (C.1) and (C.2) were too difficult to solve directly. A regular perturbation procedure is employed by first expanding the function $y(t)$ in powers of ϵ:

$$y(t) = y_0(t) + \epsilon y_1(t) + \epsilon^2 y_2(t) + \cdots = \sum_{n=0}^{N} \epsilon^n y_n(t) \quad \text{(C.3)}$$

Here, N is the number of terms desired in the series, and $y_0, y_1, y_2, \ldots, y_N$ are functions to be determined. If (C.3) is substituted into (C.1) and (C.2) and individual terms of order $n = 0, 1, 2, \ldots, N$ are collected together, an ordered set of problems is obtained, which are then solved sequentially to obtain an approximate solution to the full problem.

The square root term of (C.1) is expanded to obtain

$$\frac{1}{\sqrt{1 + \epsilon t}} = 1 - \frac{1}{2} \epsilon t + \cdots$$

and the ordered set of problems is tabulated below.

ORDER, n	DE	IC	SOLUTION
0	$\dfrac{dy_0}{dt} = 0$	$y_0(0) = 1$	$y_0(t) = 1$
1	$\dfrac{dy_1}{dt} = \dfrac{1}{2}$	$y_1(0) = 0$	$y_1(t) = \dfrac{t}{2}$
2 $\vdots$	$\dfrac{dy_2}{dt} = -\dfrac{t}{4}$	$y_2(0) = 0$	$y_2(t) = -\dfrac{t^2}{8}$

So far, the solution expressed by (C.3) is

$$y(t) = 1 + \epsilon \frac{t}{2} - \epsilon^2 \frac{t^2}{8} + \cdots \tag{C.4}$$

(C.4) gives the solution to order $n = 2$.
  This problem has the exact solution,

$$y(t) = \sqrt{1 + \epsilon t} = 1 + \epsilon \frac{t}{2} - \epsilon^2 \frac{t^2}{8} + \cdots$$

which is identical to (C.4), thus supporting the validity of the method. Each term in the series obtained by this method should be smaller than the previous one.

Nonlinear Differential Equation with a Small Disturbance

  Consider the Rayleigh bubble equation and initial conditions,

$$\text{DE:} \qquad R\ddot{R} + \frac{3}{2}\dot{R}^2 = \frac{g_0}{\rho}(P - P_\infty) \tag{C.5}$$

$$\text{IC:} \qquad t = 0, \qquad R = R_\infty(1 + \epsilon), \qquad \dot{R} = 0 \tag{C.6}$$

If the bubble contains perfect gas which undergoes adiabatic pressure changes with volume

$$P = P_\infty \left(\frac{R_\infty}{R} \right)^{3k} \tag{C.7}$$

The equilibrium bubble radius is R_∞ when its pressure P is equal to the ambient value P_∞. The equilibrium radius is disturbed by the amount ϵR_∞ in (C.6). A series expansion of $R(t)$ is assumed in the form

$$R(t) = R_0(t) + \epsilon R_1(t) + \cdots \tag{C.8}$$

If only terms up to the order ϵ are kept, the problem will be linearized, and (C.7) becomes

$$P = P_\infty \left[\frac{R_\infty}{R_0(1 + \epsilon R_1/R_0 + \cdots)} \right]^{3k} = P_\infty \left(\frac{R_\infty}{R_0} \right)^{3k} \left(1 - \epsilon 3k \frac{R_1}{R_0} + \cdots \right)$$

(C.5) and (C.6) now can be written as an ordered set:

Order n = 0

DE: $$R_0 \ddot{R}_0 + \frac{3}{2} \dot{R}_0^2 = \frac{g_0}{\rho} P_\infty \left[\left(\frac{R_\infty}{R_0} \right)^{3k} - 1 \right]$$

ICs: $$R_0(0) = R_\infty, \qquad \dot{R}_0(0) = 0$$

SOLUTION: $$R_0(t) = R_\infty \quad \text{(yields undisturbed state)}$$

Order n = 1

DE: $$R_0 \ddot{R}_1 + R_1 \overset{(0)}{\ddot{R}_0} + \frac{3}{2} (2 \overset{(0)}{\dot{R}_0} \dot{R}_1) = -\frac{g_0}{\rho} P_\infty \left(\frac{R_\infty}{R_0} \right)^{3k} 3k \frac{R_1}{R_0}$$

IC: $$R_1(0) = R_\infty, \qquad \dot{R}_1(0) = 0$$

SOLUTION: $$R_1(t) = R_\infty \cos\left(\frac{1}{R_\infty} \sqrt{\frac{3 k g_0 P_\infty}{\rho}} \, t \right)$$

The $n = 1$ solution made use of the solution for R_0 in the $n = 0$ solution. The linearized problem ($n = 1$) usually is straightforward. Sometimes higher order ($n = 2, 3, \ldots$) oscillatory solutions yield increasing amplitudes, caused by resonant drivers, which must be removed by special techniques. Texts on perturbation methods give sufficient details for handling a diversity of problems and solutions.

C.3 Vector Differentiation in Different Reference Frames

Reference frame $\textcircled{E}$ with coordinates x, y, z and unit vectors $\mathbf{n}_x$, $\mathbf{n}_y$, $\mathbf{n}_z$ in Fig. C.3 is fixed to the earth, and reference frame $\textcircled{M}$ with coordinates a, b, c and unit vectors $\mathbf{i}, \mathbf{j}, \mathbf{k}$ is translating at velocity $\mathbf{V}_m$ relative to $\textcircled{E}$ and also rotating with angular velocity

$$\boldsymbol{\omega} = \dot{\theta}_a \mathbf{i} + \dot{\theta}_b \mathbf{j} + \dot{\theta}_c \mathbf{k} \tag{C.9}$$

Point P designates a moving fluid particle or any part of a mechanical object. Position vector $\mathbf{r}$ gives the location of P relative to $\textcircled{E}$; $^m\mathbf{r}$ is the position vector of P as seen by an observer in $\textcircled{M}$; and $\mathbf{r}_m$ is the position of the origin of $\textcircled{M}$ relative to $\textcircled{E}$. Vector $\mathbf{r}$ can be expressed as

$$\mathbf{r} = \mathbf{r}_m + {}^m\mathbf{r} \tag{C.10}$$

and the time derivative or $\mathbf{r}$ relative to $\textcircled{E}$ is

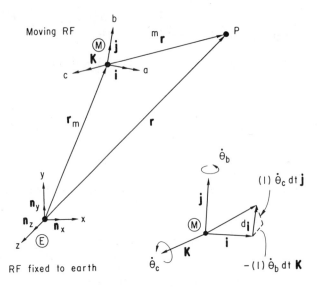

Figure C.3 Vector Differentiation in Different Reference Frames

$$\frac{d\mathbf{r}}{dt} = \frac{d\mathbf{r}_m}{dt} + \frac{d^m\mathbf{r}}{dt} \tag{C.11}$$

The vector

$$^m\mathbf{r} = a\mathbf{i} + b\mathbf{j} + c\mathbf{k} \tag{C.12}$$

is differentiated with respect to time relative to Ⓔ to get

$$\frac{d^m\mathbf{r}}{dt} = a\frac{d\mathbf{i}}{dt} + b\frac{d\mathbf{j}}{dt} + C\frac{d\mathbf{k}}{dt} + \frac{da}{dt}\mathbf{i} + \frac{db}{dt}\mathbf{j} + \frac{dc}{dt}\mathbf{k} \tag{C.13}$$

The unit vector derivatives are time rates of change relative to Ⓔ, whereas derivatives of the scalar lengths a, b, and c are the same for any reference frame. Since the last three terms of Eq. (C.13) are written with the $\mathbf{i}, \mathbf{j}, \mathbf{k}$ unit vectors, their sum is simply the time derivative of P relative to an observer in Ⓜ, $^m d\mathbf{r}^m/dt$. The terms in Eq. (C.13) which contain derivatives of unit vectors can be expressed as $\boldsymbol{\omega} \times {}^m\mathbf{r}$. This may not be obvious, but it can be seen by considering the time derivative of a single unit vector. Although unit vectors do not change in length, their orientation can change from rotation. Unit vector $\mathbf{i}$ will change its orientation in a time increment dt by an amount $d\mathbf{i}$, which corresponds to

$$d\mathbf{i} = -\dot{\theta}_b\, dt\, \mathbf{k} + \dot{\theta}_c\, dt\, \mathbf{j}$$

so that

$$\frac{d\mathbf{i}}{dt} = \dot{\theta}_c\mathbf{j} - \dot{\theta}_b\mathbf{k} = \boldsymbol{\omega} \times \mathbf{i} \tag{C.14}$$

Similarly,

$$\frac{d\mathbf{j}}{dt} = \dot{\theta}_a\mathbf{k} - \dot{\theta}_c\mathbf{i} = \boldsymbol{\omega} \times \mathbf{j} \tag{C.15}$$

$$\frac{d\mathbf{k}}{dt} = \dot{\theta}_b\mathbf{i} - \dot{\theta}_a\mathbf{j} = \boldsymbol{\omega} \times \mathbf{k} \tag{C.16}$$

It follows that Eq. (C.11) can be written as

$$\frac{d\mathbf{r}}{dt} = \boldsymbol{\omega} \times {}^m\mathbf{r} + \frac{{}^md^m\mathbf{r}}{dt} + \frac{d\mathbf{r}_m}{dt} \tag{C.17}$$

If we employ the velocity terms

$$\mathbf{V} = \frac{d\mathbf{r}}{dt}, \qquad {}^m\mathbf{V} = \frac{{}^md^m\mathbf{r}}{dt}, \qquad \mathbf{V}_m = \frac{d\mathbf{r}_m}{dt} \tag{C.18}$$

we have from Eq. (C.17)

$$\mathbf{V} = \boldsymbol{\omega} \times {}^m\mathbf{r} + {}^m\mathbf{V} + V_m \tag{C.19}$$

One more differentiation to obtain acceleration relative to $\textcircled{E}$ gives

$$\frac{d\mathbf{V}}{dt} = \boldsymbol{\omega} \times \frac{d^m\mathbf{r}}{dt} + \frac{d\boldsymbol{\omega}}{dt} \times {}^m\mathbf{r} + \frac{d^m\mathbf{V}}{dt} + \frac{d\mathbf{V}_m}{dt} \tag{C.20}$$

If we write $d^m\mathbf{r}/dt$ as before, and similarly write

$$\frac{d^m\mathbf{V}}{dt} = \frac{{}^md^m\mathbf{V}}{dt} + \boldsymbol{\omega} \times {}^m\mathbf{V} \tag{C.21}$$

we obtain

$$\frac{d\mathbf{V}}{dt} = \frac{d\mathbf{V}_m}{dt} + \frac{{}^md^m\mathbf{V}}{dt} + 2\boldsymbol{\omega} \times {}^m\mathbf{V} + \boldsymbol{\omega} \times \boldsymbol{\omega} \times {}^m\mathbf{r} + \boldsymbol{\alpha} \times {}^m\mathbf{r} \tag{C.22}$$

where $\boldsymbol{\alpha}$ is the angular acceleration, $d\boldsymbol{\omega}/dt$.

NOMENCLATURE

A	Area vector
A	Area magnitude
A_d	Drain area
A_q	Area available for heat transfer
a	Acceleration
$B(x)$	Elevation of wave channel bottom surface
b	Body force acceleration
C	Sound speed in fluid
C_{fp}	Sound speed, fluid/non-rigid pipe, Eq. (7.50)
C_d	Discharge coefficient
C_t	Thermal capacity
c	Specific heat
c_p	Specific heat at constant pressure
c_v	Specific heat at constant volume
D	Damping coefficient, bubbly mixture (10)
D_h	Hydraulic diameter
$D(x)$	Width of wave channel
E	Total storable energy
E_m	Modulus of elasticity
e	Specific internal thermal energy
erf( )	Error function (1)
F	Force vector
F	Force magnitude
F_I	Force from moving two-density interface
F_p	Force due to pressure

F_s	Surface force; shock force
F_τ	Force due to shear stress
F_f	Friction force
F_x, F_y, F_z	Force components in cartesian coordinates
f	Darcy friction factor; frequency
f_u	Unsteady friction factor
G	Mass flow rate per unit area
g	Acceleration of gravity
g_0	Newton's constant in $F = Ma/g_0$
H	Enthalpy; height dimension; liquid depth head (hydrostatic)
H_p	Pump head
H_{p0}	Pump shutoff head
$H_s(\ )$	Heaviside step function of argument ( )
h	Specific enthalpy
h_0	Stagnation enthalpy
$J(\ )$	Bessel function of the first kind
K	Spring constant
K_1, K_2, K_3	Channel constants (4)
KE	Kinetic energy
K_L	Pressure loss coefficient
k	Ratio of specific heats, c_p/c_v
L	Length dimension
L_R	Reference length
M	Mass
M_v	Virtual mass
$\dot{m}$	Mass flow rate
$\dot{m}_{fg}$	Vaporization flow rate
$\dot{m}_{gf}$	Condensation flow rate
$\mathbf{n}$	Unit vector
$\mathbf{n}_x, \mathbf{n}_y, \mathbf{n}_z$	Mutually perpendicular cartesian unit vectors
P	Pressure
PE	Gravitational potential energy
P_0	Reservoir or stagnation pressure
P_h	Heated perimeter
P_w	Wetted perimeter
Q	Amount of energy transfer by heat
q	Heat transfer rate
q'	Heat transfer rate per unit length
q''	Heat transfer rate per unit area (heat flux)
q'''	Heat generation rate per unit volume
$q_1, q_2, \ldots$	Generalized coordinates
R	Radius
R_b	Bounded pipe reaction force
R_{bd}	Blowdown (thrust) reaction force
R_g	Perfect gas constant

$\mathbf{r}$	Radial position vector
$\mathbf{r}_m$	Position vector of object m
$^m\mathbf{r}$	Position vector relative to reference frame m
r	Radial coordinate
S	Entropy;/Shock speed
s	Specific entropy
T	Temperature
t	Time
Δt	Response time
t_p	Propagation time
t_d	Disturbance time
U	Internal thermal energy
u, v, w	Velocity components in cartesian coordinates
$\mathbf{V}$	Velocity
V_p	Propagation velocity
V_R	Reference velocity
$^{cs}\mathbf{V}$	Velocity relative to control surface cs
$\mathbf{V}_{(\ )}$	Velocity of object ( )
v	Specific volume
$\mathbf{W}$	Weight vector
W	Weight magnitude
W_k	Work
x	Gas or vapor mass fraction
x, y, z	Position coordinates, cartesian
$Y(\ )$	Bessel function of the second kind
Y_L	Young's modulus, liquid
Y_p	Young's modulus, pipe
y	Elevation of liquid surface
y_d	Elevation of drain (1)
y_L	Liquid depth (1)
z	Length coordinate in pipe flow
$\mathscr{D}$	Diffusion coefficient
$\mathscr{D}_s$	Dissipation across shock, Eq. (5.75)
$\mathscr{E}$	Total stored energy per unit mass
$\mathscr{F}$	Radiation shape factor
$\mathscr{H}$	Convective heat transfer coefficient
$\mathscr{L}$	Left-traveling characteristic function
$\mathscr{M}$	Momentum
$\mathscr{P}$	Mechanical power
$\mathscr{P}_c$	Compressive power
$\mathscr{P}_\tau$	Shear power
$\mathscr{R}$	Right traveling characteristic function; radial coordinate in y, z plane
$\mathscr{T}$	Torque

$\mathcal{V}$	Volume
$\dot{\mathcal{V}}$	Volume flow rate
α	Thermal diffusivity; gas volume fraction; fluid jet parameter defined in Eq. (3.5); pump characteristic constant (7)
β	Volumetric expansivity (2)
Γ	Shear stress dyadic
$\Delta(\ )$	Large or small change of quantity ( )
$\nabla(\ )$	Gradient of ( )
$\nabla \times (\ )$	Curl of ( )
$\delta(\ )$	Increment of the quantity ( ), large or small
ϵ	Emissivity; small parameter
ζ	Second viscosity, Eq. (6.29)
η	Liquid wave elevation above undisturbed level
θ	Angular coordinate; angle of inclination
$\dot{\theta}_n$	Angular velocity component about n axis
κ	Thermal conductivity
λ	Wave length
$\lambda_1, \lambda_2, \lambda_3$	Characteristic constants
μ	Dynamic viscosity
ν	Kinematic viscosity, $\mu g_0 / \rho$
$\pi_1, \pi_2 \ldots$	Model coefficients, (6)
ρ	Density
σ	Surface tension; Stephan–Boltzmann constant
τ	Response time; time interval; entrance time of fluid particle (3)
τ_{ij}	Shear stress component, Eq. (6.31)
Φ	General extensive property
ϕ	General intensive property, Φ/M; velocity potential; general diffusing property (1); Pressure or velocity (4)
ω	Angular velocity or circular frequency
$(\ )_c$	Value of quantity at critical flow
$(\ )_d$	Disturbed property, hydrostatic shock (5)
$(\ )_{dis}$	Discharge property
$(\ )_f$	Saturated liquid; full size system
$(\ )_g$	Saturated vapor
$(\ )_{fg}$	Vaporization
$(\ )_i$	Initial value; ith component; interface
$(\ )_{out}$	Outflowing quantity
$(\ )_{in}$	Inflowing quantity
$(\ )_m$	Scale model system; mixture property
$(\ )_o$	Orifice property

$(\ )_\infty$	Ambient value; asymptotic condition
$đ$	Inexact differential
$(\ )_s$	Property at wall separation (5)
$(\ )_{ss}$	Steady state property
Δt_e	Equilibrium response time
$(\ )_r$	Reference value
$(\ )_u$	Undisturbed property, hydrostatic shock (5)
$(\ )_v$	Valve property
$(\ )_w$	Wall property
$(\ )_x$	Property, undisturbed side of pressure shock
$(\ )_y$	Property, disturbed side of pressure shock
$(\ )^*$	Nondimensional property

INDEX

649